Biomedical Applications of Laser-Processed Coatings

Irina Negut
https://orcid.org/0000-0003-4038-7548
National Institute for Lasers, Plasma, and Radiation Physics, Romania

Carmen Ristoscu
National Institute for Lasers, Plasma, and Radiation Physics, Romania

IGI Global Scientific Publishing
Publishing Tomorrow's Research Today

Vice President of Editorial Melissa Wagner
Director of Acquisitions Mikaela Felty
Director of Book Development Jocelynn Hessler
Production Manager Mike Brehm
Cover Design Jose Rosado

Published in the United States of America by
IGI Global Scientific Publishing
701 East Chocolate Avenue
Hershey, PA, 17033, USA
Tel: 717-533-8845 | Fax: 717-533-7115
Website: https://www.igi-global.com E-mail: cust@igi-global.com

Library of Congress Cataloging-in-Publication Data

LCCN: 2025041405 (CIP Data Pending)
ISBN13: 9798337341040
Isbn13Softcover: 9798337341057
EISBN13: 9798337341064
British Cataloguing in Publication Data
A Cataloguing in Publication record for this book is available from the British Library.

Table of Contents

Detailed Table of Contents

Bogdan Ionut Bita, National Institute for Optoelectronics, Romania
Ana Maria Iordache, National Institute for Optoelectronics, Romania
Ana Maria Florea, National Institute for Optoelectronics, Romania
Stefan Caramizoiu, National Institute for Optoelectronics, Romania
Stefan Marian Iordache, National Institute for Optoelectronics, Romania

Titanium Nitride (TiN) coatings, known for their hardness, corrosion resistance, and biocompatibility, are increasingly used in biomedical applications. Pulsed Laser Deposition (PLD) enables precise fabrication of TiN thin films with controlled microstructure, thickness, and stoichiometry. The PLD process involves ablation of a TiN target in a vacuum or nitrogen-rich atmosphere, producing dense, adherent coatings with excellent mechanical and biological performance. TiN's low friction and antimicrobial properties make it ideal for orthopedic, dental, and cardiovascular implants. Compared to other methods, PLD allows complex multilayer structures and fine control over surface properties, enhancing implant integration and longevity. This chapter explores the PLD process, laser-material interactions, and current biomedical applications of TiN coatings, emphasizing their role in improving implant durability and patient outcomes.

Ramyakrishna Jangama, Sri Sathya Sai University for Human Excellence, India
Chaya Suresh, Sri Sathya Sai University for Human Excellence, India
Chandana Korrapati, Sri Sathya Sai University for Human Excellence, India
Ramya Raghavan, Sri Sathya Sai University for Human Excellence, India

Hybrid bioactive coatings for cancer therapy offer controlled drug delivery, targeted treatment, and multimodal therapeutic approaches. In this study, we analyzed functional performance metrics and clinical study parameters. Data from multiple reports were evaluated on the therapeutic efficacy and safety profile of hybrid coatings. Results demonstrated an improvement in drug loading efficiency from 40% to 85%. The controlled release durations spanning hours to weeks, enabling sustained therapeutic payloads. Targeting efficiency demonstrated a 2- to 5-fold increase in cancer cell uptake, accompanied by enhanced selective cytotoxicity. In vivo tumor volume reduction ranged from 30% to 60%, with combination photothermal or photodynamic treatments. The hybrid bioactive coatings significantly improve cancer

The incorporation of advanced surface engineering techniques in biomedical textiles has opened new pathways for developing high performance, multifunctional textiles which can be tailored for healthcare applications. This chapter explores the role of laser-induced coatings and treatments in the surface functionalization of biomedical textiles. These technologies offer contactless, solvent-free, and localized surface modification. Such textile treatments enable the deposition of biocompatible and antimicrobial coatings on diverse textile substrates. These processes fine-tune surface properties like roughness, wettability, cell adhesion, and composition. These modifications are vital for wound dressings and coatings, offering antibacterial, bio-sensing, and tissue compatible functions. Laser-induced surface functionalization not only extends the functional limits of biomedical fabrics but also aligns with eco-friendly manufacturing goals. This chapter highlights research, challenges, and future directions, positioning laser-induced coating as key to next-generation medical textiles.

Ultrafast femtosecond laser surface functionalization offers a versatile single-step strategy to engineer medical implants with finely tuned biological responses. Using dental implants as a model system, this chapter illustrates how micro et nanoscale laser patterns enhance osseointegration, improve soft-tissue sealing and reduce bacterial colonization: three key challenges associated with implant failures. A major emerging direction is the integration of extracellular vesicles (EVs) through laser-induced nanostructures that act as passive topographical traps, enabling selective EV capture, protection, and controlled presentation at the tissue-implant interface. This approach opens promising routes toward next-generation bioactive implants capable of modulating the Healing process. Beyond dentistry, femtosecond laser texturing is applicable to various devices, including pacemakers, orthopedic screws,

Preface

INTRODUCTION

The field of biomaterials and biomedical engineering has experienced remarkable progress over the last few decades, driven by the continuous pursuit of advanced materials and surface modification technologies capable of enhancing the performance, functionality, and longevity of medical devices. Among these technologies, laser-based processing techniques have emerged as powerful and versatile tools for engineering coatings with precisely controlled structural, chemical, and biological properties. Their ability to tailor surfaces at micro- and nanometric scales has opened new possibilities for addressing longstanding challenges related to implant integration, tissue regeneration, antimicrobial protection, and targeted therapeutic delivery.

The increasing demand for biocompatible, multifunctional, and patient-specific solutions has accelerated research into laser-processed coatings across numerous biomedical domains. Advances in pulsed laser deposition, laser cladding, surface texturing, ultrafast laser processing, and laser-assisted manufacturing have enabled the development of coatings that simultaneously provide mechanical durability, corrosion resistance, bioactivity, and antibacterial functionality. Such innovations are transforming traditional approaches to medical implants, wound healing systems, biomedical textiles, and cancer therapies, while creating new opportunities for personalized and regenerative medicine.

This edited volume, Biomedical Applications of Laser-Processed Coatings, brings together contributions from internationally recognized researchers and emerging scholars whose expertise spans materials science, laser engineering, nanotechnology, biomedical sciences, and regenerative medicine. The chapters presented in this book provide both fundamental knowledge and state-of-the-art developments, offering readers a comprehensive understanding of the principles, technologies, characterization techniques, and biomedical applications that define this rapidly evolving field.

By integrating theoretical foundations with practical applications, this volume aims to serve as a valuable reference for researchers, engineers, clinicians, and graduate students interested in the future of laser-enabled biomedical technologies.

CHAPTER OVERVIEW

Chapter 1, "Characterization Techniques for Laser Processed Coatings," provides a comprehensive overview of the principal methodologies employed to evaluate laser-processed coatings. The authors discuss electrochemical analyses, structural and chemical characterization methods, infrared spectroscopy, and nanoindentation techniques. Particular emphasis is placed on understanding structure-property relationships and how surface morphology, roughness, and degradation behavior influence tissue integration and implant durability.

Chapter 2, "Fundamentals of Laser-Material Interaction for Coating Development," examines the underlying principles governing laser-material interactions and their implications for coating fabrication. Using titanium nitride coatings as an illustrative example, the chapter explores pulsed laser deposition processes, coating microstructure, and biomedical applications, highlighting how laser technologies contribute to improved implant performance and longevity.

Chapter 3, "Hybrid Coating Systems: Bioactive Molecules and Nanoparticles for Cancer Therapy," focuses on multifunctional hybrid coatings designed for cancer theranostics and targeted therapy. The authors analyze the incorporation of bioactive molecules and nanoparticles to achieve controlled drug release, enhanced therapeutic efficacy, and multimodal treatment capabilities, underscoring the translational potential of these advanced systems.

Chapter 4, "Laser Cladding of Bioactive Coatings on Biodegradable Implants," addresses the challenges associated with biodegradable metallic implants and presents laser cladding as a promising solution for enhancing corrosion resistance and bioactivity. The chapter discusses thermodynamic considerations, microstructural optimization, and the role of laser processing in promoting osteogenesis and extending implant functionality.

Chapter 5, "Laser-Assisted Strategies for Wound Healing and Tissue Engineering Scaffolds," reviews laser-based approaches for fabricating sophisticated scaffolds and regenerative systems. Techniques such as laser-induced forward transfer, laser-assisted bioprinting, and photobiomodulation are explored, demonstrating their capacity to improve vascularization, cell proliferation, and tissue repair while highlighting existing challenges related to scalability and clinical translation.

Chapter 6, "Laser-Based Coatings for Antibacterial and Antifouling Applications," examines how laser processing technologies can be used to combat biomaterial-associated infections. Various laser-assisted deposition and texturing techniques are discussed, emphasizing their ability to generate antimicrobial surfaces through controlled topography, reactive oxygen species production, and ion release mechanisms, thereby improving the safety and performance of implanted devices.

Chapter 7, "Laser-Coated Orthopedic and Dental Implants: Advances in Surface Engineering for Bioactivity and Longevity," presents recent developments in laser-assisted surface engineering for orthopedic and dental implants. The chapter highlights how bioactive coatings and micro- and nanostructured surfaces can enhance osseointegration, mechanical performance, and antibacterial properties, while also discussing future opportunities enabled by additive manufacturing and personalized medicine.

Chapter 8, "Surface Functionalization of Biomedical Textiles through Laser-Induced Coatings," explores the growing importance of laser technologies in the development of multifunctional biomedical fabrics. The authors demonstrate how laser-induced coatings can impart antibacterial, biosensing, and tissue-compatible characteristics to textile substrates, offering environmentally sustainable pathways for next-generation healthcare materials.

Chapter 9, "Ultrafast Laser Surface Processing for Medical Implants Functionalization," examines the application of femtosecond laser technologies to create highly controlled surface architectures for medical implants. The chapter discusses strategies to enhance osseointegration, improve soft tissue integration, reduce bacterial colonization, and integrate extracellular vesicles into laser-engineered nanostructures. It also considers industrial and regulatory aspects essential for successful clinical implementation.

Taken together, these chapters illustrate the remarkable breadth and interdisciplinary nature of laser-processed coatings in biomedical science. They reveal how advances in laser technologies are enabling the development of multifunctional surfaces capable of addressing complex biological, mechanical, and clinical requirements. The volume not only highlights current achievements but also identifies emerging directions that are expected to shape the next generation of biomaterials and medical devices.

CONCLUSION

As editors, we extend our sincere gratitude to all contributing authors for their dedication, scientific rigor, and willingness to share their expertise. Their collective efforts have made possible a volume that reflects the diversity and dynamism of contemporary research in laser-based biomedical engineering. We also acknowledge the valuable contributions of reviewers, collaborators, and publishing professionals whose support has been instrumental throughout the preparation of this book.

We hope that this volume will provide readers with both a solid foundation and a source of inspiration for future investigations. As healthcare technologies continue to evolve, laser-processed coatings are expected to play an increasingly important role in addressing unmet clinical needs and advancing personalized medicine. It is our belief that the interdisciplinary perspectives assembled here will stimulate new collaborations and accelerate innovation across materials science, medicine, and engineering.

Irina Negut
National Institute for Lasers, Plasma, and Radiation Physics, Romania

Carmen Ristoscu
National Institute for Laser, Plasma and Radiation Physics, Romania

Chapter 1
Characterization Techniques for Laser Processed Coatings

Diana Maria Vranceanu
http://orcid.org/0000-0002-6561-7023
National University of Science and Technology Politehnica Bucharest, Romania

Iulian Pana
http://orcid.org/0000-0003-1368-6219
National Institute of Research and Development for Optoelectronics, Romania

Mihaela Dinu
http://orcid.org/0000-0002-3149-6972
National Institute of Research and Development for Optoelectronics, Romania

Alina Vladescu (Dragomir)
http://orcid.org/0000-0001-5770-4541
National Institute of Research and Development for Optoelectronics, Romania

Catalin Vitelaru
National Institute of Research and Development for Optoelectronics, Romania

ABSTRACT

This chapter reviews the main characterization techniques for laser-processed coatings, such as electrochemical methods (electrochemical impedance spectroscopy and polarization) complemented by morphological, structural, and chemical analyses. Fourier transform infrared spectroscopy and nanoindentation are discussed as additional approaches for evaluating chemical integrity and local mechanical properties. The structure-property relationship is highlighted, as well as degra-

DOI: 10.4018/979-8-3373-4104-0.ch001

dation mechanisms, the role of porosity and surface roughness. Laser processing modifies the micro-/nano-topography and chemistry of coatings, thereby enhancing hydrophilicity and surface energy to facilitate apatite nucleation in simulated body fluid. By adjusting surface roughness and ion exchange, it controls biodegradation in phosphate-buffered solution, thereby improving tissue integration and the long-term durability of implants. Thus, optimized parameters allow for precise tuning of bifunctionality and durability.

1. INTRODUCTION

This chapter addresses laser-based surface modification and ways that can be leveraged to engineer implant interfaces with improved physico-chemical, mechanical, electrochemical, and biological performance.

Fourier Transform Infrared Spectroscopy (FTIR) spectroscopy is widely used for the chemical characterization of coatings obtained by laser-assisted deposition techniques intended for biomedical applications (Narazaki, 2021). It is particularly useful for verifying whether the chemical composition and characteristic functional groups of the deposited material are preserved after laser-assisted transfer, and for identifying the optimal processing window (Rezwan et al., 2006). Moreover, its relevance derives from the fact that laser-assisted deposition and laser surface processing may induce not only morphological and structural changes, but also alterations in local bonding environments, functional group integrity, phase composition, and interfacial chemistry (Sahadat Hossain & Ahmed, 2023). In coatings designed for implantable systems, such changes are important because the biological response is closely linked to the surface chemical composition, the presence of bioactive groups, and the stability of the deposited material in physiological media (Notingher et al., 2003). Nanoindentation provides complementary local mechanical information by probing hardness and elastic modulus at the scale of the coating or modified surface layer (Haque, 2003). Since laser-assisted deposition and laser surface processing can alter densification, defect content, and coating continuity, nanoindentation is particularly suitable for correlating microstructural changes with functional mechanical performance (Matei & Visan, 2026). Parameters derived from nanoindentation measurements, such as the H/E and H^3/E^2 ratios, may provide additional insight into elastic strain tolerance and resistance to plastic deformation, which are often associated with improved wear and contact-damage resistance (X. Wei et al., 2024). For coatings produced or modified by laser-based techniques, such parameters are useful because they can reveal whether a mechanically improved surface is also likely to be more damage-tolerant under repeated loading. In implant applications, this has direct relevance for resistance to microdamage, delamination, debris formation, and long-term surface degradation under specific service conditions (Shabir et al., 2025).

Electrochemical evaluation provides critical insight into the corrosion resistance and interfacial stability of laser-processed coatings intended for biomedical applications. Techniques such as open circuit potential (OCP), potentiodynamic polarization, and electrochemical impedance spectroscopy (EIS) enable both qualitative and quantitative assessment of passive thin film stability, corrosion kinetics, and barrier performance in simulated physiological environments (Anita et al., 2023; Berradja, 2019; Bierwagen et al., 2003; Trentin et al., 2022). Laser processing, due to its high spatial resolution and controllability, leads to modification of surface features at both micro- and nanometric scales (Córdoba-Torres, 2017; Pakuła et al., 2019; Schille et al., 2020; Sypniewska & Szkodo, 2022; Z. Xu et al., 2025). Such structural tailoring, microstructural refinement, metastable phase formation, and varying degrees of micro-porosity or microcracking strongly influence electrochemical behavior. In particular, EIS is highly sensitive to coating defects and porosity, as interconnected pores facilitate electrolyte penetration and modify capacitive behaviour (Córdoba-Torres, 2017). Correlating AFM-derived porosity parameters with electrochemical data reveals that increased pore density or interconnected porosity typically corresponds to decreased polarization resistance and reduced impedance modulus at low frequencies, indicating enhanced electrolyte ingress towards the substrate (Trentin et al., 2022). In contrast, laser-based processing techniques that promote dense, homogenized microstructures with minimized surface defects exhibit improved barrier properties and more stable passive behavior (Dhinasekaran et al., 2021). The combined application of electrochemical techniques and AFM therefore provides a multi-scale characterization approach, linking laser-induced microstructural features to functional corrosion performance and, consequently, to long term biomedical reliability.

In metallic biomaterials, such processing has been shown to enhance bioactivity and biocompatibility, supporting superior cell adhesion and proliferation. The application of laser-based surface modification to metallic biomaterials facilitates the rational tuning of micro/nano-scale topography and the associated wettability. This, in turn, has the effect of enhancing biocompatibility and bioactivity, while supporting robust cell adhesion and proliferation. The role of wettability in establishing a linkage between roughness and biomineralization is a subject of considerable interest (Gul et al., 2025; Y. L. Yang et al., 2010). The observation that an increase in multiscale roughness results in increased surface energy and a corresponding decrease in contact angle has led to the proposal of a hydrophilic interface hypothesis. This hypothesis suggests that such interfaces, characterized by their hydrophilic properties, play a pivotal role in facilitating mineral nucleation, thereby contributing to the process of biomineralization (Guo et al., 2025; Zhou et al., 2023). This hydrophilicity has been demonstrated to accelerate serum protein adsorption and integrin-mediated focal adhesion maturation, which contributes to the stabilization of early cell attachment and the activation of osteogenic signalling. Concurrently, hierarchical characteristics and higher energy surfaces provide favourable nucleation sites for apatite, thereby facilitating ion-mediated mineral deposition and promoting osseointegration. In the context of laser-deposited calcium phosphate coatings, it has been demonstrated that composition and processing parameters can independently modulate roughness and wettability (Cotrut et al., 2024; Dinu et al., 2025; Knabe, 2011; Kokubo & Takadama, 2006; Negut et al., 2023, 2025; Vardaki et al., 2018). However, a concurrent and reliable increase in biomineralization is observed when hydrophilicity is optimized. The concurrent optimization of roughness and wettability in laser-processed coatings enhance implant interfaces that consistently facilitate rapid and durable mineralized tissue integration while maintaining compatibility with metallic substrates. Consequently, these insights underscore the efficacy of laser-mediated surface engineering as a versatile platform for the rational design of next-generation biomaterial interfaces. The objective of this design is to align surface structure and wettability with target biological outcomes.

2. CHARACTERIZATION OF LASER-PROCESSED COATINGS BY FTIR AND NANOINDENTATION

2.1. Principles and Analytical Relevance of FTIR Spectroscopy and Nanoindentation for Biomedical Alloys and Coatings

Fourier transform infrared spectroscopy and nanoindentation are complementary techniques for characterizing laser-processed coatings intended for biomedical applications (Jażdżewska et al., 2023; Narazaki, 2021). Although each is based on different aspects of the deposited or modified surfaces, their combined use is especially important for establishing a direct structure-property relationship in complex thin-film systems (**Figure 1**).

Figure 1. Schematic representation of the information provided by FTIR spectroscopy and nanoindentation for laser-processed coatings

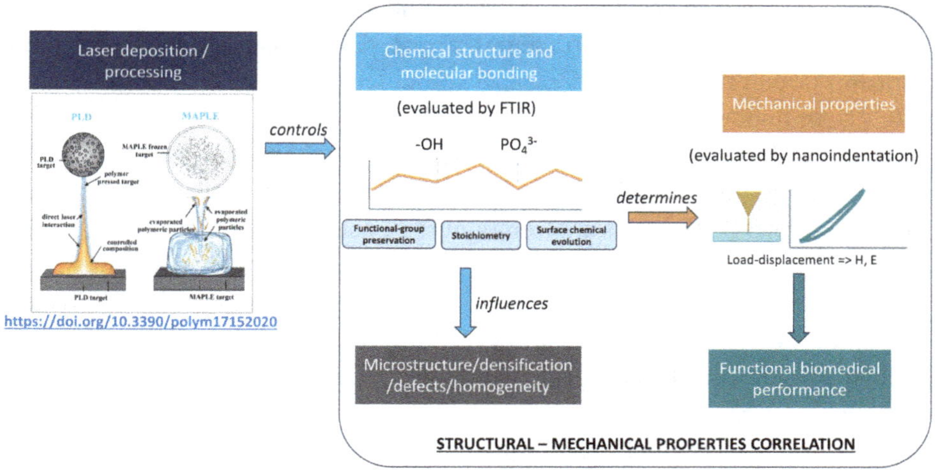

FTIR spectroscopy can provide chemical information and is particularly useful for verifying whether the transferred material maintains its relevant functional groups, stoichiometry, and bonding configuration after laser-assisted deposition/processing (Paital et al., 2009). Moreover, nanoindentation determines the local mechanical response of the coating or modified surface layer, making it suitable for assessing how laser-induced changes in densification, homogeneity, and defect distribution affect hardness, elastic modulus, and deformation tolerance (Majkowska-Marzec et al., 2020). In this sense, FTIR addresses chemical integrity, whereas nanoindentation helps determine whether the resulting surface is mechanically appropriate at the micrometer or sub-micrometer scale (Sarantopoulou et al., 2011).

In the case of laser-deposited biomedical coatings, the analytical relevance of FTIR arises from the fact that laser transfer may preserve, partially alter, or selectively reorganize the local bonding structure of the starting material. This is especially important for hybrid or composite systems containing inorganic bioactive phases alongside polymers (Rezwan et al., 2006), peptides (Bingtong et al., 2020), proteins (Al-Amin et al., 2025), growth factors (da Silva et al., 2021; Yi et al., 2026), or antibiotics (Yi et al., 2026), in which the therapeutic or biofunctional role of the coating depends strongly on chemical fidelity after deposition. In such combined systems, comparative IR studies comparing drop-cast materials with laser-processed films are frequently used to identify the optimal laser fluence and to verify that the characteristic absorption bands of all important constituents are retained after transfer. Several studies on PLA/HAp/BMP_4 and HAp/CS/FGF_2 coatings clearly show that infrared spectra and IR maps can be used not only to confirm the presence of characteristic bands, but also to evaluate the spatial distribution of components and the efficiency of material transfer across the substrate (Gherasim, Grumezescu, Grumezescu, Andronescu, et al., 2021; Grumezescu et al., 2022). An important aspect of using FTIR in this context is that it allows rapid identification of both inorganic and organic vibrational specific IR bands within the same coating. For example, in laser-processed calcium phosphate coatings, the most relevant specific IR bands are commonly associated with phosphate, hydroxyl, and carbonate groups (Vladescu et al., 2019), whereas the organic component may be identified by carbonyl, methylene, amide, N-H, or other characteristic IR bands. For instance, in HAp/CS/FGF_2 coatings, FTIR was used to identify contributions from hydroxyl, phosphate, carbonyl, and amide groups and to determine the laser fluence that best balanced efficient transfer with the preservation of chemical integrity (Grumezescu et al., 2022). Similarly, for HAp/KAN (Kanamycin) systems, characteristic methylene-, carbonyl-, and amine-related IR bands originating from the antibiotic were monitored alongside phosphate-related bands from hydroxyapatite, thereby enabling identification of the optimal fluence for uniform transfer without chemical degradation (Gherasim, Grumezescu, Grumezescu, Negut, et al., 2021). This implies that FTIR not only confirms the presence of material but also provides a direct analytical route for selecting processing conditions that preserve chemical functionality in laser-processed coatings (Al-Amin et al., 2025). In addition to verifying transfer integrity, FTIR is useful for interpreting structural changes induced by laser processing conditions (Di Siena et al., 2023). Changes in peak width, intensity, splitting, or position may reflect variations in crystallinity, local disorder, compositional substitutions, or dehydroxylation effects (Kylychbekov et al., 2026; Sahadat Hossain & Ahmed, 2023). This has a great importance in hydroxyapatite-based coatings, where broadened phosphate bands may indicate a more amorphous character, while narrower bands are commonly associated with the onset of crystallization (Zarif et al., 2024). Moreover, FTIR can be used to distinguish among different structural states of hydroxyapatite, chitosan or Sr-doped hydroxyapatite coatings, and to relate band shifts and/or splitting to crystallization and compositional modification (Zarif et al., 2024). Therefore, FTIR provides not only a compositional fingerprint, but also indirect information on structural evolution induced by processing parameters such as substrate temperature (Zarif et al., 2024), laser fluence (Radhi et al., 2024), or post-deposition treatments (Ong & Lucas, 1994). Another important reason for including FTIR in the characterization workflow for laser-processed coatings is its ability to monitor chemical evolution during interactions with physiological media (Notingher et al., 2003; Prefac et al., 2020). In the case of bioactive coatings, the functional relevance of the coating depends on how its surface chemistry changes in simulated body fluid (SBF) or related environments (Prefac et al., 2020). Therefore, FTIR is particularly effective for monitoring the appearance and growth of phosphate- and carbonate-related bands associated with apatite-like mineralization, while simultaneously verifying whether the organic component remains chemically detectable (Grunenwald et al., 2014). Also, FTIR can be used to monitor surface changes after immersion in SBF at several time periods (Dinu et al., 2025), thereby supporting the interpretation of bioactivity and mineralization behavior. The relevance of this approach is high for implant-related coatings because it links the as-deposited chemical structure to the early interfacial processes that influence osseointegration and long-term surface stability (J. Yang et al., 2024).

From the characterization point of view, FTIR is especially attractive due to its relatively rapid analysis, strong chemical sensitivity, and adaptability to both powders and thin films, whether measured in transmission mode or by attenuated total reflectivity (ATR-FTIR) (Codescu et al., 2022). In many studies, the FTIR technique is used in a complementary manner alongside IR mapping or morphological techniques such as SEM and AFM. This combination is particularly useful because chemical preservation alone is not sufficient; the same coating must also be uniformly transferred and morphologically coherent. In the case PLA/HAp/BMP4 coatings (Gherasim, Grumezescu, Grumezescu, Andronescu, et al., 2021), for example, IR analysis indicated that the optimal fluence preserved relevant absorption bands, while SEM confirmed uniform and compact transfer of the composite material. This correlation is consistent with the interpretation of IR spectral data and explains why FTIR is rarely used on its own to evaluate the multifunctionality of laser-fabricated biomedical surfaces. Moreover, FTIR provides insights into chemical composition and structural integrity, but it cannot directly assess whether the coating possesses the local mechanical characteristics required for service under repeated contact or physiological loading (**Figure 2**).

Figure 2. Schematic representation of FTIR-based analysis of laser-processed biomedical coatings

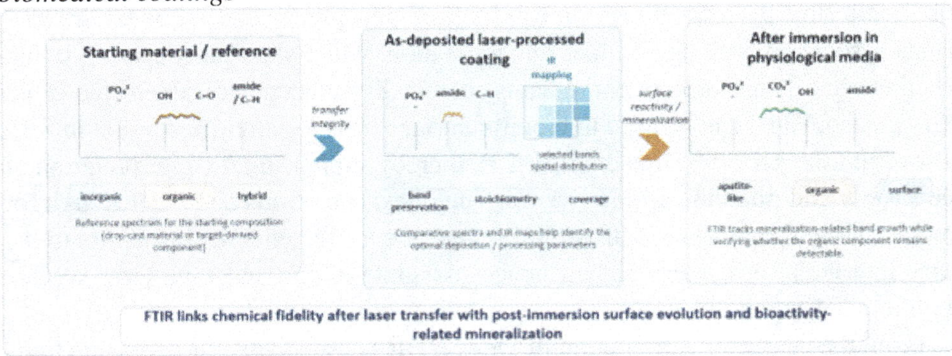

As a result, nanoindentation becomes particularly relevant for alloys within the biomedical field (Haque, 2003). Nanoindentation is a localized mechanical test at the nanoscale in which an indenter tip is pressed into the surface of a material while load and displacement are continuously recorded (Pana et al., 2017; Pshyk et al., 2025). From the resulting load-displacement curves, the hardness and elastic modulus can be determined, providing quantitative information about the material's resistance to deformation (Dinu et al., 2020). In the context of laser-processed surfaces, this is important because laser deposition and laser surface modification may affect the density, porosity, microcracking tendency, phase distribution, and coating homogeneity (Kurdi et al., 2023; Matei & Visan, 2026). Beyond hardness and elastic modulus, nanoindentation enables the use of derived mechanical indicators that are often more reliable for thin biomedical coatings than absolute values alone (Mahmoud et al., 2024). Ratios between the two parameters, such as H/E and H^3/E^2, are commonly discussed in relation to elastic strain tolerance and resistance to plastic deformation or contact damage (X. Wei et al., 2024).

For laser-modified Ti-6Al-4V, a recent study has shown that processing parameters such as scanning speed influence both molten-pool geometry and the resulting microstructure, with lower scanning speeds favoring α′ martensite formation and increasing surface hardness (Caruso et al., 2025). From a biomedical point of view, such changes are not mechanically neutral: they may improve wear resistance and local load-bearing capacity, but they may also alter stiffness gradients and how stresses are accommodated between the coating and the underlying alloy (Pshyk et al., 2025). The same considerations apply to β-type biomedical alloys such as Ti13Nb13Zr and related Ti–Nb–Zr systems, where laser processing is often employed not only to modify topography and wettability, but also to tailor local mechanical behavior (Jażdżewska et al., 2023). In the case of a Ti13Nb13Zr titanium alloy subjected to direct laser writing, nanoindentation was used to determine nanohardness and reduced modulus and to relate the results to wear susceptibility, demonstrating that the technique is well suited to evaluating the mechanical modifications induced by highly localized laser patterning (Jażdżewska et al., 2023). Earlier work on laser-surface-treated Ti13Nb13Zr with carbon-based surface modification likewise showed that laser processing can significantly alter near-surface hardness and elastic modulus, as well as corrosion-related behavior (Majkowska-Marzec et al., 2020). The mechanical perspective is equally important for coatings fabricated or processed by laser-based techniques. In PLD- and MAPLE-derived biomedical coatings, the mechanically relevant region is often only a few hundred nanometers to a few micrometers thick, which means that the indentation response is controlled not only by composition, but also by film continuity, interfacial bonding, residual porosity, crack susceptibility, and the spatial distribution of the surface features (Matei & Visan, 2026). For hydroxyapatite- and bioglass-based coatings on titanium (Dhinasekaran et al., 2021; Farooq et al., 2026), laser deposition is attractive because it can preserve bioactivity and provide good control over the film architecture (Kannan et al., 2025), yet the resulting surfaces may still differ substantially in compactness, roughness, and defect populations depending on deposition conditions. In this context, nanoindentation provides a practical way to distinguish between a compact, mechanically coherent film and a surface that is chemically promising but structurally weak due to pores, agglomerates, or poorly connected features (Pelletier et al., 2004).

The usefulness of nanoindentation in laser-fabricated coatings/alloys becomes even clearer when one considers that the most relevant mechanical parameters extend beyond hardness and elastic modulus alone (T. Wang et al., 2025; T. Z. Xu et al., 2023). Recent studies emphasize that derived parameters such as H/E and H^3/E^2 often provide a more meaningful indication of elastic strain tolerance and resistance to plastic deformation or contact damage (Drobný et al., 2021). This is highly relevant for orthopedic (Boi et al., 2019) and dental materials (Habelitz et al., 2002), where repeated microcontacts, third-body abrasion, fretting, and tribocorrosive interactions may gradually degrade the outermost surface long before any macroscopic failure occurs (Hoque et al., 2022). In practical terms, a laser-processed surface that is much harder is not necessarily the most durable; if the increase in hardness is accompanied by brittleness, pronounced heterogeneity, or excessive mismatch with the substrate, the risk of crack initiation or localized failure may remain high (Arulvel et al., 2023).

A further consideration concerns the interpretation of indentation data in systems that combine a metallic substrate with a laser-derived coating or a laser-modified surface layer. When the indentation depth becomes a substantial fraction of the film thickness, the measured hardness and modulus may represent the response of the coating/substrate system rather than the intrinsic behavior of the coating itself

(Li et al., 2024). This issue is especially important for bioactive ceramic or hybrid coatings on alloy substrates because the stiffness mismatch between the film and the substrate can be large, and the transition zone may itself be structurally graded or defect-containing (Brunello et al., 2019). Roughness, porosity, phase heterogeneity, agglomeration, and microcracks may further perturb the load–displacement response and generate substantial scatter from one indentation zone to another. For this reason, nanoindentation data in laser-engineered biomedical materials (**Figure 3**) should be interpreted together with structural and topographical information (Shabir et al., 2025).

Figure 3. Nanoindentation of thin laser-processed coatings and modified alloy surfaces

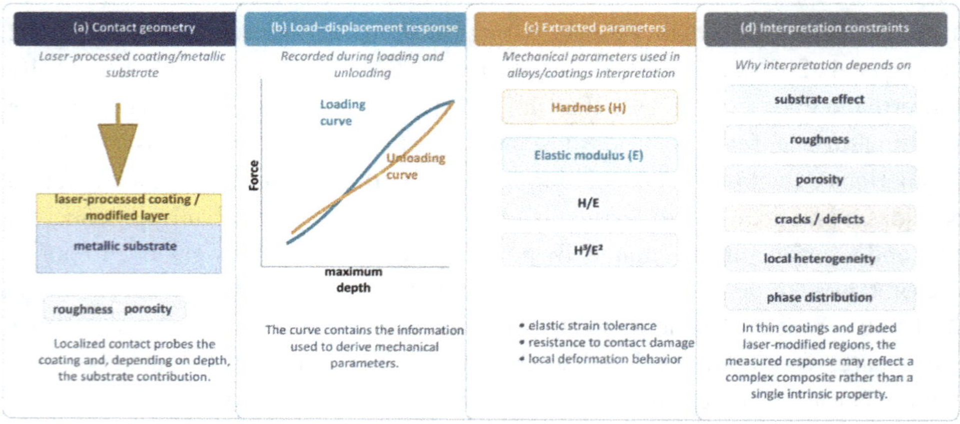

3. ELECTROCHEMICAL PERFORMANCE OF LASER PROCESSED COATINGS

3.1. Fundamentals of Electrochemical Techniques and Data Interpretation

Polarization techniques and electrochemical impedance spectroscopy (EIS) are widely used electrochemical methods for evaluating corrosion processes and the protective performance of coatings. Although both approaches provide valuable information, they differ significantly in methodology, sensitivity, and the type of data obtained (**Figure 4**).

Polarization techniques, including linear polarization resistance (LPR) and potentiodynamic polarization (e.g., Tafel analysis, potentiodynamic scanning, and cyclic polarization), are commonly employed for the rapid determination of corrosion rates. These DC-based methods use different ranges of applied potential perturbations to initiate oxidation and reduction reactions, thus providing complementary insights into corrosion mechanisms. In particular, the mentioned evaluation methods enable the direct estimation of corrosion current density (i_{corr}) through Tafel extrapolation or the Stern-Geary relationship, making them suitable for quantitative corrosion assessment (Anita et al., 2023; Berradja, 2019).

Furthermore, polarization curves provide valuable information on electrochemical phenomena such as active, passive, and transpassive behaviour, as well as key parameters including breakdown potential (Eb), passivation potential (Epp), and repassivation potential (Erp), which are essential for evaluating coating stability and susceptibility to localized corrosion (Esmailzadeh et al., 2018; Parangusan et al., 2021). In addition, open-circuit potential (OCP) measurements allow monitoring of the spontaneous electrochemical behaviour of the system without applying an external current (Trentin et al., 2022). The evolution of OCP over time can indicate processes such as the dissolution of surface oxide layers and the initiation of corrosion, or, on the contrary, the formation and stabilization of protective films (Milošev et al., 2000, 2008; Pan et al., 1996; Wolstenholme, 1973). However, OCP measurements are typically performed prior to other electrochemical tests to ensure system stabilization.

Figure 4. Overview of electrochemical techniques

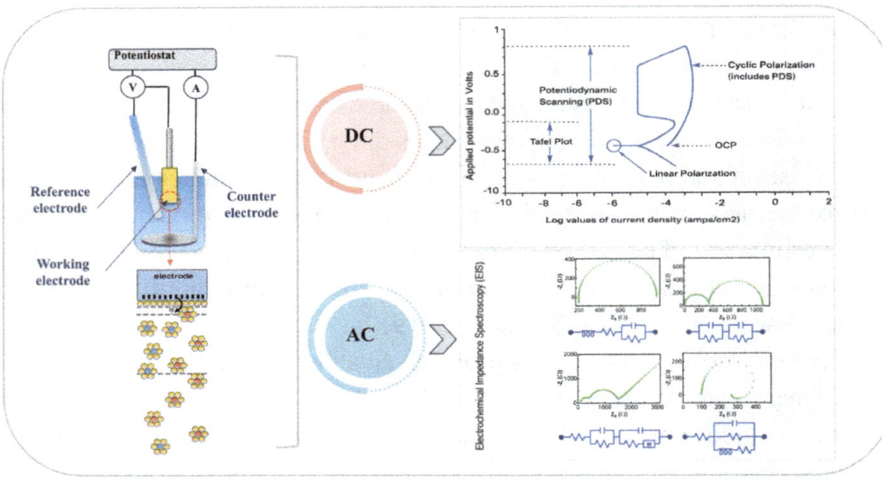

Furthermore, polarization measurements enable the separate analysis of anodic and cathodic reactions, providing insight into reaction kinetics and electrochemical mechanisms (Walsh et al., 2008). Due to their relatively simple experimental setup and short measurement time, polarization techniques are frequently used for screening corrosion inhibitors and evaluating coating performance, particularly after degradation or defect formation (Grundmeier et al., 2000; Sekine, 1997). However, these methods require significant perturbation of the system (typically ranging from ±250 mV up to ±1000 mV vs. OCP/reference electrode) and may damage or alter protective coatings, thereby limiting their applicability for long-term monitoring.

While polarization techniques provide valuable information related to corrosion kinetics under applied potential conditions, they are destructive and offer limited insight into the complex interfacial processes occurring at different time domains. In contrast, electrochemical impedance spectroscopy (EIS), an AC-based and "non-destructive" technique, enables the investigation of electrochemical systems under near-equilibrium conditions by applying a small-amplitude sinusoidal perturbation over a wide frequency range (Bierwagen et al., 2003).

One of the main advantages of EIS is its ability to provide detailed information on processes such as water uptake, pore formation, defect presence, and coating adhesion, as well as to deconvolute multiple interfacial phenomena, including charge transfer resistance, double-layer capacitance, and diffusion processes, even before visible damage (Amirudin & Thieny, 1995; Trentin et al., 2022). By analysing impedance spectra over a broad frequency range, typically represented in Nyquist and Bode magnitude plots, EIS enables a qualitatively interpretation by separating the processes occurring at different time constants: dielectric properties are observed in the high-frequency range, the coating-solution interface is associated to medium-frequency region, while the low-frequency range gives information related to corrosion processes which occur at the substrate-solution interface (Flores et al., 2006). Additionally, the presence of a broad phase angle peak (often referred to as "frequency dispersion") in the Bode phase angle plot provides important insight into surface heterogeneity and indicates the need to incorporate non-ideal elements in equivalent circuit modelling, such as constant phase elements (CPEs), to ensure physical relevance and a good correlation with experimental data (Bredar et al., 2020; Trentin et al., 2020).

For quantitative interpretation, EIS data are commonly analysed using equivalent electrical circuits (EEC), which consist of resistive elements (representing charge transfer processes) and capacitive elements describing the dielectric behaviour of the investigated material (Rammelt & Reinhard, 1992). Although the modelling procedure can be complex, the use of appropriate electrical components improves data fitting and provides a more accurate physical understanding of the corrosion mechanisms. For instance, the use of a constant phase element (CPE) instead of an ideal capacitor is often associated with deviations from ideal dielectric behaviour (Córdoba-Torres, 2017), typically arising from non-uniform current distribution along the surface (Hirschorn et al., 2010). According to the literature, such non-ideal behaviour can be attributed to several factors, including surface disorder or inhomogeneity, geometric irregularities, electrode porosity, and surface roughness (Córdoba-Torres, 2017). Nevertheless, as illustrated in **Figure 5**, coating degradation in time due to water and ion uptake leads to the formation of pores, channels, and structural defects within the coating. These processes facilitate electrolyte penetration toward the substrate and promote localized electrochemical activity at the coating-substrate interface. Based on the report published by *Trentin et al.*, this degradation is typically reflected by a decrease in impedance modulus and phase angle values at low frequencies, as well as the appearance of additional time constants in the impedance spectra (Trentin et al., 2022). Consequently, a more complex EEC, incorporating multiple resistive and non-ideal capacitive elements, is required to accurately describe the evolving interfacial processes and degradation mechanisms. Despite the valuable qualitative and quantitative information provided by EIS, the interpretation of the results and the selection of an appropriate equivalent electrical circuit must be supported by complementary morphological and structural investigations, such as SEM and AFM, to establish reliable correlations between surface features, porosity, and electrochemical performance.

Figure 5. Representative electrical equivalent circuits (EECs) as a function of electrolyte ingress over immersion time.

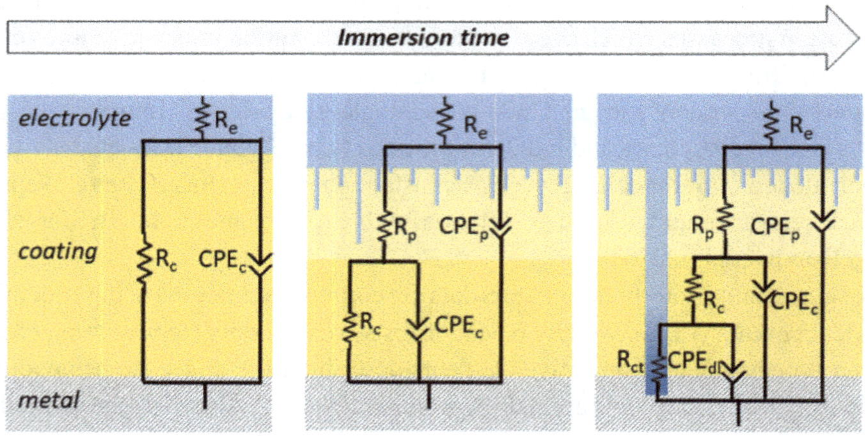

Image reprinted from ref. (Trentin et al., 2022) licensed under CC-BY.

When comparing both techniques, polarization methods are generally more suitable for quick estimation of corrosion rates, particularly in systems where the metal surface is exposed or coatings are already degraded. On the other hand, EIS is more appropriate for intact or high-impedance coating systems, where it provides superior sensitivity and deeper insight into degradation mechanisms (Berradja, 2019; Trentin et al., 2022). Since both techniques offer different advantages (**Figure 6**), multiple studies report their complementary application, where EIS characterizes coating integrity and polarization techniques quantify corrosion rates once failure initiates (Hassan et al., 2007; X. Liu et al., 2009). This combined approach enables a more detailed evaluation of coating performance under different stages of exposure. However, more advanced electrochemical methods can provide supplementary insights into system behaviour. For example, Odd Random Phase multisine Electrochemical Impedance Spectroscopy (ORP-EIS) has been used in several studies to evaluate experimental noise, non-linear distortions, and non-stationary behaviour, thereby improving the reliability of impedance measurements and helping to identify optimal experimental conditions (Alvarez-Pampliega et al., 2014; Breugelmans et al., 2010; Madelat et al., 2023). Based on statistical and signal analysis, a physically meaningful EEC models can be selected, leading to a more accurate interpretation of electrochemical processes (Dinu et al., 2017, 2024).

Figure 6. Overview of advantages of polarization methods vs. EIS

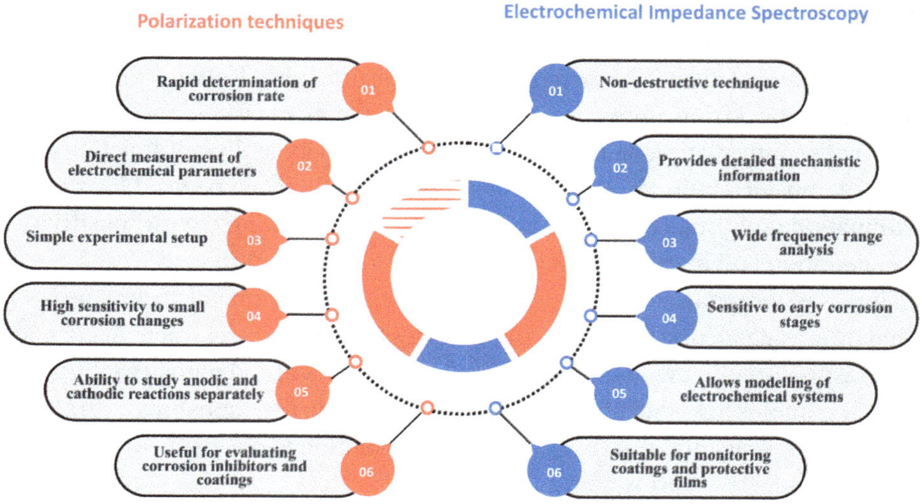

(Berradja, 2019; Herrera Hernández et al., 2020; Kendig & Scully, 1990)

Beyond conventional EIS and polarization techniques, more advanced electrochemical analyses have been often used. As previous highlighted, advanced impedance techniques combined with time-dependent measurements enable the monitoring of coating evolution over longer immersion time (Córdoba-Torres, 2017; Trentin et al., 2022). According to relevant reports, this analysis allows the identification of transitions from capacitive behaviour to diffusion-controlled processes, reflecting the development of porosity and electrolyte pathways within the coating (Berradja, 2019; Hirschorn et al., 2010). As described, Localized Electrochemical Impedance Spectroscopy (LEIS) operates on the same principle as conventional EIS, however, provides allows to evaluate local impedance from the ratio of applied AC voltage to local current density (V. M. Huang et al., 2011). However, unlike EIS, LEIS enables the investigation of coating performance in specific areas for studying substrate reactions beneath coatings, by generating 2D or 3D maps of impedance (Calado et al., 2018). This allows the visualization of defect evolution and the effectiveness of corrosion inhibitors (Kartsonakis et al., 2014) or self-healing systems (Calado et al., 2018).

Furthermore, the Scanning Vibrating Electrode Technique (SVET) is a localized electrochemical method that uses a vibrating microelectrode (typically <20 μm) to map ionic current density distributions above a corroding surface (Gonzalez-Garcia et al., 2016). By detecting potential gradients in the electrolyte, SVET provides information on anodic and cathodic activity, enabling the visualization of localized corrosion processes for active coatings also containing corrosion inhibitors or self-healing agents (Lutz et al., 2016). When combined with complementary techniques such as scanning ion-selective electrode technique (SIET), SVET enables simultaneous mapping of local current and pH variations (Jadhav & Gelling, 2019). Another advanced electrochemically technique is Scanning Kelvin Probe (SKP), a technique derived from atomic force microscopy, which measures surface potential and provides information on the electrochemical activity and work function distribution of a material (Örnek et al., 2019). Therefore, SKP can be used for studying delamination and protection mechanisms in polymer coatings, with sensitivity to ion penetration and interfacial changes (Rohwerder, 2009; Yin et al., 2022).

Thus, the integration of DC- and AC-based electrochemical techniques, supported by advanced localized methods, provides a complex assessment of corrosion mechanisms. Together with morphological, topographical and porosity information allows a complex insight into laser-processed coatings, establishing correlations between microstructural features and defect distribution to corrosion behaviour and long-term protective performance.

3.2. Electrochemical Mechanisms in Laser-Processed Coatings

Laser-based technologies have been widely explored to improve the surface properties, particularly the electrochemical performance of metallic biomaterials, by promoting stable passive film formation and controlling electrolyte-surface interactions in physiological environments (Montesissa et al., 2025; Sypniewska & Szkodo, 2022; Visan & Negut, 2025; Z. Xu et al., 2025). Among these techniques, pulsed laser deposition (PLD), matrix-assisted pulsed laser evaporation (MAPLE) and laser cladding enable the fabrication of bioactive coatings with controlled microstructure and strong adhesion to metallic substrates, as further evidenced. For instance, PLD-deposited bioactive glass and hydroxyapatite (HAp) coatings on titanium substrates form dense and homogeneous films with nanostructured surface roughness, resulting in enhanced corrosion resistance compared to uncoated substrates (Dhinasekaran et al., 2021). Despite the fact that EIS analyses reveal higher double layer capacitance for bioactive glass coatings when immersed in Hanks's solution at 37 °C, as compared to hydroxyapatite, this result was attributed to their intrinsic reactivity in physiological media, leading to dissolution and apatite layer formation (Shen et al., 2015; F. Zhang et al., 2016). In contrast, hydroxyapatite-based coatings generally exhibit a more stable electrochemical response, reflecting their enhanced barrier properties and lower interfacial reactivity (Sankar et al., 2017). This behaviour is further supported by laser cladding techniques, where SiC particle-reinforced hydroxyapatite-silver (HA-Ag) gradient coatings deposited on titanium femoral stems for artificial hip joints demonstrate enhanced corrosion resistance due to microstructural improvement and improved mechanical integrity (Deng et al., 2023). The incorporation of SiC particles increases fracture toughness through crack deflection, thus limiting defect propagation and electrolyte penetration. Electrochemical analyses showed increased corrosion potential, reduced corrosion current density, and higher impedance values, while the EIS revealed the formation of two capacitive arcs: i) the low-frequency capacitive arc is associated with ion diffusion and electrochemical reactions at the bioactive coating surface, driven by concentration gradients in the SBF (Zaludin et al., 2019; J. M. Zhang et al., 1998); ii) the high-frequency capacitive arc corresponds to surface ion adsorption phenomena and is indicative of the formation and integrity of the passive film at the coating/electrolyte interface (Jamesh et al., 2011; Kazemi et al., 2020). These findings highlight that, while bioactive glass coatings exhibit dynamic electrochemical behaviour, reinforced hydroxyapatite-based systems tend to provide more stable barrier properties.

While bioactive glass and hydroxyapatite coatings provide excellent bioactivity, their performance can be further enhanced through compositional hybridization with polymers, which improves coating stability, controls ion exchange, and enables multifunctional responses. For this purpose, MAPLE-deposited PMMA-bioactive glass (PMMA-BG61) nanocomposite coatings on 316L stainless steel exhibit homogeneous microstructures and improved electrochemical stability in SBF (37 °C), as demonstrated by EIS and potentiodynamic polarization measurements (Floroian et al., 2015). Initially, the coating exhibits a capacitive response, indicating a stable protective film. However, after 1 day of immersion, ion exchange with the electrolyte leads to interfacial reactions, reflected by decreased charge transfer resistance and increased capacitance. At longer immersion times (3-21 days), two-time constants appear, corresponding to simultaneous bioactive glass dissolution and ion adsorption processes. After 35 days, the system showed a single time constant, associated with the formation of a protective hydroxyl carbonate apatite layer. Consequently, the PMMA-BG61/316L coating ensures both bioactivity and effective corrosion protection of the metallic substrate (Floroian et al., 2015).

Moreover, further functionalization can be achieved through the incorporation of therapeutic agents, resulting in multifunctional coatings. Among these, MAPLE-derived composite coatings deposited on 316 L stainless steel substrates exhibited superior corrosion resistance, as reported for BG-PMMA-Doxy systems (Floroian et al., 2016). In addition to their electrochemical performance, these coatings benefit from the use of doxycycline, which act as broad-spectrum bacteriostatic agents by inhibiting protein synthesis in microorganisms (Gerits et al., 2016). The integration of antibiotics within bioactive glass-polymer matrices enables localized drug delivery, effectively preventing bacterial adhesion and biofilm formation on implant surfaces, avoiding implant failure (Eraković et al., 2014; Francolini & Donelli, 2010; Z. Liu et al., 2016; Patenge et al., 2012). As a result, PMMA matrix limits electrolyte penetration while the bioactive glass promotes interfacial reactivity and apatite formation (Floroian et al., 2012). Such multifunctional coatings are relevant for metallic implants, where corrosion-induced ion release (e.g., Fe, Cr, Ni, Mo from stainless steel) may lead to cytotoxic effects (Constantinescu et al., 2024; Matusiewicz, 2014).

Similar behaviour has been reported for BGT/PMMA/SS and BGB/PMMA/SS coatings deposited by MAPLE, which exhibit stable impedance responses and the progressive formation of protective layers during immersion in SBF (Floroian et al., 2017). The improved electrochemical performance is attributed to PMMA matrix, which acts as a barrier limiting electrolyte penetration, and the bioactive glass phase, which promotes apatite formation. EIS analyses reveal an initial two-time constant behaviour associated with simultaneous bioactive glass dissolution and ion adsorption. This process was followed by single capacitive response as the system becomes dominated by the protective polymer layer and the newly formed apatite layer. In addition, the incorporation of plant-derived antimicrobial agents enables controlled drug release. However, EIS analyses (in SBF, 37 °C) reported for MAPLE deposited BGN/PMMA/SS coatings indicate that surface roughness and nanoscale porosity lead to non-ideal capacitive behaviour, highlighting the influence of microstructural heterogeneity on charge transfer processes (Negut et al., 2020).

Rather than polymer-induced barrier effects, improved corrosion resistance can also be achieved through coating layering and apatite formation. In this context, multilayer coatings provide a clear design advantage over single-layer bioactive glass (BG) systems: the bottom layer can be optimized for adhesion, corrosion protection, and mechanical stability, while the top layer can be tailored for bioactivity and controlled antimicrobial or drug release. Following this approach, *Negut et al.* reported a multifunctional multilayer coating consisting of a ZnO base layer deposited by PLD and a lovastatin-functionalized bioactive glass (BG57+LOV) top layer deposited by MAPLE was developed to enhance the performance of 316L stainless steel implants (Negut et al., 2026). EIS measurements performed in SBF revealed a progressive improvement in corrosion resistance, with the multilayer coating displaying a more positive corrosion potential. This behaviour can be attributed to the synergistic contribution of the individual layers. The compact ZnO base layer acts as a barrier, limiting electrolyte ingress and stabilizing the metal-electrolyte interface, while the MAPLE-deposited BG57+0.1LOV top layer provides additional protection and bioactivity. Its moderately rough microstructure enhances surface coverage and promotes the formation of a stable surface layer in SBF.

However, PLD-derived coatings typically exhibit more dense and homogeneous microstructures, resulting in improved barrier properties and reduced ion release (Dhinasekaran et al., 2021). Laser cladding approaches further highlight the importance of microstructural refinement, as HA-based composite coatings (e.g., HA/TiO_2 and HA/ZrO_2) demonstrate significantly enhanced corrosion resistance, reflected by reduced corrosion current density and increased polarization resistance compared to pure HA coatings and the substrate (Deng et al., 2023; Kumari et al., 2024). These improvements are mainly attributed to crack-free microstructures, reduced porosity, and the incorporation of bioinert ceramic phases. Additionally, laser ablation-derived TiC coatings exhibit stable passive behaviour, further confirming the versatility of laser-based processing techniques in enhancing electrochemical performance (Mohazzab et al., 2020). Overall, these studies indicate that the electrochemical performance of laser-processed coatings is influenced by coating compactness, porosity, and surface morphology. However, porosity plays a dual role in determining coating performance: i) controlled micro- and nanoscale porosity is beneficial, as it enhances surface area and promotes osseointegration; ii) excessive or poorly distributed porosity may act as preferential pathways for electrolyte penetration, accelerating charge transfer processes and compromising corrosion resistance. Beyond porosity, surface morphology and the spatial distribution of micro- and nanoscale features also play a decisive role in controlling electrochemical behavior.

This correlation is further supported by studies on chitosan coatings deposited by PLD on NiTi alloys, where SEM investigations revealed uniform and compact surface morphologies, while AFM analyses indicated low surface roughness and a homogeneous topography, suggesting reduced porosity and restricted electrolyte ingress (Mareci et al., 2012). Based on literature data, laser-deposited coatings can exhibit even more complex porous microstructures and pronounced nanoscale topographical features. For example, borate and silicate bioactive glass coatings produced by nanosecond PLD displayed nanoscale roughness and a porous morphology, as evidenced by SEM and AFM analyses (Rau et al., 2020). Such features indicate an increased surface area and higher degree of microstructural heterogeneity, which may facilitate electrolyte penetration but also promote bioactivity and interfacial reactions.

Further evidence of the influence of surface morphology and microstructural heterogeneity on electrochemical behavior is provided by composite and multifunctional coatings obtained by PLD. For instance, hydroxyapatite-zirconia coatings deposited on biodegradable FeMnSi alloys exhibited a relatively homogeneous yet composite microstructure, where the incorporation of the ceramic phase reduced defect density and improved barrier properties, resulting in enhanced corrosion resistance (Cimpoeşu et al., 2016). Similarly, recent studies have shown that coatings with interconnected micro- and nanoscale features exhibit increased effective surface area and enhanced interfacial interactions, which significantly influence electrochemical performance, as confirmed by SEM and AFM analyses (Prodana et al., 2024).

The correlation between morphology and electrochemical behaviour is also highlighted by a recent report on a MAPLE-deposited bioactive glass embedded in a PMMA matrix and functionalized with the antibiotic ciprofloxacin (**Figure 7A**)

(Negut et al., 2022). SEM analysis confirms the formation of a continuous coating (~2 μm thickness) with localized microstructural heterogeneities, while AFM measurements reveal a textured surface with relatively high roughness (RMS= 0.21-0.28 μm), indicative of surface irregularities and localized porosity. The impedance spectra recorded over 36 hours of immersion in SBF indicate non-ideal capacitive behaviour, typically described by constant phase elements (CPE), attributed to the observed surface roughness and distributed porosity. Compared to the bare titanium substrate, the BG+CIPRO/PMMA coating exhibits significantly higher pore resistance values at the end of the test (R_{pore} of BG+CIPRO/PMMA (36 h) ~964 Ω cm^2, whereas R_{pore} for Ti (36 h) ~58 Ω cm^2), highlighting its improved barrier properties.

Figure 7. SEM, AFM, and electrochemical investigation of A: BG+CIPRO/PMMA/Ti, B: BBGi + PVP/Ti and C: BG57+VD3/Ti.

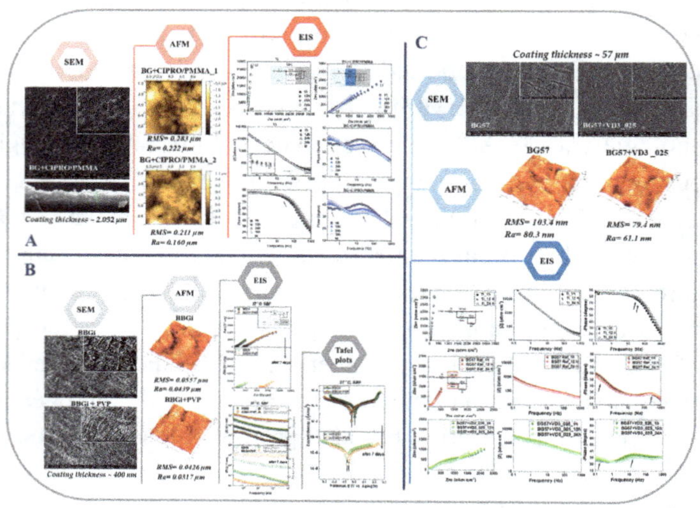

Images reprinted/adapted from ref. (Gradi teanu-Pircalabioru et al., 2025; Negut et al., 2022, 2023), licensed under CC-BY.

A similar morphology-electrochemistry correlation is observed for cerium-doped bioactive glass + polyvinylpyrrolidone (BBGi + PVP) composite coatings deposited on stainless steel by MAPLE (**Figure 7B**) (Gradi teanu-Pircalabioru et al., 2025). As reported, SEM analysis reveals a continuous morphology (~400 nm coating thickness) with moderate roughness (RMS ~ 40-50 nm), leading to non-ideal capacitive behaviour in EIS due to surface heterogeneity. Despite this, the coatings show enhanced corrosion resistance compared to the bare substrate, with higher polarization resistance (Rp ~ 51 kΩ initially, ~40 kΩ after 7 days) and lower corrosion current densities (icorr ~ 1.6-1.8 μAcm^{-2}). The addition of PVP improves electrochemical stability by delaying electrolyte penetration and stabilizing pore resistance over time, as reflected by higher low-frequency impedance values in Bode plots. Potentiodynamic measurements further confirm superior corrosion protection, with lower icorr and higher Rp compared to BBGi alone, and an increasing protection efficiency (from ~55% to ~82%) attributed to progressive pore blocking by degradation products. As observed, with increasing immersion time, a gradual reduction in the diameter of the depressed capacitive semicircles is observed, more pronounced for the BBGi + PVP coating. This evolution reflects ongoing interfacial processes and controlled degradation of the composite layer. Such behaviour can be advantageous for bone tissue engineering applications, as the progressive release of PVP may contribute to enhanced bone repair and regeneration, while supporting osseointegration (Matos et al., 2023; Yao et al., 2015). Therefore, the system enables the establishment of a degradation-time relationship, allowing the tuning of degradation rates to match the natural kinetics of bone tissue regeneration and to regulate the evolution of beneficial porosity at the coating-tissue interface (Kurakula & Koteswara Rao, 2020). In addition, PVP plays a functional role in corrosion protection. As a water-soluble and biocompatible polymer, it has been reported as an effective corrosion inhibitor, with its efficiency depending on parameters such as concentration and temperature (Al Juhaiman, 2016). Its protective behaviour is associated with adsorption at the metal/coating interface and the formation of a barrier layer that limits aggressive ion interaction. Similar effects have been observed for modified PVP systems, where high inhibition efficiencies were achieved at optimized concentrations (Nsakabwebwe et al., 2022). Owing to its biodegradability, aqueous solubility, and strong interfacial interactions, PVP contributes not only to controlled degradation but also to enhanced electrochemical stability, making it suitable for multifunctional biomedical coatings (Bejaoui et al., 2023).

In another recent report, bioactive glass matrices functionalized with therapeutic agents such as vitamin D3 (VD3), were also deposited by MAPLE (**Figure 7C**) (Negut et al., 2023). SEM analysis indicates the formation of a continuous and more homogeneous coating layer, with a thickness of ~57 μm, characterized by a finer and more uniformly distributed microstructure compared to the previous BBGi + PVP system. AFM investigations revealed reduced surface roughness with VD3 addition, with RMS values of ~ 79 nm in this case, as compared with bare BBGi where RMS was 103 nm. From electrochemical point of view, the incorporation of vitamin D3 further enhances the electrochemical performance, as reflected by increased impedance response and improved stability during immersion, which can be attributed to a more homogeneous coating structure and reduced electrolyte penetration pathways. According to *Di et al.* this result can be linked with the accumulation of corrosion products at the interface (Di et al., 2016).

Melittin-incorporated bioactive glass coatings deposited by MAPLE on titanium exhibit a structured morphology with increased roughness (RMS ~ 78.4 nm), enhancing surface area and interfacial activity (**Figure 8A**) (Dinu et al., 2025).

Figure 8. SEM, AFM, and electrochemical investigation of A: BG57+Mel/Ti; B: BG57+VD3+Mel (BAV)/Ti.

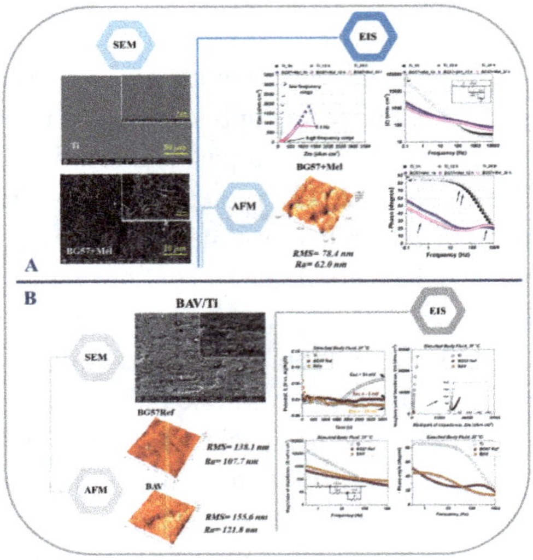

Images reprinted/adapted from ref. (Dinu et al., 2025; Negut et al., 2025), licensed under CC-BY.

EIS analysis reveals non-ideal capacitive behaviour due to surface heterogeneity and localized porosity. A significant decrease in impedance magnitude (~45%) after 12 h in SBF at 37 °C, compared to ~4% for bare titanium, indicates higher coating reactivity and dissolution, whereas titanium remains more stable due to passive oxide layer formation. This behaviour is consistent with previous studies reporting progressive impedance reduction in bioactive coatings as a result of dissolution and apatite layer formation in physiological environments (Zurita-Méndez et al., 2022). After prolonged immersion, this time-dependent response can be explained by a two-stage mechanism: (i) initial electrolyte penetration through inherent pores, facilitating mass transfer and interfacial reactions, followed by (ii) partial pore blocking by newly formed corrosion products or apatite layers, which slows further degradation and stabilizes the system (Chen et al., 2024).

The effect of melittin (providing antibacterial activity) and vitamin D3 (enhancing cell interaction and mineralization) represents a promising strategy to simultaneously address infection risks and promote osseointegration, as highlighted in *Advanced Materials Interfaces* (Negut et al., 2025). As observed (**Figure 8B**), the incorporation of VD3 and melittin led to increased coating compactness and improved barrier properties, resulting in an ~94% enhancement in protective efficiency. This improvement is associated with reduced effective porosity, enhanced interfacial stability, and the formation of a more homogeneous electrochemical interface.

Considering porosity, interfacial reactions, and protective layer formation, the role of laser processing allows a precise control over surface architecture and, consequently, the electrochemical and biological response of coatings. Laser-based strategies for biomedical surface engineering can be classified into material addition and surface modification techniques, including laser texturing (Shivakoti et al., 2021; Q. Zhang & Guan, 2023). While the former controls composition and coating integrity, the latter enables the precise tuning of surface morphology and porosity, which are critical for influencing electrochemical and biological interactions (Badiceanu et al., 2022; Matei & Visan, 2026; H. Wang et al., 2024). As a result, surface texturing can have both beneficial and detrimental effects on corrosion behaviour, depending on coating composition and application. This is because laser-induced texturing generates hierarchical surface features, including micro-cavities, pores, and periodic nanostructures, which directly influence electrolyte ingress and local electrochemical activity, as indicated by SEM analyses reported in recent studies (**Figure 9**) (Pakuła et al., 2019; Visan & Popescu-Pelin, 2024). These structures often consist micro-reservoirs, thereby modifying electrolyte transport pathways (Despres et al., 2021; Pakuła et al., 2019). At the nanoscale, the development of laser-induced periodic surface structures (LIPSS), typically with dimensions in the submicron range, further increases the effective surface area and influences local electrochemical activity (Pakuła et al., 2019a; Visan & Popescu-Pelin, 2024).

Although increased roughness and porosity may facilitate localized electrolyte ingress, controlled surface structuring can promote more uniform interfacial processes and improved stability (Despres et al., 2021; Schille et al., 2020). In addition, micro- and nano-reservoirs can trap reaction products and limit their diffusion depending on pore geometry, contributing to interfacial stabilization. Therefore, the electrochemical response of laser-processed systems should be interpreted not only in terms of coating compactness, but also as a function of multiscale surface architecture, which controls electrolyte ingress, interfacial reactions, and long-term degradation mechanisms.

Figure 9. **Representative electrical equivalent circuits (EECs) as a function of electrolyte ingress over immersion time. Image reprinted from ref. (Trentin et al., 2022) licensed under CC-BY. A: Surface textures made by ultrashort pulse laser processing under varied laser beam scanning procedures: (a) line pattern, (b) cross pattern and (c) alternating line and crosswise scanned laser beam; B: (a, b) SEM top view and (c, d) cross-sections of different laser patterning; C: Nanoripples laser-induced periodic surface structures (LIPSS) on surface cemented carbides (a) 20,000 x (zone 3), (b) 50,000 x (zone 2). Images reprinted/adapted from ref. (Despres et al., 2021; Pakuła et al., 2019; Schille et al., 2020), licensed under CC-BY.**

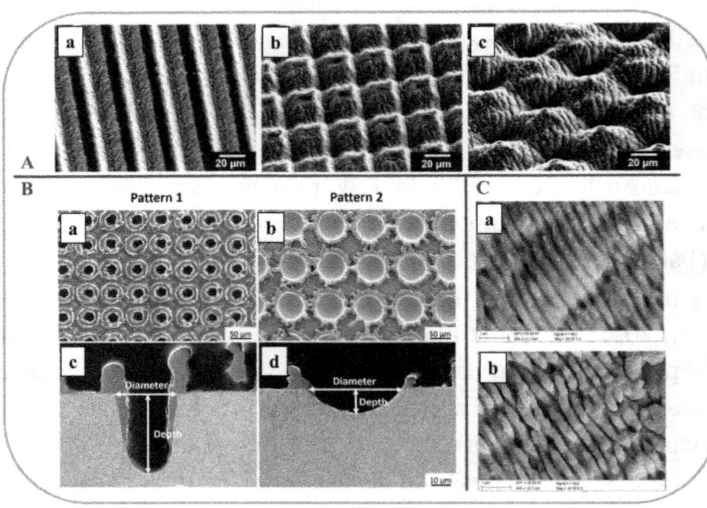

4. *IN VITRO* BIOACTIVITY ASSESSMENT OF LASER PROCESSED COATINGS

4.1. Translating Surface Engineering to Clinical Benefit: Why *in Vitro* Bioactivity and Degradation Tests are Essential

Rigorous evaluation is essential to establish the safety, efficacy, and clinical relevance of bioactive coatings. Comprehensive *in vitro* and *in vivo* assessments, supported by an extensive array of chemical, physical, and biological characterization methods, provide the fundamental evidence necessary to comprehensively understand performance across various scales, ranging from surface chemistry to tissue-level responses.

The development of bioactive coatings marks a critical milestone in biomaterials science, providing a novel approach to enhance implant performance and promote accelerated progress in the field of regenerative medicine. The transition from bioinert to bioactive materials, propelled by the innovation of bioactive glass, suggests a pivotal shift in design philosophy, transitioning from passive structural replacement to active involvement in physiological processes (Cao & Hench, 1996; Gristina, 1987; Rau, Antoniac, Cama, et al., 2016; Rau, Antoniac, Fosca, et al., 2016; Sergi et al., 2020). In this paradigm, materials are engineered not only to avoid harmful effects but also to elicit and direct beneficial biological responses (Marin et al., 2020; Webster et al., 2004; Williams, 2008; Williams & Williams, 2013).

Bioactive materials are engineered to facilitate advantageous interactions with biological tissues. In contrast to bioinert materials, which aim to minimize negative reactions, bioactive systems establish stable chemical bonds or strong interfacial interactions with host tissues. These materials are crucial in the fields of diagnostics, therapeutics, and regenerative medicine, where the regulation of cellular behaviours such as adhesion, proliferation, and differentiation influences tissue outcomes. This regulation is accomplished through compositional tuning, surface functionalization, or hybridization with inorganic phases, metals, or polymers, thereby promoting tissue repair and regeneration (Dinu et al., 2025; Gherasim, Grumezescu, Grumezescu, Andronescu, et al., 2021; Gherasim, Grumezescu, Grumezescu, Negut, et al., 2021; Grumezescu et al., 2022; Hench, 1991; Hench & Polak, 2002; Williams, 2022).

The bioactivity of materials can be modulated in accordance with the intended purpose, thereby influencing a spectrum of biological processes ranging from interfacial bonding to complex cellular mechanisms such as immunomodulation, angiogenesis, and osteoinduction. This adaptability enables coatings to fulfil clinical requirements by integrating with tissues and promoting healing. Modifications in composition, dopant chemistry, topography, surface charge, wettability, and release kinetics facilitate the regulation of biological responses, thereby aligning material performance with clinical objectives, including accelerated osseointegration, diminished infection risk, and reduced fibrotic encapsulation (Anderson et al., 2008; Hench & Polak, 2002; Hoppe et al., 2011).

Achieving clinical readiness necessitates a systematic evaluation strategy that integrates complementary *in vitro* and *in vivo* methodologies. *In vitro* assessments facilitate early examination of biocompatibility, cytotoxicity, hemocompatibility, inflammatory potential, and functional performance, supported by physicochemical and surface analyses. These assessments are complemented by measurements of wettability, surface energy, mechanical and adhesive properties, corrosion and dissolution behaviour, ion release, protein interactions, and cell-based biological responses (Di Siena et al., 2023; Gherasim, Grumezescu, Grumezescu, Negut, et al., 2021; Matei & Visan, 2026; Negut et al., 2023, 2025; Paital et al., 2009; Shabir et al., 2025; Z. Xu et al., 2025).

Established correlations between material composition, structure, and biological function enable iterative optimization of coating design and processing parameters prior to animal studies. Such methodologies facilitate rapid screening of laser-processed coatings, allowing refinement of both material formulations and processing conditions while ensuring mechanical robustness and durability. Subsequent *in vivo* evaluations contextualize material performance within living systems, examining inflammatory responses, degradation behaviour, tissue integration, vascularization, and structural remodelling through histological analysis, micro-CT, mechanical testing, and systemic toxicity assessments. Collectively, these preclinical datasets form a robust evidence base supporting clinical translation.

Within this framework, bioactive coatings exemplify a shift toward purposeful regulation of the host-material interface, reinforcing the need for standardized evaluation pathways that align physicochemical characterization with biologically relevant models to demonstrate safety and efficacy.

Figure 10 presents a workflow for *in vitro* evaluation. Biomineralization is assessed through immersion in simulated body fluid in accordance with *ISO 23317:2025 - Implants for surgery — Materials — Simulated body fluid (SBF) preparation procedure and test method to detect apatite formation in SBF for initial screening of bone-contacting implant materials*, employing either static or dynamic setups. Biodegradation is monitored in PBS, while biological performance is examined using osteogenic cell assays as per ISO 10993. Characterization involves quantifying mass evolution, surface morphology and topography (SEM, AFM), composition (EDS, XPS), crystalline phase (XRD, SAED), and chemical bonds (ATR-FTIR, Raman). Media analysis (ICP-OES/ICP-MS) is utilized to monitor ionic release and uptake, while micro-CT provides correlation with *ex-vivo* or *in vivo* observations, distinguishing surface-driven apatite nucleation from bulk dissolution.

Figure 10. Schematic workflow for in vitro assessment of bioactivity, biomineralization, and biodegradation of implant materials.

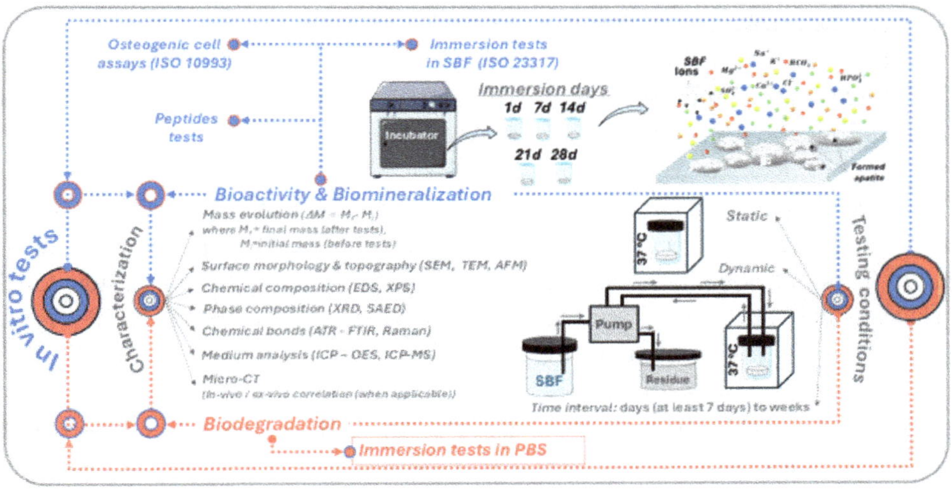

The protocol integrates standardized immersion tests in simulated body fluid (ISO 23317) and phosphate-buffered saline, osteogenic cell-based assays (ISO 10993), and multiscale physicochemical characterization to correlate surface reactivity, degradation behaviour, ionic exchange, and biological performance. Static and dynamic testing configurations enable discrimination between surface-mediated apatite formation and bulk dissolution processes.

4.2. *In Vitro* Evaluation Methods in Acelular Media of Laser-Processed Coatings

4.2.1. General Concepts

The terms bioactivity, biomineralization and biodegradation denote distinct yet interrelated phenomena within the fields of biomedical material science and biology, each possessing specific implications and functions in tissue engineering, implantology, and environmental processes.

Bioactivity refers to the capacity of a material to actively interact with living tissue when introduced into a physiological environment. Unlike inert materials, bioactive materials elicit a controlled biological response at the interface between the tissue and the material, which can result in stable bonding or a beneficial biological function. This interaction is usually characterized by the attachment of cells to the material's surface, followed by their proliferation and differentiation into specialized lineages. Concurrently, the surface chemistry of the material facilitates the formation of a biologically significant mineral coating (Kannan et al., 2025; Knabe, 2011; Kokubo & Takadama, 2006; Williams, 2022).

A common method to identify this behaviour is by observing the formation of an apatite-like layer after immersion in simulated body fluid, which replicates the initial stages of mineral formation observed *in vivo* (Cotrut et al., 2024; Dinu et al., 2025; Knabe, 2011; Kokubo & Takadama, 2006; Negut et al., 2023, 2025; Vardaki et al., 2018). When surface mineralization occurs in conjunction with favourable cellular responses, the material is more likely to integrate effectively with bone and contribute to tissue regeneration. Thus, bioactivity is best conceptualized as a functional performance at the interface - what the material accomplishes in a biological context - rather than merely its composition or structure. This is exemplified by bioactive calcium phosphate-based biomaterials that rapidly develop a uniform biomineral layer in simulated body fluid, demonstrating strong bone-bonding potential and good cell compatibility (Cotrut et al., 2024; Lei et al., 2013; Negut et al., 2025).

Biomineralization is a biologically regulated process through which living organisms produce mineralized structures, such as bones, teeth, and shells, via the controlled deposition of inorganic phases, such as calcium phosphate or calcium carbonate. This process typically occurs within an organic matrix that dictates mineral nucleation, growth, morphology, and hierarchical organization (Gajjeraman et al., 2007; Guo et al., 2025)

In biomedical contexts, biomineralization pertains to the formation of mineral layers on implant surfaces or within scaffolds, designed to mimic native tissue mineralization and thereby enhance integration with host tissues (Guo et al., 2025; Kim et al., 2004; Ma & Su, 2011). The precipitation of biologically relevant minerals, often directed by cell-derived organic components, constitutes a fundamental mechanism underpinning bone regeneration strategies, as it facilitates the formation of bone-like mineral phases essential for structural repair and functional restoration (Guo et al., 2025; Zhou et al., 2023). For example, calcium phosphate biomineralization on bioactive glasses or implant surfaces facilitates the formation of a bone-like apatite layer, enabling robust interfacial bone bonding and stimulating new bone formation (Ma & Su, 2011).

Biodegradation is the breakdown of materials through chemical and biological processes, often involving enzymes, leading to biocompatible byproducts. In biomedical implants and tissue-engineering scaffolds, it refers to materials being safely resorbed by the body without adverse effects (Easton et al., 2023; Su et al., 2016). Controlled biodegradation is crucial for temporary devices, where degradation should match tissue regeneration to provide support during healing and avoid long-term presence. This can be engineered in polymeric, inorganic, and composite systems by adjusting composition and structure. For instance, biodegradable inorganic biomaterials like nanostructured calcium phosphates dissolve into nontoxic ions that integrate into metabolic pathways, aiding tissue remodelling with minimal toxicity (Cai et al., 2020). Polymer blends and composites allow precise control of degradation to optimize tissue repair or reduce environmental impact (Easton et al., 2023).

In summary, biomineralization is a natural mineral formation process, biodegradation is the breakdown of materials, while bioactivity refers to a material's capacity to interact beneficially with biological systems, often through inducing biomineralization. While interconnected, since biodegradable bioactive implants often promote biomineralization during controlled degradation, these three terms emphasize distinct aspects of material–tissue interplay crucial for regenerative medicine and biomedical engineering. Thus, considering the points discussed earlier, it is important to highlight that the concepts of "biomineralization" and "bioactivity" are both currently utilized and acknowledged as measures to determine a biomaterial's potential to form an apatite layer on its surface.

4.2.2. Bioactivity and Biomineralization of Laser-Processed Coatings

Bioactivity/biomineralization tests carried out on laser-processed coatings, including pulsed laser deposition, matrix assisted pulsed laser evaporation and laser cladding, are designed to assess the ability of these surfaces to support and promote biological functions relevant to medical implants. Hydroxyapatite based coatings, in particular, demonstrate significant potential for enhancing biocompatibility and osteointegration of medical implants (Negut et al., 2023, 2025). It is widely recognized that these types of coatings are deposited on metallic implants, such as titanium alloys and stainless steel, through laser processing techniques to enhance osseointegration and impart beneficial biological properties.

Studies utilizing pulsed laser deposition have shown that HAp coatings deposited on stainless steel substrates exhibit favourable surface morphology, increased roughness, and strong adhesion, which facilitate osseointegration and biomineralization (Das & Shukla, 2016). Additionally, surface functionalization techniques, such as gold nanoparticle-mediated modifications promote apatite formation in simulated body fluids, indicating the potential for smart, active coatings that induce biomineralization (Vasile et al., 2012). Hydroxyapatite and dicalcium phosphate dihydrate based coatings deposited on magnesium-HAp composites have further improved corrosion resistance and biomineralization, crucial for biodegradable implants (Su et al., 2016). Nonetheless, the incorporation of dopants, bioglass or other elements, is a recognized strategy to further tailor bioactivity and mineralization potential in such coatings, aligning with ongoing research trends in biomaterials engineering

(Negut et al., 2023, 2025). Overall, these studies confirm that laser-processed HAp coatings, doped or undoped, exhibit promising biomineralization properties, supporting their application in orthopaedic or other types of implants.

Common bioactivity assays include immersion of coated samples in synthetic body fluids to observe the formation of bone-like apatite, including surface induced mineralization and potential for osteointegration. Additionally, *in vitro* cell culture assays are performed, typically using osteoblasts, pre-osteoblasts or stem cells to evaluate cell adhesion, proliferation, viability and differentiation on these coatings (De Micco et al., 2025).

Laser-processed coatings such as undoped and doped hydroxyapatite, bioglass composites, and calcium phosphate nanostructures have been subjected to biomineralization tests primarily via simulated body fluid immersion, demonstrating rapid apatite precipitation and favourable bone bioactivity.

Thus, in the studies carried out by *Dinu et al. and Negut et al.* (Dinu et al., 2025; Negut et al., 2023, 2025), multifunctional coatings consisting in a bioglass matrix (BG57; SiO_2–Na_2O–K_2O–CaO–MgO–P_2O_5) in which has been incorporated vitamin D_3 and/or melittin, a natural antimicrobial peptide derived from bee venom, were deposited on titanium by matrix-assisted pulsed laser evaporation (MAPLE) for implant applications that couple enhanced osseointegration with antibacterial protection and improved corrosion resistance. **Table 1** and **Figure 11** summarize some of the main results for the unmodified Bioglass BG57 based thin films and modified with vitamin D3 (Vit. D3) and/or melltin.

Building on the observation that contact angle (CA) varies only slightly (~50–56°) across coatings BG57 and BG57-based coatings in which it has been incorporated vitamin D_3 and melittin (denoted BAV) with markedly different roughness (Table 1, Figure 11), indicates that the BAV coatings can be framed as systems in which surface chemistry and nano-topography act to promote bioactivity. Specifically, the **Table 1** and **Figure 11.a–c** show that, relative to Ti (CA≈115.79°), all BG57-based coatings are moderately hydrophilic (BG57: CA≈50.69°; BG57+Vit.D_3: ≈54.65°; BG57+melttin: ≈55.82°; BAV: ≈53.89°) while spanning Ra from ~62 nm (Vit.D_3 or melttin alone) to ~121.8 nm for BAV.

Furthermore, the multifunctional BAV coating presents a micro/nanostructured surface (**Figure11.a**), moderate nanoscale roughness (Ra ≈ 108–122 nm; **Figure11.b**), and a distinctly hydrophilic wettability (CA ≈ 54° versus ≈ 116° for Ti; **Figure11.c**).

Taken together, these attributes are expected to promote rapid spreading of SBF medium on the coatings surface, enlarge the real contact area, and favour interfacial ion exchange (O, Ca, P), thereby reducing the kinetic barrier for heterogeneous nucleation and subsequent growth of carbonated apatite during immersion. Consistent with this mechanistic interpretation, the accompanying sub-figures substantiate the sequence of events: a near-linear, multi-week release of bioactive species (**Figure 11.d**), a time-dependent EDS enrichment in Ca and P with a concomitant rise in O (**Figure 11.e**), and the progressive formation of apatite-like agglomerates in the SEM images from day 7 to day 24 (**Figure 11.f**), collectively evidencing accelerated nucleation and growth despite the modest CA differences across the BG57 series.

Table 1. Surface parameters for different types of BioGlass (BG57) thin films deposited by MAPLE technique (Dinu et al., 2025; Negut et al., 2023, 2025)

Material	**Ti**	**BG57 (control)**	**BG57 +Vit. D_3**	**BG57 +melltin**	**BAV (BG57+Vit. D_3+melttin)**
Deposition method	-	MAPLE			
Roughness Ra (nm) RMS (nm)	-	*by AFM* 107.7 nm 138.1 nm	*by AFM* 62.1 nm 79.4 nm	*by AFM* 62.0 nm 78.4 nm	*by AFM* 121.8 nm 155.6 nm
Wettability CA (°) with SBF	115.79°	50.69°	54.65°	55.82°	53.89°
Bioactivity	Yes	Yes	-	Yes	Yes

Figure 11. **SEM images of BAV-coated Ti substrate before immersion (a). 3D AFM height map analysis and surface roughness parameter (b). Contact angle measurement for BioGlass 57 (BG57) and BAV coatings – conditions SBF, RT (22 ± 1 °C), relative humidity 42% (c). Seven-week release profile of melittin from BAV-coated Ti in SBF @ 37 °C (d). Evolution of BAV-coated surface morphology after immersion in SBF (7–24 days (e). Elemental composition (at%) of BAV coatings after SBF immersion (f)**

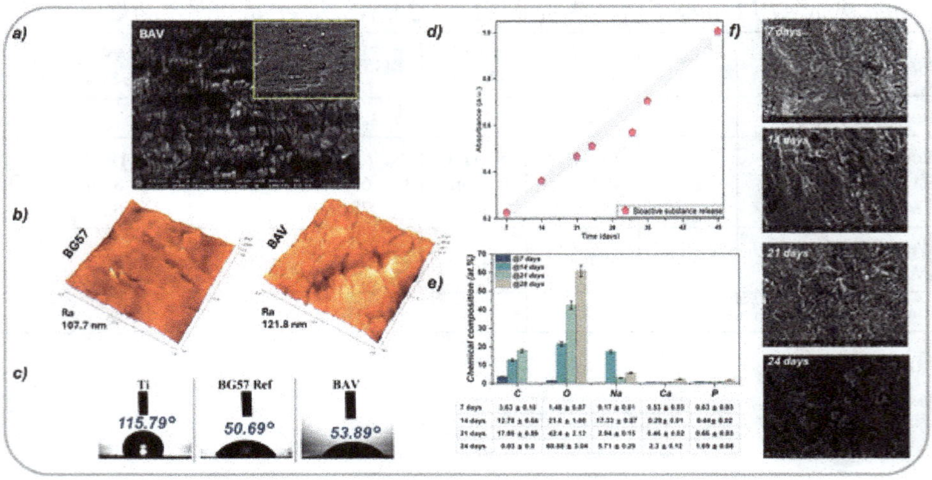

(repinted/adapted from (Dinu et al., 2025; Negut et al., 2025) - licensed under CC BY 4.0)

Thus, the coupling of bioactive mineral deposition with controlled melittin elution underscores the suitability of MAPLE-processed BAV coatings for long-term orthopaedic and dental implant performance. Overall, the MAPLE-synthesized BAV thin films offers a multifunctional platform for next-generation biomedical implants - combining enhanced cytocompatibility, osteogenic promotion, improved corrosion resistance, and durable infection control.

Other studies have shown that hydroxyapatite coatings fabricated by different laser-based deposition methods have exhibited significant biomineralization potential, assessed through immersion tests in SBF. **Table 2** summarizes some of the main findings.

Laser cladding using Nd:YAG lasers on Ti-6Al-4V substrates incorporating CaF_2 particles into hydroxyapatite has demonstrated rapid apatite precipitation on coating surfaces upon SBF immersion, indicating excellent bioactivity. The fluoridation enhanced the coating's thermostability and lowered solubility while maintaining biocompatibility (A. G. Huang et al., 2013). Surface roughness induced by laser processing creates a micro-textured topography that increases the effective surface area, facilitating protein adsorption and cell attachment, thereby enhancing osteoblast adhesion and subsequent tissue integration. The surface chemistry is modified during laser cladding by the formation of bioactive phases such as fluoridated hydroxyapatite and intermediate compounds like calcium titanate ($CaTiO_3$), which act as nucleation sites for apatite precipitation in physiological conditions. These chemical modifications also influence the surface charge and the reactivity of the coating, promoting faster and denser apatite layer formation during *in vitro* SBF immersion tests. In addition, wettability, measured by contact angle, is improved by the laser cladding process as it alters both surface roughness and chemistry. Enhanced wettability is associated with lower contact angles, which Favors the initial adsorption of proteins from body fluids and facilitates cellular responses including adhesion, proliferation, and differentiation. These effects collectively accelerate the formation of the biologically significant mineral layer that underpins successful osseointegration (Y. L. Yang et al., 2010). This biomimetic apatite layer formation is a key biomineralization indicator, showing the coating's capability to bond with natural bone through apatite nucleation. Laser cladding of fluoridated hydroxyapatite coatings on Ti-6Al-4V substrate results in a synergistic enhancement of surface roughness, chemical composition, and wettability. This creates a favourable microenvironment that facilitates rapid biomimetic apatite precipitation and strong biological interactions, which are crucial for the clinical success of implanted devices.

In a study conducted by Cheng and Ye, a low-temperature laser processing method, transmission laser coating, has been used to deposit HAp on Ti substrates while preserving the biocompatibility and enhancing coating-substrate interfacial strength (Cheng & Ye, 2009). Employing an Nd:YAG laser, this technique enables the development of multilayer coatings that integrate nanoscale hydroxyapatite (HAp) and titanium (Ti) nanoparticles. The nanoscale characteristics of HAp enhance its sinterability at low temperatures, thereby reducing thermal damage to the substrate and preserving the chemical composition and microstructure essential for biological performance. The primary advantages identified include a strong interfacial bonding strength between the coating and substrate, surpassing that of many traditional coatings, which is crucial for the durability of implants under physiological conditions. The coatings possess a porous microstructure with advantageous surface roughness, significantly contributing to cell attachment. Qualitative cell culture studies utilizing osteoblast-like cells (UMR-105) confirmed enhanced biocompatibility and cell proliferation on the coated surfaces. Additionally, immersion tests in SBF demonstrated the preservation of bioactivity through apatite precipitation, indirectly affirming efficient biomineralization behaviour. The cell culture and immersion tests in SBF showed that these coatings maintained good bioactivity and promoted cell attachment, indirectly confirming favourable biomineralization behaviour. This approach exemplifies the meticulous optimization of laser parameters and the incorporation of nanoparticle to develop coatings that integrate mechanical robustness with biological functionality. It highlights the potential of transmission laser coating to achieve stable, bioactive surface modifications that facilitate osteointegration, thereby positioning this technique as a promising method for the surface engineering of biomedical implants.

Pulsed laser deposition (PLD) of hydroxyapatite (HAp) coatings on medical-grade UNS S31254 stainless steel (254SS) substrates has been shown to enhance surface characteristics critical for orthopaedic implant applications (Das & Shukla, 2016). The deposited HAp coatings exhibited an increased average surface roughness of approximately 8 nm, as revealed by atomic force microscopy and scanning electron microscopy analyses. This nanoscale roughness is known to promote osseointegration by providing favourable topographical cues that enhance cell attachment and proliferation at the bone-implant interface. Post-deposition annealing emerged as a crucial step to improve the crystallinity and uniformity of the HAp coatings. Improved crystallinity directly impacts biomineralization by offering a more stable and bioactive apatite phase that closely mimics natural bone mineral, thus facilitating apatite nucleation and growth during implantation or *in vitro* simulated body fluid SBF immersion tests. The immersion tests conducted in SBF demonstrated rapid and dense apatite layer formation on the coating surfaces, confirming high bioactivity and indicating the coating's ability to bond effectively with host bone tissue. Furthermore, the mechanical characterization performed through tensile pull-out testing confirmed that the PLD HAp coatings form a strong bond with the 254SS substrate, indicating that these coatings provide durable bioactive surfaces capable of enduring physiological loading conditions. The observed increase in surface roughness alongside the preservation of phase purity indicates advantageous alterations in surface chemistry that are likely to enhance the biological interaction at the interface. Such improvements in surface characteristics, particularly at the nanoscale level, promote protein adsorption and subsequent cellular activities that are critical for effective and stable osseointegration. In summary, this study firmly correlates increased nanoscale surface roughness, improved crystallinity through annealing, and high adhesive strength with enhanced bioactivity of PLD hydroxyapatite coatings on 254SS. These combined surface and structural properties create a conducive microenvironment for biomineralization and stable bone-implant integration, underpinning the material's promise for orthopaedic implant applications

Laser ablation of crystalline hydroxyapatite (HAp) targets facilitates the generation of calcium phosphate aerosols, which in turn enable the deposition of porous nanostructured coatings on metallic substrates. These coatings, consisting of partially sintered microparticles embedded within nanoparticulate material, exhibit crystallinity and microstructural characteristics that promote the growth of biogenic calcium phosphate. The resultant porous nanostructures offer abundant nucleation sites and closely replicate the architecture of natural bone, thereby enhancing biomineralization. (Brown et al., 2004).

Correlating these structural characteristics to surface properties discussed in the related literature reveals a synergistic interplay between surface chemistry, nanoscale roughness, and wettability that enhances biological performance. Specifically, increased surface roughness at the nanoscale - resulting from the porous nanostructure - improves surface topography, which is well known to facilitate protein adsorption and cellular adhesion, crucial initial steps for tissue integration. Furthermore, the preservation of phase purity and crystallinity confirmed by X-ray diffraction supports a surface chemistry conducive to bioactivity by providing favorable chemical groups for calcium phosphate nucleation. These relationships align with findings reported in studies examining pulsed laser-deposited hydroxyapatite coatings, where average surface roughness values around 8 nm (Das & Shukla, 2016) were associated with improved bioactivity and osseointegration potential, largely through optimized surface energy and enhanced wettability promoting cell-material interactions. Thus, the porous nanostructured coatings produced by laser ablation not only structurally imitate bone but also generate a surface environment characterized by suitable roughness and chemical composition that collectively drive biomineralization and robust implant integration.

Thus, laser methods like Nd:YAG laser cladding (including fluoridated HAp), pulsed laser deposition, transmission laser coating, and laser aerosol synthesis have been successfully employed to fabricate bioactive coatings with strong interfacial bonding and osteoconductive properties. These coatings facilitate biomineralization crucial for their application as bone replacement structures or orthopaedic implants. Meanwhile, bioglass coatings on polymeric sutures also show robust biomineralization behavior, indicating broad potential for clinical applications (Boccaccini et al., 2003; Cheng & Ye, 2009; Das & Shukla, 2016; A. G. Huang et al., 2013). Within this context surface roughness, wettability and chemical composition are analysed in parallel since these parameters influence bioactivity. As it can be observed in **Table 2** some studies also combine antibacterial and mechanical testing with bioactivity to ensure multifunctionality.

Table 2. Summary of laser-assisted hydroxyapatite-based coatings on metallic substrates, highlighting processing methods, biomineralization outcomes, and associated mechanical/biological performance

Coating Type	Laser Processing Method	Biomineralization Findings	Mechanical / Biological Attributes	Ref.
Fluoridated Hydroxyapatite (FHA)	Nd:YAG Laser Cladding on Ti-6Al-4V	Rapid apatite formation on SBF immersion; uniform FHA crystallization; high bonding at coating-substrate interface; lower solubility and improved thermostability.	High bonding strength; comparable bioactivity and biocompatibility; improved acid resistance.	(A. G. Huang et al., 2013; Lin et al., 2017)
Hydroxyapatite (Undoped)	Transmission Laser Coating (TLC)	Maintained biocompatibility and bioactivity at low processing temperatures; pull-out tests showed enhanced coating-substrate interfacial strength; cell culture confirms good cell attachment post-SBF immersion.	Low-temperature processing preserves HAp phase; improved adhesion and biocompatibility.	(Cheng & Ye, 2009)
Pulsed Laser Deposited HAp on Stainless Steel 254SS	Pulsed Laser Deposition (PLD)	Formation of uniform bioactive apatite after SBF immersion; increased average surface roughness (~8 nm) facilitating osseointegration; annealing enhances crystallinity and phase quality.	High adhesion strength; antibacterial activity confirmed; improved mechanical robustness.	(Das & Shukla, 2016)
Cerium-Doped HAp/CNT Nanocomposites	Electrochemical deposition + Laser	Sustained ion release promotes continuous apatite nucleation in SBF; enhanced crystallinity and biomineralization potential compared to pure HAp; antibacterial effects against *E. coli* and *S. aureus*.	High hardness (354 HV) and adhesion (52 MPa); nanoscale grain refinement enhances mechanical properties and biofunction.	(Phogat et al., 2025)

In recent years, the impact of surface topography has become a particularly significant and enduring focus of research in both implant dentistry and orthopaedics. This is primarily because the nano -, micro- and macro-scale features of an implant surface can directly influence the biological processes that determine integration and long-term performance.

In clinical practice, various implant surface textures aim to enhance host response and stability. The tissue response to an implant is influenced by surface texture and morphology, affecting early cell interactions and tissue adaptation. Endosseous dental implants, available in different surface configurations, exploit bone's ability to interlock with surface irregularities, especially those around 1–100 μm. Modifying implant surface topography to optimize these micro-scale features can enhance stability and support predictable osseointegration and clinical success (Wennerberg & Albrektsson, 2009). Since the late 1980s, it was demonstrated that the rough implant surfaces stimulated the bone growth due to the high bone-implant interface (Buser et al., 1991; de Lange & Donath, 1989; Gotfredson et al., 1995; Puleo & Nanci, 1999).

In medical applications, a precise criterion for defining implant surfaces as either "smooth" or "rough" remains unestablished. According to the standards EN ISO 4287 and ISO 1071-3:2005, the surface roughness should not surpass 0.5 μm. Research assessing the roughness of dental and orthopaedic implants corroborates this standard, indicating that surfaces with average roughness values below 0.5 μm are associated with improved biological performance and mechanical stability (Fröjd et al., 2010; Ranjan et al., 2023).

Surface roughness, particularly at the micro- and nanoscale, significantly enhances biomineralization by increasing the available surface area and providing nucleation sites for apatite crystallization. Laser processing methods often modify the implant surfaces or coatings to induce controlled roughness that mimics natural bone architecture, as already mentioned and presented above. Studies indicate that increased surface roughness promotes hydroxyapatite deposition because mineral ions in SBF can more readily nucleate and grow on rough surfaces than smooth ones. For example, radio-frequency magnetron sputtered hydroxyapatite coatings on titanium show improved nanohardness, wear resistance, and enhanced wettability correlated with nano/microscale roughness variation produced by sandblasting and acid etching of the substrate before deposition (Grubova et al., 2016). Laser modification of Ti13Nb13Zr alloy increased surface roughness, which was found to improve the adhesion and homogeneity of subsequently deposited hydroxyapatite coatings, aiding apatite nucleation during biomineralization (Jażdżewska & Majkowska-Marzec, 2017). Similarly, transmission laser coating techniques on titanium substrates preserve or enhance roughness while maintaining good interfacial strength, enabling better apatite precipitation and cellular responses (Cheng & Ye, 2009).

Wettability, defined by the contact angle of biological fluids on the coating surface, governs the surface energy and thus protein adsorption, cell adhesion, and apatite nucleation. Hydrophilic surfaces (characterized by low contact angles) increase interaction with SBF and body fluids, allowing ions to easily penetrate surface pores and initiate apatite formation more readily than hydrophobic surfaces (Paital, He, et al., 2010; J. Wei et al., 2009). Hydroxyapatite-doped coatings generally exhibit higher surface energy and hydrophilicity compared to bare metallic implants, which are often hydrophobic. The enhanced wettability facilitates improved cellular adhesion and proliferation, which are crucial for the initial phase of osseointegration (Gul et al., 2025). Laser cladding of calcium phosphate (CaP) coatings with silicon dioxide (SiO_2) additives increases surface energy, reducing contact angles with SBF and promoting faster and more abundant hydroxyapatite precipitation in vitro compared to pure CaP coatings. The presence of SiO_2 also refines grain size and surface roughness, synergistically enhancing biomineralization (Dinu et al., 2013; Y. Yang et al., 2010; Y. L. Yang et al., 2010).

Thermal treatment of hydroxyapatite coatings modifies surface wettability through changes in crystallinity and roughness. Thus, annealed coatings demonstrate increased roughness and lower contact angles (higher wettability), correlating with superior corrosion resistance and bioactivity (Dinu et al., 2013). Conversely, composite polymer/hydroxyapatite coatings deposited on stainless steel showed increased hydrophobicity as hydroxyapatite content increased, suggesting that surface chemistry alongside roughness critically influences wettability and subsequent biomineralization (Dhafer et al., 2022).

The synergistic effects of surface roughness and wettability create an optimal environment that favours the rapid formation of bone-like apatite. Specifically, porous surfaces enhance wettability by allowing fluids to penetrate their surface features, thereby facilitating ion exchange at the interface. This enhanced wettability subsequently supports protein adsorption and cell spreading, which promotes the nucleation of mineralization sites. Consequently, empirical studies have demonstrated that coatings characterized by high surface roughness and increased hydrophilicity (i.e., low contact angles) result in more rapid and intense hydroxyapatite precipitation during SBF immersion, thereby indicating a superior biomineralization capacity (Gul et al., 2025; Y. L. Yang et al., 2010).

In biomineralization studies, quantitative measurement of the interplay between surface roughness and wettability combines precise characterization of surface topography (using AFM or profilometry) with wettability assessment based on contact angles. Modelling approaches such as the Wenzel and Cassie-Baxter models relate surface roughness measurements to wettability outcomes, while surface free energy calculations reveal the impact of surface chemistry. Advanced characterization methods such as NMR-based wettability indices (Kwak et al., 2022). extend applicability to complex porous structures. Integrating these methods with statistical analyses enables robust quantification of how laser processing parameters or material modifications drive the dual control of roughness and wettability, ultimately influencing biomineralization efficacy on implant coatings and related biomedical materials.

A study on laser micro-textured calcium phosphate (Ca-P) coatings on Ti-6Al-4V investigated two-line spacing patterns (100 μm and 200 μm) created by laser coating. The 100 μm spaced coatings exhibited higher surface energy and improved hydrophilicity, which correlated with enhanced apatite mineralization in SBF and better osteoblast adhesion and proliferation. This highlights the role of controlled laser texturing in tuning surface energy and wettability to promote biomineralization (Paital, Cao, et al., 2010).

Femtosecond laser processing has been demonstrated as a highly efficient and precise technique for fabricating microchannels and functional surfaces on polymers such as medical-grade polydimethylsiloxane (PDMS). This ultrafast laser processing enables fine manipulation at micro- and nanoscale levels, allowing controlled modification of surface roughness and topography. These structural changes directly influence surface wettability, facilitating spatially selective patterning of hydrophilic and hydrophobic regions with significant contact angle contrasts. The resultant surfaces not only exhibit tailored wettability but also improved bioactivity by promoting favourable interactions with biological fluids and cells, making this method highly promising for biomedical applications involving microfluidics and tissue engineering. This precision and versatility underscore the capability of femtosecond laser processing to engineer polymeric surfaces with multifunctional properties optimized for specific biomedical uses (Atanasov et al., 2015).

A comprehensive overview on laser-induced micro/nano functional surfaces on metals elaborates on techniques such as selective laser melting (SLM), laser surface melting, and femtosecond laser-induced periodic surface structures (LIPSS) on titanium alloys that create hierarchical surface roughness and chemically altered surfaces, has been presented by Arthanari et al. (Arthanari et al., 2021). These modifications significantly enhance bioactivity by promoting biomineralization and cell proliferation, improving antibacterial properties, and guiding cell attachment. Relatedly, laser texturing of metallic alloys (Co-based, Ti, NiTi) showed that laser parameters like scanning speed and pattern design drastically influence microstructure and wettability. Contact angle measurements and surface energy calculations combined with SEM and XPS analyses confirmed that laser-induced roughness and chemistry modifications synergistically tailor hydrophilicity/hydrophobicity for biomedical interfaces (Kasman & Ozan, 2025; C.-J. Yang et al., 2016; D. Zhang et al., 2019)

Collectively, these studies offer both quantitative and qualitative insights that establish a connection between laser processing parameters, the resulting surface roughness, and wettability, with biological outcomes such as biomineralization, cell adhesion, and antibacterial activity. They illustrate that multi-scale laser-induced topographies, when combined with controlled surface chemistry, represent a highly efficient toolset for optimizing bioactive implant surfaces.

One other important aspect which should be considered for the calcium phosphate-based coatings is the Ca/P variation which is influenced by *1)* coating properties: the chemical composition, crystallinity, the addition of other elements, and *2)* chemical composition of the testing medium. With respect to the aforementioned factors the newly precipitated apatite can be either enriched in calcium (Ca-enriched apatite, Ca/P ratio higher than 1.67 – results obtained in Dulbecco's Modified Eagle Medium medium (DMEM) and SBF) or deficient in calcium (Ca-deficient apatite, Ca/P ratio lower than 1.67 – results obtained in PBS). By using different types of medium a clearer overview of the biomaterials/coating behaviour can be pointed out (Drevet et al., 2011). Overall, testing calcium phosphate-based coatings in conditions closer to those found in vivo can provide valuable information regarding their integrity and the catalytic-reactive properties of their surfaces towards precipitation processes.

For bioactivity immersion tests, SBF and DMEM serve complementary but distinct purposes due to their differing compositions and effects on apatite formation and biological interactions. In **Table 3** are presented the chemical compositions of the most used simulated testing media.

SBF is an inorganic ion-rich solution designed to mimic human plasma (**Table 3**), making it well-suited for assessing the intrinsic bioactivity of materials through apatite nucleation and growth on surfaces. Studies consistently show rapid and uniform apatite precipitation within a few days of immersion in SBF, reflecting the material's ability to bond to bone-like mineral phases (Clèries et al., 2000; Kim et al., 2004; Kokubo et al., 1990; Kokubo & Takadama, 2006; Rohanová et al., 2014; Stuart et al., 2015; Theodorou et al., 2011). Because SBF doesn't present proteins, it isolates the mineralization process, providing a clear measure of surface bioactivity free from biological modulation.

In contrast, DMEM is a nutrient- and protein-containing cell culture medium that better approximates the biochemical environment cells experience in vivo. While less optimized for direct apatite precipitation, DMEM supports cellular viability, proliferation, and differentiation, enabling evaluation of biological responses to biomaterials. The inclusion of proteins in DMEM can modulate or even inhibit apatite formation compared to SBF, as protein adsorption alters surface chemistry and mineralization kinetics (Magyari et al., 2014). This characteristic makes DMEM particularly relevant for studies on how materials influence osteoblast behaviour, nodule formation, and bone development.

Table 3. Chemical composition of the most used simulated acellular media.(Faure et al., 2009; Kokubo & Takadama, 2006, 2008; Rohanová et al., 2014; Vladescu et al., 2019; Vranceanu et al., 2019)

Chemical composition		**Human Blood Plasma**	**Acellular Media**		
			SBF	**DMEM**	**PBS**
Ion Concentration (mM)	Na^+ K^+ Mg^{2+} Ca^{2+} Cl^- HCO_3^- HPO_4^- SO_4^{2-}	142.0 5.0 1.5 2.5 103.0 27.0 1.0 0.5	142.0 5.0 1.5 2.5 148.8 4.2 1.0 –	154.5 5.4 0.8 1.8 120.544.0 0.9 0.8	137 2.7 – – – – 10.0 –
Buffer			TRIS and 1 M HCl	No	$NaHCO_3$
pH		7.2–7.4	7.4	7.4	7.4

continued on following page

Table 3. Continued

Chemical composition		Human Blood Plasma	Acellular Media		
			SBF	**DMEM**	**PBS**
Amino Acids		N/A	No	0.084 g/L L-Arginine ·HCl 0.042 g/L L-Histidine · HCl · H_2O 0.105 g/L L-Isoleucine 0.105 g/L L-Leucine 0.146 g/L L-Lysine · HCl 0.03 g/L L-Methionine 0.066 g/L L-Phenylalanine 0.042 g/L L-Serine 0.095 g/L L-Threonine 0.016 g/L L-Tryptophan 0.10379 g/L L-Tyrosine, 2Na · $2H_2O$ 0.094 g/L L-Valine	No
Vitamins		N/A	No	0.004 g/L Choline Chloride 0.004 g/L Folic Acid 0.0072 g/L Myo-Inositol 0.004 g/L Niacinamide 0.004 g/L D-Pantothenic Acid · ½Ca 0.004 g/L Pyridoxal · HC 0.0004 g/L Riboflavin 0.004 g/L Thiamine · HCl	No
Other Components		N/A	No	1.0 g/L D-Glucose 0.0159 g/L Phenol Red · Na 0.11 g/L Pyruvic Acid · Na	No

According to Mandel and Tas, DMEM solution was found to have the ability of transforming brushite powders into octacalcium phosphate (OCP), which is a biological precursor for HAp (Mandel & Tas, 2010). It was also shown that the DMEM can be regarded as a feasible alternative to using SBF solutions for *in vitro* bioactivity testing of synthetic biomaterials for hydroxyapatite analysed at 36.5°C within 1 week (Mandel & Tas, 2010).

In DMEM, amino acids can influence HAp precipitation mainly by (i) weakly complexing Ca^{2+}, thereby changing the "free" calcium activity available for nucleation, and (ii) adsorbing onto surfaces or early calcium–phosphate clusters, which can promote or inhibit nucleation and redirect crystal growth, affecting mineral rate, morphology, and crystallinity. However, this is only a partial model of *in vivo* biomineralization, because in the human body, mineral formation is strongly governed by proteins (e.g., collagen and non-collagenous proteins), enzymes (e.g., alkaline phosphatase), cellular activity, and dynamic fluid flow, which are not replicated by DMEM alone.

Amino acids, especially polar and charged residues prevalent in non-collagenous proteins (NCPs) and within characteristic regions of collagen, play a critical regulatory role in hydroxyapatite (HAp) biomineralization by modulating ion availability and interfacial nucleation and growth processes (Tavafoghi & Cerruti, 2016). *In vitro* studies reviewed in *Tavafoghi and Cerruti* emphasize the importance of amino acid mobility in their functional roles: dissolved amino acids, particularly negatively charged ones like aspartic acid, glutamic acid, and phosphoserine, as well as positively charged residues such as arginine and lysine, tend to inhibit HAp precipitation through chelation of calcium and phosphate ions or by binding to nascent calcium phosphate nuclei, thus blocking further growth. In contrast, when immobilized on mineral or organic surfaces, these amino acids can enhance mineralization by locally concentrating calcium and phosphate ions, thereby increasing supersaturation and facilitating heterogeneous nucleation and crystal growth (Tavafoghi & Cerruti, 2016; Z. Wang et al., 2015).

However, the authors caution against directly equating *in vitro* free amino acid behavior to *in vivo* mineralization mechanisms since free amino acids are simpler models of complex protein domains within the extracellular matrix. The process of biomineralization is governed by a dynamic interaction among non-collagenous proteins (NCPs), cells, and matrix components. Physiological amino acid concentrations exhibit significant variation across different compartments, with notably higher levels present in the extracellular fluid of cartilage compared to blood plasma. Furthermore, dietary amino acid effects on bone mineral density predominantly act indirectly via anabolic signaling pathways and enzymatic processes rather than through free amino acids acting solely as mineral nucleators or inhibitors (Tavafoghi & Cerruti, 2016). This nuanced perspective highlights that while amino acids are fundamental regulators of HAp formation, their biological roles are embedded within a complex matrix milieu that modulates their functional impact during bone biomineralization. Thus, amino-acid effects observed in DMEM are best interpreted as simplified physicochemical contributions rather than a complete representation of physiological mineralization.

In light of these considerations, it is appropriate to assert that SBF is more suitable for evaluating the mineralization potential and inorganic bioactivity of biomaterials, while DMEM offers a milieu that is more physiologically relevant for assessing biological responses, including cell-material interactions and protein effects. Employing both media in complementary tests allows a comprehensive understanding of a biomaterial's performance, from apatite formation capability to cellular biocompatibility and osteogenic potential. Therefore, the optimal choice depends on the research focus: use SBF to measure apatite formation as a marker of bioactivity, and DMEM when investigating cell-mediated responses and protein interactions related to bone regeneration (Magyari et al., 2014; Negut et al., 2023).

4.2.3. Biodegradation of Laser-Processed Coatings

Biodegradation of laser-processed coatings in phosphate-buffered saline (PBS) is a critical factor for their biomedical performance, particularly in implant and tissue engineering contexts. PBS is a widely accepted *in vitro* medium mimicking physiological conditions (**Table 3**), providing a controlled environment to evaluate degradation kinetics and mechanisms of biomaterial coatings. Laser processing techniques, such as laser-assisted biomimetic coating during immersion in super-saturated calcium phosphate solutions, allow rapid and localized deposition of bioactive coatings that can influence subsequent degradation profiles and biological interactions (Oyane, 2019).

Compared with SBF and DMEM, PBS medium lacks in Ca^{2+} ions and other ions and does not allow precipitation of new apatite, favouring so the material degradation by releasing Ca^{2+} from the HAp based coatings, for example. Through these degradation mechanisms, the coating surface is negatively charged and at the same time the ionic exchange in the media occurs. Furthermore, the dissolution rate is determined by the intrinsic physicochemical properties of the material and by the media (Levard et al., 2012).

The coatings dissolution is controlled by their crystallinity, presence of secondary phases, porosity in the coating, and solution pH whereas their precipitation is dominated by ion concentration (Mohammadi et al., 2008; Vahabzadeh et al., 2015; Q. Zhang et al., 2003), and by the diffusion phenomenon caused by crystal defects such as impurities, vacancies, or dislocations (van Raemdonck et al., 1984; H. Wang et al., 2003). Some studies have indicated that HAp coatings with lower crystallinity tend to release more Ca^{2+} (Frayssinet et al., 1992; Maxian et al., 1993), indicating that the HAp coating can be slowly degraded *in vitro*. Nonetheless, it is known that a too fast degradation rate can lead to poor mechanical properties, structural failure and even chronic inflammatory reactions (W. Wang & Yeung, 2017).

One important aspect of biodegradation in PBS relates to the coating composition and structure. For instance, phosphate-based glass thin-film coatings deposited by radio-frequency magnetron sputtering exhibit rapid initial degradation manifested by pitting corrosion, then transition to slower linear degradation due to formation of an amorphous, adherent metaphosphate protective layer on the surface. This diffusion barrier effectively slows further dissolution, indicating that the initial surface structure and post-degradation surface transformations critically affect degradation behaviour in PBS (Stuart et al., 2015). Similarly, polycaprolactone (PCL) coatings deposited by plasma polymerization degrade gradually in PBS at body temperature, with degradation rate modulated by deposition parameters such as applied power and precursor flow rate, illustrating that laser-processed polymer coatings can be tailored to adjust biodegradability according to desired application timelines (Tynan et al., 2009).

When focusing on metallic implant coatings like HAp nanostructured layers deposited on magnesium alloys, laser or plasma-based deposition yields coatings that significantly moderate the rapid corrosion and degradation of magnesium substrates in PBS. These nanostructured HAp coatings exhibit improved resistance to biodegradation compared to uncoated magnesium, thereby prolonging implant stability in physiological fluids. The interaction between coating and substrate is crucial. Coatings act as physical barriers limiting exposure to PBS and reduce metallic ion release and hydrogen evolution, a byproduct of magnesium corrosion, thereby mitigating rapid biodegradation (Iskandar et al., 2011; Mukhametkaliyev et al., 2017; L. Wei & Gao, 2023). Moreover, surface modification techniques like sol-gel spin coating of calcium phosphate with fluorine doping can further enhance degradation resistance and promote biomineralization on magnesium alloys, emphasizing the importance of coating chemistry and microstructure in tuning biodegradation behavior in PBS (Su et al., 2018). Overall, the dissolution rate is determined by the intrinsic physicochemical properties of the material and by the media (Levard et al., 2012). Thus, materials with a fine-tune degradation rate need to be developed.

From a mechanistic standpoint, biodegradation in PBS often leads to changes in surface ultrastructure and crystallinity. For example, enzymatic degradation studies illustrate that amorphous polymer components tend to dissolve, increasing crystallinity and mechanical stiffness of polymer films during degradation. Surface lamellar structures can emerge, disappear, or fragment as degradation progresses. These remodelling influences subsequent material stability and biological performance (Zhuikov et al., 2017). In coatings formed by laser-assisted processes, controlled deposition allows immobilization of bioactive molecules, which can impact both biodegradation rate and cellular responses, demonstrating that laser processing provides a versatile approach to fabricate coatings with tailored degradation kinetics in PBS (Oyane, 2019).

In summary, biodegradation of laser-processed coatings in PBS is governed by coating composition, microstructure, and processing parameters. Polymer coatings such as plasma-polymerized PCL show degradation that can be tuned by deposition conditions. Phosphate-based glass coatings undergo complex surface chemistry changes that modulate degradation rates, while nanostructured calcium phosphate coatings on magnesium alloys effectively control rapid metal biodegradation while promoting cellular compatibility. Laser-assisted techniques enable precise control over coating properties, facilitating the design of coatings with optimal degradation profiles for biomedical applications where PBS serves as an important model physiological environment.

5. CONCLUSION

The characterization of laser-processed coatings cannot rely on a single analytical perspective. In multifunctional biomedical films, chemical preservation after laser transfer is only one part of the problem; the same coating must also develop a coherent microstructure and a sufficiently stable local mechanical response. Within this context, FTIR spectroscopy provides the molecular-level evidence of compositional integrity and interfacial evolution, whereas nanoindentation offers a route for translating these structural features into mechanically meaningful parameters. Their combined use therefore establishes a direct connection between molecular bonding, coating architecture, and expected in-service performance, forming an essential basis for the rational design of next-generation laser-fabricated biomedical surfaces.

Through electrochemical techniques, particularly polarization methods and electrochemical impedance spectroscopy (EIS), a complex evaluation of corrosion behaviour and degradation mechanisms is made. The electrochemical performance of laser-processed coatings is depending on microstructural features such as porosity, roughness, and coating morphology. While controlled micro- and nanoscale porosity can enhance bioactivity and interfacial interactions, excessive or poorly distributed porosity leads to electrolyte ingress and an enhanced corrosion process. Therefore, optimizing surface roughness is essential to balance corrosion resistance and biological functionality.

Laser-processed coatings for medical applications exhibit significantly enhanced bioactivity, biomineralization in simulated body fluid (SBF), and controlled biodegradation in phosphate-buffered saline (PBS), primarily driven by the fine-tuning of surface properties such as roughness, wettability, and surface chemistry. Laser techniques allow precise manipulation of these surface characteristics, which synergistically promote apatite nucleation and growth, optimize ion exchange, and regulate degradation kinetics essential for successful tissue integration and implant performance. The method of laser processing critically influences these biological behaviours by enabling controlled creation of micro- and nanoscale features that enhance hydrophilicity and surface energy while tailoring chemical composition. Consequently, optimizing laser parameters and process conditions facilitates the design of coatings with superior biofunctional responses, promoting rapid biomineralization and balanced biodegradation aligned with physiological environments. This fine control over laser deposition thus represents an effective strategy to achieve coatings with improved biological activity and durability suited to demanding medical implant applications.

ACKNOWLEDGEMENT

The work performed by INOE team was supported by the Core Program within the National Research Development and Innovation Plan 2022–2027, carried out with the support of MCID, project no. PN 23 05 (id: PN11N-03-01-2023).

REFERENCES

Al-Amin, K., Kawsar, M., Mamun, M., & Sahadat Hossain, M. (2025). Fourier transform infrared spectroscopic technique for analysis of inorganic materials: A review. *Nanoscale Advances, 7*(21), 6677–6702. DOI: 10.1039/D5NA00522A PMID: 40949700

Al Juhaiman, L. A. (2016). Polyvinyl pyrrolidone as a Corrosion Inhibitor for Carbon Steel in HCl. *International Journal of Electrochemical Science, 11*(3), 2247–2262. DOI: 10.1016/S1452-3981(23)16098-6

Alvarez-Pampliega, A., Hauffman, T., Petrova, M., Breugelmans, T., Muselle, T., Van den Bergh, K., De Strycker, J., Terryn, H., & Hubin, A. (2014). Corrosion study on Al-rich metal-coated steel by odd random phase multisine electrochemical impedance spectroscopy. *Electrochimica Acta, 124*, 165–175. DOI: 10.1016/j.electacta.2013.09.159

Amirudin, A., & Thieny, D. (1995). Application of electrochemical impedance spectroscopy to study the degradation of polymer-coated metals. *Progress in Organic Coatings, 26*(1), 1–28. DOI: 10.1016/0300-9440(95)00581-1

Anderson, J. M., Rodriguez, A., & Chang, D. T. (2008). Foreign body reaction to biomaterials. In *Seminars in Immunology* (Vol. 20, Number 2, pp. 86–100). DOI: 10.1016/j.smim.2007.11.004

Anita, N., Joany, R. M., Dorothy, R., Aslam, J., Rajendran, S., Subramania, A., Singh, G., & Verma, C. (2023). Linear polarization resistance (LPR) technique for corrosion measurements. In *Electrochemical and Analytical Techniques for Sustainable Corrosion Monitoring* (pp. 59–80). Elsevier. DOI: 10.1016/B978-0-443-15783-7.00005-0

Arthanari, S., Zhang, J., Xue, X., Li, Y., & Guan, Y. (2021). *Laser Induced Micro/Nano Functional Surfaces on Metals for Biomedical Applications* (pp. 439–464). DOI: 10.1007/978-981-33-6252-9_15

Arulvel, S., Dsilva Winfred Rufuss, D., Akshat, J., Kandasamy, J., & Singhal, M. (2023). Laser processing techniques for surface property enhancement: Focus on material advancement. *Surfaces and Interfaces*, *42*, 103293. DOI: 10.1016/j.surfin.2023.103293

Atanasov, P. A., Stankova, N. E., Nedyalkov, N. N., Fukata, N., Hirsch, D., Rauschenbach, B., Amoruso, S., Wang, X., Kolev, K. N., Valova, E. I., Georgieva, J. S., & Armyanov, St. A. (2015). Fs-laser processing of medical grade polydimethylsiloxane (PDMS). *Applied Surface Science*, *374*, 229–234. Advance online publication. DOI: 10.1016/j.apsusc.2015.11.175

Badiceanu, M., Anghel, S., Mihailescu, N., Visan, A. I., Mihailescu, C. N., & Mihailescu, I. N. (2022). Coatings Functionalization via Laser versus Other Deposition Techniques for Medical Applications: A Comparative Review. *Coatings*, *12*(1), 71. DOI: 10.3390/coatings12010071

Bejaoui, M., Galai, H., Touati, F., & Kouass, S. (2023). Multifunctional Roles of PVP as a Versatile Biomaterial in Solid State. In *Dosage Forms - Innovation and Future Perspectives*. IntechOpen., DOI: 10.5772/intechopen.99431

Berradja, A. (2019). Electrochemical Techniques for Corrosion and Tribocorrosion Monitoring: Methods for the Assessment of Corrosion Rates. In *Corrosion Inhibitors*. IntechOpen. DOI: 10.5772/intechopen.86743

Bierwagen, G., Tallman, D., Li, J., He, L., & Jeffcoate, C. (2003). EIS studies of coated metals in accelerated exposure. *Progress in Organic Coatings*, *46*(2), 149–158. DOI: 10.1016/S0300-9440(02)00222-9

Bingtong, L., Yongliang, Z., & Liping, S. (2020). Identification and characterization of the peptides with calcium-binding capacity from tilapia (Oreochromis niloticus) skin gelatin enzymatic hydrolysates. *Journal of Food Science*, *85*(1), 114–122. DOI: 10.1111/1750-3841.14975 PMID: 31869867

Boccaccini, A. R., Stamboulis, A. G., Rashid, A., & Roether, J. A. (2003). Composite surgical sutures with bioactive glass coating. *Journal of Biomedical Materials Research. Part B, Applied Biomaterials*, *67*(1), 618–626. Advance online publication. DOI: 10.1002/jbm.b.10047 PMID: 14528459

Boi, M., Marchiori, G., Berni, M., Gambardella, A., Salamanna, F., Visani, A., Bianchi, M., Fini, M., & Filardo, G. (2019). Nanoindentation: An advanced procedure to investigate osteochondral engineered tissues. *Journal of the Mechanical Behavior of Biomedical Materials*, *96*, 79–87. DOI: 10.1016/j.jmbbm.2019.04.042 PMID: 31029997

Bredar, A. R. C., Chown, A. L., Burton, A. R., & Farnum, B. H. (2020). Electrochemical Impedance Spectroscopy of Metal Oxide Electrodes for Energy Applications. *ACS Applied Energy Materials*, *3*(1), 66–98. DOI: 10.1021/acsaem.9b01965

Breugelmans, T., Tourwé, E., Van Ingelgem, Y., Wielant, J., Hauffman, T., Hausbrand, R., Pintelon, R., & Hubin, A. (2010). Odd random phase multisine EIS as a detection method for the onset of corrosion of coated steel. *Electrochemistry Communications*, *12*(1), 2–5. DOI: 10.1016/j.elecom.2009.10.008

Brown, S., Kim, H., & Camata, R. P. (2004). A Hybrid Laser/Aerosol Method for the Synthesis of Porous Nanostructured Calcium Phosphate Materials for Bone Tissue Engineering Applications. *MRS Proceedings, 845*. DOI: 10.1557/PROC-845-AA5.35

Brunello, G., Elsayed, H., & Biasetto, L. (2019). Bioactive Glass and Silicate-Based Ceramic Coatings on Metallic Implants: Open Challenge or Outdated Topic? *Materials (Basel)*, *12*(18), 2929. DOI: 10.3390/ma12182929 PMID: 31510062

Buser, D., Schenk, R. K., Steinemann, S., Fiorellini, J. P., Fox, C. H., & Stich, H. (1991). Influence of surface characteristics on bone integration of titanium implants. A histomorphometric study in miniature pigs. *Journal of Biomedical Materials Research*, *25*(7), 889–902. Advance online publication. DOI: 10.1002/jbm.820250708 PMID: 1918105

Cai, A., Zhu, Y., & Qi, C. (2020). Biodegradable Inorganic Nanostructured Biomaterials for Drug Delivery. *Advanced Materials Interfaces*, *7*(20), 2000819. Advance online publication. DOI: 10.1002/admi.202000819

Calado, L. M., Taryba, M. G., Carmezim, M. J., & Montemor, M. F. (2018). Self-healing ceria-modified coating for corrosion protection of AZ31 magnesium alloy. *Corrosion Science*, *142*, 12–21. DOI: 10.1016/j.corsci.2018.06.013

Cao, W., & Hench, L. L. (1996). Bioactive materials. *Ceramics International*, *22*(6), 493–507. DOI: 10.1016/0272-8842(95)00126-3

Caruso, S., Saffioti, M. R., Del Prete, A., & Rotella, G. (2025). Laser surface treatment of Ti6Al4V alloy: Finite element analysis for predicting molten pool geometry and microstructure modifications. *International Journal of Advanced Manufacturing Technology*, *137*(7–8), 3967–3976. DOI: 10.1007/s00170-025-15403-3

Chen, H., Shi, Q., & Zheng, K. (2024). Mathematical modeling of bioactive glass degradation. *Journal of Non-Crystalline Solids*, *646*, 123265. DOI: 10.1016/j.jnoncrysol.2024.123265

Cheng, G. J., & Ye, C. (2009). Experiment, thermal simulation, and characterizations on transmission laser coating of hydroxyapatite on metal implant. *Journal of Biomedical Materials Research. Part A*, *92A*(1), 70–79. Advance online publication. DOI: 10.1002/jbm.a.32325 PMID: 19165793

Cimpoeşu, N., Trincă, L. C., Dascălu, G., Stanciu, S., Gurlui, S. O., & Mareci, D. (2016). Electrochemical Characterization of a New Biodegradable FeMnSi Alloy Coated with Hydroxyapatite-Zirconia by PLD Technique. *Journal of Chemistry*, *2016*, 1–9. DOI: 10.1155/2016/9520972

Clèries, L., Fernández-Pradas, J. M., & Morenza, J. L. (2000). Behavior in simulated body fluid of calcium phosphate coatings obtained by laser ablation. *Biomaterials*, *21*(18), 1861–1865. DOI: 10.1016/S0142-9612(00)00060-0 PMID: 10919689

Codescu, M. M., Vladescu, A., Geanta, V., Voiculescu, I., Pana, I., Dinu, M., Kiss, A. E., Braic, V., Patroi, D., Marinescu, V. E., & Iordoc, M. (2022). Zn based hydroxyapatite based coatings deposited on a novel FeMoTaTiZr high entropy alloy used for bone implants. *Surfaces and Interfaces*, *28*, 101591. DOI: 10.1016/j.surfin.2021.101591

Constantinescu, S., Gherasim, O., Dorcioman, G., Grumezescu, V., Iosub, G., Niculescu, A.-G., Moldoveanu, E.-T., Rădulescu, D. M., & Grumezescu, A. M. (2024). MAPLE-prepared graphene oxide-based coatings for improved orthopedic screws used in knee interventions. *Romanian Journal of Morphology and Embryology = Revue Roumaine de Morphologie et Embryologie*, *65*(3), 433–442. DOI: 10.47162/RJME.65.3.05 PMID: 39529336

Córdoba-Torres, P. (2017). Relationship between constant-phase element (CPE) parameters and physical properties of films with a distributed resistivity. *Electrochimica Acta*, *225*, 592–604. DOI: 10.1016/j.electacta.2016.12.087

Cotrut, C. M., Blidisel, A., & Vranceanu, D. M., Vladescu (Dragomir), A., Ungureanu, E., Pana, I., Dinu, M., Vitelaru, C., Parau, A. C., Pruna, V., Magurean, M. S., & Titorencu, I. (. (2024). Evaluation of the In Vitro Behavior of Electrochemically Deposited Plate-like Crystal Hydroxyapatite Coatings. *Biomimetics*, *9*(11), 704. DOI: 10.3390/biomimetics9110704 PMID: 39590276

da Silva, M. J., Alves, W., Graeff, C. F. de O., & D'Alpino, P. H. P. (2021). Modified Synthesis and Physicochemical Characterization of a Bioglass-Based Composite for Guided Bone Regeneration. *TheScientificWorldJournal*, *2021*, 1–9. DOI: 10.1155/2021/4295433 PMID: 34899084

Das, A., & Shukla, M. (2016). Surface morphology, bioactivity, and antibacterial studies of pulsed laser deposited hydroxyapatite coatings on Stainless Steel 254 for orthopedic implant applications. *Proceedings of the Institution of Mechanical Engineers, Part L: Journal of Materials: Design and Applications, 233*. DOI: 10.1177/1464420716663029

de Lange, G. L., & Donath, K. (1989). Interface between bone tissue and implants of solid hydroxyapatite or hydroxyapatite-coated titanium implants. *Biomaterials, 10*(2), 121–125. Advance online publication. DOI: 10.1016/0142-9612(89)90044-6 PMID: 2706299

De Micco, S., Bellucci, D., & Cannillo, V. (2025). The Assessment of Bioactivity and Biological Responsiveness in Bioactive Glasses and Ceramics: A Review of Available Techniques. *Materials (Basel), 18*(18), 4393. DOI: 10.3390/ma18184393 PMID: 41010235

Deng, Z., Liu, D., Liu, G., Xiong, Y., Xin, S., Li, S., Li, C., & Chen, T. (2023). Study on the corrosion resistance of SiC particle-reinforced hydroxyapatite-silver gradient bioactive ceramic coatings prepared by laser cladding. *Surface and Coatings Technology, 467*, 129734. DOI: 10.1016/j.surfcoat.2023.129734

Despres, L., Costil, S., Cormier, J., Villechaise, P., & Cariou, R. (2021). Impact of Laser Texturing on Ni-Based Single Crystal Superalloys. *Metals, 11*(11), 1737. DOI: 10.3390/met11111737

Dhafer, G., Al-Shroofy, M. N., & Al-Kaisy, H. A. (2022). Electrostatic Deposition of Bio-Composite Polymer/Hydroxyapatite Coatings on 316L Stainless Steel. *Key Engineering Materials, 911*, 40–45. Advance online publication. DOI: 10.4028/p-40btz7

Dhinasekaran, D., Kaliaraj, G. S., Jagannathan, M., Rajendran, A. R., Prakasarao, A., Ganesan, S., & Subramanian, B. (2021). Pulsed laser deposition of nanostructured bioactive glass and hydroxyapatite coatings: Microstructural and electrochemical characterization. *Materials Science and Engineering C, 130*(April), 112459. DOI: 10.1016/j.msec.2021.112459 PMID: 34702534

Di, H., Yu, Z., Ma, Y., Zhang, C., Li, F., Lv, L., Pan, Y., Shi, H., & He, Y. (2016). Corrosion-resistant hybrid coatings based on graphene oxide–zirconia dioxide/epoxy system. *Journal of the Taiwan Institute of Chemical Engineers, 67*, 511–520. DOI: 10.1016/j.jtice.2016.08.008

Di Siena, M., Genna, S., Moretti, P., Ponticelli, G. S., Venettacci, S., & Russo, P. (2023). Study of the laser-material interaction for innovative hybrid structures: Thermo-mechanical characterization of polyethylene-based polymers. *Polymer Testing*, *120*, 107947. DOI: 10.1016/j.polymertesting.2023.107947

Dinu, M., Bita, B., Parau, A. C., Ristoscu, C., & Negut, I. (2025). Bioactive Glass and Melittin Thin Films Deposited by MAPLE for Titanium Implant Functionalization. *Materials (Basel)*, *18*(10), 2410. DOI: 10.3390/ma18102410 PMID: 40429147

Dinu, M., Cotrut, C. M., Petreus, T., Hauffman, T., & Baciu, F., & Vladescu (Dragomir), A. (. (2024). Structural, electrochemical, biological, and mechanical assessment of functionally graded Cr-based multilayers for enhanced metal-ceramic bond strength in dental restorations. *Materials Chemistry and Physics*, *312*, 128655. DOI: 10.1016/j.matchemphys.2023.128655

Dinu, M., Hauffman, T., Cordioli, C., Vladescu, A., Braic, M., Hubin, A., & Cotrut, C. M. (2017). Protective performance of Zr and Cr based silico-oxynitrides used for dental applications by means of potentiodynamic polarization and odd random phase multisine electrochemical impedance spectroscopy. *Corrosion Science*, *115*, 118–128. DOI: 10.1016/j.corsci.2016.11.018

Dinu, M., Kiss, A. E., Parau, A. C., Braic, V., Vitelaru, C., Braic, M., Pana, I., Balaceanu, M., & Vladescu, A. (2013). Influence of Thermal Treatment on the Roughness, Corrosion Resistance and Wettability of Hydroxyapatite Films Deposited by RF Magnetron Sputtering. *Key Engineering Materials*, *587*, 297–302. Advance online publication. . DOI: 10.4028/www.scientific.net/KEM.587.297

Dinu, M., Pana, I., Scripca, P., Sandu, I. G., Vitelaru, C., & Vladescu, A. (2020). Improvement of CoCr Alloy Characteristics by Ti-Based Carbonitride Coatings Used in Orthopedic Applications. *Coatings*, *10*(5), 495. DOI: 10.3390/coatings10050495

Drevet, R., Velard, F., Potiron, S., Laurent-Maquin, D., & Benhayoune, H. (2011). In vitro dissolution and corrosion study of calcium phosphate coatings elaborated by pulsed electrodeposition current on Ti6Al4V substrate. *Journal of Materials Science. Materials in Medicine*, *22*(4), 753–761. Advance online publication. DOI: 10.1007/s10856-011-4251-5 PMID: 21290169

Drobný, P., Mercier, D., Koula, V., Škrobáková, S. I., Čaplovič, Ľ., & Sahul, M. (2021). Evaluation of Adhesion Properties of Hard Coatings by Means of Indentation and Acoustic Emission. *Coatings*, *11*(8), 919. DOI: 10.3390/coatings11080919

Easton, Z. H. W., Essink, M. A. J., Rodriguez Comas, L., Wurm, F. R., & Gojzewski, H. (2023). Acceleration of Biodegradation Using Polymer Blends and Composites. *Macromolecular Chemistry and Physics*, *224*(6), 2200421. Advance online publication. DOI: 10.1002/macp.202200421

Eraković, S., Janković, A., Ristoscu, C., Duta, L., Serban, N., Visan, A., Mihailescu, I. N., Stan, G. E., Socol, M., Iordache, O., Dumitrescu, I., Luculescu, C. R., Janaćković, Dj., & Mišković-Stanković, V. (2014). Antifungal activity of Ag:hydroxyapatite thin films synthesized by pulsed laser deposition on Ti and Ti modified by TiO2 nanotubes substrates. *Applied Surface Science*, *293*, 37–45. DOI: 10.1016/j.apsusc.2013.12.029

Esmailzadeh, S., Aliofkhazraei, M., & Sarlak, H. (2018). Interpretation of Cyclic Potentiodynamic Polarization Test Results for Study of Corrosion Behavior of Metals: A Review. *Protection of Metals and Physical Chemistry of Surfaces*, *54*(5), 976–989. DOI: 10.1134/S207020511805026X

Farooq, R. S., Yaqoob, K., Liaqat, U., Ahmed, R., & Baig, M. A. (2026). Pulsed laser deposition of hydroxyapatite coatings on titanium substrates with varying surface topographies. *Optics & Laser Technology*, *195*, 114557. DOI: 10.1016/j.optlastec.2025.114557

Faure, J., Balamurugan, A., Benhayoune, H., Torres, P., Balossier, G., & Ferreira, J. M. F. (2009). Morphological and chemical characterisation of biomimetic bone like apatite formation on alkali treated Ti6Al4V titanium alloy. *Materials Science and Engineering C*, *29*(4), 1252–1257. DOI: 10.1016/j.msec.2008.09.047

Flores, J. F., Olaya, J. J., Colás, R., Rodil, S. E., Valdez, B. S., & Fuente, I. G. (2006). Corrosion behaviour of TaN thin PVD films on steels. *Corrosion Engineering, Science and Technology*, *41*(2), 168–176. DOI: 10.1179/174327806X107941

Floroian, L., Florescu, M., Sima, F., Popescu-Pelin, G., Ristoscu, C., & Mihailescu, I. N. (2012). Synthesis of biomaterial thin films by pulsed laser technologies: Electrochemical evaluation of bioactive glass-based nanocomposite coatings for biomedical applications. *Materials Science and Engineering C*, *32*(5), 1152–1157. DOI: 10.1016/j.msec.2012.03.001

Floroian, L., Ristoscu, C., Candiani, G., Pastori, N., Moscatelli, M., Mihailescu, N., Negut, I., Badea, M., Gilca, M., Chiesa, R., & Mihailescu, I. N. (2017). Antimicrobial thin films based on ayurvedic plants extracts embedded in a bioactive glass matrix. *Applied Surface Science*, *417*, 224–233. DOI: 10.1016/j.apsusc.2017.02.197

Floroian, L., Ristoscu, C., Mihailescu, N., Negut, I., Badea, M., Ursutiu, D., Chifiriuc, M., Urzica, I., Dyia, H., Bleotu, C., & Mihailescu, I. (2016). Functionalized Antimicrobial Composite Thin Films Printing for Stainless Steel Implant Coatings. *Molecules (Basel, Switzerland), 21*(6), 740. DOI: 10.3390/molecules21060740 PMID: 27294895

Floroian, L., Samoila, C., Badea, M., Munteanu, D., Ristoscu, C., Sima, F., Negut, I., Chifiriuc, M. C., & Mihailescu, I. N. (2015). Stainless steel surface biofunctionalization with PMMA-bioglass coatings: Compositional, electrochemical corrosion studies and microbiological assay. *Journal of Materials Science. Materials in Medicine, 26*(6), 195. DOI: 10.1007/s10856-015-5527-y PMID: 26085116

Francolini, I., & Donelli, G. (2010). Prevention and control of biofilm-based medical-device-related infections. *FEMS Immunology and Medical Microbiology, 59*(3), 227–238. DOI: 10.1111/j.1574-695X.2010.00665.x PMID: 20412300

Frayssinet, P., Tourenne, F., Rouquet, N., Giammara, B., & Bonel, G. (1992). Biological influence of crystallinity of hydroxyapatite coatings for orthopaedic use on cell growth. In B. C (Ed.), *Thermal Spray: International Advances in Coatings Technology Conference Proceedings* (pp. 477–486). ASM International, Material Park, Ohio.

Fröjd, V., Wennerberg, A., & Franke Stenport, V. (2010). Importance of Ca2+ Modifications for Osseointegration of Smooth and Moderately Rough Anodized Titanium Implants – A Removal Torque and Histological Evaluation in Rabbit. *Clinical Implant Dentistry and Related Research, 14*(5), 737–745. Advance online publication. DOI: 10.1111/j.1708-8208.2010.00315.x PMID: 20977616

Gajjeraman, S., Narayanan, K., Hao, J., Qin, C., & George, A. (2007). Matrix Macromolecules in Hard Tissues Control the Nucleation and Hierarchical Assembly of Hydroxyapatite. *The Journal of Biological Chemistry, 282*(2), 1193–1204. Advance online publication. DOI: 10.1074/jbc.M604732200 PMID: 17052984

Gerits, E., Kucharíková, S., Van Dijck, P., Erdtmann, M., Krona, A., Lövenklev, M., Fröhlich, M., Dovgan, B., Impellizzeri, F., Braem, A., Vleugels, J., Robijns, S. C. A., Steenackers, H. P., Vanderleyden, J., De Brucker, K., Thevissen, K., Cammue, B. P. A., Fauvart, M., Verstraeten, N., & Michiels, J. (2016). Antibacterial activity of a new broad-spectrum antibiotic covalently bound to titanium surfaces. *Journal of Orthopaedic Research, 34*(12), 2191–2198. DOI: 10.1002/jor.23238 PMID: 27003909

Gherasim, O., Grumezescu, A. M., Grumezescu, V., Andronescu, E., Negut, I., Bîrcă, A. C., Gălă eanu, B., & Hudi ă, A. (2021). Bioactive Coatings Loaded with Osteogenic Protein for Metallic Implants. *Polymers, 13*(24), 4303. DOI: 10.3390/polym13244303 PMID: 34960852

Gherasim, O., Grumezescu, A. M., Grumezescu, V., Negut, I., Dumitrescu, M. F., Stan, M. S., Nica, I. C., Holban, A. M., Socol, G., & Andronescu, E. (2021). Bioactive Coatings Based on Hydroxyapatite, Kanamycin, and Growth Factor for Biofilm Modulation. *Antibiotics (Basel, Switzerland), 10*(2), 160. DOI: 10.3390/antibiotics10020160 PMID: 33562515

Gonzalez-Garcia, Y., Garcia, S. J., & Mol, J. M. C. (2016). *Electrochemical Techniques for the Study of Self Healing Coatings.*, DOI: 10.1007/978-94-017-7540-3_9

Gotfredson, K., Wennerberg, A., Johansson, C., Skovgaard, L. T., & Hjørting-Hansen, E. (1995). Anchorage of TiO2-blasted, HA-coated, and machined implants: An experimental study with rabbits. *Journal of Biomedical Materials Research, 29*(10), 1223–1231. Advance online publication. DOI: 10.1002/jbm.820291009 PMID: 8557724

Gradi teanu-Pircalabioru, G., Negut, I., Dinu, M., Parau, A. C., Bita, B., Duta, L., Ristoscu, C., & Sava, B. (2025). Enhancing orthopaedic implant efficacy: The development of cerium-doped bioactive glass and polyvinylpyrrolidone composite coatings via MAPLE technique. *Biomedical Materials (Bristol, England), 20*(1), 015019. DOI: 10.1088/1748-605X/ad98d5 PMID: 39612575

Gristina, A. G. (1987). Biomaterial-Centered Infection: Microbial Adhesion Versus Tissue Integration. *Science, 237*(4822), 1588–1595. DOI: 10.1126/science.3629258 PMID: 3629258

Grubova, I., Priamushko, T., Chudinova, E., Surmeneva, M., Korneva, O., Epple, M., Prymak, O., Shulepov, I., & Surmenev, R. (2016). Formation and Characterization of Crystalline Hydroxyapatite Coating with the (002) Texture. *IOP Conference Series. Materials Science and Engineering, 116*, 012016. Advance online publication. DOI: 10.1088/1757-899X/116/1/012016

Grumezescu, V., Grumezescu, A., Ficai, A., Negut, I., Vasile, B., Gălă eanu, B., & Hudi ă, A. (2022). Composite Coatings for Osteoblast Growth Attachment Fabricated by Matrix-Assisted Pulsed Laser Evaporation. *Polymers, 14*(14), 2934. DOI: 10.3390/polym14142934 PMID: 35890714

Grundmeier, G., Schmidt, W., & Stratmann, M. (2000). Corrosion protection by organic coatings: Electrochemical mechanism and novel methods of investigation. *Electrochimica Acta, 45*(15–16), 2515–2533. DOI: 10.1016/S0013-4686(00)00348-0

Grunenwald, A., Keyser, C., Sautereau, A. M., Crubézy, E., Ludes, B., & Drouet, C. (2014). Revisiting carbonate quantification in apatite (bio)minerals: A validated FTIR methodology. *Journal of Archaeological Science*, *49*, 134–141. DOI: 10.1016/j.jas.2014.05.004

Gul, I. A., Abdul-Rani, A. M., Ahmad, A., Al-Amin, M., Aliyu, A. A., & Garba, E. (2025). Surface Wettability Response of Hydroxyapatite-Doped Coatings on Metallic Biomaterials: A Concise Review. *Diffusion and Defect Data, Solid State Data. Part B, Solid State Phenomena*, *370*, 61–72. Advance online publication. DOI: 10.4028/p-c4DEVS

Guo, J., Gao, F., Zhang, F., Zhao, X., & Zhao, R. (2025). Biomineralization-Driven Advances in Materials Science and Biomedical Engineering. *JACS Au*, *5*(9), 4134–4154. DOI: 10.1021/jacsau.5c00669 PMID: 41001633

Habelitz, S., Marshall, G. W.Jr, Balooch, M., & Marshall, S. J. (2002). Nanoindentation and storage of teeth. *Journal of Biomechanics*, *35*(7), 995–998. DOI: 10.1016/S0021-9290(02)00039-8 PMID: 12052404

Haque, F. (2003). Application of Nanoindentation Development of Biomedical to Materials. *Surface Engineering*, *19*(4), 255–268. DOI: 10.1179/026708403322499173

Hassan, H. H., Abdelghani, E., & Amin, M. A. (2007). Inhibition of mild steel corrosion in hydrochloric acid solution by triazole derivatives. *Electrochimica Acta*, *52*(22), 6359–6366. DOI: 10.1016/j.electacta.2007.04.046

Hench, L. L. (1991). Bioceramics: From Concept to Clinic. *Journal of the American Ceramic Society*, *74*(7), 1487–1510. DOI: 10.1111/j.1151-2916.1991.tb07132.x

Hench, L. L., & Polak, J. M. (2002). Third-Generation Biomedical Materials. *Science*, *295*(5557), 1014–1017. DOI: 10.1126/science.1067404 PMID: 11834817

Herrera Hernández, H., & Ruiz Reynoso, M. A., C. Trinidad González, J., O. González Morán, C., G. Miranda Hernández, J., Mandujano Ruiz, A., Morales Hernández, J., & Orozco Cruz, R. (2020). Electrochemical Impedance Spectroscopy (EIS): A Review Study of Basic Aspects of the Corrosion Mechanism Applied to Steels. In *Electrochemical Impedance Spectroscopy*. IntechOpen. DOI: 10.5772/intechopen.94470

Hirschorn, B., Orazem, M. E., Tribollet, B., Vivier, V., Frateur, I., & Musiani, M. (2010). Determination of effective capacitance and film thickness from constant-phase-element parameters. *Electrochimica Acta*, *55*(21), 6218–6227. DOI: 10.1016/j.electacta.2009.10.065

Hoppe, A., Güldal, N. S., & Boccaccini, A. R. (2011). A review of the biological response to ionic dissolution products from bioactive glasses and glass-ceramics. In *Biomaterials* (Vol. 32, Number 11, pp. 2757–2774). DOI: 10.1016/j.biomaterials.2011.01.004

Hoque, M. A., Yao, C.-W., Khanal, M., & Lian, I. (2022). Tribocorrosion Behavior of Micro/Nanoscale Surface Coatings. *Sensors (Basel)*, *22*(24), 9974. DOI: 10.3390/s22249974 PMID: 36560343

Huang, A. G., Liu, Y., Hu, S. H., Wang, Q. H., & Zhen, L. L. (2013). Laser Cladding of Fluoridated Hydroxyapatite Coatings on Titanium Alloy for Bone Replacement Structures. *Advanced Materials Research*, *815*, 540–546. Advance online publication. . DOI: 10.4028/www.scientific.net/AMR.815.540

Huang, V. M., Wu, S.-L., Orazem, M. E., Pébère, N., Tribollet, B., & Vivier, V. (2011). Local electrochemical impedance spectroscopy: A review and some recent developments. *Electrochimica Acta*. Advance online publication. DOI: 10.1016/j.electacta.2011.03.018

Iskandar, M. E., Lock, J. Y., Aslani, A., & Liu, H. (2011). Controlling the Biodegradation of Magnesium Implants Through Nanostructured Coatings. *Volume 2: Biomedical and Biotechnology Engineering; Nanoengineering for Medicine and Biology*. DOI: 10.1115/IMECE2011-65901

Jadhav, N., & Gelling, V. J. (2019). Review—The Use of Localized Electrochemical Techniques for Corrosion Studies. *Journal of the Electrochemical Society*, *166*(11), C3461–C3476. DOI: 10.1149/2.0541911jes

Jamesh, M., Kumar, S., & Sankara Narayanan, T. S. N. (2011). Corrosion behavior of commercially pure Mg and ZM21 Mg alloy in Ringer's solution – Long term evaluation by EIS. *Corrosion Science*, *53*(2), 645–654. DOI: 10.1016/j.corsci.2010.10.011

Jażdżewska, M., & Majkowska-Marzec, B. (2017). Hydroxyapatite deposition on the laser modified Ti13Nb13Zr alloy. *Advances in Materials Science*, *17*(4), 5–13. Advance online publication. DOI: 10.1515/adms-2017-0017

Jażdżewska, M., Majkowska-Marzec, B., Zieliński, A., Ostrowski, R., Frączek, A., Karwowska, G., & Olive, J.-M. (2023). Mechanical Properties and Wear Susceptibility Determined by Nanoindentation Technique of Ti13Nb13Zr Titanium Alloy after "Direct Laser Writing.". *Materials (Basel)*, *16*(13), 4834. DOI: 10.3390/ma16134834 PMID: 37445148

Kannan, S., Sivashanmugam, P., Shivalingam, C., & Thayumanavan, P. (2025). Tailored bioactive glass coating on titanium implants to elevate bioactivity and longevity. *Journal of Sol-Gel Science and Technology*, *115*(1), 63–72. DOI: 10.1007/s10971-025-06780-y

Kartsonakis, I. A., Athanasopoulou, E., Snihirova, D., Martins, B., Koklioti, M. A., Montemor, M. F., Kordas, G., & Charitidis, C. A. (2014). Multifunctional epoxy coatings combining a mixture of traps and inhibitor loaded nanocontainers for corrosion protection of AA2024-T3. *Corrosion Science*, *85*, 147–159. DOI: 10.1016/j.corsci.2014.04.009

Kasman, Ş., & Ozan, S. (2025). Marine Shell-Inspired Laser Surface Texturing: Characterizing the Surface Properties of Co-Based Alloy. *Journal of Bionic Engineering*, *23*(1), 341–353. Advance online publication. DOI: 10.1007/s42235-025-00712-w

Kazemi, M., Ahangarani, S., Esmailian, M., & Shanaghi, A. (2020). Investigation on the corrosion behavior and biocompatibility of Ti-6Al-4V implant coated with HA/TiN dual layer for medical applications. *Surface and Coatings Technology*, *397*, 126044. DOI: 10.1016/j.surfcoat.2020.126044

Kendig, M., & Scully, J. (1990). Basic Aspects of Electrochemical Impedance Application for the Life Prediction of Organic Coatings on Metals. *Corrosion*, *46*(1), 22–29. DOI: 10.5006/1.3585061

Kim, H.-M., Himeno, T., Kawashita, M., Kokubo, T., & Nakamura, T. (2004). The mechanism of biomineralization of bone-like apatite on synthetic hydroxyapatite: An in vitro assessment. *Journal of the Royal Society, Interface*, *1*(1), 17–22. DOI: 10.1098/rsif.2004.0003 PMID: 16849149

Knabe, C. (2011). Bioactivity : Mechanisms. In *Comprehensive Biomaterials*. Elsevier Ltd., DOI: 10.1016/B978-0-08-055294-1.00026-X

Kokubo, T., Kushitani, H., Sakka, S., Kitsugi, T., & Yamamuro, T. (1990). Solutions able to reproduce in vivo surface-structure changes in bioactive glass-ceramic A-W3. *Journal of Biomedical Materials Research*, *24*(6), 721–734. DOI: 10.1002/jbm.820240607 PMID: 2361964

Kokubo, T., & Takadama, H. (2006). How useful is SBF in predicting in vivo bone bioactivity? *Biomaterials*, *27*(15), 2907–2915. DOI: 10.1016/j.biomaterials.2006.01.017 PMID: 16448693

Kokubo, T., & Takadama, H. (2008). Simulated Body Fluid (SBF) as a Standard Tool to Test the Bioactivity of Implants. *Handbook of Biomineralization: Biological Aspects and Structure Formation, 3*, 97–109. DOI: 10.1002/9783527619443.ch51

Kumari, R., Kumar, S., Das, A. K., Murmu, A. M., & Kanchan, K. (2024). Microstructural characterization and corrosion analysis of HA/TiO2 and HA/ZrO2 composite coating on Ti- alloy by laser cladding. *Applied Surface Science Advances, 24*, 100655. DOI: 10.1016/j.apsadv.2024.100655

Kurakula, M., & Koteswara Rao, G. S. N. (2020). Moving polyvinyl pyrrolidone electrospun nanofibers and bioprinted scaffolds toward multidisciplinary biomedical applications. *European Polymer Journal, 136*, 109919. DOI: 10.1016/j.eurpolymj.2020.109919

Kurdi, A., Tabbakh, T., & Basak, A. K. (2023). Microstructural and Nanoindentation Investigation on the Laser Powder Bed Fusion Stainless Steel 316L. *Materials (Basel), 16*(17), 5933. DOI: 10.3390/ma16175933 PMID: 37687627

Kwak, H., Gao, J., An, Y., Kasule, M., Doyle, P., Kleinhammes, A., & Wu, Y. (2022, February 21). Connection Between Wettability and NMR-Detected Isotherms. *International Petroleum Technology Conference*. DOI: 10.2523/IPTC-22408-EA

Kylychbekov, S., Abdisatarov, B., Allamyradov, Y., Annamuradov, B., Duta, L., & Er, A. O. (2026). The mechanism of hydroxyapatite coatings degradation at high substrate temperatures. *Journal of Applied Physics, 139*(6), 065304. Advance online publication. DOI: 10.1063/5.0297660

Lei, B., Wang, L., Chen, X., & Chae, S.-K. (2013). Biomimetic and molecular level-based silicate bioactive glass-gelatin hybrid implants for loading-bearing bone fixation and repair. *Journal of Materials Chemistry. B, 1*(38), 5153. Advance online publication. DOI: 10.1039/c3tb20941e PMID: 32261107

Levard, C., Hotze, E. M., Lowry, G. V., & Brown, G. E. (2012). Environmental transformations of silver nanoparticles: Impact on stability and toxicity. In *Environmental Science and Technology*. DOI: 10.1021/es2037405

Li, X., Wang, M., Xu, L., Xu, T., Wu, W., Pan, S., Wang, C., Zhang, W., & Cai, X. (2024). Progress in indentation test for material characterization: A systematic review. *Results in Surfaces and Interfaces, 17*, 100358. DOI: 10.1016/j.rsurfi.2024.100358

Lin, J.-S., Tsai, T.-B., Say, W.-C., Chiu, C., & Chen, S.-H. (2017). In vitro study of electrodeposited fluoridated hydroxyapatite coating on G-II titanium with a nanostructured TiO2 interlayer. *Biomedical Materials (Bristol, England), 12*(2), 025018. Advance online publication. DOI: 10.1088/1748-605X/aa6264 PMID: 28374679

Liu, X., Xiong, J., Lv, Y., & Zuo, Y. (2009). Study on corrosion electrochemical behavior of several different coating systems by EIS. *Progress in Organic Coatings, 64*(4), 497–503. DOI: 10.1016/j.porgcoat.2008.08.012

Liu, Z., Ma, S., Duan, S., Xuliang, D., Sun, Y., Zhang, X., Xu, X., Guan, B., Wang, C., Hu, M., Qi, X., Zhang, X., & Gao, P. (2016). Modification of Titanium Substrates with Chimeric Peptides Comprising Antimicrobial and Titanium-Binding Motifs Connected by Linkers To Inhibit Biofilm Formation. *ACS Applied Materials & Interfaces*, *8*(8), 5124–5136. DOI: 10.1021/acsami.5b11949 PMID: 26863404

Lutz, A., van den Berg, O., Wielant, J., De Graeve, I., & Terryn, H. (2016). A Multiple-Action Self-Healing Coating. *Frontiers in Materials*, *2*. Advance online publication. DOI: 10.3389/fmats.2015.00073

Ma, M.-G., & Su, R.-C. (2011). Biomineralization and Biomimetic Synthesis of Biomineral and Nanomaterials. In *Advances in Biomimetics*. InTech., DOI: 10.5772/13838

Madelat, N., Wouters, B., Jiryaeisharahi, Z., Hubin, A., Terryn, H., & Hauffman, T. (2023). An ORP-EIS approach to distinguish the contribution of the buried interface to the electrochemical behaviour of coated aluminium. *Electrochimica Acta*, *455*, 142400. DOI: 10.1016/j.electacta.2023.142400

Magyari, K., Baia, L., Vulpoi, A., Simon, S., Popescu, O., & Simon, V. (2014). Bioactivity evolution of the surface functionalized bioactive glasses. *Journal of Biomedical Materials Research. Part B, Applied Biomaterials*, *103*(2), 261–272. Advance online publication. DOI: 10.1002/jbm.b.33203 PMID: 24820252

Mahmoud, Z. H., AL-Salman, H. N. K., & Kianfar, E. (2024). Nanoindentation: Introduction and applications of a non-destructive analysis. *Nano TransMed*, *3*, 100057. DOI: 10.1016/j.ntm.2024.100057

Majkowska-Marzec, B., Tęczar, P., Bartmański, M., Bartosewicz, B., & Jankiewicz, B. J. (2020). Mechanical and Corrosion Properties of Laser Surface-Treated Ti13Nb13Zr Alloy with MWCNTs Coatings. *Materials (Basel)*, *13*(18), 3991. DOI: 10.3390/ma13183991 PMID: 32916961

Mandel, S., & Tas, A. C. (2010). Brushite (CaHPO4 ·2H2O) to octacalcium phosphate (Ca 8 (HPO4) 2 (PO4)4·5H2O) transformation in DMEM solutions at 36.5 °C. *Materials Science and Engineering C*, *30*(2), 245–254. DOI: 10.1016/j.msec.2009.10.009 PMID: 30011614

Mareci, D., Cimpoesu, N., & Popa, M. I. (2012). Electrochemical and SEM characterization of NiTi alloy coated with chitosan by PLD technique. *Materials and Corrosion*, *63*(11), 985–991. DOI: 10.1002/maco.201206501

Marin, E., Boschetto, F., & Pezzotti, G. (2020). Biomaterials and biocompatibility: An historical overview. *Journal of Biomedical Materials Research. Part A*, *108*(8), 1617–1633. DOI: 10.1002/jbm.a.36930 PMID: 32196949

Matei, A. T., & Visan, A. I. (2026). Biocompatible Thin Films Deposited by Laser Techniques. *Materials (Basel)*, *19*(5), 925. DOI: 10.3390/ma19050925 PMID: 41828192

Matos, R. J. R., Silva, J. C., Soares, P. I. P., & Borges, J. P. (2023). Polyvinylpyrrolidone Nanofibers Incorporating Mesoporous Bioactive Glass for Bone Tissue Engineering. *Biomimetics*, *8*(2), 206. DOI: 10.3390/biomimetics8020206 PMID: 37218792

Matusiewicz, H. (2014). Potential release of in vivo trace metals from metallic medical implants in the human body: From ions to nanoparticles – A systematic analytical review. *Acta Biomaterialia*, *10*(6), 2379–2403. DOI: 10.1016/j.actbio.2014.02.027 PMID: 24565531

Maxian, S. H., Zawadsky, J. P., & Dunn, M. (1993). In vitro evaluation of amorphous calcium phosphate and poorly crystallized hydroxyapatite coatings on titanium implants. *Journal of Biomedical Materials Research*, *27*(1), 111–117. DOI: 10.1002/jbm.820270114 PMID: 8380594

Milošev, I., Kosec, T., & Strehblow, H.-H. (2008). XPS and EIS study of the passive film formed on orthopaedic Ti–6Al–7Nb alloy in Hank's physiological solution. *Electrochimica Acta*, *53*(9), 3547–3558. DOI: 10.1016/j.electacta.2007.12.041

Milošev, I., Metikoš-Huković, M., & Strehblow, H.-H. (2000). Passive film on orthopaedic TiAlV alloy formed in physiological solution investigated by X-ray photoelectron spectroscopy. *Biomaterials*, *21*(20), 2103–2113. DOI: 10.1016/S0142-9612(00)00145-9 PMID: 10966021

Mohammadi, Z., Ziaei-Moayyed, A. A., & Sheikh-Mehdi Mesgar, A. (2008). In vitro dissolution of plasma-sprayed hydroxyapatite coatings with different characteristics: Experimental study and modeling. *Biomedical Materials (Bristol, England)*, *3*(1), 015006. Advance online publication. DOI: 10.1088/1748-6041/3/1/015006 PMID: 18458493

Mohazzab, B. F., Jaleh, B., Fattah-alhosseini, A., Mahmoudi, F., & Momeni, A. (2020). Laser surface treatment of pure titanium: Microstructural analysis, wear properties, and corrosion behavior of titanium carbide coatings in Hank's physiological solution. *Surfaces and Interfaces*, *20*, 100597. DOI: 10.1016/j.surfin.2020.100597

Montesissa, M., Tommasini, V., Rubini, K., Boi, M., Baldini, N., & Boanini, E. (2025). State of Art and Perspective of Calcium Phosphate-Based Coatings Coupled with Bioactive Compounds for Orthopedic Applications. *Nanomaterials (Basel, Switzerland)*, *15*(15), 1199. DOI: 10.3390/nano15151199 PMID: 40801737

Mukhametkaliyev, T. M., Surmeneva, M. A., Vladescu, A., Cotrut, C. M., Braic, M., Dinu, M., Vranceanu, M. D., Pana, I., Mueller, M., & Surmenev, R. A. (2017). A biodegradable AZ91 magnesium alloy coated with a thin nanostructured hydroxyapatite for improving the corrosion resistance. *Materials Science and Engineering C, 75*, 95–103. DOI: 10.1016/j.msec.2017.02.033 PMID: 28415551

Narazaki, A. (2021). Characterization of Laser-Processed Samples. In *Handbook of Laser Micro- and Nano-Engineering* (pp. 1–38). Springer International Publishing., DOI: 10.1007/978-3-319-69537-2_64-1

Negut, I., Bita, B., Dinu, M., Parau, A. C., Tozar, T., Gherasim, O., Grumezescu, V., Ristoscu, C., Marutescu, L., & Gradisteanu Pircalabioru, G. (2025). Multifunctional Bioactive Coatings Based on Bioglass, Vitamin D3, and Melittin Deposited via MAPLE for Enhanced Osseointegration, Antibacterial Activity, and Corrosion Resistance of Titanium Implants. *Advanced Materials Interfaces, 12*(22), e00505. Advance online publication. DOI: 10.1002/admi.202500505

Negut, I., Bita, B., Dorcioman, G., Dinu, M., Parau, A. C., Ristoscu, C., & Gradisteanu-Pircalabioru, G. (2026). Laser-Engineered Multilayer Coatings Based on Zinc Oxide and Lovastatin-Functionalized Bioactive Glasses for Corrosion-Resistant and Antimicrobial Stainless Steel Implants. *Biomimetics, 11*(4), 227. DOI: 10.3390/biomimetics11040227 PMID: 42041457

Negut, I., Floroian, L., Ristoscu, C., Mihailescu, C. N., Mirza Rosca, J. C., Tozar, T., Badea, M., Grumezescu, V., Hapenciuc, C., & Mihailescu, I. N. (2020). Functional Bioglass—Biopolymer Double Nanostructure for Natural Antimicrobial Drug Extracts Delivery. *Nanomaterials (Basel, Switzerland), 10*(2), 385. DOI: 10.3390/nano10020385 PMID: 32098412

Negut, I., Gradisteanu-Pircalabioru, G., Dinu, M., Bita, B., Parau, A. C., Grumezescu, V., Ristoscu, C., & Chifiriuc, M. C. (2023). Bioglass and Vitamin D3 Coatings for Titanium Implants: Osseointegration and Corrosion Protection. *Biomedicines, 11*(10), 2772. DOI: 10.3390/biomedicines11102772 PMID: 37893145

Negut, I., Ristoscu, C., Tozar, T., Dinu, M., Parau, A. C., Grumezescu, V., Hapenciuc, C., Popa, M., Stan, M. S., Marutescu, L., Mihailescu, I. N., & Chifiriuc, M. C. (2022). Implant Surfaces Containing Bioglasses and Ciprofloxacin as Platforms for Bone Repair and Improved Resistance to Microbial Colonization. *Pharmaceutics, 14*(6), 1175. DOI: 10.3390/pharmaceutics14061175 PMID: 35745748

Notingher, I., Jones, J. R., Verrier, S., Bisson, I., Embanga, P., Edwards, P., Polak, J. M., & Hench, L. L. (2003). Application of FTIR and Raman Spectroscopy to Characterisation of Bioactive Materials and Living Cells. *Journal of Spectroscopy*, *17*(2–3), 275–288. DOI: 10.1155/2003/893584

Nsakabwebwe, C., Makhatha, M. E., Tsoeunyane, G. M., & Baruwa, A. D. (2022). Corrosion Inhibition Efficiency of Polyvinylpyrrolidone-Cysteine on Mild Steel in 1.0 M HCl Solution. *Journal of Bio- and Tribo-Corrosion*, *8*(2), 48. DOI: 10.1007/s40735-022-00647-3

Ong, J. L., & Lucas, L. C. (1994). Post-deposition heat treatments for ion beam sputter deposited calcium phosphate coatings. *Biomaterials*, *15*(5), 337–341. DOI: 10.1016/0142-9612(94)90245-3 PMID: 8061124

Örnek, C., Leygraf, C., & Pan, J. (2019). On the Volta potential measured by SKPFM – fundamental and practical aspects with relevance to corrosion science. *Corrosion Engineering, Science and Technology*, *54*(3), 185–198. DOI: 10.1080/1478422X.2019.1583436

Oyane, A. (2019). Calcium Phosphate Coating Using Supersaturated Solutions for Biomedical Applications. *Vacuum and Surface Science*, *62*(4), 217–222. Advance online publication. DOI: 10.1380/vss.62.217

Paital, S. R., Balani, K., Agarwal, A., & Dahotre, N. B. (2009). Fabrication and evaluation of a pulse laser-induced Ca–P coating on a Ti alloy for bioapplication. *Biomedical Materials (Bristol, England)*, *4*(1), 015009. DOI: 10.1088/1748-6041/4/1/015009 PMID: 19020344

Paital, S. R., Cao, Z., He, W., & Dahotre, N. B. (2010). Wetting effects on in vitro bioactivity and in vitro biocompatibility of laser micro-textured Ca-P coating. *Biofabrication*, *2*(2), 025001. Advance online publication. DOI: 10.1088/1758-5082/2/2/025001 PMID: 20811129

Paital, S. R., He, W., & Dahotre, N. B. (2010). Laser pulse dependent micro textured calcium phosphate coatings for improved wettability and cell compatibility. *Journal of Materials Science. Materials in Medicine*, *21*(7), 2187–2200. Advance online publication. DOI: 10.1007/s10856-010-4085-6 PMID: 20464459

Pakuła, D., Staszuk, M., Dziekońska, M., Kožmín, P., & Čermák, A. (2019). Laser Micro-Texturing of Sintered Tool Materials Surface. *Materials (Basel)*, *12*(19), 3152. DOI: 10.3390/ma12193152 PMID: 31561577

Pan, J., Thierry, D., & Leygraf, C. (1996). Electrochemical impedance spectroscopy study of the passive oxide film on titanium for implant application. *Electrochimica Acta, 41*(7–8), 1143–1153. DOI: 10.1016/0013-4686(95)00465-3

Pana, I., Vitelaru, C., Kiss, A., Zoita, N. C., Dinu, M., & Braic, M. (2017). Design, fabrication and characterization of TiO 2 -SiO 2 multilayer with tailored color glazing for thermal solar collectors. *Materials & Design, 130*, 275–284. DOI: 10.1016/j.matdes.2017.05.063

Parangusan, H., Bhadra, J., & Al-Thani, N. (2021). A review of passivity breakdown on metal surfaces: Influence of chloride- and sulfide-ion concentrations, temperature, and pH. *Emergent Materials, 4*(5), 1187–1203. DOI: 10.1007/s42247-021-00194-6

Patenge, N., Arndt, K., Eggert, T., Zietz, C., Kreikemeyer, B., Bader, R., Nebe, B., Stranak, V., Hippler, R., & Podbielski, A. (2012). Evaluation of antimicrobial effects of novel implant materials by testing the prevention of biofilm formation using a simple small scale medium-throughput growth inhibition assay. *Biofouling, 28*(3), 267–277. DOI: 10.1080/08927014.2012.671305 PMID: 22435853

Pelletier, H., Nelea, V., Mille, P., & Muller, D. (2004). Determination of mechanical properties of pulsed laser-deposited hydroxyapatite thin film implanted at high energy with N+ and Ar+ ions, using nanoindentation. *Journal of Materials Science, 39*(11), 3605–3611. DOI: 10.1023/B:JMSC.0000030712.81704.b0

Phogat, D., Rani, P., Biswas, A., Balani, K., & Awasthi, S. (2025). Architecturally Refined Cerium-Integrated Hydroxyapatite/CNT Nanocomposite Coatings: Enhanced Mechanics and Biofunction for Orthopaedic Implantation. *Macromolecular Bioscience, 25*(12), 70106. Advance online publication. DOI: 10.1002/mabi.70106 PMID: 41059707

Prefac, G.-A., Milea, M.-L., Vadureanu, A.-M., Muraru, S., Dobrin, D.-I., Isopencu, G.-O., Jinga, S.-I., Raileanu, M., Bacalum, M., & Busuioc, C. (2020). CeO2 Containing Thin Films as Bioactive Coatings for Orthopaedic Implants. *Coatings, 10*(7), 642. DOI: 10.3390/coatings10070642

Prodana, M., Stoian, A. B., Ionita, D., Brajnicov, S., Boerasu, I., Enachescu, M., & Burnei, C. (2024). In-Depth Characterization of Two Bioactive Coatings Obtained Using MAPLE on TiTaZrAg. *Materials (Basel), 17*(12), 2989. DOI: 10.3390/ma17122989 PMID: 38930358

Pshyk, O., Nakonechna, O., & Coy, E. (2025). Current View of Nanoindentation: Recent Developments and Application in Material Characterization. *ACS Applied Materials & Interfaces, 17*(25), 36275–36300. DOI: 10.1021/acsami.5c05434 PMID: 40499095

Puleo, D. A., & Nanci, A. (1999). Understanding and controlling the bone-implant interface. In *Biomaterials*. DOI: 10.1016/S0142-9612(99)00160-X

Radhi, A. B., Khashan, K. S., & Sulaiman, G. M. (2024). Antibacterial Activity of Gold Nanoparticles Produced by One-Step Pulsed Laser Ablation in Liquid. *Plasmonics, 19*(3), 1173–1185. DOI: 10.1007/s11468-023-02081-8

Rammelt, U., & Reinhard, G. (1992). Application of electrochemical impedance spectroscopy (EIS) for characterizing the corrosion-protective performance of organic coatings on metals. *Progress in Organic Coatings, 21*(2–3), 205–226. DOI: 10.1016/0033-0655(92)87005-U

Ranjan, A., Shetty, P., Despande, R., Biradar, A., Khan, W., & Kulshrestha, R. (2023). A comparative surface evaluation of orthodontic mini-implants before and after en masse retraction-A SEM study. *Journal of Orthodontic Science, 12*(1). Advance online publication. DOI: 10.4103/jos.jos_166_21 PMID: 37351404

Rau, J. V., Antoniac, I., Cama, G., Komlev, V. S., & Ravaglioli, A. (2016). Bioactive Materials for Bone Tissue Engineering. *BioMed Research International, 1–3*, 1–3. Advance online publication. DOI: 10.1155/2016/3741428 PMID: 26942191

Rau, J. V., Antoniac, I., Fosca, M., De Bonis, A., Blajan, A. I., Cotrut, C., Graziani, V., Curcio, M., Cricenti, A., Niculescu, M., Ortenzi, M., & Teghil, R. (2016). Glass-ceramic coated Mg-Ca alloys for biomedical implant applications. *Materials Science and Engineering C, 64*, 362–369. Advance online publication. DOI: 10.1016/j.msec.2016.03.100 PMID: 27127065

Rau, J. V., De Bonis, A., Curcio, M., Schuhladen, K., Barbaro, K., De Bellis, G., Teghil, R., & Boccaccini, A. R. (2020). Borate and Silicate Bioactive Glass Coatings Prepared by Nanosecond Pulsed Laser Deposition. *Coatings, 10*(11), 1105. DOI: 10.3390/coatings10111105

Rezwan, K., Chen, Q. Z., Blaker, J. J., & Boccaccini, A. R. (2006). Biodegradable and bioactive porous polymer/inorganic composite scaffolds for bone tissue engineering. *Biomaterials, 27*(18), 3413–3431. DOI: 10.1016/j.biomaterials.2006.01.039 PMID: 16504284

Rohanová, D., Boccaccini, A. R., Horkavcová, D., Bozděchová, P., Bezdička, P., & Častorálová, M. (2014). Is non-buffered DMEM solution a suitable medium for *in vitro* bioactivity tests? *Journal of Materials Chemistry. B, 2*(31), 5068–5076. DOI: 10.1039/C4TB00187G PMID: 32261840

Rohwerder, M. (2009). Conducting polymers for corrosion protection: A review. *International Journal of Materials Research*, *100*(10), 1331–1342. DOI: 10.3139/146.110205

Sahadat Hossain, M., & Ahmed, S. (2023). FTIR spectrum analysis to predict the crystalline and amorphous phases of hydroxyapatite: A comparison of vibrational motion to reflection. *RSC Advances*, *13*(21), 14625–14630. DOI: 10.1039/D3RA02580B PMID: 37197675

Sankar, M., Suwas, S., Balasubramanian, S., & Manivasagam, G. (2017). Comparison of electrochemical behavior of hydroxyapatite coated onto WE43 Mg alloy by electrophoretic and pulsed laser deposition. *Surface and Coatings Technology*, *309*, 840–848. DOI: 10.1016/j.surfcoat.2016.10.077

Sarantopoulou, E., Petrou, P. S., Kollia, Z., Palles, D., Spyropoulos-Antonakakis, N., Kakabakos, S., & Cefalas, A.-C. (2011). Protein immobilization and detection on laser processed polystyrene surfaces. *Journal of Applied Physics*, *110*(6), 064309. Advance online publication. DOI: 10.1063/1.3627160

Schille, J., Schneider, L., Mauersberger, S., Szokup, S., Höhn, S., Pötschke, J., Reiß, F., Leidich, E., & Löschner, U. (2020). High-Rate Laser Surface Texturing for Advanced Tribological Functionality. *Lubricants (Basel, Switzerland)*, *8*(3), 33. DOI: 10.3390/lubricants8030033

Sekine, I. (1997). Recent evaluation of corrosion protective paint films by electrochemical methods. *Progress in Organic Coatings*, *31*(1–2), 73–80. DOI: 10.1016/S0300-9440(97)00020-9

Sergi, R., Bellucci, D., & Cannillo, V. (2020). A Comprehensive Review of Bioactive Glass Coatings: State of the Art, Challenges and Future Perspectives. *Coatings*, *10*(8), 757. DOI: 10.3390/coatings10080757

Shabir, S., Khan, N. Z., & Siddiquee, A. N. (2025). Laser-induced surface modifications in biomedical materials: A comprehensive review. *Surface Topography : Metrology and Properties*, *13*(2), 023002. DOI: 10.1088/2051-672X/addc85

Shen, S., Cai, S., Xu, G., Zhao, H., Niu, S., & Zhang, R. (2015). Influence of heat treatment on bond strength and corrosion resistance of sol–gel derived bioglass–ceramic coatings on magnesium alloy. *Journal of the Mechanical Behavior of Biomedical Materials*, *45*, 166–174. DOI: 10.1016/j.jmbbm.2015.02.005 PMID: 25728582

Shivakoti, I., Kibria, G., Cep, R., Pradhan, B. B., & Sharma, A. (2021). Laser Surface Texturing for Biomedical Applications: A Review. *Coatings*, *11*(2), 124. DOI: 10.3390/coatings11020124

Stuart, B. W., Gimeno-Fabra, M., Segal, J., Ahmed, I., & Grant, D. M. (2015). Degradation and Characterization of Resorbable Phosphate-Based Glass Thin-Film Coatings Applied by Radio-Frequency Magnetron Sputtering. *ACS Applied Materials & Interfaces*, *7*(49), 27362–27372. DOI: 10.1021/acsami.5b08957 PMID: 26523618

Su, Y., Li, D., Su, Y., Lu, C., Niu, L., Lian, J., & Li, G. (2016). Improvement of the Biodegradation Property and Biomineralization Ability of Magnesium-Hydroxyapatite Composites with Dicalcium Phosphate Dihydrate and Hydroxyapatite Coatings. *ACS Biomaterials Science &amp. ACS Biomaterials Science & Engineering*, *2*(5), 818–828. Advance online publication. DOI: 10.1021/acsbiomaterials.6b00013 PMID: 33440579

Su, Y., Lu, C., Hu, X., Guo, Y., Xun, X., Zhang, Z., Li, G., Lian, J., & Ren, L. (2018). Improving the Degradation Resistance and Surface Biomineralization Ability of Calcium Phosphate Coatings on a Biodegradable Magnesium Alloy via a Sol-Gel Spin Coating Method. *Journal of The Electrochemical Society, 165*. https://doi.org/ DOI: 10.1149/2.0901803jes

Sypniewska, J., & Szkodo, M. (2022). Influence of Laser Modification on the Surface Character of Biomaterials: Titanium and Its Alloys—A Review. *Coatings*, *12*(10), 1371. DOI: 10.3390/coatings12101371

Tavafoghi, M., & Cerruti, M. (2016). The role of amino acids in hydroxyapatite mineralization. In *Journal of the Royal Society Interface*. DOI: 10.1098/rsif.2016.0462

Theodorou, G., Goudouri, O. M., Kontonasaki, E., Chatzistavrou, X., Papadopoulou, L., Kantiranis, N., & Paraskevopoulos, K. M. (2011). Comparative Bioactivity Study of 45S5 and 58S Bioglasses in Organic and Inorganic Environment. *Bioceramics Development and Applications*, *1*, 1–4. DOI: 10.4303/bda/D110154

Trentin, A., de L. Gasparini, A., Faria, F. A., Harb, S. V., dos Santos, F. C., Pulcinelli, S. H., Santilli, C. V., & Hammer, P. (2020). Barrier properties of high performance PMMA-silica anticorrosion coatings. *Progress in Organic Coatings*, *138*, 105398. DOI: 10.1016/j.porgcoat.2019.105398

Trentin, A., Pakseresht, A., Duran, A., Castro, Y., & Galusek, D. (2022). Electrochemical Characterization of Polymeric Coatings for Corrosion Protection: A Review of Advances and Perspectives. *Polymers*, *14*(12), 2306. DOI: 10.3390/polym14122306 PMID: 35745882

Tynan, J., Ward, P., Byrne, G., & Dowling, D. P. (2009). Deposition of Biodegradable Polycaprolactone Coatings Using an In-line Atmospheric Pressure Plasma System. *Plasma Processes and Polymers*, *6*(S1). Advance online publication. DOI: 10.1002/ppap.200930310

Vahabzadeh, S., Roy, M., Bandyopadhyay, A., & Bose, S. (2015). Phase stability and biological property evaluation of plasma sprayed hydroxyapatite coatings for orthopedic and dental applications. *Acta Biomaterialia*, *17*, 47–55. Advance online publication. DOI: 10.1016/j.actbio.2015.01.022 PMID: 25638672

van Raemdonck, W., Ducheyne, P., & de Meester, P. (1984). Auger Electron Spectroscopic Analysis of Hydroxylapatite Coatings on Titanium. *Journal of the American Ceramic Society*, *67*(6), 381–384. Advance online publication. DOI: 10.1111/j.1151-2916.1984.tb19721.x

Vardaki, M., Prodana, M., Stoian, A. B., & Ionita, D. (2018). Corrosion and bioactivity of a bioinspired coating on tizr alloys. *UPB Scientific Bulletin, Series B. Chemistry and Materials Science*, *80*(1), 209–218.

Vasile, E., Serafim, A., Dragusin, D.-M., Petrea, C., Iovu, H., & Stancu, I.-C. (2012). Apatite formation on active nanostructured coating based on functionalized gold nanoparticles. *Journal of Nanoparticle Research*, *14*(6), 918. Advance online publication. DOI: 10.1007/s11051-012-0918-1

Visan, A. I., & Negut, I. (2025). Polymeric Composite Thin Films Deposited by Laser Techniques for Antimicrobial Applications—A Short Overview. *Polymers*, *17*(15), 2020. DOI: 10.3390/polym17152020 PMID: 40808071

Visan, A. I., & Popescu-Pelin, G. F. (2024). Advanced Laser Techniques for the Development of Nature-Inspired Biomimetic Surfaces Applied in the Medical Field. *Coatings*, *14*(10), 1290. DOI: 10.3390/coatings14101290

Vladescu, A., Pârâu, A., Pana, I., Cotrut, C. M., Constantin, L. R., Braic, V., & Vrânceanu, D. M. (2019). In vitro activity assays of sputtered HAp coatings with SiC addition in various simulated biological fluids. *Coatings*, *9*(6), 389. Advance online publication. DOI: 10.3390/coatings9060389

Vranceanu, D. M., Parau, A. C., Cotrut, C. M., Kiss, A. E., Constantin, L. R., Braic, V., & Vladescu, A. (2019). In vitro evaluation of Ag doped hydroxyapatite coatings in acellular media. *Ceramics International*, *45*(8), 11050–11061. DOI: 10.1016/j.ceramint.2019.02.191

Walsh, F. C., Ponce de León, C., Kerr, C., Court, S., & Barker, B. D. (2008). Electrochemical characterisation of the porosity and corrosion resistance of electrochemically deposited metal coatings. *Surface and Coatings Technology, 202*(21), 5092–5102. DOI: 10.1016/j.surfcoat.2008.05.008

Wang, H., Deng, D., Zhai, Z., & Yao, Y. (2024). Laser-processed functional surface structures for multi-functional applications-a review. *Journal of Manufacturing Processes, 116*, 247–283. DOI: 10.1016/j.jmapro.2024.02.062

Wang, H., Lee, J.-K., Moursi, A., & Lannutti, J. (2003). Ca/P ratio effects on the degradation of hydroxyapatite in vitro. *Journal of Biomedical Materials Research. Part A, 67*(2), 599–608. DOI: 10.1002/jbm.a.10538 PMID: 14566803

Wang, T., Li, Q., Huo, F., Chen, H., & Xu, T. (2025). Comparison of Microstructure and Properties of CoCrMo Coatings Prepared by High-Speed and Conventional Laser Cladding. *Coatings, 15*(10), 1200. DOI: 10.3390/coatings15101200

Wang, W., & Yeung, K. W. K. (2017). Bone grafts and biomaterials substitutes for bone defect repair: A review. In *Bioactive Materials*. DOI: 10.1016/j.bioactmat.2017.05.007

Wang, Z., Xu, Z., Zhao, W., & Sahai, N. (2015). A potential mechanism for amino acid-controlled crystal growth of hydroxyapatite. *Journal of Materials Chemistry. B, 3*(47), 9157–9167. Advance online publication. DOI: 10.1039/C5TB01036E PMID: 32263130

Webster, T. J., Massa-Schlueter, E. A., Smith, J. L., & Slamovich, E. B. (2004). Osteoblast response to hydroxyapatite doped with divalent and trivalent cations. *Biomaterials, 25*(11), 2111–2121. DOI: 10.1016/j.biomaterials.2003.09.001 PMID: 14741626

Wei, J., Igarashi, T., Okumori, N., Igarashi, T., Maetani, T., Liu, B., & Yoshinari, M. (2009). Influence of surface wettability on competitive protein adsorption and initial attachment of osteoblasts. *Biomedical Materials (Bristol, England), 4*(4), 045002. Advance online publication. DOI: 10.1088/1748-6041/4/4/045002 PMID: 19525576

Wei, L., & Gao, Z. (2023). Recent research advances on corrosion mechanism and protection, and novel coating materials of magnesium alloys: A review. *RSC Advances, 13*(12), 8427–8463. DOI: 10.1039/D2RA07829E PMID: 36926015

Wei, X., Zheng, S., An, Y., Yu, S., Liu, G., & Pu, J. (2024). Microstructure and tribocorrosion properties of Fe-based amorphous composite coatings fabricated by laser cladding. *Journal of Materials Science, 59*(9), 4039–4058. DOI: 10.1007/s10853-024-09436-4

Wennerberg, A., & Albrektsson, T. (2009). On implant surfaces: A review of current knowledge and opinions. *The International Journal of Oral & Maxillofacial Implants*, *25*(1), 63–74. DOI: 10.1111/j.1600-051X.2008.01321.x PMID: 20209188

Williams, D. F. (2008). On the mechanisms of biocompatibility. *Biomaterials*, *29*(20), 2941–2953. DOI: 10.1016/j.biomaterials.2008.04.023 PMID: 18440630

Williams, D. F. (2022). Biocompatibility pathways and mechanisms for bioactive materials: The bioactivity zone. *Bioactive Materials*, *10*, 306–322. DOI: 10.1016/j.bioactmat.2021.08.014 PMID: 34901548

Williams, D. F., & Williams, R. L. (2013). Degradative Effects of the Biological Environment on Metals and Ceramics. *Biomaterials Science*, (1998), 728–738. DOI: 10.1016/B978-0-08-087780-8.00062-0

Wolstenholme, J. (1973). Electrochemical methods of assessing the corrosion of painted metals—A review. *Corrosion Science*, *13*(7), 521–530. DOI: 10.1016/S0010-938X(73)80002-2

Xu, T. Z., Zhang, S., Wang, L., Du, Y., Wu, C. L., Zhang, C. H., Sun, X. Y., Chen, H. T., & Chen, J. (2023). Influence of scanning speed on the microstructure, nanoindentation characteristics and tribological behavior of novel maraging steel coatings by laser cladding. *Materials Characterization*, *205*, 113335. DOI: 10.1016/j.matchar.2023.113335

Xu, Z., Wang, Y. A., Ng, V., Yin, H., & Xu, S. (2025). Advancements in Laser-Processed Functional Surfaces for Medical Devices: A Current Review. *Nanomaterials (Basel, Switzerland)*, *15*(13), 999. DOI: 10.3390/nano15130999 PMID: 40648706

Yang, C.-J., Mei, X.-S., Tian, Y.-L., Zhang, D.-W., Li, Y., & Liu, X.-P. (2016). Modification of wettability property of titanium by laser texturing. *International Journal of Advanced Manufacturing Technology*, *87*(5-8), 1663–1670. Advance online publication. DOI: 10.1007/s00170-016-8601-9

Yang, J., Zhang, W., Lin, B., Mao, S., Liu, G., Tan, K., & Tang, J. (2024). Enhancement of Local Osseointegration and Implant Stability of Titanium Implant in Osteoporotic Rats by Biomimetic Multilayered Structures Containing Catalpol. *ACS Omega*, *9*(27), 29544–29556. DOI: 10.1021/acsomega.4c02322 PMID: 39005760

Yang, Y., Paital, S. R., & Dahotre, N. B. (2010). Effects of SiO2 substitution on wettability of laser deposited Ca-P biocoating on Ti-6Al-4V. *Journal of Materials Science. Materials in Medicine*, *21*(9), 2511–2521. Advance online publication. DOI: 10.1007/s10856-010-4105-6 PMID: 20549311

Yang, Y. L., Paital, S. R., & Dahotre, N. B. (2010). Wetting and in vitro bioactivity of laser processed CaP coating with presence and variation of SiO2 on Ti–6Al–4V. *Materials Technology*, *25*(3-4), 137–142. Advance online publication. DOI: 10.1179/175355510X12723642365287

Yao, Q., Li, W., Yu, S., Ma, L., Jin, D., Boccaccini, A. R., & Liu, Y. (2015). Multifunctional chitosan/polyvinyl pyrrolidone/45S5 Bioglass® scaffolds for MC3T3-E1 cell stimulation and drug release. *Materials Science and Engineering C*, *56*, 473–480. DOI: 10.1016/j.msec.2015.06.046 PMID: 26249617

Yi, G., Teong, S. P., Zhang, Y., & Ying, J. Y. (2026). Inorganic Antimicrobial Materials. *ACS Nano*, *20*(15), 11495–11523. Advance online publication. DOI: 10.1021/acsnano.5c21363 PMID: 41945705

Yin, Y., Prabhakar, M., Ebbinghaus, P., Corrêa da Silva, C., & Rohwerder, M. (2022). Neutral inhibitor molecules entrapped into polypyrrole network for corrosion protection. *Chemical Engineering Journal*, *440*, 135739. DOI: 10.1016/j.cej.2022.135739

Zaludin, M. A. F., Jamal, Z. A. Z., Derman, M. N., & Kasmuin, M. Z. (2019). Corrosion analysis of hydroxyapatite coated magnesium in simulated body fluid (SBF), phosphate buffered saline (PBS), and Ringer's solution. *Materials Today: Proceedings*, *16*, 1686–1691. DOI: 10.1016/j.matpr.2019.06.037

Zarif, M. E., Bita, B., Yehia-Alexe, S. A., Negut, I., Gradisteanu Pircalabioru, G., Andronescu, E., & Groza, A. (2024). Biological and Physicochemical Analysis of Sr-Doped Hydroxyapatite/Chitosan Composite Layers. *Polymers*, *16*(13), 1922. DOI: 10.3390/polym16131922 PMID: 39000777

Zhang, D., Wang, M., Zhao, Y., Yang, C., & Yang, Z. (2019). Modification of Wettability Property of NITI Alloy by Laser Texturing and Carbon Ion Implantation. *2019 IEEE International Conference on Manipulation, Manufacturing and Measurement on the Nanoscale (3M-NANO)*, 224–227. DOI: 10.1109/3M-NANO46308.2019.8947423

Zhang, F., Cai, S., Xu, G., Shen, S., Li, Y., Zhang, M., & Wu, X. (2016). Corrosion behavior of mesoporous bioglass-ceramic coated magnesium alloy under applied forces. *Journal of the Mechanical Behavior of Biomedical Materials*, *56*, 146–155. DOI: 10.1016/j.jmbbm.2015.11.029 PMID: 26703229

Zhang, J. M., Lin, C. J., Feng, Z. D., & Tian, Z. W. (1998). Mechanistic studies of electrodeposition for bioceramic coatings of calcium phosphates by an in situ pH-microsensor technique. *Journal of Electroanalytical Chemistry (Lausanne, Switzerland)*, *452*(2), 235–240. DOI: 10.1016/S0022-0728(98)00107-7

Zhang, Q., Chen, J., Feng, J., Cao, Y., Deng, C., & Zhang, X. (2003). Dissolution and mineralization behaviors of HA coatings. *Biomaterials, 24*(26), 4741–4748. Advance online publication. DOI: 10.1016/S0142-9612(03)00371-5 PMID: 14530071

Zhang, Q., & Guan, Y. (2023). Laser induced periodic surface structure on titanium alloy and its effect on microstructure refinement and corrosion behavior. *Journal of Alloys and Compounds, 936*, 168188. DOI: 10.1016/j.jallcom.2022.168188

Zhou, Y., Liu, K., & Zhang, H. (2023). Biomimetic Mineralization: From Microscopic to Macroscopic Materials and Their Biomedical Applications. *ACS Applied Bio Materials, 6*(9), 3516–3531. DOI: 10.1021/acsabm.3c00109 PMID: 36944024

Zhuikov, V. A., Bonartsev, A. P., Bagrov, D. V., Yakovlev, S. G., Myshkina, V. L., Makhina, T. K., Bessonov, I. V., Kopitsyna, M. N., Morozov, A. S., Rusakov, A. A., Useinov, A. S., Shaitan, K. V., & Bonartseva, G. A. (2017). Mechanics and surface ultrastructure changes of poly(3-hydroxybutyrate) films during enzymatic degradation in pancreatic lipase solution. *Molecular Crystals and Liquid Crystals (Philadelphia, Pa.), 648*(1), 236–243. Advance online publication. DOI: 10.1080/15421406.2017.1302580

Zurita-Méndez, N. N., Carbajal-De la Torre, G., Flores-Merino, M. V., & Espinosa-Medina, M. A. (2022). Development of Bioactive Glass-Collagen-Hyaluronic Acid-Polycaprolactone Scaffolds for Tissue Engineering Applications. *Frontiers in Bioengineering and Biotechnology, 10*, 825903. Advance online publication. DOI: 10.3389/fbioe.2022.825903 PMID: 35252134

Chapter 2
Fundamentals of Laser–Material Interaction for Coating Development

Bogdan Ionut Bita
http://orcid.org/0000-0002-4102-635X
National Institute for Optoelectronics, Romania

Ana Maria Iordache
National Institute for Optoelectronics, Romania

Ana Maria Florea
http://orcid.org/0009-0008-2571-8242
National Institute for Optoelectronics, Romania

Stefan Caramizoiu
National Institute for Optoelectronics, Romania

Stefan Marian Iordache
National Institute for Optoelectronics, Romania

ABSTRACT

Titanium Nitride (TiN) coatings, known for their hardness, corrosion resistance, and biocompatibility, are increasingly used in biomedical applications. Pulsed Laser Deposition (PLD) enables precise fabrication of TiN thin films with controlled microstructure, thickness, and stoichiometry. The PLD process involves ablation of a TiN target in a vacuum or nitrogen-rich atmosphere, producing dense, adherent coatings with excellent mechanical and biological performance. TiN's low friction and antimicrobial properties make it ideal for orthopedic, dental, and cardiovascular implants. Compared to other methods, PLD allows complex multilayer

DOI: 10.4018/979-8-3373-4104-0.ch002

structures and fine control over surface properties, enhancing implant integration and longevity. This chapter explores the PLD process, laser-material interactions, and current biomedical applications of TiN coatings, emphasizing their role in improving implant durability and patient outcomes.

INTRODUCTION

Laser-based technologies such as PLD are especially suited to deposit thin films of TiN with tailored morphology, thickness, and crystallinity. The process begins with a high-energy pulsed laser beam typically an excimer focused onto a solid TiN target inside a vacuum chamber (Brown et al., 2017). Each laser pulse delivers a burst of energy that rapidly heats and vaporizes the target's surface layer, forming a dense plasma plume composed of energetic ions, atoms, and clusters of TiN. This plasma expands rapidly toward a substrate, often heated and positioned directly opposite the target. Upon arrival, the high-energy species condense into a thin film, replicating the target's composition and forming a coating with exceptional adhesion and uniformity (Dilova et al., 2019).

One of the defining advantages of PLD lies in its control over key deposition variables such as laser fluence, repetition rate, ambient gas composition, and substrate temperature. By fine-tuning these parameters, researchers can customize TiN coatings to exhibit different grain sizes, porosity levels, phase structures, and mechanical performance. For instance, adjusting the nitrogen partial pressure during deposition allows for the formation of stoichiometrically accurate TiN or even sub-stoichiometric variants (TiNx) with modified electrical and antimicrobial properties (Mahamood, 2017). Furthermore, PLD enables the deposition of ultra-thin coatings ranging from a few nanometers to several microns making it ideal for medical applications where coating thickness must be balanced with surface functionality and underlying material integrity (Shen et al., 2023).

Compared to traditional deposition methods like physical vapor deposition (PVD) (Mahamood, 2017; Shen et al., 2023), chemical vapor deposition (CVD) ((Mahamood, 2017; Shen et al., 2023) or sputtering (Sattar et al., 2023), PLD offers several unique benefits for biomedical applications (Lu et al., 2024). Most notably, PLD allows the fabrication of coatings with complex architectures such as multilayers, compositional gradients, and nanostructured morphologies. In the case of TiN, multilayer coatings can be developed by alternating it with other biocompatible materials such as titanium oxide, zirconia, or hydroxyapatite to combine mechanical toughness with biological performance. Such designs can produce gradient coatings that mimic natural tissue transitions for example, from soft tissue to bone enhancing implant integration and reducing stress concentration at the interface. From a biological perspective, TiN coatings have demonstrated outstanding biocompatibility in both in vitro and in vivo studies. Cell culture experiments consistently show that osteoblasts, fibroblasts, and endothelial cells adhere well to TiN-coated surfaces, exhibiting good proliferation and metabolic activity. Furthermore, TiN's smooth, non-porous surface limits bacterial adhesion and biofilm formation, an essential property in reducing the risk of postoperative infections a growing concern in implant related surgeries. When optimized, TiN coatings also reduce platelet activation and thrombus formation, making them suitable for cardiovascular implants such as stents and pacemaker leads (Irimiciuc et al., 2019). PLD deposited TiN coatings also contribute to improved mechanical reliability of medical devices. With a hardness typically in the range of 20–30 GPa and high resistance to fatigue and wear, TiN provides a durable surface that significantly extends the lifespan of implants subjected to cyclic mechanical loads (Eason, 2007). The deposition process can also be tailored to manage residual stress levels within the coating, improving adhesion and mitigating risks of delamination. Importantly, PLD enables the fabrication of defect-free, pinhole-free coatings crucial in applications such as dental abutments or joint articulating surfaces where corrosion and wear resistance are paramount. An essential segment of the chapter delves into the scientific understanding of laser-material interactions that govern the formation of TiN coatings via PLD. These interactions depend on laser pulse duration (ranging from nanoseconds to femtoseconds), fluence, wavelength, and repetition rate. Depending on these parameters, the ablation mechanism may be predominantly thermal or photomechanical, influencing both the kinetic energy of the ablated species and the quality of the resulting film (Harilal et al., 2018). For instance, short-pulse lasers minimize thermal damage to the substrate while promoting fine-grained film growth. Additionally, PLD offers the capability to conduct reactive depositions in nitrogen-rich environments, further enhancing the nitridation process and improving film purity. As PLD transitions from laboratory settings to industrial and clinical environments, advances in deposition system design are playing a crucial role (Cai et al., 2025). High-throughput PLD systems now incorporate multiple target holders, automated substrate motion, in-situ monitoring, and post-deposition annealing capabilities (Figure 1). These improvements ensure reproducibility and consistency key factors for medical device manufacturing (Amaral et al., 2024). TiN coatings produced via PLD are already under commercial exploration for applications in dental prosthetics and orthopedic implants, with growing interest in expanding their use to biosensors, wound closure devices, and endovascular tools.

Figure 1. PLD design for multuple application

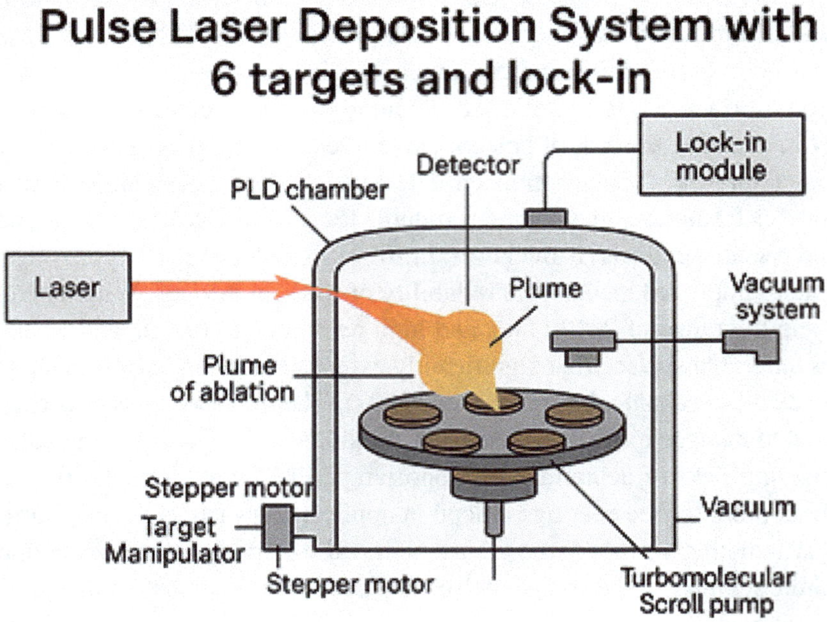

The future of TiN coatings deposited by PLD is promising and multifaceted. Emerging directions include doping TiN with antibacterial agents such as silver or copper to create dual-functional surfaces that resist microbial colonization while maintaining mechanical integrity. Researchers are also investigating stimuli-responsive TiN coatings that can modulate their properties in response to temperature (Balcerak-Woźniak et al., 2024), pH (Lang et al., 2024), or light (Bril et al., 2022), offering dynamic functionality for drug delivery or tissue regeneration platforms (Municoy et al., 2020). In parallel, innovations in laser technology, such as ultrafast lasers, beam shaping, and multi-beam systems, are enhancing the precision and scalability of PLD processes, opening new frontiers in the engineering of complex biomedical surfaces (Visan & Popescu-Pelin, 2024).

THE INTEGRATION OF SURFACE ENGINEERING IN BIOMEDICAL TECHNOLOGY

The integration of surface engineering into biomedical technology has dramatically reshaped the functionality and performance of medical devices. In particular, surface coatings have become critical in determining the biological performance of implants and devices, influencing parameters such as corrosion resistance, osseointegration, antimicrobial efficacy, and mechanical durability (Vakulov et al., 2021). The ability to precisely tailor surface properties at micro- and nano-scales is a driving force behind innovations in biomaterials. Within this context, laser based processing has emerged as one of the most promising technologies for the development and modification of biomedical coatings. Lasers offer a unique set of capabilities that are unmatched by conventional surface modification techniques (Sypniewska & Szkodo, 2022). These include localized energy delivery, high spatial resolution, non-contact operation, and excellent controllability of process parameters. As a result, lasers can be employed to achieve a wide range of effects from surface texturing and microstructuring to the deposition of biocompatible or bioactive films. The versatility of laser-material interaction mechanisms enables the formation of coatings with highly customized physical (Matei et al., 2025), chemical (Chowdhury et al., 2022), and biological properties (Yang et al., 2019), specifically suited for applications in orthopedics (Zhu et al., 2024), cardiology (Huang et al., 2020), dentistry (Zhu et al., 2024), and regenerative medicine (Ahsan et al., 2011).

The field of laser-processed coatings in biomedical applications lies at the intersection of materials science, laser physics, and bioengineering. A comprehensive understanding of laser-material interactions is essential for optimizing coating processes and ensuring their functionality under physiological conditions (Park et al., 2024). These interactions are governed by a range of parameters including laser wavelength, fluence, pulse duration, and scanning strategies, as well as material-specific properties such as absorptivity, thermal conductivity, and phase transition behavior. Laser-material interactions occur across several regimes thermal, photothermal, photochemical, and mechanical depending on the nature of the laser pulse and the material's response (Theerthagiri et al., 2022). For instance, pulsed lasers with nanosecond, picosecond, or femtosecond durations can induce rapid melting, vaporization, or plasma formation, allowing for precise ablation or deposition of thin films without excessive thermal damage to the underlying substrate. Continuous-wave (CW) lasers, on the other hand, are often used for bulk heating, sintering, or cladding processes where a higher thermal budget is acceptable or even desirable. In the context of coating development, lasers can be utilized in a variety of modalities: Pulsed Laser Deposition (PLD) and related methods allow for the fabrication of ultra-thin, stoichiometrically accurate films, often used for titanium nitride or composite coatings in orthopedic and dental implants. Moreover, the biomedical environment presents additional challenges that laser-processed coatings must meet (Xu et al., 2025). These include long term biocompatibility, resistance to body fluids, the ability to resist bacterial colonization, and in some cases, the ability to promote specific cellular responses. Laser processing offers the added advantage of enabling in-situ functionalization, where coatings can be designed to be multifunctional combining, for example, mechanical durability with antimicrobial and osteogenic properties (Nikolova & Apostolova, 2022).

THE PHYSICS BEHIND PLD

Pulsed Laser Deposition represents a sophisticated physical vapor deposition technique that harnesses the power of high-energy laser pulses to ablate material from a target surface, creating a plasma plume that subsequently deposits onto a substrate (Gao et al., 2025). This sudden energy transfer results in explosive evaporation, creating a characteristic plasma plume containing atoms, ions, molecules, and even small clusters of the target material. The plasma plume, often described as a "laser-induced plasma," expands perpendicular to the target surface at velocities approaching 10^6 cm/s (Yu et al., 2023). This expansion is governed by complex gas dynamics and electromagnetic interactions, which significantly influence the final film properties. The energetic species within the plume typically possess kinetic energies ranging from 1–100 eV, providing sufficient mobility upon reaching the substrate to form dense, well-adhered films (Papavlu et al., 2017).

PULSED LASER DEPOSITION: A REVOLUTIONARY TOOL FOR BIOMEDICAL COATINGS

In the ever-evolving landscape of biomedical engineering, the development of functional coatings on medical implants has emerged as a cornerstone of innovation. Among the various techniques available, Pulsed Laser Deposition (PLD) stands out as a versatile and cutting-edge method, capable of fabricating ultra-thin, stoichiometrically accurate films (Gaudiuso, 2023). This technology, with its roots in laser physics and material science, has become a powerful ally in the creation of titanium nitride and composite coatings for orthopedic and dental implants. These coatings, designed to enhance durability and biointegration, represent a fascinating interplay of science, engineering, and medicine. What is it about PLD that makes it so effective in coating development? At its core, Pulsed Laser Deposition relies on the interaction between a high-energy laser beam and a target material. Imagine a focused laser beam, pulsing with incredible intensity, striking a solid target in a vacuum chamber. In an instant, the surface of the target material absorbs the laser energy, heating up to extreme temperatures often exceeding thousands of degrees Celsius in just nanoseconds (Soloviev et al., 2017). This rapid energy transfer causes the material to evaporate violently, forming a plasma plume that contains atoms, ions, and molecules of the target material. This plume, glowing with energy, expands rapidly toward a carefully positioned substrate, where it condenses into a thin film (Kim et al., 2024).

The beauty of PLD lies in its precision. By controlling the laser's energy, wavelength, and pulse duration, as well as environmental factors like substrate temperature and gas pressure, researchers can fine-tune the properties of the deposited film. This level of control enables the fabrication of coatings that are not only ultra-thin but also chemically accurate-qualities that are essential for biomedical applications such as titanium nitride coatings for implants (AbuAlia et al., 2024).

TITANIUM NITRIDE: ENGINEERING A FUNCTIONAL INTERFACE

Titanium nitride (TiN), a ceramic-like compound known for its exceptional hardness, corrosion resistance, and biocompatibility, has become a preferred choice for coating orthopedic and dental implants. It is especially valued for its ability to reduce wear, inhibit bacterial colonization, and enhance the lifespan of medical devices. PLD offers unique advantages for depositing titanium nitride coatings (Chen et al., 2017). Unlike conventional methods, PLD can produce films with high purity and excellent crystallinity, ensuring that the coatings perform reliably under physiological conditions. Furthermore, the versatility of PLD allows researchers to control film thickness, crystal structure, and surface morphology attributes that directly influence the implant's performance in the body. Developing titanium nitride coatings using PLD is both an art and a science. The choice of laser parameters plays a critical role in determining the quality of the film (Atta et al., 2025). For instance, excimer lasers are commonly used for TiN deposition. The energy density must be carefully calibrated to ensure efficient ablation of the target material. Even the duration of each laser pulse, measured in nanoseconds or femtoseconds, can affect the dynamics of the plasma plume and, consequently, the properties of the deposited film. The environment inside the deposition chamber often nitrogen rich is critical to ensure the correct stoichiometry and phase of titanium nitride (Lapitskaya et al., 2023).

Beyond its biomedical relevance, TiN is an archetypal engineering material with a cubic rocksalt (B1–TiN) structure that provides a remarkable combination of hardness, metallic conductivity, and thermal stability. Its properties stem from partially filled Ti3d orbitals hybridized with N2p states, resulting in strong metallic bonding and a high melting point. The microstructure defined by grain size, texture, and density plays a decisive role in determining TiN's electronic and optical performance. PLD-grown TiN films are often dense and stoichiometric, exhibiting superior crystallinity and controlled morphology compared to sputtered or CVD counterparts, leading to lower resistivity and enhanced mechanical integrity. From a materials-science standpoint, the optical and electronic behavior of TiN is particularly notable. The dielectric response of TiN can be described by a Drude–Lorentz model combining intraband (free-electron) and interband (N–p → Ti–d) transitions. This dual contribution produces a characteristic screened plasma energy (Eps ≈ 2.5–2.9 eV), a key indicator of stoichiometric quality and metallic character. The unscreened plasma energy (Epu > 6.9 eV for conductive TiN) reflects the high density of conduction electrons and correlates directly with low resistivity and high carrier mobility. Moreover, the mean free path of conduction electrons in dense TiN films is typically below 50 nm, highlighting the importance of nanoscale control over crystallite boundaries during deposition. Such precise control of stoichiometry and microstructure through PLD allows TiN to serve as a platform for advanced functional interfaces. Its high thermal and mechanical stability, combined with tunable electronic and optical parameters, makes TiN a model system for studying the relationship between structure, bonding, and performance in transition metal nitrides. The ability to tailor these characteristics through laser fluence, nitrogen partial pressure, and substrate temperature provides an exceptional level of design freedom for applications ranging from protective coatings to plasmonic and microelectronic devices.

BEYOND TITANIUM NITRIDE: EXPLORING ADVANCED COATING ARCHITECTURES

While titanium nitride is widely used, PLD's capabilities extend far beyond a single material. Researchers are now exploring more complex coating architectures, leveraging PLD's precision to create multilayered and nanostructured films. Imagine an orthopedic implant with a multilayer coating system designed to optimize performance at every level. The innermost layer might consist of titanium oxide for adhesion, followed by a tough, wear-resistant TiN layer, and an outermost functional layer engineered to combat microbial threats or deliver localized therapeutic agents (Ahmed et al., 2025). Recent developments in pulsed laser deposition have demonstrated the potential of graded, composite, and heterostructured coatings, where each layer is engineered to perform a specific role in mechanical, chemical, or biological functionality. The sequential ablation of multiple targets allows for atomic-scale modulation of composition and thickness, enabling the fabrication of nanolaminates such as TiN/AlN, TiN/TiO_2, or TiN/ZrN that exhibit enhanced hardness, thermal stability, and diffusion barrier performance (Shepelin et al., 2023; Kolaklieva et al., 2019). Such nanolayered systems often exploit strain engineering and interface confinement effects, in which periodic variations in lattice parameters and bonding character give rise to superlattice strengthening and improved crack resistance. The resulting interfaces not only hinder dislocation motion but also suppress oxidation and corrosion under physiological or high-temperature conditions (Theerthagiri et al., 2022).

Furthermore, PLD enables in situ control of phase composition and crystallographic orientation, allowing the integration of metastable or non-equilibrium phases that cannot be achieved through conventional deposition techniques. By tuning laser fluence, repetition rate, and background gas pressure, the degree of ionization and kinetic energy in the plasma plume can be precisely regulated, thereby tailoring adatom mobility and film density at the nanoscale (Mahajan et al., 2024). This flexibility enables the creation of graded stoichiometry coatings such as TiN_XOy or TiAlN systems that exhibit optimized combinations of hardness, electrical conductivity, and corrosion resistance (Chen et al., 2022). In addition to multilayer systems, the fabrication of nanostructured and textured surfaces via PLD has opened new opportunities for enhancing interfacial performance. Controlled growth kinetics and shadowing effects during deposition can generate columnar, nanoporous, or hierarchical morphologies, which improve adhesion and reduce residual stress accumulation. Nanostructured TiN-based coatings, for example, exhibit improved load-bearing capacity and mechanical resilience owing to their defect-tolerant microstructure (Albu et al., 2022). Moreover, nanoscale texturing can be engineered to biological architecture senhancing wettability, mechanical anchorage, and protein adsorption, while promoting cell attachment and tissue integration (Li et al., 2021).

Overall, these advances mark a transition from conventional monolithic coatings toward smart, multifunctional surface systems capable of dynamic responses to environmental stimuli. Future trends include the integration of stimuli-responsive oxynitride layers, ion-doped antimicrobial nanocomposites, and bioactive multilayers that combine the mechanical robustness of TiN with the chemical functionality of oxides, nitrides, or phosphates (Negut et al., 2024). By uniting these design strategies, PLD is poised to play a central role in the next generation of adaptive and durable surface architectures for biomedical, electronic, and energy-related applications.

BRINGING IT ALL TOGETHER: BIOLOGICAL PERFORMANCE

The true test of any biomedical coating lies in its biological performance. Once deposited, coatings must undergo rigorous in vitro and in vivo testing to evaluate cytocompatibility, antibacterial effectiveness, and mechanical stability in a physiological environment. In the case of PLD-derived titanium nitride coatings, a growing body of in vitro evidence confirms their strong potential for biomedical use (Shepelin et al., 2023; Kolaklieva et al., 2019). Notably, TiN surfaces have been shown to support osteoblast adhesion and proliferation, resist bacterial colonization, and maintain structural integrity under cyclic loading without delamination or excessive wear Kolaklieva et al., 2019). For example, Jang et al. demonstrated that TiN and TiAlN coatings improved osteoblast responses relative to bare titanium substrates by providing favorable nanoscale roughness and chemistry (Theerthagiri et al., 2022). Evaluated mixed TiN–ZrN coatings and observed both high antibacterial activity and osteoblastic viability in vitro (Mahajan et al., 2024).

To characterize these biological materials interactions, a suite of surface-analytical and imaging techniques is indispensable. Scanning electron microscopy (SEM) often combined with energy dispersive X-ray spectroscopy (EDS) is used to examine surface morphology, coating uniformity, grain boundaries, defects, and cell-exposed surfaces post-culture (e.g., bacterial biofilms). Atomic force microscopy (AFM) is employed to quantify nanoscale topography (roughness metrics), map mechanical properties (e.g. stiffness via nanoindentation or force modulation), and sometimes monitor friction or adhesion forces in situ (Mahajan et al., 2024; Chen et al., 2022). For instance, the review by Smith et al. outlines many applications of SEM and AFM in metal coatings e.g. assessing film uniformity, cracking, and surface texture (Chen et al., 2022). More advanced techniques, such as correlative AFM in SEM, enable simultaneous morphological and mechanical mapping on the same region, thereby linking topography to local mechanical response (Albu et al., 2022). Beyond morphology, X-ray diffraction (XRD) (and grazing incidence XRD) provides insights into the crystal structure, phase purity, texture or preferred orientation, and residual stress state of the coating. In addition, X-ray photoelectron spectroscopy (XPS), Auger electron spectroscopy (AES), time-of-flight secondary ion mass spectrometry (TOF-SIMS), and Fourier transform infrared spectroscopy (FTIR) furnish chemical composition, bonding states (e.g. TiN vs TiOx), and possible contamination or oxidation. These complementary tools (Figures 2, 3 and 4) together allow researchers to correlate coating microstructure, chemical stability, and mechanical integrity with observed biological responses (e.g. cell spreading, antimicrobial action, protein adsorption) (Negut et al., 2024; Duta et al., 2016).

Figure 2. Plan-view SEM of PLD TiN on 410 SS showing a dense, defect-free morphology (Duta et al., 2016)

Figure 3. Cross-sectional SEM confirming compact TiN architecture and uniform thickness (Duta et al., 2016)

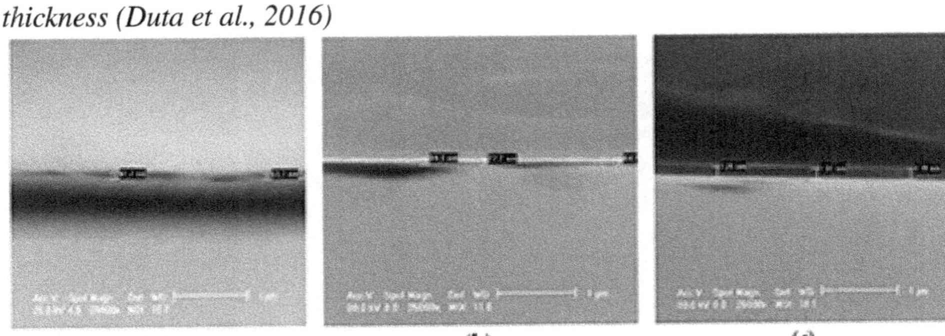

Figure 4. AFM 3D topography (2 × 2 μm²) of PLD TiN illustrating low Rq/Ra and homogeneous nanoroughness (Duta et al., 2016)

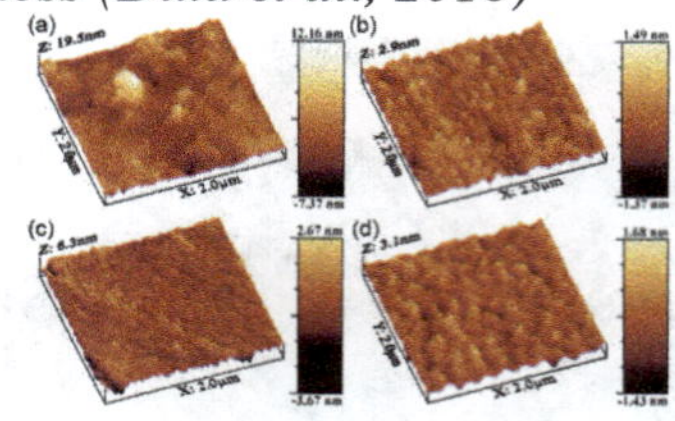

Figure 5. XRD (θ–θ) and GI-XRD patterns confirming cubic TiN B1 phase with (200) orientation; crystallite size estimated by Scherrer broadening (Duta et al., 2016)

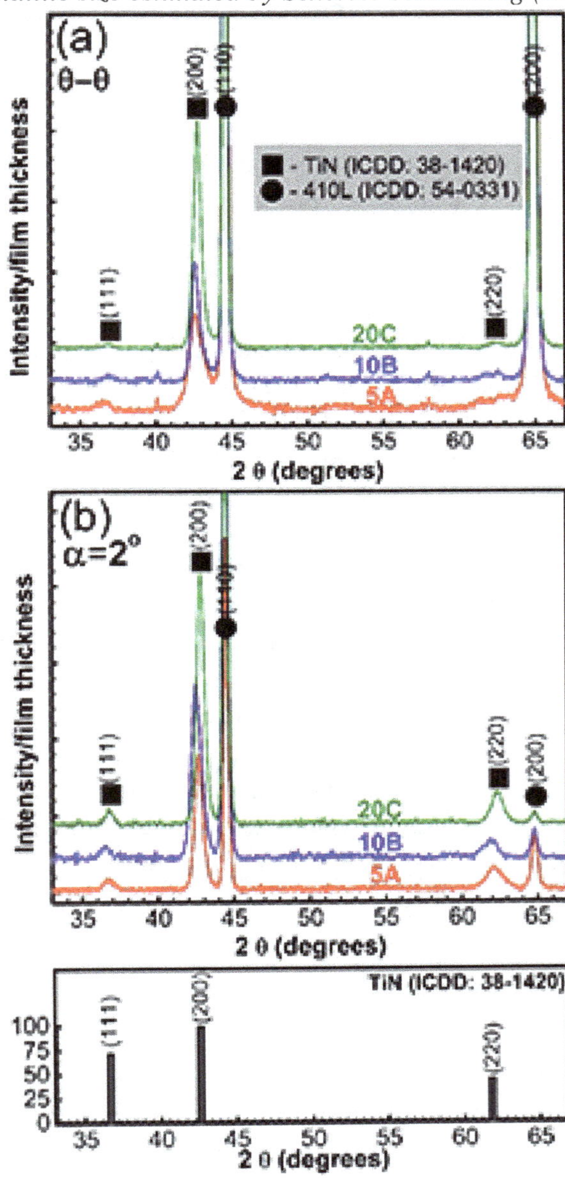

The true test of any biomedical coating lies in its biological performance. Once deposited, coatings must undergo rigorous in vitro and, where feasible, in vivo evaluation to confirm cytocompatibility, antibacterial activity, and mechanical stability under physiologically relevant loads. For PLD derived TiN, multiple studies report high osteoblast adhesion/proliferation, suppressed bacterial colonization, and robust interfacial integrity during wear or scratch testing (Shepelin et al., 2023; Negut et al., 2024). A representative PLD study showed that thickness-tuned, dense TiN layers with cubic B1 structure and low roughness (Rq $\lesssim$ 1 nm over 2×2 μm^2 scans) remained adherent and cytocompatible, linking film stoichiometry, texture, and nanoroughness to cell response (Duta et al., 2016). Complementary characterization provides the structure property bridge: SEM (plan-view and cross-section) confirms densification, thickness uniformity, and absence of microcracks; AFM quantifies nm-scale topography linked to protein adsorption and cell spreading; XRD/ GI XRD establishes phase purity, preferred orientation (often TiN(200)), and crystallite size evolution with processing (Shepelin et al., 2023; Duta et al., 2016). In parallel, XPS or TOF SIMS validates surface chemistry (e.g., TiN vs. Ti OxNy), while standardized cell assays (MTS/AlamarBlue, LDH, live/dead) and tribocorrosion tests in simulated body fluids assess bioperformance durability. Collectively, this integrated workflow ensures that PLD TiN and advanced multilayers meet stringent biomedical standards before translation (Negut et al., 2024; Duta et al., 2016; Song et al., 2020).

Finally, in vitro assays such as live/dead fluorescent staining, Alamar Blue or MTT for cell viability, fluorescence or luminescence-based bacterial viability tests, and shear or scratch tests under simulated physiological loads are crucial to validate that the engineered coatings function as intended. Long term tests in simulated body fluids (SBF) or under cyclic mechanical loads help assess durability, ion leaching, and fatigue behavior (Al Jabbari et al., 2012). In vivo animal studies remain the gold standard for evaluating osseointegration, biofilm resistance, and immune response. By integrating these analytical methods and biological assays under a unified framework, researchers can rigorously validate advanced PLD coatings (e.g. multilayer, nanostructured, doped systems) as safe, functional, and robust interfaces for implantable devices.

FROM LABORATORY TO OPERATING ROOM: SCALING UP PLD

Although Pulsed Laser Deposition (PLD) has demonstrated remarkable precision and versatility in laboratory settings, translating this technology into large-scale industrial production remains a complex and multifaceted challenge (Molaei et al., 2025). The inherent precision of PLD requiring meticulous control over parameters such as laser fluence, pulse frequency, target substrate distance, and substrate temperature poses significant difficulties when reproducing uniform results across larger surface areas or batch productions. The process's sensitivity to even minor variations in these parameters can affect film thickness, stoichiometry, and microstructure, which in turn influence the coating's mechanical, chemical, and biological performance. Consequently, the transition from laboratory-scale experimentation to industrial scale fabrication necessitates significant advancements in process automation, system design, and quality assurance protocols. To address these challenges, modern PLD systems have evolved to incorporate sophisticated engineering solutions (Molaei et al., 2025). Advanced multi-target configurations now allow sequential or simultaneous ablation of different materials, facilitating the deposition of multilayered or compositionally graded coatings (Thapliyal et al., 2025). Automated substrate manipulation systems enable precise motion control, ensuring uniform film deposition across complex geometries typical of biomedical implants. Real-time monitoring and feedback mechanisms such as in situ optical emission spectroscopy, quartz crystal microbalance (QCM) sensors, and laser interferometry allow dynamic adjustment of process parameters to maintain coating consistency and repeatability (Schneider et al., 2023). Additionally, innovations in high-power laser technology and large-area plume control have significantly improved deposition rates, bringing PLD closer to meeting industrial throughput requirements. From a manufacturing perspective, scalability also involves considerations beyond technical reproducibility. Factors such as cost-effectiveness, energy efficiency, and compatibility with existing production lines play crucial roles in determining the commercial viability of PLD based coating systems (Bi et al., 2023). Collaborative efforts between academic research institutions, equipment manufacturers, and medical device companies have therefore focused on developing hybrid deposition platforms and modular PLD systems capable of continuous operation under cleanroom conditions. These efforts have led to the emergence of semi-industrial PLD systems capable of producing high quality coatings with tailored properties on orthopedic and dental implants at near-commercial scales (Schmidt et al., 2024).

The successful industrialization of PLD represents a key step toward bridging the gap between laboratory innovation and clinical application. With ongoing advances in system design, automation, and process standardization, PLD is poised to become a cornerstone technology for the scalable production of next-generation biomedical coatings combining scientific precision with industrial efficiency to meet the growing demands of the medical device industry.

THE FUTURE OF PLD IN BIOMEDICAL APPLICATIONS

As we look to the future, the potential of Pulsed Laser Deposition (PLD) in biomedical applications continues to broaden, driven by rapid advances in materials science, laser technology, and surface engineering. Researchers are increasingly investigating novel coating compositions and architectures that transcend traditional functionalities, moving toward smart and responsive surfaces capable of interacting dynamically with the biological environment (Zuo et al., 2025). Emerging research directions include the development of drug-eluting coatings for controlled therapeutic release, antibacterial and antifouling surfaces that actively prevent microbial colonization, and stimuli-responsive materials that adapt their physicochemical behavior in response to changes in pH (Lorusso et al., 2024), temperature (Chen et al., 2024), or biochemical (Depablos-Rivera et al., 2019). Such innovations are expected to significantly enhance implant integration, reduce postoperative complications, and improve overall patient outcomes. Advances in ultrafast laser systems including femtosecond and picosecond laser technologies are poised to further revolutionize PLD by providing unparalleled control over energy delivery, plume dynamics, and film growth mechanisms (Liguori et al., 2019). The use of beam shaping and spatial light modulation techniques allows for precise tailoring of laser target interactions, enabling deposition of films with highly uniform thickness, controlled crystallinity, and nanostructured morphologies. These capabilities will facilitate the fabrication of coatings with complex compositional gradients and hierarchical architectures that mimic the multi-scale organization of natural tissues (Yu et al., 2024). Moreover, the integration of PLD with complementary thin-film technologies, such as magnetron sputtering, atomic layer deposition (ALD), and pulsed electron deposition, is anticipated to yield hybrid fabrication platforms capable of producing multifunctional and customizable biomedical coatings (Cai et al., 2025). In parallel, computational modeling and machine learning assisted process optimization are emerging as powerful tools for predicting film properties, optimizing deposition parameters, and accelerating materials discovery. The application of in silico approaches to PLD process design can help bridge the gap between fundamental research and industrial manufacturing by enabling predictive control and process standardization (Ryou et al., 2025). Furthermore, as regulatory frameworks and clinical validation protocols continue to evolve, PLD's reproducibility and precision make it particularly attractive for medical-grade surface engineering applications that demand stringent quality assurance.

Ultimately, the convergence of advanced laser technology, intelligent process control, and biomaterial innovation will position PLD as a cornerstone technique for the next generation of biomedical coatings. Its unique capacity to fabricate tailored, high purity, and functionally graded surfaces offers immense promise for the realization of smart implants and medical devices that not only restore biological function but also actively participate in the healing process.

CONCLUSION

In conclusion, this chapter emphasizes the transformative role of Pulsed Laser Deposition (PLD) in the development of advanced titanium nitride (TiN) and related coatings for biomedical applications. Through its distinctive capability to combine atomic-scale precision, compositional purity, and structural versatility, PLD stands apart from conventional deposition technologies. The technique's ability to manipulate key parameters laser fluence, pulse duration, substrate temperature, and background atmosphere allows the creation of thin films with tailored crystallinity, surface morphology, and stoichiometry. These advantages have made PLD deposited TiN coatings a benchmark in the pursuit of biointegrative, wear-resistant, and infection-resistant surfaces for orthopedic, dental, and cardiovascular devices. Beyond its practical successes, PLD represents a convergence of physics, chemistry, and biology in materials engineering. The controlled ablation of a solid target and the subsequent formation of a dynamic plasma plume enable the growth of films with nanometer precision, allowing researchers to explore novel interface phenomena that directly influence cellular behavior. TiN coatings, when optimized for stoichiometric balance and surface roughness, have consistently demonstrated improved osteoblast adhesion, balanced wettability, and enhanced mechanical anchorage at the implant tissue interface. Importantly, these effects emerge not solely from chemical inertness but from the interplay between surface energy, grain orientation, and nanoscale topography parameters that PLD can adjust with exceptional fidelity.

The flexibility of PLD extends naturally into the design of multifunctional and hybrid architectures. Multilayer systems such as TiN/TiO_2, TiN/ZrN, or TiN/AlN capitalize on interface engineering to combine hardness, corrosion resistance, and biocompatibility within a single coating. The capability to alternate or co-ablate multiple targets within one deposition chamber encourages the creation of gradient and nanocomposite coatings, bridging mechanical robustness with biological activity. Moreover, the integration of PLD with complementary methods such as atomic layer deposition (ALD), magnetron sputtering, or pulsed electron deposition (PED) is expanding the available material palette and enabling hierarchically organized coatings that mimic the functional gradation of natural tissues. Equally significant is the role of advanced characterization in correlating structure and performance. Modern surface analysis techniques ranging from X-ray diffraction (XRD) and X-ray photoelectron spectroscopy (XPS) to atomic force microscopy (AFM) and electron microscopy have revealed how nanoscale variations in roughness, grain size, and oxide incorporation affect cell adhesion and protein adsorption. Combined with mechanical and electrochemical testing, these insights establish a feedback loop between synthesis and performance, reinforcing PLD's reputation as both a scientific tool and a manufacturing route. Looking ahead, the ongoing miniaturization of laser systems and the emergence of ultrafast femtosecond and picosecond regimes promise to refine energy delivery and plasma dynamics, fostering smoother films with reduced particulate contamination. The introduction of beam shaping and spatial light modulation further broadens control, enabling the deposition of coatings with spatially resolved properties and complex geometries suited to next-generation implants. At the same time, data-driven approaches, including machine learning assisted process optimization and predictive modeling of plasma substrate interactions, are poised to revolutionize how PLD processes are designed, monitored, and scaled. Such computational frameworks will promote reproducibility and accelerate the transition of PLD from laboratory demonstrations to standardized biomedical manufacturing.

Ultimately, PLD embodies the evolution of thin-film science from a process of passive coating toward active interface engineering. Its capacity to synthesize biocompatible, durable, and functionally graded surfaces will be instrumental in developing medical devices that not only restore but also enhance biological performance. As materials science continues to merge with digital design, robotics, and personalized medicine, PLD is expected to remain a cornerstone technique uniting precision engineering with biomedical innovation. Through sustained interdisciplinary collaboration, the promise of PLD-fabricated TiN and composite coatings may soon translate into safer, longer-lasting, and smarter implant technologies that define the future of regenerative and restorative medicine.

REFERENCES

AbuAlia, M., Fullam, S., Cinotti, F., Manninen, N., & Wimmer, M. A. (2024). Titanium nitride coatings on CoCrMo and Ti6Al4V alloys: Effects on wear and ion release. *Lubricants (Basel, Switzerland), 12*(3), 96. Advance online publication. DOI: 10.3390/lubricants12030096 PMID: 41743082

Ahmed, Z., Wang, Z., Adil, M., Bhatti, I. A., & Cao, H. (2025). Bright antimicrobial surfaces based on Schottky interfaces: From light illumination to bacterial charging. *Smart Materials in Medicine, 6*(1), 67–94. Advance online publication. DOI: 10.1016/j.smaim.2025.02.001

Ahsan, S., Ahmed, F., Kim, Y. G., Lee, M. S., & Jun, M. B. (2011). Colorizing stainless steel surface by femtosecond laser induced micro/nano-structures. *Applied Surface Science, 257*(17), 7771–7777. DOI: 10.1016/j.apsusc.2011.04.027

Al Jabbari, Y. S., Fehrman, J., Barnes, A. C., Zapf, A. M., Zinelis, S., & Berzins, D. W. (2012). Titanium nitride and nitrogen ion implanted coated dental materials. *Coatings, 2*(3), 160–178. DOI: 10.3390/coatings2030160

Albu, D. F., Caraman, M., Cotorobai, V., Radu, A., Cobianu, C., & Gurlui, S. (2022). Thin film fabrication by pulsed laser deposition from TiO_2: Growth atmosphere effects on structure and optoelectronic properties. *Coatings, 12*(3), 293. DOI: 10.3390/coatings12030293

Amaral, C., Paiva, M., Rodrigues, A. R., Veiga, F., & Bell, V. (2024). Global regulatory challenges for medical devices: Impact on innovation and market access. *Applied Sciences (Basel, Switzerland), 14*(20), 9304. Advance online publication. DOI: 10.3390/app14209304

Atta, S., Dash, S., Mathew, A. T., & NarendraKumar, U. (2025). Recent developments on titanium based mono and multilayer nitride films deposited through HiPIMS. *Next Materials, 9*, 100983. Advance online publication. DOI: 10.1016/j.nxmate.2025.100983

Balcerak-Woźniak, A., Dzwonkowska-Zarzycka, M., & Kabatc-Borcz, J. (2024). A comprehensive review of stimuli-responsive smart polymer materials—Recent advances and future perspectives. *Materials (Basel), 17*(17), 4255. Advance online publication. DOI: 10.3390/ma17174255 PMID: 39274645

Bi, J., Li, H., Ren, W., Wang, Z., & Li, C. (2023). Beam shaping technology and its application in metal laser processing. *Journal of Manufacturing Processes, 93*, 88–108. DOI: 10.1016/j.jmapro.2023.08.038

Bril, M., Fredrich, S., & Kurniawan, N. A. (2022). Stimuli-responsive materials: A smart way to study dynamic cell responses. *Smart Materials in Medicine, 3*, 257–273. DOI: 10.1016/j.smaim.2022.01.010

. Brown, J. J., Smith, A. B., & Johnson, C. D. (2017). Pulsed laser deposition of thin films: A review. Journal of Vacuum Science & Technology A, 35(1), Article 01A101. https://doi.org/DOI: 10.1116/1.4971234

Cai, J., Li, F., Zhang, X., Wang, J., Yu, Z., Feng, B., & Li, Y. (2025). Application of pulsed laser deposition (PLD) technology in the preparation of two-dimensional (2D) film materials. *Materials (Basel), 18*(13), 2999. Advance online publication. DOI: 10.3390/ma18132999 PMID: 40649486

Cai, J., Yang, F., & Xu, H.. (2025). Application of pulsed laser deposition technology in the preparation of 2D film materials. *Materials (Basel), 18*(13), 2999. Advance online publication. DOI: 10.3390/ma18132999 PMID: 40649486

Chen, L., Wang, Z., Wang, L., & Li, B. (2024). Hydrodynamic expansion and plume splitting of the ultrafast laser-induced plasma of an Al-Cu alloy. *Journal of Applied Physics, 135*(8), 083101. Advance online publication. DOI: 10.1063/1.3565061

Chen, N., Tian, L., Patil, A. C., Peng, S., Yang, I. H., Thakor, N. V., & Ramakrishna, S. (2017). Neural interfaces engineered via micro- and nanostructured coatings. *Nano Today, 14*, 59–83. DOI: 10.1016/j.nantod.2017.04.007

Chen, Z., Zhen, Y.-H., Huang, Y., Gao, Z.-C., Sheng, H.-P., Mayrhofer, P. H., & Zhang, Z.-L. (2022). Atomic-scale understanding of the structural evolution of TiN/AlN superlattices during nanoindentation—Part I: Deformation. *Acta Materialia, 234*, 118008. DOI: 10.1016/j.actamat.2022.118008

Chowdhury, S., Yadaiah, N., Prakash, C., Ramakrishna, S., Dixit, S., Gupta, L. R., & Buddhi, D. (2022). Laser powder bed fusion: A state-of-the-art review of the technology, materials, properties & defects, and numerical modelling. *Journal of Materials Research and Technology, 20*, 2109–2172. DOI: 10.1016/j.jmrt.2022.07.121

Depablos-Rivera, O., Pedraza-Hernández, R., Martínez, E., Valdivia-Barrientos, R., & Rangel, R. H.Jr. (2019). Hybrid magnetron sputtering and pulsed laser ablation for the synthesis of ZnO–Au composite thin films. *Thin Solid Films, 687*, 137518. Advance online publication. DOI: 10.1016/j.tsf.2019.04.032

Dilova, T., Atanasova, G., Dikovska, A., & Nedyalkov, N. (2019). The effect of light irradiation on the gas-sensing properties of nanocomposites based on ZnO and Ag nanoparticles. *Applied Surface Science, 505*, 144625. Advance online publication. DOI: 10.1016/j.apsusc.2019.144625

Duta, L., Stan, G. E., Popa, A. C., Husanu, M. A., Moga, S., Socol, M., Zgura, I., Miculescu, F., Urzica, I., Popescu, A. C., & Mihailescu, I. (2016). Thickness influence on in vitro biocompatibility of titanium nitride thin films synthesized by pulsed laser deposition. *Materials (Basel), 9*(1), 38. Advance online publication. DOI: 10.3390/ma9010038 PMID: 28787846

Eason, R. (Ed.). (2007). *Pulsed laser deposition of thin films: Applications-led growth of functional materials*. Wiley.

Gao, H., Wang, Y., Ye, J., Du, B., Wang, D., Li, S., Cui, Q., Wang, S., & Zhang, T. (2025). Study on the ablation behavior of high-intensity lasers in vacuum. *Applied Sciences (Basel, Switzerland), 15*(2), 848. Advance online publication. DOI: 10.3390/app15020848

Gaudiuso, R. (2023). Pulsed laser deposition of carbon-based materials: A focused review of methods and results. *Processes (Basel, Switzerland), 11*(8), 2373. Advance online publication. DOI: 10.3390/pr11082373

Harilal, S. S., Brumfield, B. E., LaHaye, N. L., Hartig, K. C., & Phillips, M. C. (2018). Optical spectroscopy of laser-produced plasmas for standoff isotopic analysis. *Applied Physics Reviews, 5*(2), 021301. Advance online publication. DOI: 10.1063/1.5016053

Huang, Z., Tsui, G. C.-P., Deng, Y., & Tang, C.-Y. (2020). Two-photon polymerization nanolithography technology for fabrication of stimulus-responsive micro/nanostructures for biomedical applications. *Nanotechnology Reviews, 9*(1), 1118–1136. DOI: 10.1515/ntrev-2020-0073

Irimiciuc, S. A., Nica, P., Agop, M., & Focsa, C. (2019). Target properties – Plasma dynamics relationship in laser ablation of metals: Common trends for fs, ps and ns irradiation regimes. *Applied Surface Science, 506*, 144926. Advance online publication. DOI: 10.1016/j.apsusc.2019.144926

Kim, Y., Choi, Y., Lee, S. A., Choi, W. S., & Kang, K. T. (2024). Complex oxide thin films: A review on pulsed laser epitaxy growth. *Current Applied Physics, 68*, 113–130. DOI: 10.1016/j.cap.2024.09.001

Kolaklieva, L., Grozdanov, I., Zafirova, K., Boev, V., Srebrenkoska, V., Mitrovic, M., & Krumme, A. (2019). Pulsed laser deposition of aluminum nitride films: Correlation between mechanical, optical and structural properties. *Coatings, 9*(3), 195. DOI: 10.3390/coatings9030195

. Lang, F., Pang, J., & Bu, X. (2024). Stimuli-responsive coordination polymers toward next-generation smart materials and devices. eScience, 4(3), Article 100231. DOI: 10.1016/j.esci.2024.100231

Lapitskaya, V., Nikolaev, A., Khabarava, A., Sadyrin, E., Antipov, P., Abdulvakhidov, K., Aizikovich, S., & Chizhik, S. (2023). The influence of nitrogen flow on the stoichiometric composition, structure, mechanical, and microtribological properties of TiN coatings. *Materials (Basel), 17*(1), 120. Advance online publication. DOI: 10.3390/ma17010120 PMID: 38203974

Li, J., Zhou, P., Attarilar, S., & Shi, H. (2021). Innovative surface modification procedures to achieve micro/nano-graded Ti-based biomedical alloys and implants. *Coatings, 11*(6), 647. Advance online publication. DOI: 10.3390/coatings11060647

Liguori, A., Capasso, L., & Di Benedetto, A.. (2019). The pulsed electron deposition technique for biomedical applications: A review. *Coatings, 9*(1), 16. Advance online publication. DOI: 10.3390/coatings10010016

Lorusso, A., Campanella, D., Di Fonzo, F., Loconsole, A., Girolami, M., & Rea, M.. (2024). The potential of spot size control in shaping the thickness distribution of PLD films. *Materials (Basel), 17*(11), 2712. Advance online publication. DOI: 10.3390/ma17112712 PMID: 38893975

Lu, X., Chang, L., Tran, M. A., Komljenovic, T., Bowers, J. E., & Srinivasan, K. (2024). Emerging integrated laser technologies in the visible and short near-infrared regimes. *Nature Photonics, 18*(10), 1010–1023. DOI: 10.1038/s41566-024-01529-5

Mahajan, U., Bhatia, H., Singh, A., Gupta, R., & Mahapatra, P. S. (2024). Titanium nitride as a promising alternative to plasmonic metals: Properties, processing, and applications. *Materials Advances, 5*(7), 3642–3667.

Mahamood, R. M. (2017). *Laser metal deposition process of metals, alloys, and composite materials*. Springer., DOI: 10.1007/978-3-319-64985-6

Matei, A. T., Visan, A. I., & Negut, I. (2025). Laser-fabricated micro/nanostructures: Mechanisms, fabrication techniques, and applications. *Micromachines, 16*(5), 573. Advance online publication. DOI: 10.3390/mi16050573 PMID: 40428699

Molaei, M., Al-Mangour, B., Alhawiti, A., Alidrisi, H., & Manladan, S. M. (2025). Plasma electrolytic oxidation (PEO) coatings for biomedical implants: A review on enhancing antibacterial efficacy through controlled antibiotic release. *Coatings, 15*(8), 925. Advance online publication. DOI: 10.3390/coatings15080925

Municoy, S., Echazú, M. I. Á., Antezana, P. E., Galdopórpora, J. M., Olivetti, C., Mebert, A. M., Foglia, M. L., Tuttolomondo, M. V., Alvarez, G. S., Hardy, J. G., & Desimone, M. F. (2020). Stimuli-responsive materials for tissue engineering and drug delivery. *International Journal of Molecular Sciences*, *21*(13), 4724. Advance online publication. DOI: 10.3390/ijms21134724 PMID: 32630690

Negut, I., Chiriac, A. P., & Grumezescu, A. M. (2024). Advances in antimicrobial coatings for preventing infections in implantable and wearable medical devices. *Coatings*, *14*(3), 256. DOI: 10.3390/coatings14030256

Nikolova, M. P., & Apostolova, M. D. (2022). Advances in multifunctional bioactive coatings for metallic bone implants. *Materials (Basel)*, *16*(1), 183. Advance online publication. DOI: 10.3390/ma16010183 PMID: 36614523

Papavlu, A. P., Dinca, V., Filipescu, M., & Dinescu, M. (2017). Matrix-assisted pulsed laser evaporation of organic thin films: Applications in biology and chemical sensors. In InTech eBooks. DOI: 10.5772/intechopen.70676

Park, J. H., Pattipaka, S., Hwang, G., Park, M., Woo, Y. M., Kim, Y. B., Lee, H. E., Jeong, C. K., Zhang, T., Min, Y., Park, K., Lee, K. J., & Ryu, J. (2024). Light–material interactions using laser and flash sources for energy conversion and storage applications. *Nano-Micro Letters*, *16*(1), 314. Advance online publication. DOI: 10.1007/s40820-024-01483-5 PMID: 39186184

Ryou, S., Xie, M., & Lee, J.. (2025). Machine-learning-enabled on-the-fly analysis of RHEED during thin-film growth. *Journal of Vacuum Science & Technology. A, Vacuum, Surfaces, and Films*, *43*(3), 032702. Advance online publication. DOI: 10.1116/6.0004493

Sattar, H., Hu, Z., Zheng, W., Zhou, J., Shahzad, M. K., Rehman, F., Guo, L., & Luo, W. (2023). Exploring the potential and recent advancement in laser opto-ultrasonic detection for material characterization: A state-of-the-art review. *Optics & Laser Technology*, *171*, 110316. Advance online publication. DOI: 10.1016/j.optlastec.2023.110316

Schmidt, M., Hassel, T., & Vollertsen, F.. (2024). Dynamic beam shaping—Improving laser materials processing by tailored energy deposition. *CIRP Annals*, *73*(2), 501–524. DOI: 10.1016/j.cirp.2024.05.005

Schneider, C. W., Lippert, T., & Wokaun, A. (2023). PLD plasma plume analysis: A summary of the PSI contributions. *Applied Sciences (Basel, Switzerland)*, *13*(2), 1186. Advance online publication. DOI: 10.3390/app13021186

Shen, D., Zhang, X., & Zhu, L. (2023). Laser processing for electricity generators: Physics, methods and applications. *Nano Energy*, *120*, 109182. Advance online publication. DOI: 10.1016/j.nanoen.2023.109182

Shepelin, N. A., Tehrani, Z. P., Ohannessian, N., Schneider, C. W., Pergolesi, D., & Lippert, T. (2023). A practical guide to pulsed laser deposition: From fundamentals to advanced functional thin films. *Chemical Society Reviews*, *52*(7), 2294–2321. DOI: 10.1039/D2CS00938B PMID: 36916771

Soloviev, A., Burdonov, K., Chen, S. N., Eremeev, A., Korzhimanov, A., Pokrovskiy, G. V., Pikuz, T. A., Revet, G., Sladkov, A., Ginzburg, V., Khazanov, E., Kuzmin, A., Osmanov, R., Shaikin, I., Shaykin, A., Yakovlev, I., Pikuz, S., Starodubtsev, M., & Fuchs, J. (2017). Experimental evidence for short-pulse laser heating of solid-density target to high bulk temperatures. *Scientific Reports*, *7*(1), 11675. Advance online publication. DOI: 10.1038/s41598-017-11675-2 PMID: 28939883

Song, S. H., Lee, S.-W., Kim, Y.-K., Lee, Y., Ryu, J.-J., Heo, S.-J., & Lee, J.-H. (2020). Application of a novel CVD TiN coating on a biomedical Co–Cr alloy: An evaluation of coating layer and substrate characteristics. *Materials (Basel)*, *13*(5), 1145. Advance online publication. DOI: 10.3390/ma13051145 PMID: 32150803

Sypniewska, J., & Szkodo, M. (2022). Influence of laser modification on the surface character of biomaterials: Titanium and its alloys—A review. *Coatings*, *12*(10), 1371. Advance online publication. DOI: 10.3390/coatings12101371

Thapliyal, D., Kumar, P., & Saxena, A. (2025). Nanoparticle-doped antibacterial and antifungal coatings: A review. *Polymers*, *17*(2), 247. Advance online publication. DOI: 10.3390/polym17020247 PMID: 39861318

Theerthagiri, J., Karuppasamy, K., Lee, S. J., Shwetharani, R., Kim, H., Pasha, S. K. K., Ashokkumar, M., & Choi, M. Y. (2022). Fundamentals and comprehensive insights on pulsed laser synthesis of advanced materials for diverse photo- and electrocatalytic applications. *Light, Science & Applications*, *11*(1), 210. Advance online publication. DOI: 10.1038/s41377-022-00904-7 PMID: 35945216

Theerthagiri, J., Senthil, R. A., Madhavan, J., Arunachalam, P., Ashokkumar, M., Choi, M. Y., & Kim, H.-S. (2022). Fundamentals and comprehensive insights on pulsed laser–based nanomaterial fabrication. *Light, Science & Applications*, *11*, 210. DOI: 10.1038/s41377-022-00904-7

Vakulov, Z., Khakhulin, D., Zamburg, E., Mikhaylichenko, A., Smirnov, V. A., Tominov, R., Klimin, V. S., & Ageev, O. A. (2021). Towards scalable large-area pulsed laser deposition. *Materials (Basel)*, *14*(17), 4854. Advance online publication. DOI: 10.3390/ma14174854 PMID: 34500943

Visan, A. I., & Popescu-Pelin, G. F. (2024). Advanced laser techniques for the development of nature-inspired biomimetic surfaces applied in the medical field. *Coatings*, *14*(10), 1290. Advance online publication. DOI: 10.3390/coatings14101290

Xu, Z., Wang, Y. A., Ng, V., Yin, H., & Xu, S. (2025). Advancements in laser-processed functional surfaces for medical devices: A current review. *Nanomaterials (Basel, Switzerland)*, *15*(13), 999. Advance online publication. DOI: 10.3390/nano15130999 PMID: 40648706

Yang, L., Wei, J., Ma, Z., Song, P., Ma, J., Zhao, Y., Huang, Z., Zhang, M., Yang, F., & Wang, X. (2019). The fabrication of micro/nano structures by laser machining. *Nanomaterials (Basel, Switzerland)*, *9*(12), 1789. DOI: 10.3390/nano9121789 PMID: 31888222

Yu, C., Ye, J., Chang, H., Zhou, W., Han, X., Li, M., & Gao, H. (2023). Experimental research on characteristics of impulse coupling and plasma plume generated by laser irradiating copper target with nanosecond pulsed laser propulsion. *Aerospace (Basel, Switzerland)*, *10*(6), 544. Advance online publication. DOI: 10.3390/aerospace10060544

Yu, J., Huang, W., & Li, J.. (2024). Recent advances on pulsed laser deposition of large-area 2D materials. *Small Methods*, *8*(10), 2301282. Advance online publication. DOI: 10.1002/smtd.202301282 PMID: 39082065

Zhu, D., Zuo, P., Li, F., Tian, H., Liu, T., Hu, L., Huang, H., Liu, J., & Qian, X. (2024). Fabrication and applications of surface micro/nanostructures by femtosecond laser. *Colloid and Interface Science Communications*, *59*, 100700. DOI: 10.1016/j.colcom.2024.100770

Zhu, D., Zuo, P., Li, F., Tian, H., Liu, T., Hu, L., Huang, H., Liu, J., & Qian, X. (2024). Fabrication and applications of surface micro/nanostructures by femtosecond laser. *Colloid and Interface Science Communications*, *59*, 100700. DOI: 10.1016/j.colcom.2024.100770

Zuo, S., Wang, T., & Wang, Y.. (2025). High-performance NIR laser-beam shaping and materials processing at 350 W with a spatial light modulator. *Photonics*, *12*(6), 544. Advance online publication. DOI: 10.3390/photonics12060544

Chapter 3
Hybrid Coating Systems: Bioactive Molecules and Nanoparticles for Cancer Therapy

Ramyakrishna Jangama
Sri Sathya Sai University for Human Excellence, India

Chaya Suresh
Sri Sathya Sai University for Human Excellence, India

Chandana Korrapati
Sri Sathya Sai University for Human Excellence, India

Ramya Raghavan
http://orcid.org/0000-0002-9953-543X
Sri Sathya Sai University for Human Excellence, India

ABSTRACT

Hybrid bioactive coatings for cancer therapy offer controlled drug delivery, targeted treatment, and multimodal therapeutic approaches. In this study, we analyzed functional performance metrics and clinical study parameters. Data from multiple reports were evaluated on the therapeutic efficacy and safety profile of hybrid coatings. Results demonstrated an improvement in drug loading efficiency from 40% to 85%. The controlled release durations spanning hours to weeks, enabling sustained therapeutic payloads. Targeting efficiency demonstrated a 2- to 5-fold increase in cancer cell uptake, accompanied by enhanced selective cytotoxicity. In vivo tumor volume reduction ranged from 30% to 60%, with combination photothermal or photodynamic treatments. The hybrid bioactive coatings significantly improve cancer treatment efficacy and safety, supporting their continued development and clinical

DOI: 10.4018/979-8-3373-4104-0.ch003

translation. Hybrid coatings for cancer theranostics combine diagnostic imaging, targeted drug delivery, and therapeutic modalities within a single multifunctional platform.

INTRODUCTION

Cancer is a leading cause of premature mortality globally, with economic implications projected to reach USD 25.2 trillion from 2020 to 2050. The evolution of nanomedicines in cancer treatment focuses on improving drug delivery and reducing toxicity (Pallares et al., 2025). Hybrid coating systems that integrate bioactive molecules and nanoparticles are an emerging field in cancer therapy. Hybrid coatings combine organic (bioactive molecules) and inorganic (nanoparticles) components, offering tailored biological and physicochemical properties. Laser processing, pulsed laser deposition, and laser ablation enable precise surface modification, enhanced coating adhesion, and controlled nanostructure formation. These coatings can be applied to medical devices and implants or administered as standalone therapeutic systems, offering the potential for sustained drug release, enhanced biocompatibility, and improved targeting efficiency. The integration of nanoparticles, targeting ligands, and bioactive molecules within these coatings enables precise control over drug loading, release kinetics, and therapeutic specificity. Drug-loading efficiency across hybrid bioactive coatings ranged from 40% to 85%, with higher loading capacities achieved by optimizing coating composition and nanoparticle integration. The controlled-release duration varied significantly from hours to weeks, depending on the coating design and the intended therapeutic application. This broad range enables customization for both acute and chronic treatment protocols.

Laser techniques create micro- and nanoscale surface features that improve safety profiles compared to systemic drug administration. Biocompatibility assessments showed greater than 85% cell viability in non-cancerous cells. The therapeutic applications of hybrid coating adhesion were explored, including red blood cells, platelets, cancer cells, neutrophils, macrophages, mesenchymal stem cells (MSCs), T cells, natural killer (NK) cells, and bacterial membranes (E. coli). The degradation profiles and chronic toxicity of the inorganic NP core (Au, Ag, SiO_2) remain incompletely understood. While biomimetic coatings reduce clearance, they can actively engage the immune system. Immune interactions can be therapeutic (activating T-cells) or detrimental (inducing an unpredictable cytokine storm). Long-Term effects between the synthetic core and the biological shell are a new frontier in immunotoxicology. (Lakshmaiah & Raghavan, 2026) Nanoparticles loaded with doxorubicin and indocyanine green can generate heat under near-infrared irradiation while releasing drugs in response to tumor microenvironment conditions such as low pH and high glutathione levels, resulting in effective tumor suppression with reduced systemic toxicity.

- Gold (Au) & Silver (Ag): For photothermal therapy, radiation sensitization, and imaging.
- Iron Oxide (Fe_3O_4): For magnetic hyperthermia and as MRI contrast agents.
- Silica (SiO_2): As mesoporous carriers for drug loading.
- Quantum Dots & Carbon-Based (Graphene, CNTs) biosensing and imaging.

Figure 1. Hybrid bioactive coatings offer significant advantages over conventional cancer therapy approaches. Performance ranges were extracted from reported values across multiple studies.

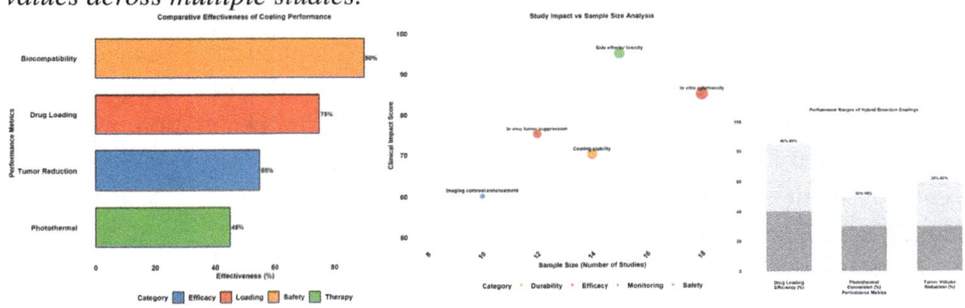

Bioactive composite coatings containing bioactive glass and antimicrobial peptides have been deposited onto titanium substrates, producing nanoscale structures that improve cellular adhesion, corrosion resistance, and bioactivity. Such coatings demonstrate potential for cancer-related orthopedic applications, including bone tumor treatment and implant integration. RBCs can be sourced from donor blood, but using patient-derived cancer cells or immune cells raises questions of cost, access, and the ethics of creating personalized, complex therapeutics. The large-scale production and eventual disposal of persistent nanomaterials (quantum dots, carbon nanotubes) require a proactive environmental risk assessment. The future of cancer nano-therapy lies in bio-inspired, multifunctional platforms. The most promising trends are those that leverage well-understood biology (RBC, leukocyte membranes) to solve clear clinical problems (long circulation, tumor targeting) (Figure 1). Novel hybrid-coated mechanisms need to address translational scalability, manufacturing, safety profiles, and patient-centered outcomes. (M et al., 2026)

Nanoparticles made of gold (Au), silver (Ag), iron oxide (Fe3O4), silica (SiO2), and quantum dots, as well as carbon-based materials (graphene, CNTs), have been reported. We found several advantages related to drug or nanoparticle distribution, including better surface wettability and cell interactions. Targeting efficiency demonstrated a 2 to 5-fold increase in cancer cell uptake compared to non-targeted systems, highlighting the effectiveness of integrated targeting ligands. The saffron derivatives with nanoparticle-based delivery systems, including liposomes and metal–organic frameworks, to maximize stability and control drug release at the tumor site (Korrapati et al., 2024). In vivo studies showed tumor volume reductions of 30% to 60% with combination therapies. Ultrafast femtosecond laser (fs-laser) surface functionalization is a single-step, contamination-free technique to engineer implant surfaces with controlled micro–nano topography, chemistry, and wettability. In dental implants, titanium or zirconia enables precise patterning that directly influences cell adhesion, osteogenesis, antimicrobial behavior, and long-term osseointegration. By incorporating both drug delivery and photothermal methods, photodynamic therapy achieves significant therapeutic outcomes. Photothermal conversion efficiency ranged from 30% to 50%, enabling effective thermal therapy when combined with appropriate laser irradiation protocols. Advancements in nanomaterials for light-mediated therapeutics that utilize near-infrared light, ultrasound, or X-ray excitation, which offer deeper tissue penetration compared to conventional UV or visible light methods. Hybrid laser coatings, laser-assisted techniques with nanoparticle systems, strategize to overcome barriers and enhance delivery efficiency (Jiménez-Boland et al., 2025).

Photothermal conversion efficiencies of 30% to 50% facilitated effective thermal ablation, while biocompatibility assays confirmed greater than 85% viability in non-cancerous cells, indicating minimized off-target toxicity. Clinical study data further supported these findings, with in vitro cancer cell death rates of 50% to 90% and notably low systemic toxicity in multiple investigations. Coating stability was maintained for weeks to months in physiological environments, a critical requirement for long-term therapeutic applications. Coating nanoparticles with cancer cell membranes or hybrid membranes by fusing cancer and immune cell membranes enhances targeting specificity (Figure 2). It prolongs circulation and triggers immunogenic cell death (ICD), thus increasing therapeutic efficacy and reducing toxicity to healthy tissues. Advances in nanotechnology, biomaterials engineering, and laser-assisted fabrication have enabled the development of hybrid coatings capable of simultaneously performing imaging, targeted drug delivery, and therapeutic functions. These multifunctional coatings are particularly relevant for implants, nanoparticles, microneedles, and biosensing platforms, where surface engineering plays a critical role in biological interaction and clinical efficacy. Hybrid coatings integrate organic, inorganic, and biomimetic components, producing smart interfaces that can interact dynamically with biological systems while delivering therapeutic agents and enabling real-time monitoring. (Korrapati et al., 2026)

Figure 2. A Diverse range of membrane sources was employed to engineer biomimetic nanoparticles. The major study outcomes and types of cell membranes tested, along with nanoparticle sizes in nanometers, are provided.

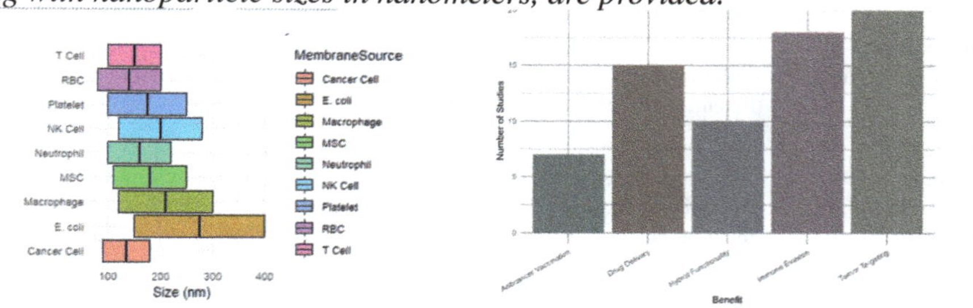

NEED FOR LASER COATED HYBRID COATED BIOMIMETIC OR NANOPARTICLE SYSTEMS

The foundation of any discussion on novel cancer therapies is the inadequacy of conventional approaches. Chemotherapy and radiotherapy are often plagued by systemic toxicity. Non-specific biodistribution harms healthy cells, dividing them and causing side effects (neuropathy, cardiotoxicity, myelosuppression) that diminish a patient's quality of life. Polyethylene Glycol (PEG) was the "gold standard" for creating a hydrophilic stealth layer, thereby reducing opsonization and clearance by the Mononuclear Phagocyte System (MPS) and extending circulation half-life. Poly(lactic-co-glycolic acid) (PLGA) is a biodegradable polymer used for controlled drug release, and it demonstrated proof of concept that nano-encapsulation can improve drug solubility, reduce off-target effects to some degree, and enable passive targeting via the Enhanced Permeability and Retention (EPR) effect. Limited efficacy due to poor solubility and rapid clearance of drugs, coupled with the inability to penetrate deep into tumor tissue and the emergence of multi-drug resistance, limits therapeutic success and impacts survival rates (Table 1).

Table 1. Traditional Chemotherapy Challenges and The Nanoparticle Solution

Conventional Therapy	Nanoparticle Bioactive Molecules Solution	Nano carriers	Examples
Non-Specific Biodistribution of Chemotherapy Drugs	Passive or Active Targeting Enhanced Permeability and Retention (EPR) effect	Iron Oxide (Fe_3O_4), Silica (SiO_2), Gold (Au), Liposomes, Polymer NPs.	Liposomal Doxorubicin, Albumin-bound Paclitaxel (Abraxane),
Potent drugs are poorly soluble or degrade rapidly in the bloodstream.	The nanoparticle core acts as a stable, solubilizing carrier for hydrophobic or fragile payloads.	Lipids (LNPs), Polymers (PLGA), Silica (SiO_2), Carbon-based.	Antibody-Drug Conjugates (ADCs): Lipid Nanoparticles (LNPs) for mRNA cancer vaccines
Multidrug Resistance (MDR), chemotherapeutic drugs reduce efficacy	Multi-Modal & Bypass Strategies, Co-delivery of drugs with MDR inhibitors, hyperthermia bypass pump-based resistance.	Gold (Au) for photothermal, Iron Oxide (Fe_3O_4) for magnetic hyperthermia.	Hybrid Bioactive Coatings, Gold Nanoshells (AuroShell)
Diagnosis and treatment are often separate, sequential processes.	Theranostics integrates imaging and therapy into a single agent for real-time treatment monitoring.	Quantum Dots (imaging), Gold NPs (photoacoustic), Iron Oxide (MRI + therapy)	Iron Oxide Nanoparticles, synthetic nanoparticles (Au, SiO_2, PLGA) with natural cell membranes (RBC, platelet, leukocyte, cancer cell) or biomimetic polymers

Immune Suppression therapies inadvertently suppress the immune system, hindering the body's natural ability to combat cancer. The EPR effect is highly heterogeneous in humans, limiting tumor accumulation. PEGylation can induce anti-PEG antibodies, leading to accelerated blood clearance upon repeated dosing. Most critically, these synthetic coatings are still recognized as "foreign" by the immune system, leading to eventual MPS uptake in the liver and spleen, and a failure to interact with the complex tumor microenvironment (TME) actively. (M et al., 2026) Hybrid coatings refer to multicomponent surface architectures that combine two or more functional materials to achieve synergistic biomedical effects. These coatings may incorporate nanoparticles, polymers, biomolecules, lipids, or cell membranes, forming layered or composite structures with both diagnostic and therapeutic capabilities. The goal is to produce materials that can detect disease biomarkers, deliver drugs in a controlled manner, and monitor treatment outcomes.

Antibody-Drug Conjugates (ADCs)

A monoclonal antibody (targeting a cancer-specific antigen) is chemically linked to a potent cytotoxic drug. The ADC binds to the cancer cell, is internalized, causing cell death with high specificity. Proven to improve survival in specific cancers (Trastuzumab emtansine (T-DM1) in HER2+ breast cancer). Multiple ADCs are

FDA-approved. The trend is toward new targets, more stable linkers. ADCs reduce off-target toxicity compared to conventional chemotherapy, though unique side effects like ocular toxicity or neuropathy can occur. Cancer cell membrane–coated nanoparticles have demonstrated homologous targeting ability, meaning they preferentially bind to tumor cells originating from the same tissue type. This approach improves tumor localization while reducing immune clearance, thereby enhancing therapeutic efficiency. Such biomimetic coatings also help nanoparticles evade immune detection and extend circulation time in the bloodstream. (P. et al., 2025) Laser-based fabrication techniques are increasingly employed to produce precise and functionalized hybrid coatings. Technologies such as pulsed laser deposition (PLD) and matrix-assisted pulsed laser evaporation (MAPLE) enable controlled deposition of biomaterials, polymers, and bioactive molecules onto implant surfaces. Composite coatings consisting of bioactive glass and antimicrobial peptides have been deposited onto titanium implants using laser-assisted techniques. These coatings generate nanoscale roughness and improve hydrophilicity, thereby enhancing cell adhesion and corrosion resistance. Laser-assisted deposition can also embed antimicrobial nanoparticles or therapeutic drugs within polymer matrices, enabling coatings that actively prevent bacterial colonization and biofilm formation. These advanced coatings may release antimicrobial agents in response to environmental triggers, thereby addressing the growing challenge of antibiotic-resistant infections in biomedical devices.(Korrapati et al., 2026)

Future research is expected to focus on integrating artificial intelligence, advanced manufacturing techniques, and biomimetic design principles to optimize hybrid coatings for personalized medicine. Smart coatings capable of responding to disease-specific signals or releasing therapeutics on demand may revolutionize the management of cancer, infections, and chronic diseases (Figure 3).

Figure 3. Hybrid Nanomaterial Applications

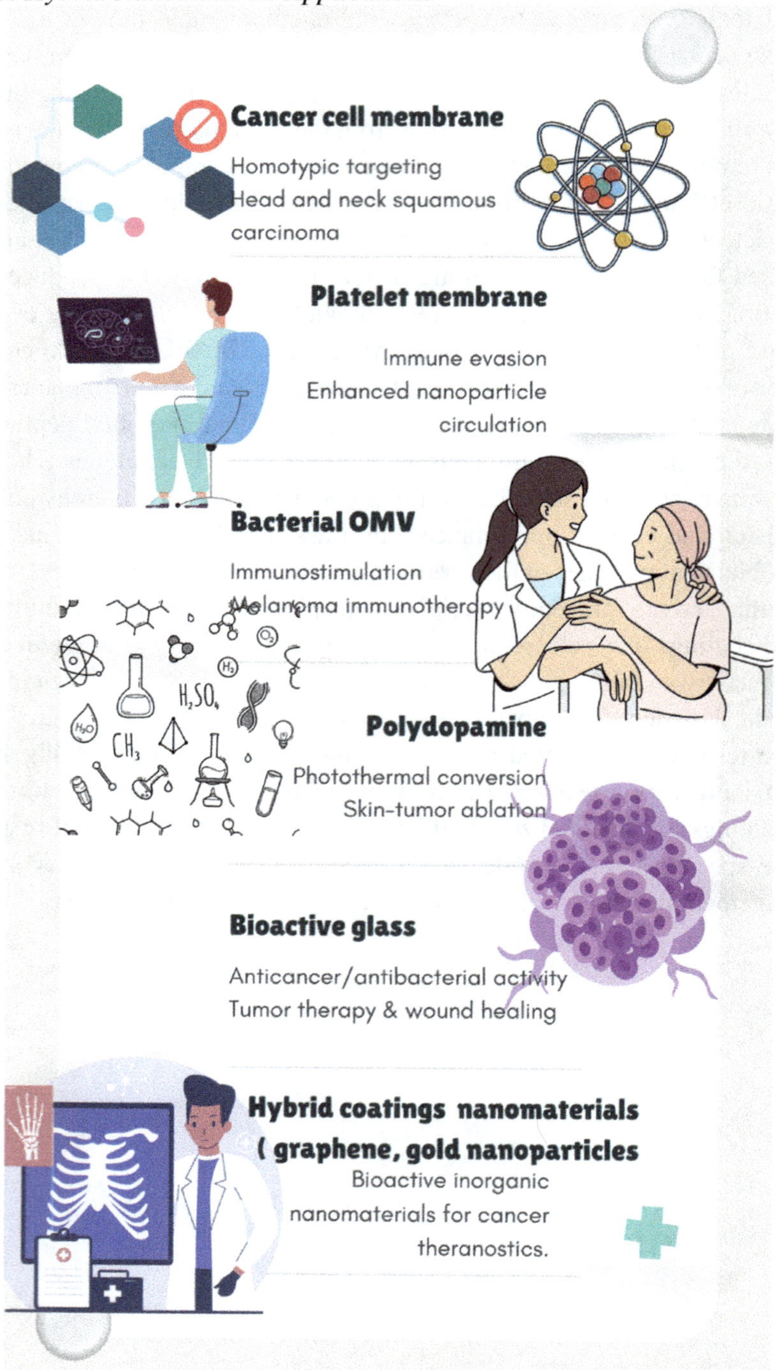

Lipid Nanoparticles (LNPs) for mRNA Cancer Vaccines

Nanoparticle-based coatings are central to theranostic technology because they enable the integration of drug-delivery systems with imaging capabilities.

Gold nanoparticles, for instance, exhibit strong plasmonic properties that can convert light into heat, enabling photothermal therapy for cancer treatment. In addition to their therapeutic activity, these nanoparticles can serve as imaging contrast agents, enabling clinicians to visualize tumors and monitor treatment response. (Nanjundeshwara et al., 2025) LNPs protect and deliver mRNA encoding for tumor-specific antigens (neoantigens) into host cells. The host cells then produce the antigen, triggering a potent, personalized adaptive immune response against the cancer. Demonstrated remarkable success in clinical trials for melanoma (Moderna/Merck mRNA-4157), significantly reducing the risk of recurrence and death—generally manageable (transient flu-like symptoms and injection-site pain), representing a major improvement over chemotherapy. As an outpatient therapy, it minimizes hospital stays and long-term organ damage. The sequencing of the patient's tumor and the manufacturing of a custom vaccine using mRNA-LNPs require ultra-cold storage and pose distribution challenges. (Korrapati et al., 2026) Hydrogel-based coatings represent another important class of hybrid theranostic systems. Hydrogels are highly hydrated polymer networks capable of encapsulating therapeutic agents and releasing them in a controlled manner. Nanoparticles carrying doxorubicin and indocyanine green were coated with polydopamine and attached to probiotic bacteria for targeted delivery to tumor microenvironments. Laser-coated hybrid system enabled combined chemotherapy and photothermal therapy, producing sustained tumor suppression with reduced systemic toxicity.

CRISPR-Based Gene Editing for Cancer Therapy

Based on viral or non-viral (nanoparticle) delivery, CRISPR systems can be used to edit oncogenes in vivo. Evidence shows that laser-treated titanium surfaces exhibit reduced adhesion of gram-positive and gram-negative bacteria in gene therapies for tumor-suppressor genes or in engineering a patient's own T-cells ex vivo (CAR-T). CRISPR systems aim for durable remissions, but off-target editing and cytokine release syndrome (CRS) are also known risks in CAR-T. Ex vivo CAR-T therapy is a clinical reality. However, safe, efficient, and tumor-specific in vivo gene editing remains a formidable challenge. Somatic cell editing is tightly controlled. Any germline editing is currently considered ethically impermissible. Long-term patient monitoring is essential. (Pallavi et al., 2026) Among the various technologies supporting theranostics, hybrid coatings have gained significant attention for their ability to combine multiple functional materials to enhance drug delivery, imaging, and therapeutic performance. Next-generation implants are moving toward biomimetic surface design. The Shark-skin–like textures have antimicrobial properties. Plant-based Lotus-leaf structures have been shown to have superhydrophobic surfaces. Trabecular bone patterns enable osteogenic signaling. Femtosecond lasers allow direct writing of biomimetic nanoarchitectures without chemical coatings.

Table 2. Cell-Derived Membranes for Creating Bio-Hybrid Systems

Source	Mechanism of Action	Translational Challenges
Red Blood Cells	Extended circulation time (superior to PEG) CD47 inhibits phagocytosis Active targeting of membrane vehicles	Abundant source, well-understood biology
Leukocytes Neutrophils	Improved navigation to inflamed tumor sites Ability to cross biological barriers. Inherits adhesion molecules (integrins) for margination and trafficking.	Active targeting and penetration. Sourcing and complexity are higher.
Platelets	Enhanced adhesion to damaged vasculature and circulating tumor cells CTCs preventing metastasis. Express P-selectin, which binds to CD44 on CTCs and tumor cells.	Promising for metastatic cancers. Complex role in coagulation requires careful safety evaluation.
Cancer Cells	Homotypic targeting Adhesion proteins for homologous binding.	Personalized therapy, Patient-derived tumors may have risk of oncogenic signaling
Macrophages	Deep tumor penetration and potential for immune modulation. Inherits chemokine receptors to follow tumor homing signals	Potential for modulating the immunosuppressive TME, Control over M1/M2 polarization is a major hurdle.

Biomimetic Nanomaterials

Biomimetic nanomaterials, when fabricated via laser processing techniques, produce multifunctional coatings. Nanomaterials objectives include developing multifunctional polymeric micelles for drug delivery, modulating tumor microenvironments. They help identify biomarkers for patient selection and improving immunotherapy efficacy. The delivery of bioactive nanoparticles in oncological therapeutics, target drug release, with reduced systemic toxicity. For example, metastatic colorectal cancer (mCRC) urgent need is for the 95% of patients with mismatch repair-proficient (pMMR) or microsatellite instability-low (MSI-low) tumors who do not benefit from immune checkpoint blockade (ICB). The development of new delivery concepts for mRNA and siRNA therapies, utilizing (bio) hybrid carrier materials and microbubble-based formulations, overcomes challenges in in vivo delivery to tumors. However, the efficiency of nanoparticle delivery or hybrid carrier systems remains a critical challenge, due to biological barriers that limit penetration.

Microchannel Formation

Hybrid laser coatings primarily enhance nanoparticle delivery through Microchannel Formation. They can be engineered for targeted drug delivery, enhanced imaging, and improved compatibility with biological tissues. Ablative fractional lasers create channels in tissue, increasing permeability and enabling deeper penetration of nanoparticles (Visan & Negut, 2025). The diameter of these channels should exceed the size of delivered particles for optimal efficiency. The most robust quantitative data indicate an 11-fold increase in drug delivery using hybrid AuNP-lipid nanoparticles with pulsed laser activation compared to conventional carriers (Mundekkad & Mallya, 2025). Dual pulsed laser deposition of Ag nanoparticles on calcium phosphate coatings has been shown to work for biomedical applications. Dual pulsed laser deposition and material shaping fabricated antimicrobial silver-doped carbon structures via combinatorial pulsed laser deposition.Hydrogel particles can carry anticancer drugs, proteins, or imaging probes, enabling both therapeutic delivery and diagnostic monitoring.

Table 3. Bioactive Molecules such as anticancer drugs (doxorubicin, paclitaxel), peptides (RGD), growth factors, and antibodies promote selective cytotoxicity, enhance adhesion, or target specific cancer cell markers.

Bioactive Molecule	Role in Cancer Therapy	Compatibility Laser Processing	Examples
Chemo therapeutic Drugs	Direct cytotoxicity to cancer cells	Sensitive To Heat	Doxorubicin, Paclitaxel, Cisplatin
Targeting Peptides	Enhance selective binding to cancer cell markers	Small Molecules	RGD peptides, Folic acid derivatives, Transferrin
Antibodies	Specific targeting and immune modulation	May Require Gentle Processing	Anti-HER2, Anti-EGFR, Anti-CD20
Growth Factors	Promote tissue integration or modulate microenvironment	Risk Of Denaturation	VEGF, BMPs
Photo sensitizers	Enable photodynamic therapy	Stable Under Laser Conditions	Porphyrins, Chlorins

Stimuli-Responsive Release

Gold nanoparticles (AuNPs) are highly promising in cancer therapy due to their low toxicity and unique photothermal properties. Hybrid nanoparticles containing gold (Au) clusters can be triggered by nanosecond pulsed lasers to release drugs precisely at the tumor site, increasing intracellular drug concentration up to 11-fold compared to conventional carriers (Amaral et al., 2025). Comparative Delivery Efficiency Metrics indicate Laser-triggered hybrid nanoparticles deliver up to 11 times more drug into target cells than conventional lipid nanoparticles. Laser-assisted drug delivery (LADD) increases drug penetration, especially when channel diameter exceeds nanoparticle size and low-viscosity formulations are used. Light-mediated therapies employing advanced nanosystems show enhanced tissue penetration and therapeutic efficacy in deep-seated tumors (Tsang & Zhang, 2024).

- Laser-textured titanium implants with antibacterial nanostructures that reduce biofilm formation while enhancing osseointegration.
- Laser-processed gold or graphene-based coatings used for photothermal cancer therapy and imaging applications.
- Laser-engineered hydroxyapatite or bioactive glass coatings that combine regenerative capability with drug-delivery potential.
- Laser-fabricated plasmonic nanostructures enabling biosensing and localized therapeutic activation.

Gold nanoparticle (AuNP)-mediated photothermal therapy (PTT) is a promising cancer treatment method that offers high precision, minimal invasiveness, and selective tumor ablation through localized hyperthermia. The integration of AuNPs with other therapeutic modalities, such as chemotherapy, radiotherapy, and immunotherapy, is explored to achieve synergistic treatment outcomes (Table 4).

Antitumoral Strategies

Hybrid laser coatings leverage the precision of laser energy to modify tissue or carrier properties. Nanotechnological devices are designed for the topical and transdermal delivery of bioactive compounds. Exposome research reveals how specific exposures drive genetic and epigenetic changes, identifying new molecular targets for therapy (P. et al., 2025).

The mechanisms that enable these devices to deliver nutritive molecules effectively to the dermis. Transdermal delivery strategies promote systemic distribution, facilitating deeper and more controlled penetration of nanoparticles (Ananthakumar, Nanjundeshwara, et al., 2025). Recent studies have explored various approaches, including ablative fractional lasers to create microchannels in tissue, plasmonic nanoparticles activated by pulsed lasers for site-specific drug release, and light-mediated therapies that exploit improved tissue penetration (Tsang & Zhang, 2024).

Table 4. Current Status of Clinical Trials: Gold Nanoparticles as Drug Delivery Carriers

NCT Trial Number	Study Title	Phase	Cancer Type	AuNP Application
NCT03020084	Photoimmunotherapy with AuroLase® Therapy for Solid Tumors	Phase I	Solid Tumors	Photothermal Ablation Therapy
NCT01679069	Gold Nanoparticles for Radiation Dose Enhancement in Solid Tumors	Phase I	Solid Tumors	Radiation Dose Enhancer
NCT03610034	Tumor-Targeting AuNPs for Pancreatic Cancer Therapy	Phase I/II	Pancreatic Cancer	Radiosensitizer

THERANOSTIC FUNCTIONS

Bioinspired materials, such as nanocomposites and polymers, integrate theranostics combining therapy and real-time imaging. Exploiting tumor-specific triggers for on-demand drug release discusses their applications in targeted drug delivery, regenerative medicine.

Some of the applications are

- Chemotherapy: Improved drug stability, controlled release, and reduced systemic toxicity.
- Targeted photothermal therapy (PTT): Antibody-mediated targeting and light-to-heat conversion for localized cell death.
- Photodynamic therapy (PDT): Generation of reactive oxygen species (ROS) upon light activation to induce cell death.
- Tissue regeneration: Promotes cell adhesion and growth, and enhances tissue integration with implants.
- Angiogenesis and osteogenesis: Stimulate blood vessel and bone tissue formation.

A common approach involves targeting malignant cell receptors that drive cell proliferation and differentiation. Anticancer drugs such as doxorubicin, cisplatin, denosumab, bisphosphonates, and paclitaxel are frequently employed to inhibit cancer cell growth. These strategies are especially relevant in cases where bone metastases result in osteolytic lesions, increasing fracture risk, and necessitating surgical intervention with metallic prostheses or fixation. Hybrid coatings with bioprinting to respond to biomarkers or microenvironmental changes (pH, enzymes) (Byalahunasishetter et al., 2025). Spatial omics data on environmental exposures, enabling site-specific, controlled drug release.

The use of 3D printed metallic porous implants is highlighted for their ability to enhance bone tissue ingrowth due to favorable osseointegration properties. However, the inert nature of metal surfaces can impede recovery, making functional coatings critical for improved outcomes in cancer patients undergoing bone surgeries.

Table 5. Comparative Delivery Efficiency Metrics

Approach/Technique	Delivery Efficiency Improvement	Key Mechanism
Hybrid AuNP-Lipid Nanoparticles + Pulsed Laser	11-fold increase in intracellular drug concentration	Plasmonic resonance-triggered release
Ablative Fractional Laser-Assisted Drug Delivery (LADD)	Increased penetration (qualitative)	Microchannel formation in tissue
Light-Mediated Nanoparticle Systems (NIR/Ultrasound/X-ray)	Enhanced deep tissue penetration (qualitative)	Improved excitation depth

MICROFLUIDICS-BASED DROPLET IN GENE AND DRUG DELIVERY

Traditional liposome production methods are inadequate in achieving precise control over liposome dimensions and uniformity, leading to inefficient use of reagents and resources.

Microfluidic technology offers a solution by enabling liposome synthesis on a centimeter-level chip, which improves manufacturing efficiency and reduces costs. Despite the advantages of microfluidic technologies, the manufacturing of microfluidic devices is often hindered by high costs and time requirements. 3D printing presents a promising and cost-effective method for producing microfluidic devices tailored for liposome production, with significant potential for biomedical applications (Biradar, Suresh, et al., 2025). Microfluidics in lipid nanoparticles and polymeric microparticles allows precise control, optimal mixing, and prevention of cross-contamination. Bioequivalence is defined by the FDA as the equality in the rate and extent of the active pharmaceutical ingredient's availability at the drug action sites.

Hydrogel particles are crucial in the synthesis of various dimensional materials (0D, 1D, 2D, 3D) for diagnosis and drug delivery. In disease therapy, their ability to retain water encapsulates active molecules, with sizes ranging from nanometers to micrometers, based on their application and preparation method. These particles can take on various morphologies, such as spherical, rod-shaped, and core-shell structures, each offering specific advantages for drug delivery, including improved circulation time and targeted delivery. Hydrogel particles can be engineered to respond to different stimuli (temperature, pH, light, etc.), making them suitable for advanced applications like gene therapy, tissue regeneration, and cell therapies. The use of microfluidics in hydrogel particle production allows for precise control over their characteristics, leading to efficient and reproducible manufacturing while minimizing reagent waste and optimizing drug encapsulation (Hinojosa-Ventura et al., 2025). The methacrylated chitosan-coated liposomes (CSMA-liposomes) as a novel ocular drug delivery system designed to enhance drug retention in the eye. The liposomes, synthesized using microfluidic techniques, showed improved mucoadhesion and prolonged residence time, with in vitro and ex vivo results indicating significant enhancements in drug retention and controlled release profiles (Mirkani et al., 2025).

ADVANCED IMMUNE CELL MEMBRANE COATINGS

T cell and NK cell membrane-coated nanoparticles represent emerging approaches for targeted immunotherapy delivery. While showing promise in preclinical studies, these platforms face challenges related to stability and manufacturing complexity. methacrylated chitosan-coated liposomes (CSMA-liposomes) as a novel ocular drug delivery system aimed at addressing the rapid removal of drugs from the eye. Membrane components derived from immune cells (dendritic cells) can directly activate immune responses against tumors.

CSMA-liposomes were synthesized using microfluidic techniques, resulting in enhanced mucoadhesion and prolonged residence time due to electrostatic and covalent interactions with mucin.

Optimization of the liposomes showed that uncoated liposomes had a size of 154 nm and a PDI of 0.12, which increased to 208 nm after CSMA coating, with a significant change in zeta potential from -8.3 to +36 mV. In vitro studies demonstrated that CSMA-liposomes had significant mucoadhesion and improved retention on bovine conjunctival tissues, along with a sustained, biphasic release profile of dexamethasone sodium phosphate, indicating enhanced ocular bioavailability and therapeutic potential (Mirkani et al., 2025).

Figure 4. Hybrid Nanomaterial Functionality and Applications

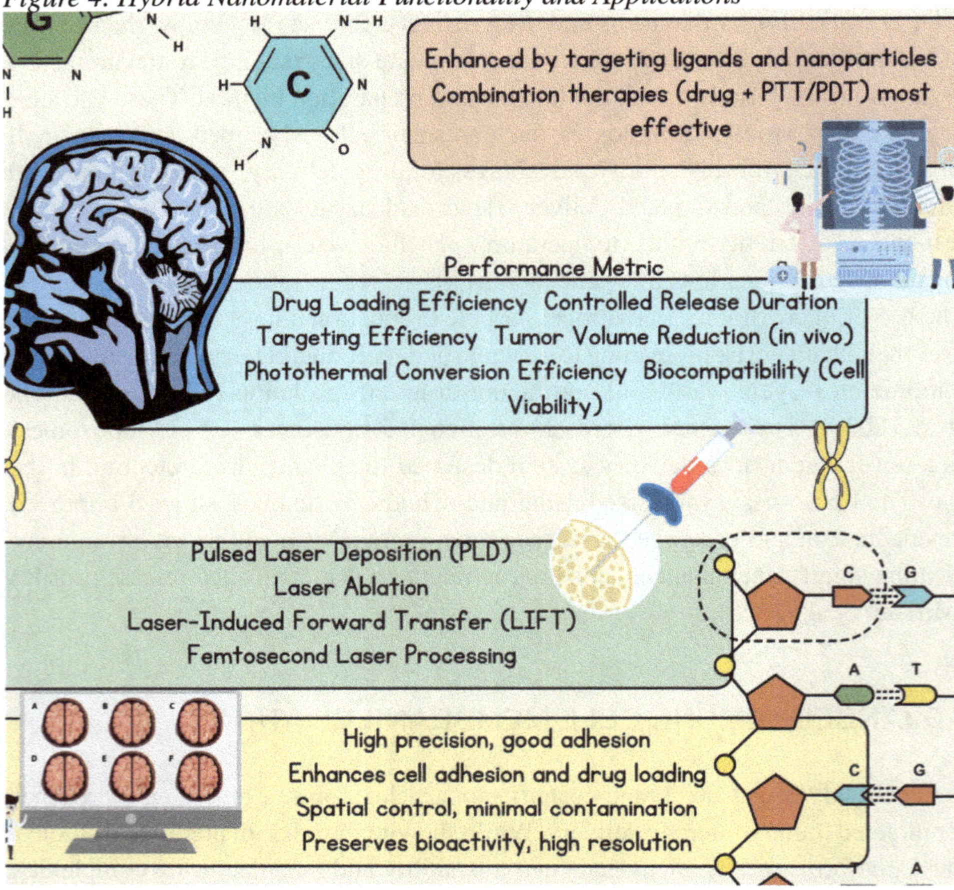

Calcium Hybrid Coating Applications

Hydroxyapatite (HA) via Micro Arc Oxidation with Paclitaxel demonstrates toxic effects against human cervical carcinoma cells. HA via Matrix Assisted Pulsed Laser Evaporation (MAPLE) with Carboplatin and Quercetin cytotoxic effects against osteosarcoma cells during MAPLE deposition are reported (Maskowicz et al., 2024). HA via Plasma Spray with Curcumin and Epigallocatechin shows cytotoxic effects against osteosarcoma cells using a drop-casting method. HA via Plasma Spray with Curcumin and Vitamin D3 is cytotoxic against osteosarcoma cells through absorption. Low-viscosity formulations consistently outperform high-viscosity ones in filling laser-created channels and accelerating drug delivery. HA via Micro-Arc Oxidation with Selenium enhances the versatility of calcium phosphate (CaP) coatings when combined with various bioactive compounds to target different cancer cell types (Montesissa et al., 2025).

SCAFFOLD COATING FABRICATION TECHNOLOGIES

The integration of stem cell and gene therapies, and the potential use of artificial intelligence in scaffold design for personalized medicine. In orthopedic oncology, particularly for bone sarcomas and metastatic cancers. A range of coating fabrication technologies, which can be broadly categorized into physical and chemical techniques (Montesissa et al., 2025).

Physical Techniques

Physical techniques are noted for producing high-quality coatings with superior adhesion. These methods are based on the physical ablation or vaporization of the source material, followed by condensation as a thin film on the substrate. Hydroxyapatite (HAp) is extensively researched for artificial implant development, focusing on improving coating production and biocompatibility. The study successfully deposited HAp microparticles on Ti6Al4V alloy substrates using matrix-assisted pulsed laser evaporation (MAPLE), resulting in uniform coatings without additional treatment. A tunable pulsed laser was employed to optimize deposition speed and quality by adjusting the wavelength to match selected solvents, including deionized water, isopropyl alcohol, and a mixture of isopropanol and acetonitrile. The resulting HAp coatings were approximately 200 nm thick for water and isopropanol matrices and about 320 nm for the isopropanol: acetonitrile matrix, achieving a 37% increase in deposition rate and meeting criteria for further biocompatibility testing and material modification (Maskowicz et al., 2024).

Table 6. Key Parameters Influencing Hybrid Laser Coating Efficiency

Parameter	Recommended Setting/Effect	Impact on Delivery Efficiency
Channel Diameter	> Particle diameter	Improves penetration
Formulation Viscosity	Low (aqueous solutions/gels)	Accelerates channel filling/delivery
Laser Fluence	Optimized per application	Controls the depth and extent of channels
Pulsed Laser Wavelength	527 nm (for AuNP resonance)	Maximizes plasmonic effect
Incubation Time	Optimized for formulation	Influences uptake

Key physical techniques mentioned include:

- Plasma Spray (PS)
- Radiofrequency Magnetron Sputtering (RF-MS)
- Pulsed Laser Deposition (PLD)
- Ion Beam Assisted Deposition (IBAD)
- Pulsed Electron Deposition (PED)
- Matrix-Assisted Pulsed Laser Evaporation (MAPLE)
- Ionized Jet Deposition (IJD)
- Cathodic Arc Deposition
- High Power Impulse Magnetron Sputtering (HiPIMS)

These physical techniques offer improved control over microstructure, chemical composition, and mechanical properties compared to chemical processes, although they tend to be more complex and costly. Coatings produced by these methods are generally less crystalline and more amorphous, but post-deposition thermal treatments can enhance crystallinity.

Key parameters influencing delivery efficiency include Laser Parameters: Fluence, channel depth, density surface area, spot size, and number of passes. Low-viscosity aqueous formulations fill laser channels more efficiently (Labadie et al., 2022). Hybrid Nanoparticle Design: Incorporating gold clusters within lipid layers enables precise light-triggered release.

Chemical Techniques

The applications of coatings for cancer treatment, particularly in the context of orthopedic implants and bone tissue engineering. Two main chemical techniques are highlighted for ceramic coatings:

Electrophoretic Deposition (EPD) utilizes electrically charged ceramic particles suspended in a liquid medium. Under an applied electric field, these particles migrate toward a conductive substrate, forming a uniform coating. EPD is valued for its simplicity, low cost, and ability to coat complex geometries with thicknesses ranging from less than 1 μm to over 100 μm.

Electrochemical Deposition (ECD) relies on electric fields to promote the formation of ceramic films on substrates. The development of tumor-targeted NCPs, specifically the OX/SN38 NCP, which combines a hydrophilic oxaliplatin prodrug and a hydrophobic SN38 prodrug, demonstrates potent anticancer effects and synergy with anti-PD-L1 antibodies for enhanced chemo-immunotherapy. The proposed studies aim to optimize a core-shell NCP, OX/CDN/Chol-D, for the co-delivery of oxaliplatin and a cholesterol-conjugated drug, enhancing drug delivery to tumors through LDLR-mediated endocytosis and disrupting tumor vasculature to improve accumulation.

Pulsed Laser Deposition (PLD)

Personalized, regenerative approaches use hybrid coatings for patient-specific therapy and tissue repair. The bioactive inorganic nanomaterials with hybrid coatings have assistive technologies. (S. et al., 2025). Photothermal therapy and polydopamine or iron oxide cores absorb near-infrared light, generating heat to ablate tumor cells. Calcium phosphate coatings enhance the osteoinductive and osseointegrative properties of implanted devices, with pain relief for assistive augmentative applications. (Raghavan et al., 2025). The incorporation of bioactive molecules into calcium phosphate coatings can improve their osteointegrative, antibacterial, and antitumor properties. Laser irradiation induces localized heating around AuNP clusters, causing vaporization or thermo-mechanical decomposition of lipid layers. Predictive pain solutions facilitate nanoparticle entry into cancer cells without compromising drug integrity. (Ananthakumar, Nagaraja, et al., 2025). The PDD depends on several factors for clinical translation. Current challenges include

Optimize Laser Parameters: Ensure channel diameter exceeds nanoparticle size and use low-viscosity formulations for maximal efficiency.

Employ Stimuli-Responsive Nanoparticles: Design carriers that respond specifically to clinically safe laser wavelengths for controlled release.

Photothermal Therapy (PTT)

Photothermal therapy (PTT) is presented as a promising adjunctive treatment that enhances surgical effectiveness and helps prevent tumor recurrence through hyperthermal therapeutic effects. Various advanced biomaterials, including nanomaterials, bioceramics, polymers, and hydrogels, are being developed for their photothermal capabilities to address malignant bone tumors and perform unique biological functions. Gold nanoparticle-mediated photothermal therapy (PTT) is a promising cancer treatment method that offers high precision and minimal invasiveness through localized hyperthermia. (Amaral et al., 2025).

The efficacy of MR fusion imaging technology combined with nanoparticle-directed laser irradiation for the focal ablation of prostate tissue in men with localized prostate cancer. Efficacy will be assessed through contrast-enhanced MRI 48-96 hours post-treatment to observe coagulative necrosis, and through MRI with ultrasound-guided target biopsies at 6 months and 1 year after the laser treatment. AuroLase Therapy in the treatment of subjects with primary and/or metastatic tumors of the lung where there is airway obstruction. In this study, patients will be given a systemic IV infusion of particles and a subsequent escalating dose of laser radiation delivered by optical fiber via bronchoscopy.

Matrix-Assisted Pulsed Laser Evaporation (MAPLE) Membrane-Coated Nanoparticles

Microbial biofilms pose a significant threat to human health by causing persistent infections and increasing antimicrobial resistance, often leading to the failure of traditional therapies. Antimicrobial nano-material-based coatings are explored as a revolutionary approach to combat biofilm infections by disrupting the formation of established biofilm communities. Advanced nanocoatings have the potential to not only prevent biofilm formation but also respond dynamically to infection environments, releasing antimicrobial agents on demand and adapting to specific bacterial strains. Various types of nanocoatings, including metal-based, polymer-based, and ceramic-based materials, and their applications in medical devices such as implants, contact lenses, and wound dressings, highlight the importance of nanotechnology in improving global healthcare. Bacterial membrane coatings, primarily using E. coli outer membranes, provide strong adjuvant properties for cancer immunotherapy. The use of matrix-assisted pulsed laser evaporation to print amphotericin B onto polyglycolic acid microneedles addresses the challenge of its poor solubility. The amphotericin B-loaded microneedles exhibited concentration-dependent activity against the yeast Candida albicans, indicating the potential of this method for delivering drugs with complex solubility issues. However, safety concerns and regulatory challenges limit their clinical development (Sachan et al., 2025). It highlights the advantages of these methods in controlling film properties and embedding antimicrobial agents, resulting in enhanced efficacy against various pathogens, including resistant strains. The potent immune activation through PAMPs and vaccine delivery capabilities are novel immunotherapy approaches.

Nanoparticle-Based Bioactive Coatings for Cancer Therapy

The development of composite thin films made from bioactive glass and melittin for titanium implants aims to enhance osseointegration and implant longevity. The coatings demonstrated improved biocompatibility, corrosion resistance, and bioactivity, with favorable surface characteristics for cell adhesion and the formation of a carbonated apatite layer. (Kantaros et al., 2025). Hybrid systems that trigger release only upon laser activation minimize off-target toxicity. Future research directions are outlined, focusing on the guanidium-functionalized gold nanoparticles with phospholipid bilayers. Artificial intelligence integration of advanced computational techniques (Lakshmaiah et al., 2025). The interdisciplinary collaboration with neuromorphic computing could improve the design and application of biomimetic nanomaterials. (GVS et al., 2025).

Table 7. Silica Nanoparticles for Photothermal Therapy, Drug Delivery, and Imaging

NCT Trial Number	Study Title	Phase	Cancer Type
NCT02588521	Mesoporous Silica Nanoparticle-Based Formulation of Doxorubicin in Hepatocellular Carcinoma	Phase I	Hepatocellular Carcinoma
NCT04452140	SiNP-based photothermal therapy for Breast Cancer	Phase I	Breast Cancer
NCT02102120	Silica-coated gold nanoparticles for imaging and therapy in Prostate Cancer	Phase I/II	Prostate Cancer

Silica Nanoparticles (SNPs)

Mesoporous silica nanoparticles (MSNs) are excellent drug delivery systems due to their high surface area and modifiable pore size, which allows for high drug loading capacity. They are often used in combination therapies to improve the effectiveness of traditional chemotherapy. Synthetic biology methods are essential for understanding biomimetic materials at the nanoscale. Biomimetic nanomaterials are engineered at the nanoscale to replicate the intricate structures and oncogenesis (Biradar, Korrapati, et al., 2025). Researchers study how organisms achieve remarkable properties such as self-healing, self-assembly, and hierarchical organization. By emulating these features, scientists develop innovative nanomaterials with enhanced functionalities across a broad range of fields. The challenges are scalability, biological compatibility, and safety (Mundekkad & Mallya, 2025).

Iron Oxide Nanoparticles (IONPs)

Iron oxide nanoparticles (IONPs) are primarily used in cancer therapy for their magnetic properties. They can be utilized in magnetic fluid hyperthermia, where an alternating magnetic field heats the nanoparticles to kill cancer cells. They are also effective as contrast agents for magnetic resonance imaging (MRI).

TITANIUM

The environment affects the wetting properties of metals like copper, nickel, and tin during nanosecond laser processing. The optimized laser processing (at 32 W) resulted in a unique hierarchical mesoporous surface with cauliflower-like nanostructures, uniform silver distribution, prolonged superhydrophilicity (40 days), and antibacterial efficacy (12 days) against common bacteria. Soft exoskeleton biomimetics have the ability to induce oxidative stress and apoptosis in cancerous cells. To monitor safety rigorously, safety and ethical protocols prevent inadvertent systemic exposure during LADD procedures. Advance soft robotics studies should focus on direct comparative metrics and mechanistic insights into cellular uptake pathways. (Chandra et al., 2025).

Furthermore, biocompatible and antimicrobial polymer coatings developed via pulsed laser deposition and MAPLE techniques could be adapted for localized cancer therapy, implantable devices, or drug delivery systems aimed at cancer treatment.

The changes in wetting characteristics are primarily due to the re-deposition of ablation products that create a nanoporous layer. It enhances wicking capabilities and can lead to hydrophobization of materials. Titanium-based implants develop antibacterial surfaces and maintain the mechanical properties of the implants. Emerging technologies, such as electronic noses, gas sensors, and smart packaging, leverage microbial volatiles to detect infection and contamination. Scavenger receptor class B member 1 is a multiligand receptor involved in various functions, including the uptake of cholesteryl esters from high-density lipoproteins and the transport of micronutrients like fat-soluble vitamins and carotenoids across cell membranes. An integrated approach combining functional foods and nanoparticle-based delivery systems is explored. Silver nanoparticle coating and selective laser-assisted surface alloying were employed on titanium, alumina, and vanadium surfaces to enhance antibacterial properties. These hybrid systems and microbial volatiles technologies are advancing both cancer therapy (Nagashri et al., 2024). The findings suggest potential methods for developing materials with biphilic and anisotropic wetting properties by controlling the nanoporous layer's thickness.

POLYMERS

Polymeric composite thin films are effective antimicrobial materials developed in response to increasing antibiotic resistance applications in biomedical devices, implants, wound dressings, and surfaces in healthcare. Laser-based deposition techniques, such as pulsed laser deposition and matrix-assisted pulsed laser evaporation, allow for precise control over the films' composition, structure, and thickness. The incorporation of antimicrobial agents like metal nanoparticles, antibiotics, and natural compounds into polymeric matrices enhances the antimicrobial properties of the coatings against various pathogens, including antibiotic-resistant strains. The factors affecting antimicrobial efficacy include film morphology, composition, and substrate interactions. Polymersomes are nanostructures with a hollow aqueous compartment and a coating of amphiphilic block copolymers, providing them with higher mechanical stability compared to liposomes.

The tunable properties of polymersome membranes make them promising carriers for drug delivery, with applications ranging from protein therapy to medical imaging. (Lai, 2025).

SELF-ASSEMBLY POLYMERS

Self-assembly and hierarchical structures of polyelectrolytes can self-assemble under external electric fields, enabling controlled ionic transport in biomimetic nanochannels.

The strength and Adhesion of Spider silk serve as a model for materials with high strength-to-weight ratios. The adhesive mechanisms of gecko feet inspire medical adhesives and shoes with superior grip. The lotus leaf's hydrophobic surface structure has led to the development of self-cleaning materials for applications such as window panes and coatings. Adaptive Networks from Swarm intelligence, as seen in ants and bees, have inspired adaptive nanorobotic networks capable of real-time communication and collective environmental monitoring (Mundekkad & Mallya, 2025). Tumor-targeted nanoscale coordination polymers as chemo-immunotherapy in metastatic colorectal cancer (mCRC) facilitate nanoparticle entry into cancer cells without compromising drug integrity. Laser irradiation induces localized heating around AuNP clusters, causing vaporization or thermo-mechanical decomposition of lipid layers. The patients with MMR-proficient or MSI-low tumors currently have limited treatment options. The project focuses on creating a core-shell NCP, OX/CDN/Chol-D, to co-deliver oxaliplatin and a cholesterol-conjugated drug, promoting immunogenic cell death and activating the STING pathway to improve T-cell responses and synergize with immune checkpoint blockade therapies. This innovative approach seeks to transform "cold" tumors into "hot" ones, ultimately improving outcomes for mCRC patients with poor prognoses.

Despite advancements in AuNP-based PTT, challenges such as biodistribution, long-term safety, and regulatory approval pose significant barriers to its clinical translation. A multifunctional hydrogel coating for poly (aryl ether nitrile ketone) implants, designed to address tumor recurrence, bacterial infection, and bone loss in cancer patients with bone defects. The coating, which incorporates black phosphorus nanoparticles, enables drug delivery, antibacterial action, and promotes osteointegration through a combination of photothermal and photodynamic therapies.

SYNTHESIS AND CHARACTERIZATION PROTOCOLS

Membrane Isolation: Differential centrifugation (800xg for 10min, 20,000xg for 45min) followed by sucrose density gradient purification yields 98% pure membrane fractions. The research optimized a cell membrane isolation protocol using U87-MG human GBM cells, identifying the Heidolph homogenizer as an effective method for producing pure and scalable cell membrane fractions. G-NPs demonstrated improved uptake by GBM cells and increased BBB permeability (Jiménez-Boland et al., 2025).

Core Fabrication: Poly(lactic-co-glycolic acid) (PLGA) cores (153±12nm) are synthesized via nanoprecipitation, loaded with 12.7% w/w doxorubicin through emulsion-diffusion. Hybrid-coated nanoparticles exhibit superior targeting abilities due to the presence of specific surface adhesion molecules derived from source cells. For instance, platelet–cancer stem cell hybrid membrane-coated iron oxide nanoparticles demonstrate both immune evasion and active homotypic targeting. This duality enables efficient delivery to tumors while minimizing clearance by the immune system, thereby increasing therapeutic efficacy.

Membrane Coating: Extrusion through 400nm polycarbonate membranes at 42°C achieves 89% coating efficiency, verified by fluorescence resonance energy transfer between membrane dyes and core labels.

Functional Validation: Flow cytometry demonstrates 93% retention of original cell surface markers (EpCAM, HER2), while quartz crystal microbalance measurements confirm 5.6×10^6 antibodies/μm^2 surface density.

Advanced characterization techniques include Synthetic cells (SCs) that can be engineered to mimic natural cell functions and perform specialized tasks such as drug delivery, biosensing, bioproduction, vaccine development, and environmental remediation. Cryo-electron tomography visualizes a 7.2nm thick membrane bilayer enveloping cores. Recent advancements in synthetic biology, biomaterials, and microfluidics have led to the creation of more sophisticated synthetic cells. (Yandrapalli, 2025). Surface plasmon resonance: Quantifies binding affinity (KD=18nM) to homologous cancer cells. A thorough understanding of design principles, materials, and fabrication techniques is essential for transitioning SCs from proof-of-concept to practical applications.

Table 8. Ongoing Clinical Trials: Applications of Iron Nanoparticles for Magnetic Fluid Hyperthermia

NCT Trial Number	Study Title	Phase	Cancer Type
NCT03892705	Phase 1 Study of Ferromagnetic Nanoparticle Hyperthermia for Glioblastoma	Phase I	Glioblastoma
NCT01957591	Nanoparticle-Mediated Hyperthermia for Prostate Cancer	Phase I	Prostate Cancer
NCT03348169	Iron Nanoparticles with Chemotherapy for Glioblastoma Multiforme	Phase II	Glioblastoma

CASE STUDIES OF THERAPEUTIC APPLICATIONS

The advanced biomaterials in photothermal therapy (PTT) for malignant bone tumors highlight their potential to enhance surgical treatment effectiveness and reduce recurrence.

Solid tumors (breast, lung, ovarian cancer): Doxorubicin and Paclitaxel are potent chemotherapy drugs with poor water solubility and significant side effects. Encapsulating them in hybrid nanoparticles improves their bioavailability and allows for targeted delivery to tumor sites, minimizing damage to healthy tissues.

HER2-positive breast cancer, EGFR-positive cancers: By conjugating anti-HER2 or anti-EGFR antibodies to gold nanorods, the nanoparticles can specifically bind to cancer cells that overexpress these receptors. When exposed to near-infrared (NIR) light, the nanorods generate intense heat, selectively destroying the cancer cells while sparing healthy ones.

Cancer: Porphyrins and Chlorins are effective photosensitizers. However, they can be unstable or have poor water solubility. Encapsulating them within the pores of MSNs enhances their stability and dispersibility. The MSNs can also be functionalized with targeting ligands to deliver the photosensitizers directly to cancer cells for more precise and effective PDT.

Cardiovascular diseases, Orthopedic injuries: RGD peptides are a common motif in the extracellular matrix (ECM) that promotes cell adhesion. Coating medical implants (cardiovascular stents, bone implants) with RGD-functionalized hydrogel nanoparticles improves the integration of the implant with surrounding host tissue, reducing the risk of implant rejection or failure.

Bone defects, Orthopedic implants: VEGF promotes the formation of new blood vessels (angiogenesis), which is crucial for nutrient supply and waste removal in new bone. BMPs are powerful inducers of bone formation (osteogenesis). Hybrid coatings with these growth factors on orthopedic implants accelerate bone healing and improve implant stability.

MELANOMA

Cancer metabolism and CD8+ T cell exhaustion in local metastatic niches occur in melanoma metastases. CD8+ T cells, exhausted in tumors, lead to increased cancer progression and resistance to treatment. The importance of cellular metabolism in this process shows

nutrient availability in the tumor microenvironment on T cell exhaustion, with a specific emphasis on differences between primary tumors and metastases. Disruption of tumor-stroma crosstalk via nanoparticle-based photodynamic therapy increases tumor permeability, allowing nanoparticle delivery to be more effective. LADD is considered safe across skin types and immunosuppressed patients.

The integration of artificial intelligence and advanced in vitro test platforms, such as 3D tumor spheroids and bioprinted vascularized tumors. Bioactive glass nanoparticles and hybrid membrane-coated nanoparticles demonstrate selective toxicity toward cancer cells while supporting tissue regeneration or immune activation. To optimize drug delivery research, improve early tumor detection, and reduce reliance on animal experiments, align with the 3R principles.

BONE CANCERS

The primary bone cancers are expected to see nearly 4000 new cases diagnosed in the U.S. in 2024. 3D-printed biological scaffolds made from various biomaterials are used in treating bone tumors and promoting tissue regeneration. Various physicochemical analysis methods, including scanning electron microscopy, atomic force microscopy, and electrochemical impedance spectroscopy, were employed to characterize the coatings on titanium substrates.

Composite thin films made of bioactive glass and melittin demonstrated a significant reduction in hydrophobicity, shifting from ~95° for bare titanium to ~62°, indicating enhanced biocompatibility favorable for cell adhesion. The coatings showed high bioactivity with the formation of a carbonated apatite layer after 7 days in simulated body fluid and exhibited improved corrosion resistance. Bioactive glass and melittin enhance osseointegration and longevity of coatings with a ~45% decrease in impedance magnitude after 12 hours of immersion compared to only 4% for bare titanium. (Dinu et al., 2025).

The treatment of bone cancer is complicated due to its rarity and diversity. Laser-triggered release ensures reducing systemic exposure and toxicity issues related to surgical interventions, metastatic spread, and post-operative complications. Integrating biomimetic principles into additive manufacturing (AM) that goes beyond mere morphological mimicry, emphasizing functional adaptation and sustainability. Innovative therapeutic use of biomaterials, scaffolds, and gene therapy, to promote bone regeneration and address tumor recurrence, utilizing technologies like 3D bioprinting (Khobragade et al., 2025). By emulating biological systems such as nacre, spider silk, and bone, AM can achieve multifunctionality, responsiveness, and resource efficiency through traditional geometric replication. Recent advancements in 4D printing, soft robotics, and self-morphing hydrogel particles retain water and encapsulate active molecules, making them ideal for drug delivery (Hinojosa-Ventura et al., 2025). It emphasizes the transformative role of microfluidics in the precise creation of these particles, enhancing their dimensions, morphology, and stability for improved pharmaceutical applications. The advantages of 3D printing technology include the ability to create adjustable shapes, porosity, pore sizes, and mechanical properties, which surpass the limitations of natural and synthetic materials. Various types of 3D-printed biological scaffolds, such as bioceramic, metal alloy, and nano-scaffolds, illustrate the potential for engineering time-dependent behaviors and environmental adaptability using bioinspired material architectures (Kantaros et al., 2025).

METASTATIC BREAST CANCER

A comprehensive nanomedicine approach improves the treatment of metastatic breast cancer, addressing its complexity and heterogeneity. Key objectives include creating polymeric micelles that can co-load multiple drugs and an imaging agent, and using physical and pharmacological methods to enhance tumor vasculature and microenvironment. The focus is on identifying biomarkers for patient selection through imaging and biopsy, which is essential for effective clinical application of cancer nanomedicines. The development of bioactive coatings, ceramics, biopolymers, and nanomaterials promotes bone integration, tissue regeneration, and antimicrobial effects. Bioactive hybrid nanoparticles co-deliver doxorubicin (DOX) and indocyanine green (ICG) for targeted chemotherapy and photothermal therapy in breast cancer treatment. The synthesized nanoparticles, with an average size of 113.58 ± 2.14 nm, are modified with polydopamine (PDA) and attached to the anaerobic probiotic Bifidobacterium infantis, enhancing their ability to target tumor hypoxic regions. The photothermal conversion efficiency of approximately 38.04% utilizes the tumor microenvironment (TME) conditions (low pH and high

GSH levels) to control the release of DOX and ICG. The combination of therapies leads to apoptosis of tumor cells and immunogenic cell death (ICD. It is an effective antitumor response with minimal systemic toxicity, positioning the hybrid as a promising nanodrug for solid cancer treatment (Lu et al., 2023). The research intends to enhance immunotherapy efficacy by using multifunctional micelles, with specific strategies to induce immunogenic cell death while minimizing side effects associated with drug resistance.

Limitations

The research highlights the need for interdisciplinary collaboration to address challenges in scalable fabrication, dynamic material programming, and true functional emulation, aiming to develop sustainable, high-performance solutions for various applications, including aerospace and biomedical fields (Kantaros et al., 2025). Optimization of composition and structure: The impact of nanoparticle size, morphology, and membrane composition on efficacy needs further clarification for each cancer type. Future research will likely focus on refining hybrid coating strategies for personalized medicine, integrating computational design to optimize formulations, and advancing clinical translation. Biomimicry at the nanoscale contributes significantly to sustainability and addresses ethical considerations by leveraging nature's designs and processes to create innovative, environmentally conscious solutions.

FUTURE DIRECTIONS

Environmentally Friendly and Resource-Efficient Solutions by nanoscale biomimicry aims to create solutions that are environmentally friendly and resource-efficient by drawing inspiration from nature's designs and processes, which have evolved over millions of years to be highly efficient and sustainable. Efficient Energy Systems leads to the development of more efficient nanostructured materials for solar panels and energy storage systems, mimicking natural processes like photosynthesis. This reduces reliance on fossil fuels and minimizes the ecological footprint associated with energy production.

Eco-Conscious Materials for nanoscale biomimicry supports the creation of sustainable, eco-conscious nanomaterials that offer efficient functionality and superior performance. This can include materials with improved strength, light-weight properties, self-healing capabilities, or self-cleaning and self-adhesive textiles and coatings, inspired by natural examples like lotus leaves and spider silk. Biomimetic nanomaterials are crucial for efficient environmental remediation, including water purification, pollution control, and waste management. For instance, nanoparticles modeled after algae or plants can efficiently absorb toxins or catalyze the breakdown of harmful substances. The nano field offers adaptive solutions to reduce carbon footprints and manage climate-related challenges, such as designing buildings inspired by termite mounds for energy-efficient ventilation. Ongoing research is focused on developing sustainable and environmentally friendly materials, including biodegradable nanocomposites, which offer alternatives to drastically reduce environmental impact and support global sustainable development goals. Advancements in biomimetic materials, like self-healing concrete or adaptive building materials, can enhance durability and ensure the longevity of structures, revolutionizing sectors like construction.

HYBRID MEMBRANE COATING INNOVATIONS

These systems, often involving the combination of multiple membrane types or functional nanomaterials, improve tumor specificity, prolong circulation time, evade immune clearance, and can integrate multiple treatment modalities such as photothermal therapy, immunotherapy, and drug delivery Transdermal Drug Delivery Systems (TDDSs) provide non-invasive drug administration and sustained release, but their effectiveness is limited by the skin's barrier, especially for high molecular weight and hydrophilic substances. Hybrid nanophotonic graphene systems utilize localized photothermal effects to create microchannels in the skin, while graphene improves permeability due to its electrical and thermal conductivity. These hybrid systems include various advanced materials such as photonic crystals, plasmonic nanoparticles, metamaterials, quantum dots, nanowires, fiber optic nanosensors, and nanoantennas, enabling precise control and real-time monitoring for medical applications. The research highlights the design, effectiveness, biocompatibility, and clinical implications of hybrid nanophotonic graphene systems, emphasizing their potential to enhance transdermal drug delivery and transform personalized medicine, with patentable advancements in the field.

Hybrid systems combine membranes from different cell types (cancer cells, immune cells, platelets) to endow nanoparticles with multiple functionalities. LADD is considered safe across skin types and immunosuppressed patients; however, unpredictable systemic effects may occur if inadvertent systemic delivery happens. For example, a hybrid membrane coating can provide both tumor-targeting and immune-modulating properties, or combine homotypic targeting with prolonged circulation. Laser-assisted printing technologies can deposit drugs directly onto microneedle surfaces, enabling targeted transdermal delivery for antifungal or anticancer treatments. The design, fabrication, and application of polymersomes as carriers for oral drug administration, highlighting their mechanical stability and tunable properties compared to other nanostructures (Lai, 2025).

Multifunctional Platforms for Theranostics

Hybrid CCM-NPs are being developed as multifunctional platforms for both therapy and diagnostics (theranostics). These systems can be engineered to carry imaging agents alongside drugs, enabling real-time monitoring of drug delivery and therapeutic response. The bottom-up development of synthetic cells (SCs) that mimic natural cell functions and can be used for various applications, including drug delivery and environmental remediation. It highlights recent advancements in synthetic biology and related fields (Ananthakumar et al., 2024). The versatility of hybrid systems opens new avenues in composite coatings produced using laser deposition techniques have shown improved osteointegration and antibacterial properties when applied to titanium implants. These coatings combine structural reinforcement with therapeutic functionality.

ETHICAL CONSIDERATIONS

Hybrid coatings face several challenges that must be addressed before widespread clinical translation. These include issues related to scalability, reproducibility, long-term stability, and regulatory approval. Additionally, understanding the interactions between complex coating materials and biological systems remains an ongoing research challenge. The intersection of nanoscale biomimicry with sustainability requires a careful consideration of ethical aspects. The challenges alongside innovations enhance scientific credibility, and accelerate the development of truly transformative cancer therapies. Material-Specific Hurdles:

- Gold (Au), Silver (Ag): Long-term biodistribution and potential for ion leaching/toxicity
- Iron Oxide (Fe_3O_4): Clinical use (Ferumoxytol). Challenges are more about functionalizing and targeting rather than core safety.
- Quantum Dots: Heavy metal toxicity (Cd, Se) limiting to diagnostic applications.
- Carbon-based (CNTs, Graphene): Biopersistent aspect ratio of CNTs can cause asbestos-like pathogenicity

The use of nanotechnology raises concerns about potential environmental impacts. The lifecycle of nanomaterials, particularly their disposal, could introduce unforeseen risks if not managed responsibly. There are concerns regarding human health, specifically the potential for unexpected biological interactions (like abrupt immune responses or cytotoxicity) when biomimetic nanoparticles interact with biological systems. Assessing long-term biocompatibility is crucial, especially when chronic exposure could lead to accumulation and potential toxicity. Social Equity is important to ensure equitable access to these technologies to prevent disparities from widening if advancements in nanoscale biomimicry only benefit certain populations. The development and use of biomimetic nanoparticles face stringent regulatory requirements, which can be complex and vary by region. Rigorous testing for safety and efficacy, along with a current lack of standardized tests and protocols, complicates regulatory approval processes and can delay market introduction. Fostering an ethical framework that prioritizes transparency, inclusivity, and ecological stewardship is essential. By accommodating ethical aspects into research and applying these technologies responsibly, advancements can address environmental challenges while also promoting social good, respecting both the natural world and dependent communities. Issues related to large-scale manufacturing, reproducibility, long-term stability, and regulatory approval must be addressed. Furthermore, understanding the complex interactions between hybrid materials and biological systems is essential to ensure safety and effectiveness. Future research is expected to focus on integrating artificial intelligence, biomimetic design principles, and advanced nanofabrication techniques to develop next-generation theranostic coatings capable of personalized cancer treatment.

CONCLUSIONS

Hybrid coatings represent a transformative strategy for cancer theranostics by combining diagnostic imaging, targeted drug delivery, and therapeutic modalities within a single multifunctional platform. By integrating nanomaterials, biomimetic membranes, polymers, and laser-engineered structures, these coatings offer improved tumor targeting, enhanced therapeutic efficacy, and reduced systemic toxicity. As research in nanotechnology and biomaterials continues to evolve, hybrid coatings are poised to play a central role in the development of precision oncology and next-generation cancer therapeutics. Nanomedicine remains one of the most challenging areas in modern medicine, with traditional pharmaceutical treatment modalities often limited by poor drug targeting, systemic toxicity, and inadequate therapeutic efficacy (Bhuvana et al., 2025). Hybrid bioactive coatings have emerged as a promising therapeutic platform that combines multiple functional components to enhance cancer treatment outcomes while minimizing adverse effects. These advanced coating systems integrate drug delivery capabilities, targeting mechanisms, and therapeutic modalities such as photothermal therapy (PTT) and photodynamic therapy (PDT) into a single platform. Hybrid coating systems combining bioactive molecules and nanoparticles, particularly when fabricated by laser processing, represent a promising direction for advanced cancer therapies. Our meta-analysis indicates enhanced targeting, controlled drug release, and multifunctional theranostics. While challenges remain regarding scalability and regulatory approval, ongoing research and technological advancements continue to push this field toward clinical application. In summary, hybrid coating systems based on bioactive nanoparticles are transforming cancer therapy by integrating advanced targeting, immune modulation, and multimodal treatment capabilities within a single platform. Their adaptability and multifunctionality position them at the forefront of next-generation oncologic nanomedicine. Nanomaterials, biomimetic membranes, polymers, and laser-engineered coatings can enhance targeting accuracy, improve drug delivery efficiency, and enable real-time monitoring of therapeutic outcomes. As interdisciplinary collaboration continues to expand across nanotechnology, biomaterials science, and biomedical engineering, hybrid coatings are likely to play a central role in the next generation of precision medicine and smart biomedical devices.

REFERENCES

Amaral, M. N., Kumar, P., Faísca, P., Ferreira, H. A., Coelho, J. M. P., Gaspar, M. M., & Reis, C. P. (2025). Gold nanoparticle-mediated photothermal therapy: Expanding the frontiers of cancer treatment and theragnostics. *Biomedicine and Pharmacotherapy*, *190*, 118399. DOI: 10.1016/j.biopha.2025.118399 PMID: 40784123

Ananthakumar, B., Nagaraja, B. G., Umesh, S. M., & Raghavan, R. (2025). Predictive Medicine. *Predictive Algorithms for Rehabilitation and Assistive Systems*, 291–314. DOI: 10.4018/979-8-3373-0194-5.ch011

Ananthakumar, B., Nanjundeshwara, N., Lakshmaiah, A., & Raghavan, R. (2024). Synthetic Gene Circuits as a Promising Approach in Cancer Immunotherapy. *Critical Developments in Cancer Immunotherapy*, 391–418. DOI: 10.4018/979-8-3693-3976-3.ch011

Ananthakumar, B., Nanjundeshwara, N., Lakshmaiah, A., & Raghavan, R. (2025). Nanocarriers for Delivering Nucleic Acids in Plants. *Nanocarriers in Plant Science and Agriculture*, 207–234. DOI: 10.4018/979-8-3693-9869-2.ch008

Bhuvana, G. N., Umesh, S. M., & Raghavan, R. (2025). Expanding the frontiers of Schistosomiasis Treatment: The promise of Nanoparticle-Based Therapies. *Journal of Vector Borne Diseases, 10*(4103/jvbd.jvbd_180_23). DOI: 10.4103/jvbd.jvbd_180_23

Biradar, S., Korrapati, C., Krishna, R. J., & Nanjundeshwara, N. (2025). Spatial Omics and Gene Circuits: The Future in Cancer Biology. *Spatially Variable Genes in Cancer: Development, Progression, and Treatment Response*, 383–408. DOI: 10.4018/979-8-3693-7728-4.ch014

Biradar, S., Suresh, C., Nanjundeshwara, N., & Raghavan, R. (2025). Spatially Variable Genes. *Spatially Variable Genes in Cancer*, 1–28. DOI: 10.4018/979-8-3693-7728-4.ch001

Byalahunasishetter, B. I., N., V., P., S., C., M., S., S., & Raghavan, R. (2025). Spatial Omics and Bioprinting. *Innovations in Precision Medicine and Genomics*, 85–110. DOI: 10.4018/979-8-3693-5787-3.ch003

Chandra, G. V. S. A., Biradar, S., & Raghavan, R. (2025). Recent Trends in Soft Robotics for Assistive Technologies. *Assistive Technology Solutions for Aging Adults and Individuals With Disabilities*, 119–144. DOI: 10.4018/979-8-3693-6308-9.ch006

Dinu, M., Bita, B., Parau, A. C., Ristoscu, C., & Negut, I. (2025). Bioactive Glass and Melittin Thin Films Deposited by MAPLE for Titanium Implant Functionalization. *Materials (Basel)*, *18*(10), 2410. DOI: 10.3390/ma18102410 PMID: 40429147

GVS, A. C., Ananthakumar, B., & Raghavan, R. (2025). Enhancing Assistive Technologies With Neuromorphic Computing. *Revolutionizing AI with Brain-Inspired Technology: Neuromorphic Computing*, 207–232. DOI: 10.4018/979-8-3693-6303-4.ch009

Hinojosa-Ventura, G., Acosta-Cuevas, J. M., Velázquez-Carriles, C. A., Navarro-López, D. E., López-Alvarez, M. Á., Ortega-de la Rosa, N. D., & Silva-Jara, J. M. (2025). From Basic to Breakthroughs: The Journey of Microfluidic Devices in Hydrogel Droplet Generation. *Gels (Basel, Switzerland)*, *11*(5), 309. DOI: 10.3390/gels11050309 PMID: 40422329

Jiménez-Boland, D., Robles-Fernández, A., Martín-Rodríguez, A., Cuadros, M. Á., Traverso, J. Á., Sánchez-Moreno, P., & Bramini, M. (2025). Breaking Barriers in Glioblastoma Targeting through Advanced Nanoparticle Cell Membrane Coating. *ACS Applied Materials & Interfaces*, *17*(24), 35288–35303. DOI: 10.1021/acsami.5c07306 PMID: 40488690

Kantaros, A., Ganetsos, T., Pallis, E., & Papoutsidakis, M. (2025). Biomimetic Additive Manufacturing: Engineering Complexity Inspired by Nature's Simplicity. *Biomimetics*, *10*(7), 453. DOI: 10.3390/biomimetics10070453 PMID: 40710266

Khobragade, S. S., Deshmukh, M., Vyas, U., & Ingle, R. G. (2025). Innovative Approaches in Bone Tissue Engineering: Strategies for Cancer Treatment and Recovery. *International Journal of Molecular Sciences*, *26*(9), 3937. DOI: 10.3390/ijms26093937 PMID: 40362178

Korrapati, C., Chowdareddy, S., Suresh, C., & Raghavan, R. (2026). Cancer Cell Membrane-Coated Nanoparticles: Therapeutic Advancements and Hybrid Systems. *Applications of Nanomaterials for Oncology*, 131–160. DOI: 10.4018/979-8-3373-3765-4.ch005

Korrapati, C., Yaraladdi, A. B., Vaddi, H., & Raghavan, R. (2024). Applications of the Saffron European Kitchen Spice Crocus sativus. *Therapeutic and Pharmacological Applications of Ethnobotany*, 98–127. DOI: 10.4018/979-8-3693-1986-4.ch005

Lai, W.-F. (2025). Design and Applications of Polymersomes for Oral Drug Administration. *ACS Applied Materials & Interfaces*, *17*(21), 30423–30435. DOI: 10.1021/acsami.5c04658 PMID: 40370090

Lakshmaiah, A., Korrapati, C., Suresh, C., & Raghavan, R. (2025). Synthetic Biology and AI. *Spatially Variable Genes in Cancer*, 265–290. DOI: 10.4018/979-8-3693-7728-4.ch010

Lakshmaiah, A., & Raghavan, R. (2026). Cardiovascular Protector Activity of Medicinal Mushrooms. *Medicinal Mushrooms: Conservation and Bioprospecting*, 209–218. DOI: 10.1007/978-1-0716-4993-0_28

Lu, Y., Chen, L., Wu, Z., Zhou, P., Dai, J., Li, J., Wen, Q., Fan, Y., Zeng, F., Chen, Y., & Fu, S. (2023). Self-driven bioactive hybrids co-deliver doxorubicin and indocyanine green nanoparticles for chemo/photothermal therapy of breast cancer. *Biomedicine and Pharmacotherapy*, *169*, 115846. DOI: 10.1016/j.biopha.2023.115846 PMID: 37944443

M, M. K., P, D. A., J, S. P., Moger, S. S., & Raghavan, R. (2026). Spatial Multi-Omics in Cancers: Integrating Molecular Landscapes for Precision Oncology. *Spatially Resolved Single-Cell and Multiple Omics in Human Diseases*, 217–248. DOI: 10.4018/979-8-3373-4849-0.ch007

Maskowicz, D., Maroszek, K., Jendrzejewski, R., & Sawczak, M. (2024). Hydroxyapatite Nanocoatings Deposited by Means of Resonant Matrix-Assisted Pulsed Laser Evaporation. *Materials (Basel)*, *17*(23), 5778. DOI: 10.3390/ma17235778 PMID: 39685214

Mirkani, A., Nabid, M. R., & Pakian, S. (2025). Microfluidics-Facilitated Synthesis of Methacrylated Chitosan-Coated Liposomes: An Innovative Platform for Prolonged Ocular Retention and Controlled Drug Release. *Biomacromolecules*, *26*(8), 5336–5352. DOI: 10.1021/acs.biomac.5c00849 PMID: 40665681

Montesissa, M., Tommasini, V., Rubini, K., Boi, M., Baldini, N., & Boanini, E. (2025). State of Art and Perspective of Calcium Phosphate-Based Coatings Coupled with Bioactive Compounds for Orthopedic Applications. *Nanomaterials (Basel, Switzerland)*, *15*(15), 1199. DOI: 10.3390/nano15151199 PMID: 40801737

Mundekkad, D., & Mallya, A. R. (2025). Biomimicry at the nanoscale—A review of nanomaterials inspired by nature. *Nano Trends*, *10*, 100119. DOI: 10.1016/j.nwnano.2025.100119

Nagashri, N., Archana, L., & Raghavan, R. (2024). Microbial Volatile Compounds (MVOCs) in Food Industries and Food Safety Applications. *Innovations, Securities, and Case Studies Across Healthcare, Business, and Technology*, 75–96. DOI: 10.4018/979-8-3693-1906-2.ch005

Nanjundeshwara, N., Lakshmaiah, A., & Raghavan, R. (2025). Aflatoxigenic Fungi. *Aflatoxigenic Fungi*, 225–239. DOI: 10.1201/9781003487043-29

P., A., D., K. M., M., J., P., S., & Raghavan, R. (2025). The Cancer Exposome. *Cancer Exposomics and Environmental Influences on Carcinogenesis*, 161–192. DOI: 10.4018/979-8-3373-2165-3.ch004

Pallares, R. M., Barmin, R. A., Wang, A., Kiessling, F., & Lammers, T. (2025). Clinical cancer nanomedicines. *Journal of Controlled Release*, *385*, 113991. DOI: 10.1016/j.jconrel.2025.113991 PMID: 40582644

Pallavi, V. N., Chowdareddy, S., Byalahunasishetter, B. I., & Raghavan, R. (2026). CRISPR-Cas9 Technology for Precision Cancer Medicine: Target Discovery, Disease Modeling, and Therapeutics. *CRISPR-Cas9 Technology for Precision Medicine*, 35–72. DOI: 10.4018/979-8-3373-4857-5.ch002

Raghavan, R., Moger, S. S., Umesh, S., & Bhuvana, G. N. (2025). Revolutionizing Pain Management. *Unveiling Technological Advancements and Interdisciplinary Solutions for Pain Care*, 129–164. DOI: 10.4018/979-8-3693-9501-1.ch005

S., G. V., C., A., Biradar, S., & Raghavan, R. (2025). Neuroscience-Driven Solutions for Speech Impairment. *Advancing Medical Research Through Neuroscience*, 119–146. DOI: 10.4018/979-8-3693-5464-3.ch006

Sachan, R., Jaipan, P., Zhang, J. Y., Degan, S., Erdmann, D., Tedesco, J., Vanderwal, L., Stafslien, S. J., Negut, I., Visan, A., Dorcioman, G., Socol, G., Cristescu, R., Chrisey, D. B., & Narayan, R. J. (2025). Printing amphotericin B on microneedles using matrixassisted pulsed laser evaporation. *International Journal of Bioprinting*, *3*(2), 147–157. DOI: 10.18063/IJB.2017.02.004 PMID: 33094188

Tsang, C. Y., & Zhang, Y. (2024). Nanomaterials for light-mediated therapeutics in deep tissue. *Chemical Society Reviews*, *53*(6), 2898–2931. DOI: 10.1039/D3CS00862B PMID: 38265834

Visan, A. I., & Negut, I. (2025). Polymeric Composite Thin Films Deposited by Laser Techniques for Antimicrobial Applications—A Short Overview. *Polymers*, *17*(15), 2020. DOI: 10.3390/polym17152020 PMID: 40808071

Yandrapalli, N. (2025). Bottom-up development of lipid-based synthetic cells for practical applications. *Trends in Biotechnology*, *0*(0), 2150–2169. Advance online publication. DOI: 10.1016/j.tibtech.2025.03.009 PMID: 40263003

Chapter 4
Laser Cladding of Bioactive Coatings on Biodegradable Implants

Muhammed Enes İlgazi
http://orcid.org/0000-0002-5763-6427
Bitlis Eren University, Turkey

Sercan Basit
http://orcid.org/0000-0002-2368-6362
Kırşehir Ahi Evran University, Turkey

Yusuf Uzunoğlu
Bitlis Eren University, Turkey

ABSTRACT

Biodegradable metals like magnesium and zinc offer temporary mechanical support for implants, preventing stress shielding and secondary extraction surgeries. However, their clinical viability is severely limited by rapid, uncontrolled degradation and cytotoxic hydrogen gas accumulation. To overcome these barriers, this chapter presents laser cladding as a transformative surface engineering strategy. Unlike traditional coatings with weak physical adhesion, laser processing creates a durable, true metallurgical bond between heat-sensitive biodegradable substrates and bioactive ceramics like hydroxyapatite. This review explores the thermodynamic challenges of processing low-boiling-point alloys, focusing on controlling laser energy density to mitigate porosity and cracking. Furthermore, it evaluates clinical performance, demonstrating how these microstructural architectures exponentially

DOI: 10.4018/979-8-3373-4104-0.ch004

slow electrochemical corrosion while simultaneously stimulating osteoblast adhesion and bone regeneration, paving the way for reliable next-generation transient medical devices.

1. BIODEGRADABLE IMPLANTS AND THE NEED FOR SURFACE ENGINEERING

1.1. The Paradigm Shift in Medical Implants: From Permanent to Transient

Orthopedic trauma fixation and cardiovascular surgeries have traditionally relied heavily on permanent metallic alloys, such as 316L stainless steel, titanium, and cobalt-chromium (Marin & Lanzutti, 2025). These materials offer exceptional mechanical strength and resist corrosion beautifully inside the human body (Verma et al., 2025). However, their permanence is not always a clinical advantage (Aziz et al., 2025). Bone tissue requires physical mechanical stress to maintain its density and strength—a physiological concept known as Wolff's Law (Ruff et al., 2006). When a rigid titanium plate takes over the entire mechanical load, the surrounding natural bone is deprived of this necessary stress (Mellon & Tanner, 2012). This phenomenon, known as stress shielding, inevitably leads to bone weakening (osteopenia) and eventual implant loosening (Baskerville, 2022).

Beyond biomechanical mismatch, leaving a foreign metal inside the body long after a fracture has healed poses significant risks, including chronic inflammation and late-stage infections (Moriarty et al., 2022). For many patients, especially children and active adults, this necessitates a second surgical procedure simply to remove the hardware (Boulos et al., 2017). Recognizing these physical and economic burdens, the biomedical community has begun shifting its focus toward "biodegradable" or "transient" implants (Fernandes & Taurino, 2022). The concept is straightforward but revolutionary: an implant should provide robust mechanical support while the tissue heals, and then gradually dissolve into harmless by-products, leaving behind only healthy, fully integrated natural tissue (Xia et al., 2025).

1.2. The Triumvirate of Biodegradable Metals: Magnesium, Zinc, and Iron

When designing these temporary structures, researchers generally focus on three primary metallic systems: iron, zinc, and magnesium (Prasadh et al., 2022). Each brings a unique set of metallurgical properties and biological responses to the table. Iron is structurally robust and mimics the mechanical behavior of traditional steels, making it an attractive candidate for load-bearing applications (Salama et al., 2022). The problem with iron, however, is that it degrades far too slowly in the body (Abbaspour et al., 2014). Its corrosion products, primarily bulky iron oxides, tend to accumulate at the implantation site, potentially disrupting local cellular functions rather than being safely absorbed (Szczęsny et al., 2025).

Zinc serves as an interesting middle ground. Its degradation rate perfectly matches the healing timeline of cardiovascular tissues, making it a favorite for temporary stents (Prasadh et al., 2022). The major drawback of pure zinc is its inherent mechanical weakness; it lacks the strength and ductility required for heavy orthopedic loads without extensive and complex alloying (Kong et al., 2023).

Magnesium stands out as the most biologically and biomechanically compatible option. Its elastic modulus (around 41-45 GPa) is remarkably close to that of human cortical bone, which essentially eliminates the risk of stress shielding (Kanaginahal et al., 2024). Furthermore, magnesium is a vital trace element in the human body (Pethő et al., 2024). As a magnesium implant degrades, the released ions actively stimulate osteoblasts, accelerating new bone formation (Bai et al., 2023). Yet, this biological harmony comes with a massive engineering trade-off: magnesium is fiercely reactive (Mubeen et al., 2025).

1.3. The Clinical Dilemma: Uncontrolled Degradation Kinetics

The greatest hurdle in utilizing magnesium and its alloys clinically is their rapid, almost aggressive degradation when exposed to chloride-rich body fluids like blood plasma (Wu et al., 2026). When a magnesium implant contacts an aqueous physiological environment, it undergoes rapid anodic dissolution (Mousavian & Bautin, 2026). This electrochemical reaction releases magnesium ions while simultaneously reducing water to produce hydroxide ions and hydrogen gas (Liang et al., 2024). If the hydrogen gas evolves faster than the surrounding tissue can absorb it, gas pockets form beneath the skin (Tanaka et al., 2022). These pockets physically separate the implant from the bone bed, preventing proper tissue integration (Alrmali et al., 2026).

Simultaneously, the localized spike in hydroxide ions drives the pH of the surrounding tissue to highly alkaline levels, which can be toxic to nearby cells (Pearson, 2025). Most critically, this rapid corrosion means the implant loses its structural integrity far too quickly (Bender et al., 2022). An orthopedic plate might literally crumble before the bone has had the 12 to 24 weeks it needs to heal (Graeser, 2023). Attempts to solve this purely by altering the bulk material—such as adding heavy alloying elements—often backfire (Raabe, 2023). Introducing different metals into the magnesium matrix can create internal micro-galvanic cells, which paradoxically accelerate the corrosion process rather than slowing it down (Ikeuba et al., 2024).

1.4. The Imperative for Surface Engineering

Since modifying the bulk alloy has its limitations, engineers have increasingly turned to surface engineering as the primary defense mechanism (L. Wang et al., 2025). The rationale is to separate the mechanical duties of the bulk metal from the chemical reactions happening at the surface (Miu et al., 2022). By modifying just the outermost layer of the implant, we can construct a temporary kinetic barrier (Yang et al., 2022). This engineered surface must perform a dual role: it needs to shield the highly reactive underlying metal from the aggressive physiological fluids during the critical early stages of healing, while simultaneously offering a "bioactive" interface that encourages proteins and bone cells to attach and proliferate (Nikolova & Apostolova, 2022).

1.5. Limitations of Conventional Coating Technologies

Historically, researchers have tried to protect these biodegradable metals using a variety of conventional coating techniques, including dip coating, sol-gel deposition, and polymer wrapping (De Riccardis, 2023). While these methods can successfully deposit bioactive materials like hydroxyapatite onto the implant, they share a fatal flaw: poor interfacial adhesion (M. F. Wang et al., 2025). Most traditional coatings rely on weak physical interlocking or minimal chemical bonding (Li et al., 2022).

Orthopedic implants are not static; they bend, twist, and compress with every movement the patient makes (Thakur, 2022). Under these dynamic physiological loads, weakly bonded coatings simply crack and peel off (Kang & Sakaguchi, 2025). Once a piece of the coating delaminates, the exposed magnesium becomes a highly concentrated anode against the remaining coated area (Shapagina & Dushik, 2025). This localized exposure triggers severe pitting corrosion, which hollows out the metal and destroys the implant structurally much faster than if it had never been coated at all (Daniyan et al., 2023).

1.6. Transitioning to Advanced Manufacturing: The Promise of Laser Cladding

To permanently solve the adhesion problem, the biomedical field requires a manufacturing method that fuses the coating directly into the atomic structure of the base metal (Saba et al., 2024). This is precisely where laser cladding demonstrates its transformative potential (Zhao et al., 2025). Instead of merely laying a material on top of the substrate, a high-energy laser beam is used to simultaneously melt the deposited bioactive powder and a microscopic top layer of the underlying biodegradable metal (Badiceanu et al., 2022).

As the laser moves, the localized melt pool cools almost instantaneously, forming a true metallurgical bond between the coating and the implant (Alontseva et al., 2023). This continuous atomic interface provides exceptional shear strength, ensuring that the coating remains firmly attached even under severe physiological bending (Wen et al., 2024). By mastering laser-assisted surface engineering, we can finally bridge the gap between metallurgy and biology—harnessing the regenerative benefits of materials like magnesium and zinc while strictly dictating their lifespan inside the human body (Singh et al., 2024).

2. FUNDAMENTALS OF LASER CLADDING FOR HEAT-SENSITIVE BIODEGRADABLE METALS

2.1. The Physics of Laser-Matter Interaction in Coating Processes

Before diving into the specific challenges of biodegradable implants, we need to look at what actually happens when a laser beam strikes a metal surface (Ahmed et al., 2025). At its core, laser cladding is a specialized additive manufacturing technique (Cheng et al., 2022). It relies on a highly focused laser beam to melt a coating material—usually fed as a powder or wire—directly onto a target substrate (Biswas, 2026).

When the laser hits metals like magnesium or zinc, the free electrons in the metallic lattice immediately absorb the incoming photons (Mostafa, 2025). This electromagnetic energy turns into localized heat almost instantly through electron-phonon collisions (H. Zhang et al., 2023). The catch here is that magnesium and zinc are highly reflective. If we use traditional CO2 lasers (which operate at a 10.6 mikrometre wavelength), these metals will bounce back over 90% of the beam, making the process incredibly inefficient (Sarkar, 2022).

Because of this high reflectivity, modern surface engineering leans heavily on solid-state or fiber lasers (typically operating between 1.03 mikrometre and 1.07 mikrometre) (Bülbül Tatbul, 2022). These shorter wavelengths provide much better energy absorption and help maintain a stable melt pool (Zhao et al., 2022). Once the absorbed heat overcomes the metal's melting point, a tiny pool of liquid metal forms (G. Lu et al., 2024). As we inject powder into this pool under a protective argon or helium gas shield, the laser moves along its path, and the molten material cools rapidly (Grohol, 2024). The end result is a solid, continuous coating track that is metallurgically fused to the base metal (C. J. Li et al., 2022).

2.2. Overcoming the Thermodynamic Hurdles of Mg and Zn

We have successfully used laser cladding on tough, high-temperature materials like titanium and stainless steel for decades (John et al., 2022). However, applying this exact same technology to biodegradable metals is a completely different thermodynamic challenge (Kim et al., 2023). Iron behaves reasonably well under a laser, but magnesium and zinc are notoriously sensitive to heat (Bîrsan & Susac, 2026).

The main issue stems from their incredibly low melting and boiling points (Mao et al., 2024). Pure magnesium melts around 650°C and boils at 1090°C, while zinc melts at just 419°C and boils at 907°C (Gracheva et al., 2025). Considering that a laser-induced melt pool can easily shoot past 1500°C in a fraction of a second, the risk is obvious: if we don't control the heat input perfectly, these metals will literally boil away (Blutinger, 2022).

This vaporization causes two major manufacturing headaches (Saxena & Shukla, 2023). First, the expanding metal vapor creates recoil pressure that blows molten droplets out of the pool—a defect known as spatter (Shi et al., 2024). Second, gas bubbles that fail to escape before the metal freezes get trapped as internal pores (Thiévenaz et al., 2025). For a medical implant, a porous coating is a critical failure point because it allows corrosive body fluids to bypass the protective layer and attack the base metal directly (Daniyan et al., 2023). On top of that, magnesium has a massive chemical affinity for oxygen. Without flawless inert gas shielding, it can oxidize rapidly or even ignite, leaving behind brittle magnesium oxide (MgO) inclusions that severely reduce the implant's fatigue life under physiological loads (D. Kumar & Ansari, 2024).

2.3. Fine-Tuning Laser Parameters for Heat-Sensitive Alloys

Because of these severe thermal risks, getting laser cladding to work on magnesium and zinc requires meticulous parameter tuning (Jeyaprakash et al., 2025).

The entire process hinges on finding the perfect balance between laser power (P), scanning speed (v), beam spot diameter (D), and powder feed rate (f). In materials science, we typically use the concept of energy density to track and manage this heat input. The linear energy density (El) and area energy density (Ea) are calculated using the following formulas:

$$E_l = \frac{P}{v}(J/mm)$$

$$E_a = \frac{P}{v \cdot D}(J/mm^2)$$

When dealing with heat-sensitive biomaterials, the golden rule is to keep the energy density (Ea) as low as practically possible (T. Li et al., 2025).

Too much heat leads to excessive penetration into the substrate, heavy vaporization, and even the thermal distortion of delicate implant structures like cardiovascular stents (Duan et al., 2024).

On the flip side, if the energy density is too low, the injected powder absorbs all the laser's energy (M. Li et al., 2023). The substrate never melts, resulting in a "lack of fusion" defect (Subraveti et al., 2024). The coating basically just sits on the surface without forming a chemical bond, meaning it will easily peel off once implanted in the body (Saba et al., 2024). The powder feed rate (f) actually acts as a helpful thermal buffer here; since the powder requires energy to melt, injecting more powder can help keep the underlying substrate relatively cool (Hsu et al., 2026).

2.4. Controlling Melt Pool Convection and Dilution

The melt pool isn't just a stagnant puddle; it's a violently turbulent micro-reactor. The fluid flow inside is driven primarily by thermocapillary forces, commonly known as Marangoni convection (Khudair et al., n.d.). Because the laser's energy is highly concentrated in the center of the beam, the middle of the melt pool is significantly hotter than its edges (Khan et al., 2022). This temperature difference creates a surface tension gradient that constantly pulls liquid metal outward from the center, generating strong convective currents (X. D. Zhang, 2024).

We can quantify the intensity of this flow using the dimensionless Marangoni number (M_a):

$$Ma = \frac{-\frac{d\gamma}{dT} \cdot \Delta T \cdot L}{\mu \cdot \alpha}$$

In this equation, $d\gamma/dT$ represents how the surface tension changes with temperature, ΔT is the temperature gap, L is the characteristic length of the pool, μ is viscosity, and α is thermal diffusivity.

For biodegradable implants, managing this Marangoni flow is the key to controlling "dilution" (Griffis et al., 2025). We actually need a slight mix (typically between 5% and 15%) of the substrate and the coating to ensure a strong metallurgical bond (X. Liu, 2025). However, if the Marangoni convection is too aggressive, too much of the highly reactive Mg or Zn substrate gets churned into our carefully formulated bioactive coating (Madhuri et al., 2025). This over-dilution can form brittle intermetallic phases and completely destroy the coating's intended bioactivity and corrosion resistance (P. Kumar et al., 2026).

2.5. The Impact of Rapid Solidification on Microstructure

What truly sets laser cladding apart from older, conventional coating methods is the rapid solidification process (Zhao et al., 2025). As soon as the laser beam moves away, the surrounding cold metal acts like a massive heat sink, quenching the tiny melt pool almost instantaneously (Vedani, 2025).

The cooling rates ( ) in laser processing are extreme, often hitting anywhere from 10^3 to 10^6 K/s (Bähr, 2025).

This cooling rate is directly tied to the temperature gradient (G) and the solidification growth rate (R):

$$\dot{T} = G \cdot R$$

This lightning-fast cooling forces the metal to solidify under non-equilibrium conditions, which brings several massive structural benefits for biodegradable implants (Shuai et al., 2025). First, we achieve extreme grain refinement. The rapid cooling creates ultra-fine, sometimes nano-scale grains that significantly boost the implant's surface hardness and yield strength (Gholamzadeh et al., 2024). Second, it traps alloying elements evenly throughout the structure rather than allowing them to segregate at the grain boundaries (Hofmann, 2025). This chemical homogenization minimizes the micro-galvanic cells that usually cause magnesium and zinc to corrode erratically, resulting in a much smoother, more predictable degradation profile in physiological fluids (Imani, 2024).

2.6. Managing the Heat-Affected Zone (HAZ)

Unfortunately, the intense heating and cooling cycle isn't entirely beneficial for the whole part (J. Lu et al., 2025). The area immediately beneath the melt pool—the Heat-Affected Zone (HAZ)—takes a severe thermal beating without actually melting (Lam et al., 2022). Many magnesium and zinc alloys are carefully rolled or extruded during manufacturing to create fine, strong grain structures (Leśniak et al., 2024).

When the laser's heat hits the HAZ, it can cause those carefully engineered grains to grow and recrystallize (Y. Wang et al., 2024). When grains coarsen, that specific local area loses mechanical strength (H. Wang et al., 2024). If you place that implant in a load-bearing environment, like a healing bone fracture, this weakened HAZ becomes a prime location for fatigue failure and localized pitting corrosion (Das et al., 2022).

To mitigate this, engineers often switch to pulsed lasers instead of continuous wave lasers (Haider et al., 2022). By delivering the energy in rapid, short bursts with distinct "off" times, we can lower the overall heat input and keep the HAZ as narrow as possible (Meng et al., 2026). Researchers are also exploring auxiliary cooling methods, like using liquid nitrogen to chill the fixture, to protect the bulk integrity of the implant during coating (Mohammed et al., 2022).

2.7. Epitaxial Growth and True Metallurgical Bonding

Ultimately, the primary reason we use lasers for this application is to create an unbreakable bond between the bioactive coating and the biodegradable metal underneath (Park et al., 2023). Because the laser slightly melts the top layer of the substrate, the newly deposited coating literally grows out of the partially melted substrate grains—a process known as epitaxial growth (M. Li et al., 2023).

As the liquid coating cools, it wets the solid base perfectly, aligning its crystal lattice directly with the crystal lattice of the substrate (Butt, 2022). This continuous atomic arrangement completely eliminates the gaps, mechanical interlocking dependencies, and weak physical bonds that plague older coating methods like sol-gel or dip coating (Pastrafidou et al., 2025).

The resulting metallurgical bond is incredibly robust, often possessing a shear strength that exceeds the base metal itself (Wei et al., 2025). Therefore, when the implant bends or flexes inside the human body, the coating stays firmly attached (Van den Borre et al., 2022). It doesn't flake off or crack, ensuring it can effectively shield the reactive metal underneath and allow the implant to degrade predictably, perfectly synchronized with the patient's tissue healing process (Bose et al., 2025).

3. COATING MATERIALS AND MICROSTRUCTURAL EVOLUTION AT THE INTERFACE

3.1. Selection of Bioactive Coating Materials for Laser Processing

The fundamental purpose of surface engineering on biodegradable metals is to construct a barrier that temporarily dictates the degradation kinetics while actively communicating with the surrounding biological environment (Ahmed et al., 2025). To achieve this dual functionality, the selection of the coating material is just as critical as the laser parameters themselves (Xia et al., 2026).

In the realm of orthopedic and dental implants, the preferred materials are bioceramics, specifically those belonging to the calcium phosphate family (Dorozhkin, 2022). Hydroxyapatite (HA), with the chemical formula $Ca10(PO4)6(OH)2$, remains the undisputed gold standard for bioactive coatings (Beqain, 2023). The rationale behind this preference lies in its crystallographic and chemical similarity to the inorganic mineral phase of human bone (B. Wang et al., 2023). Hydroxyapatite exhibits a calcium-to-phosphorus (Ca/P) atomic ratio of approximately 1.67, which is remarkably close to that of natural bone apatite (Ahmed & Omer, 2024).

When introduced into a physiological environment, HA surfaces possess a unique ability to adsorb extracellular matrix proteins, which in turn attract osteoprogenitor cells (Tang et al., 2024). These cells attach, proliferate, and eventually differentiate into mature osteoblasts, laying down new bone tissue directly onto the implant surface without any intervening fibrous capsule (X. Lu et al., 2025).

However, pure hydroxyapatite is not the only option. Depending on the desired resorption rate of the coating itself, researchers often turn to other calcium phosphate phases, such as β-tricalcium phosphate (β-TCP) (Oladele et al., 2024). β-TCP is highly resorbable compared to the relatively stable HA (Ni et al., 2025). By physically blending HA and β-TCP powders before laser cladding, engineers can create Biphasic Calcium Phosphate (BCP) coatings (Tzagiollari, 2024). The ratio of HA to TCP within the BCP blend allows for precise tailoring of the coating's dissolution rate, perfectly synchronizing it with the degradation profile of the underlying magnesium or zinc substrate (J. Wang, 2023).

Bioactive glasses, composed primarily of SiO2, Na2O, CaO, and P2O5, represent another highly attractive class of coating materials (Raghavendra et al., 2025). Unlike crystalline calcium phosphates, bioactive glasses are amorphous (Degli Esposti et al., 2024). When exposed to body fluids, they undergo a rapid surface ion exchange, releasing soluble silica and calcium ions that actively upregulate the genetic expression of osteoblasts (Bai et al., 2023). Integrating these bioactive glasses into laser-processed surface layers provides a powerful chemical stimulus for early-stage bone healing, essentially transforming a degrading metal piece into an active pharmaceutical-like delivery system (Sheriff, 2026).

3.2. Managing the Ceramic-Metal Thermal Mismatch

While bioceramics offer unparalleled biological benefits, attempting to fuse them directly onto highly reactive, low-melting-point metals using a high-energy laser beam introduces a massive thermomechanical conflict (Mohammed et al., 2026). The physical properties of ceramics and metals are fundamentally opposed, and this opposition becomes starkly apparent during the rapid heating and cooling cycles of laser cladding (Cui et al., 2025).

The primary engineering obstacle is the severe disparity in the coefficients of thermal expansion (CTE) between the calcium phosphate coating and the biodegradable substrate (Dorozhkin, 2023). Magnesium alloys generally exhibit a CTE of around 26 μm/(m·°C), while zinc alloys sit even higher, near 30 μm/(m·°C) (Seetharaman et al., 2025). In stark contrast, hydroxyapatite has a CTE of approximately 11 to 14 μm/(m·°C) (Smith et al., 2023).

When the laser melts the ceramic powder and the uppermost layer of the metal substrate, they temporarily exist as a liquid mixture (B. Gao et al., 2022). However, as the laser heat source moves away, the melt pool solidifies and begins to cool down to room temperature (Khan et al., 2022). During this cooling phase, the metal substrate attempts to shrink significantly more than the ceramic coating (C. J. Li et al., 2022). Because the coating is metallurgically bonded to the substrate, it cannot freely contract (Song & Li, 2022). This restricted movement generates immense residual tensile stresses within the brittle ceramic layer (Tabatabaeian et al., 2022).

If these internal stresses exceed the fracture toughness of the hydroxyapatite, the coating will inevitably develop extensive networks of transverse and longitudinal micro-cracks (Nikam et al., 2026). In severe cases, macroscopic delamination occurs, causing entire sections of the coating to violently peel away from the base metal (Buchman, 2025).

To circumvent this thermal mismatch, engineers utilize the concept of Functionally Graded Materials (FGMs) (K. Li et al., 2025). Instead of applying a pure ceramic layer directly onto the pure metal, the coating is designed with a compositional gradient (Arun et al., 2023). The initial powder layers injected into the laser beam consist primarily of the substrate metal powder mixed with a very small volume fraction of hydroxyapatite (Mocanu et al., 2022). As the cladding process builds the coating upwards, layer by layer, the ratio of ceramic to metal in the powder feedstock is gradually increased (Verma, Vasudev, & Singh, 2025). The final, outermost layer consists entirely of the bioactive ceramic (K. Lu et al., 2025). This gradual transition effectively smooths out the CTE mismatch across the thickness of the coating, distributing the thermal stresses over a wider volume and drastically reducing the incidence of interfacial cracking (Bogdan & Peter, 2024).

3.3. High-Temperature Phase Transformations of Hydroxyapatite

Beyond thermal expansion issues, the sheer temperature of the laser melt pool threatens the chemical stability of the bioactive coating itself (S. Kumar et al., 2025). Laser cladding is a high-energy-density process, and temperatures within the interaction zone routinely exceed 1500°C (Zhao et al., 2025).

Hydroxyapatite is a complex molecule that behaves unpredictably under extreme thermal loads (Cherqaoui et al., 2026). When exposed to temperatures above 800°C to 1000°C, hydroxyapatite begins to undergo dehydroxylation—it loses its vital hydroxyl (OH) groups, transforming into oxyhydroxyapatite (Kosińska et al., 2024). As the temperature climbs beyond 1200°C, which is well within the operating range of a typical laser melt pool, the crystalline structure of HA completely breaks down (Omigbodun et al., 2026). It thermally decomposes into secondary calcium phosphate phases, primarily α-tricalcium phosphate (α-TCP) and tetracalcium phosphate (TTCP) (Kwiecień et al., 2026).

If the heating is intense enough, the decomposition pathway leads to the formation of calcium oxide (CaO) (Boning, 2024). From a clinical standpoint, the presence of calcium oxide within the coating is catastrophic (Z. Cao et al., 2024). When the implanted device comes into contact with human blood plasma or interstitial fluids, any localized CaO rapidly reacts with water to form calcium hydroxide (Z. Wang et al., 2025). This reaction is highly exothermic and causes a localized, violent spike in the pH level, making the microenvironment severely alkaline (Boning, 2024). This sudden alkalinization is highly cytotoxic; it kills the very osteoblast cells the coating was designed to attract and triggers severe tissue necrosis around the implant (Du et al., 2024).

Preventing this thermal decomposition requires strict optimization of the laser processing parameters (Murzin, 2024a). Engineers must operate at the lowest possible linear energy density that still allows for adequate melting and interfacial bonding (Fu et al., 2022). Additionally, utilizing carrier gases enriched with water vapor, rather than dry argon, has been shown to suppress the dehydroxylation of HA in the melt pool (Heimann, 2024). By maintaining a high partial pressure of water vapor directly above the melt pool, the thermodynamic equilibrium is shifted, forcing the hydroxyl groups to remain within the apatite lattice during the rapid solidification phase (Dorozhkin, 2023).

3.4. Solidification Kinetics and Microstructural Evolution

Understanding the final properties of the coating requires a deep dive into the solidification kinetics of the melt pool (P. Kumar et al., 2026). The microstructural architecture of a laser-clad layer is never uniform; it varies significantly from the interface up to the free surface, dictated entirely by the local thermal gradients and solidification velocities (Zhao et al., 2025).

The morphology of the solidifying front is governed by the ratio of the temperature gradient (G) to the solidification growth rate (R), denoted as G/R (J. Wu et al., 2025). At the exact interface where the molten pool meets the cold, solid metal substrate, the temperature gradient is extremely steep, while the growth rate is initially zero (Gu et al., 2023). This high G/R ratio promotes planar solidification, ensuring a continuous, defect-free atomic bond between the substrate and the coating (J. Cao et al., 2026).

As the solidification front moves upward away from the substrate and toward the surface of the coating, the temperature gradient decreases because the substrate is warming up, and the growth rate increases rapidly (Ma et al., 2023). This shift causes the planar front to break down, transitioning first into a cellular structure and subsequently into highly directional columnar dendrites (Abou-Khalil et al., 2022). These columnar dendrites grow parallel to the direction of heat extraction—which is straight down into the substrate (Maheshwaran et al., 2026).

Finally, near the very top surface of the coating, the melt pool is exposed to the shielding gas and the ambient atmosphere, leading to multi-directional heat loss (J. Li et al., 2025). Here, the cooling rate (the product of G and R) reaches its absolute maximum (Han et al., 2024). This extreme chilling effect often triggers massive nucleation events, resulting in a top layer composed of ultra-fine equiaxed grains (Y. Zhou et al., 2022).

This naturally occurring microstructural gradient is highly advantageous for biodegradable implants (Xia et al., 2025). The fine equiaxed grains at the surface provide high hardness and excellent resistance to initial mechanical wear during surgical insertion (Saha & Roy, 2022). Meanwhile, the robust columnar structures extending downward act as a dense physical barrier, effectively blocking the penetration pathways of corrosive chloride ions and delaying the onset of substrate degradation (X. Liu, 2025).

3.5. Interfacial Characterization and Elemental Diffusion

The true success of laser cladding lies hidden within the microscopic transition zone between the biodegradable metal and the bioactive coating (Nikolova & Apostolova, 2022). This interface is not a sharp, distinct line, but rather a complex metallurgical region characterized by intense elemental diffusion and the formation of new phases (Yadav et al., 2026).

When analyzing this interface using techniques like Scanning Electron Microscopy equipped with Energy Dispersive X-ray Spectroscopy (SEM-EDS), researchers observe a distinct diffusion band (Matsui et al., 2022). During the brief period when both materials are molten, the intense Marangoni convection currents actively mix the elements (Qin, 2026). Calcium and phosphorus from the molten hydroxyapatite migrate downward into the partially melted substrate (Dorozhkin, 2023). Conversely, metal atoms—whether magnesium, zinc, or iron—diffuse upward into the ceramic coating matrix (Muresan, 2023).

In magnesium-based implants, this bidirectional diffusion creates a highly unique and beneficial transition zone (Mubeen et al., 2025). Magnesium has a strong chemical affinity for phosphorus (Y. Liu et al., 2024). As magnesium atoms diffuse into the decomposing calcium phosphate melt, they frequently react to form complex magnesium phosphate compounds, such as magnesium-substituted tricalcium phosphate or even farringtonite phases (Lopera & Godoy, 2025).

Unlike the brittle intermetallic compounds that typically form when welding dissimilar metals, these newly synthesized magnesium phosphates are actually highly biocompatible (Jadav et al., 2022). They act as a biological bridge, structurally anchoring the hydroxyapatite layer to the magnesium substrate while simultaneously providing a secondary source of osteoinductive magnesium ions (D. Li et al., 2023). This specific interfacial chemistry is a primary reason why laser cladding provides vastly superior coating adhesion compared to mechanical deposition methods, ensuring the structural integrity of the implant remains intact under dynamic bodily loads (Amirtharaj Mosas et al., 2022).

3.6. Managing Manufacturing Defects: Porosity and Micro-Cracking

Despite its immense potential, laser processing of metal-ceramic systems is inherently prone to manufacturing defects that must be rigorously managed to ensure clinical safety (Abdullah et al., 2026). The two most critical defects encountered are internal porosity and extensive micro-cracking (Gomah et al., 2024).

Porosity in these systems generally arises from two distinct sources (T. Li et al., 2022). The first is gas entrapment. As the powder particles are injected into the melt pool, small pockets of the shielding gas can be carried beneath the surface (Z. Zhou et al., 2025). Due to the incredibly rapid cooling rates associated with laser processing, the melt pool often freezes before these gas bubbles have the necessary time to rise to the surface and escape, leaving behind spherical voids (M. Chen et al., 2025).

The second source is far more problematic: metallurgical vaporization (Shi et al., 2024). As discussed previously, elements like zinc and magnesium vaporize easily under the laser beam (Shi et al., 2024). If the energy density is too high, the substrate literally boils at the interface, creating irregular, keyhole-like pores that severely compromise the bond strength (Zhu & Fuh, 2026).

Micro-cracking, on the other hand, is almost exclusively driven by the relief of thermal residual stresses caused by the CTE mismatch mentioned earlier (Ferguson, 2023). While employing a functionally graded material approach helps mitigate macroscopic delamination, microscopic cracks often still propagate within the pure ceramic layers at the surface (Sai Charan et al., 2022).

To combat these defects, engineers employ a combination of pre-processing and in-situ control strategies (J. Zhang et al., 2025). Pre-heating the substrate prior to laser cladding is a common metallurgical technique to reduce the temperature gradient, thereby lowering the cooling rate and minimizing thermal stress (Gupta et al., 2025). However, pre-heating reactive biodegradable metals like magnesium in an open atmosphere is dangerous due to the risk of severe oxidation. Therefore, pre-heating must be conducted within a strictly controlled, inert gas chamber (Pesode & Barve, 2023).

Alternatively, researchers are exploring the use of ductile metallic buffer layers (Pramesti et al., 2025). By depositing a very thin layer of a pure, highly ductile metal (such as ultra-pure zinc on a harder zinc-alloy substrate) before applying the ceramic coating, the buffer layer acts as a mechanical shock absorber (Meng & Du, 2024). It plastically deforms to accommodate the thermal contraction of the ceramic, effectively absorbing the residual stress and preventing the propagation of cracks into the bioactive surface (K. Lu et al., 2025).

By mastering these defect-mitigation techniques, the production of flawless, highly controlled, and bioactive biodegradable implants via laser surface engineering transitions from a theoretical concept to a reliable clinical reality (S. Kumar et al., 2025).

4. CLINICAL PERFORMANCE – DEGRADATION KINETICS AND IN VITRO/IN VIVO BIOCOMPATIBILITY

4.1. Transitioning From the Laboratory to the Physiological Environment

Once a flawless, metallurgically bonded coating is achieved via laser processing, the evaluation paradigm must shift entirely (Xia et al., 2026). A surface that appears structurally sound under an electron microscope might still fail catastrophically when exposed to the harsh, dynamic conditions of the human body (Mokhtari et al., 2025). The true measure of a biodegradable implant's success is its clinical performance, which is dictated by how the engineered surface interacts with aggressive biological fluids, proteins, and living cells over time (Alam et al., 2024).

To bridge the gap between materials science and clinical application, researchers rely heavily on in vitro degradation testing (Pagar et al., 2022). Instead of standard industrial corrosive media, implants are submerged in Simulated Body Fluids (SBF), Hank's Balanced Salt Solution (HBSS), or Dulbecco's Modified Eagle Medium (DMEM) (Imani, 2024). These solutions carefully replicate the ion concentrations, chloride levels, and buffering capacity of human blood plasma at a strict physiological temperature of 37°C (Watson et al., 2023). By observing the laser-clad materials in these environments, scientists can accurately predict how the degradation kinetics will unfold once the device is surgically placed into a patient's bone defect or vascular system (Kumar Yadav et al., 2023).

4.2. Electrochemical Evaluation of Degradation Kinetics

The most precise method for quantifying the protective capability of a laser-processed coating is electrochemical testing (H. Liu et al., 2022). Because the degradation of metals like magnesium and zinc in an aqueous environment is fundamentally an electrochemical corrosion process, tools like Potentiodynamic Polarization (PDP) and Electrochemical Impedance Spectroscopy (EIS) are indispensable (Bijapur et al., 2023).

When conducting a potentiodynamic polarization scan on a bare, uncoated magnesium alloy in SBF, the resulting Tafel plot typically reveals a highly negative corrosion potential (Ecorr) and a massive corrosion current density (Icorr) (van Gaalen, 2023).

This profile mathematically confirms an aggressive, rapid dissolution rate. However, when the same test is applied to a magnesium implant featuring a dense, laser-clad hydroxyapatite layer, the electrochemical landscape changes dramatically (Qazi, 2025).

The ceramic coating acts as a robust physical and dielectric barrier, severely restricting the flow of electrons and blocking chloride ions from reaching the underlying reactive metal (X. Liu, 2025). Consequently, the Tafel curve undergoes a drastic shift: the corrosion potential moves significantly toward the noble (positive) direction, and the corrosion current density drops by several orders of magnitude (van Ede & Angst, 2024).

Electrochemical Impedance Spectroscopy provides an even deeper diagnostic view (Vivier & Orazem, 2022). By applying a small alternating current across a spectrum of frequencies, researchers can map the resistive and capacitive behavior of the coating-substrate interface using Nyquist and Bode plots (Ajayan et al., 2024).

A successful laser cladding process yields a massive increase in polarization resistance (Rp). Unlike chemically deposited coatings that often show a rapid drop in resistance within hours due to fluid seeping through structural micropores, laser-clad surfaces maintain high impedance values for weeks (K. Lu et al., 2022).

The rapid solidification of the laser melt pool creates a highly dense, impermeable columnar microstructure that forces the corrosive physiological fluids to take an exceptionally long and tortuous path before they can initiate any localized pitting on the substrate (L. Wang et al., 2026).

4.3. Controlling Hydrogen Evolution and pH Modulation

While electrochemical data provides instantaneous corrosion rates, the physical byproducts of degradation pose the most immediate threat to clinical healing (F. M. Li et al., 2022). For magnesium-based implants, the anodic dissolution reaction forces the reduction of water, generating significant volumes of hydrogen gas and releasing hydroxide ions into the surrounding tissue bed (Abraham & Subramani, 2023).

If an uncoated magnesium screw is implanted into a bone fracture, the rapid burst of hydrogen gas accumulates under the periosteum faster than the local blood supply can clear it (Espiritu, 2025). These expanding gas cavities physically separate the bone from the implant, effectively destroying any chance of osseointegration (Pandey et al., 2022). Simultaneously, the massive release of hydroxide ions can drive the local pH of the tissue from a healthy 7.4 up to highly cytotoxic levels exceeding 10 or 11, causing severe localized tissue necrosis (Hajjar & Zhou, 2023).

A properly engineered laser-clad bioactive coating elegantly neutralizes both of these threats (Diwakar et al., 2024). By drastically reducing the corrosion current density, the coating limits the hydrogen evolution rate to a level that the surrounding vascular network can easily absorb and diffuse (F. M. Li et al., 2022). Volume displacement tests consistently show that laser-clad samples produce a fraction of the hydrogen gas compared to their bare counterparts (Bunaziv et al., 2023).

Furthermore, as the hydroxyapatite or calcium phosphate coating slowly dissolves in the body, it acts as a dynamic chemical buffer (Drevet et al., 2023). The dissolution products help neutralize the hydroxide ions generated by any minor substrate corrosion, stabilizing the local pH near the physiological ideal (F. M. Li et al., 2022). This controlled microenvironment ensures that the delicate cells migrating to the surgical site survive the initial inflammatory phase of wound healing (Z. Wang et al., 2022).

4.4. In Vitro Cellular Adhesion and Proliferation Dynamics

Biocompatibility is not merely the absence of toxicity; it is the active promotion of tissue regeneration (Saikia, 2024). When a medical device is implanted, cells do not interact with the bare material directly (Luu et al., 2024). Within milliseconds of contact with blood, the implant surface is covered by a complex layer of water molecules and proteins, such as fibronectin, vitronectin, and albumin (Nascimento et al., 2023). The composition and orientation of this protein layer dictate whether bone-forming cells will recognize the implant as a friendly scaffold or a foreign body (Aslam Khan et al., 2024).

Laser cladding offers a unique advantage here (K. Wang et al., 2023). The rapid thermal cycling inevitably leaves behind a specific micro-topography—a subtle surface roughness dictated by the overlapping laser tracks and powder particles (Grumezescu & Duta, 2026). Unlike perfectly polished metal, this laser-induced texture provides an immense surface area that is highly favorable for protein adsorption (Shabir et al., 2025).

When human osteoblasts or mesenchymal stem cells are seeded onto these laser-clad surfaces in a laboratory incubator, the biological response is striking (L. Chen et al., 2026). Fluorescence microscopy reveals that cells do not simply sit round and dormant. Instead, they actively recognize the bioactive ceramic chemistry and the micro-roughness (W. Xie et al., 2023). They flatten out and extend long, finger-like projections called filopodia and lamellipodia, searching for anchorage points (Pratiwi et al., 2024). Through transmembrane proteins called integrins, the cells form strong focal adhesions with the coating (Aman & Margadant, 2023). Standard colorimetric assays, such as MTT or CCK-8 tests, routinely confirm that cell viability and proliferation rates on laser-clad hydroxyapatite surfaces are significantly higher than on bare zinc or magnesium, proving that the surface is highly cytocompatible and actively encourages cellular colonization (Homa et al., 2024).

4.5. Osteogenic Differentiation and the Healing Cascade

Surviving and multiplying is only the first step; for an orthopedic implant to succeed, the attached cells must actively build new bone tissue—a process known as osteogenic differentiation (Tajabadi et al., 2026). The genius of utilizing calcium phosphate-based coatings on biodegradable metals lies in their active pharmacological role during the later stages of healing (Montesissa et al., 2025). As the laser-clad coating slowly dissolves over several months, it releases a localized cocktail of calcium and phosphate ions (S. Saha, 2023). If the underlying substrate is magnesium or zinc, these essential trace elements are also gradually released at safe, controlled therapeutic doses (Y. Xie et al., 2022).

This specific ionic environment serves as a powerful biochemical signal. It triggers the osteoblasts to upregulate the expression of specific osteogenic marker genes, including Runx2, alkaline phosphatase (ALP), and osteocalcin (Ariffin et al., 2022). Laboratory assays tracking ALP activity typically show a massive spike when cells are cultured on these degrading bioactive surfaces (De Micco et al., 2025). Eventually, the cells begin to secrete their own extracellular matrix, primarily composed of collagen type I (Bandzerewicz & Gadomska-Gajadhur, 2022). Over time, they mineralize this matrix, depositing naturally formed biological apatite directly onto the dissolving implant (Hong et al., 2022). This active, ion-driven tissue regeneration ensures that as the metal hardware slowly loses its mechanical integrity, the newly formed bone is already taking over the load-bearing responsibilities, seamlessly completing the transient healing cycle (Shrivastava et al., 2024).

4.6. In Vivo Translation and Future Perspectives

The ultimate validation of any biomaterial occurs within complex living systems. In vivo animal models, typically involving the implantation of coated pins or screws into the femoral condyles or tibias of rats and rabbits, provide the final proof of concept (Varut et al., 2025). Histological sections taken weeks or months post-surgery offer a definitive visual narrative (Garcia-Ortega, 2024). Bare biodegradable metals often show a distinct gap between the implant and the host bone, filled with fibrous scar tissue—a clear sign of poor integration and rapid gas evolution (Lata et al., 2025).

In stark contrast, histological staining of laser-clad implants consistently demonstrates direct Bone-to-Implant Contact (BIC) (Artemchuk et al., 2025). The newly formed woven bone can be seen growing directly into the micro-pores and dissolving regions of the laser coating, eventually maturing into dense, load-bearing lamellar bone without any intervening fibrous capsule (Gatto, 2022).

Moving forward, the clinical translation of laser-processed biodegradable implants relies on mastering the scalability and reproducibility of the manufacturing process (Rahman et al., 2026). Regulatory bodies like the FDA and the European MDR demand strict quality control, and the inherent thermal complexities of laser cladding require advanced monitoring systems, such as real-time melt pool imaging and AI-driven predictive modeling (Murzin, 2024).

As the technology matures, the future points toward personalized medicine: utilizing tailored laser parameters and patient-specific powder blends to create biodegradable implants with degradation profiles specifically tuned to an individual's unique bone density, age, and healing capacity (B. Gao et al., 2025).

ACKNOWLEDGEMENT

Artificial intelligence tools were used as an assistant during the drafting and language editing processes of this article, in accordance with ethical guidelines.

REFERENCES

Abbaspour, N., Hurrell, R., & Kelishadi, R. (2014). Review on iron and its importance for human health. *Journal of Research in Medical Sciences : the Official Journal of Isfahan University of Medical Sciences, 19*(2), 164. PMID: 24778671

Abdullah, F. R., Mafri, T. A., & Shen, Y. K. (2026). Principles, current applications, and future perspectives of selective laser melting in prosthodontics: A review. *Metals, 16*(4), 415. DOI: 10.3390/met16040415

Abou-Khalil, L., da Cruz, K. S., Reinhart, G., Mangelinck-Noël, N., & Nguyen-Thi, H. (2022). Influence of growth velocity on fragmentation during directional solidification of Al–14 wt.% Sn alloy studied by in-situ synchrotron X-radiography. *Acta Materialia, 241*, 118370. DOI: 10.1016/j.actamat.2022.118370

Abraham, A. M., & Subramani, V. (2023). Effect of magnesium as biomaterial in biodegradation. *Materials Today: Proceedings.*

Ahmed, L. O., & Omer, R. A. (2024). Hydroxyapatite biomaterials: A comprehensive review of their properties, structures, clinical applications, and producing techniques. *Reviews in Inorganic Chemistry, 44*(4), 599–618. DOI: 10.1515/revic-2024-0018

Ahmed, Y., Ankah, N. K., Ogunlakin, N., Toor, I. U., & Farooq, W. (2025). Exploring the future of metallic implants: A review of biodegradable and non-biodegradable solutions. *Corrosion Reviews, 43*(5), 495–521. DOI: 10.1515/corrrev-2024-0125

Ajayan, A., Joseph, V., Paul, A., & Joseph, A. (2024). Coating characterization with electrochemical techniques. In *Encapsulated corrosion inhibitors for eco-benign smart coatings* (pp. 205–224). CRC Press. DOI: 10.1201/9781032677255-12

Alam, F., Ashfaq Ahmed, M., Jalal, A. H., Siddiquee, I., Adury, R. Z., Hossain, G. M., & Pala, N. (2024). Recent progress and challenges of implantable biodegradable biosensors. *Micromachines, 15*(4), 475. DOI: 10.3390/mi15040475 PMID: 38675286

Alontseva, D., Azamatov, B., Safarova, Y., Voinarovych, S., & Nazenova, G. (2023). A brief review of current trends in the additive manufacturing of orthopedic implants with thermal plasma-sprayed coatings to improve the implant surface biocompatibility. *Coatings, 13*(7), 1175. DOI: 10.3390/coatings13071175

Alrmali, A. E., Misch, J., Gomez-Meda, R., Saleh, M. H., & Wang, H. L. (2026). The MICH concept for long-term implant stability. *The International Journal of Oral Implantology*, *19*(1), 13–29. PMID: 41823002

Aman, J., & Margadant, C. (2023). Integrin-dependent cell–matrix adhesion in endothelial health and disease. *Circulation Research*, *132*(3), 355–378. DOI: 10.1161/CIRCRESAHA.122.322332 PMID: 36730379

Amirtharaj Mosas, K. K., Chandrasekar, A. R., Dasan, A., Pakseresht, A., & Galusek, D. (2022). Recent advancements in materials and coatings for biomedical implants. *Gels (Basel, Switzerland)*, *8*(5), 323. DOI: 10.3390/gels8050323 PMID: 35621621

Ariffin, S. H. Z., Lim, K. W., Wahab, R. M. A., Ariffin, Z. Z., Din, R. D. R., Shahidan, M. A., & Abidin, I. Z. Z. (2022). Gene expression profiles for in vitro human stem cell differentiation into osteoblasts and osteoclasts: A systematic review. *PeerJ*, *10*, e14174. DOI: 10.7717/peerj.14174 PMID: 36275474

Artemchuk, A., Osmanov, B., & Chepurnyi, Y. (2025). Evaluation of bone-to-implant contact area in patients with reduced residual bone height in the posterior maxilla: A virtual simulation study. *Journal of Oral and Maxillofacial Surgery, Medicine, and Pathology*.

Arun, K. L., Udhayakumar, M., & Radhika, N. (2023). A comprehensive review on various ceramic nanomaterial coatings over metallic substrates: Applications, challenges and future trends. *Journal of Bio- and Tribo-Corrosion*, *9*(1), 11. DOI: 10.1007/s40735-022-00717-6

Aslam Khan, M. U., Aslam, M. A., Bin Abdullah, M. F., & Stojanović, G. M. (2024). Current perspectives of protein in bone tissue engineering: Bone structure, ideal scaffolds, fabrication techniques, applications, scopes, and future advances. *ACS Applied Bio Materials*, *7*(8), 5082–5106. DOI: 10.1021/acsabm.4c00362 PMID: 39007509

Aziz, U., McAfee, M., Manolakis, I., Timmons, N., & Tormey, D. (2025). A review of optimization of additively manufactured 316/316L stainless steel process parameters, post-processing strategies, and defect mitigation. *Materials (Basel)*, *18*(12), 2870. DOI: 10.3390/ma18122870 PMID: 40573001

Badiceanu, M., Anghel, S., Mihailescu, N., Visan, A. I., Mihailescu, C. N., & Mihailescu, I. N. (2022). Coatings functionalization via laser versus other deposition techniques for medical applications: A comparative review. *Coatings*, *12*(1), 71. DOI: 10.3390/coatings12010071

Bähr, S. H. (2025). Process gas influences during the powder bed fusion of metals using a laser beam [Doctoral dissertation, Technische Universität München].

Bai, L., Song, P., & Su, J. (2023). Bioactive elements manipulate bone regeneration. *Biomaterials Translational*, *4*(4), 248. PMID: 38282709

Bandzerewicz, A., & Gadomska-Gajadhur, A. (2022). Into the tissues: Extracellular matrix and its artificial substitutes: Cell signalling mechanisms. *Cells*, *11*(5), 914. DOI: 10.3390/cells11050914 PMID: 35269536

Baskerville, S. J. J. (2022). Effects of orthopedic implant and host bone properties on stress-shielding induced osteopenia [Unpublished data or report].

Bender, R., Féron, D., Mills, D., Ritter, S., Bäßler, R., Bettge, D., De Graeve, I., Dugstad, A., Grassini, S., Hack, T., Halama, M., Han, E.-H., Harder, T., Hinds, G., Kittel, J., Krieg, R., Leygraf, C., Martinelli, L., Mol, A., & Zheludkevich, M. (2022). Corrosion challenges towards a sustainable society. *Materials and Corrosion*, *73*(11), 1730–1751. DOI: 10.1002/maco.202213140

Beqain, M. D. Y. S. (2023). Novel substituted hydroxyapatites [Doctoral dissertation, The University of Waikato].

Bijapur, K., Molahalli, V., Shetty, A., Toghan, A., De Padova, P., & Hegde, G. (2023). Recent trends and progress in corrosion inhibitors and electrochemical evaluation. *Applied Sciences (Basel, Switzerland)*, *13*(18), 10107. DOI: 10.3390/app131810107

Bîrsan, D. C., & Susac, F. (2026). Recent advances and future perspectives in friction stir welding and processing. *RE:view*.

Biswas, D. J. (2026). Shaping with light: The many faces of laser material processing. In A beginner's guide to lasers and their applications, Part 2: The laser's impact on science and humanity (pp. 297-327). Springer Nature Switzerland.

Blutinger, J. D. (2022). Digital cuisine: Food printing and laser cooking [Doctoral dissertation, Columbia University].

Bogdan, M., & Peter, I. (2024). A comprehensive understanding of thermal barrier coatings (TBCs): Applications, materials, coating design and failure mechanisms. *Metals*, *14*(5), 575. DOI: 10.3390/met14050575

Boning, R. (2024). Review on thermal properties and reaction kinetics of Ca (OH) 2/CaO thermochemical energy storage materials. *Electrical Materials and Applications*, *1*(1), e12007. DOI: 10.1049/ema3.12007

Bose, S., Dahiya, A., & Bandyopadhyay, A. (2025). Of biomaterials. *Additive Manufacturing*, 169.

Boulos, A., DeFroda, S. F., Kleiner, J. E., Thomas, N., Gil, J. A., & Cruz, A. I.Jr. (2017). Inpatient orthopaedic hardware removal in children: A cross-sectional study. *Journal of Clinical Orthopaedics and Trauma*, *8*(3), 270–275. DOI: 10.1016/j.jcot.2017.06.020 PMID: 28951646

Buchman, A. (2025). Failure cases in adhesive joints and coatings. *Progress in Adhesion and Adhesives*, *9*, 317–367. DOI: 10.1002/9781394315086.ch9

Bülbül Tatbul, B. E. S. N. A. (2022). 3D-microstructuring of silicon induced by nanosecond pulsed infrared fiber laser for potential solar cell applications [Master's thesis, Middle East Technical University].

Bunaziv, I., Ren, X., Hagen, A. B., Hovig, E. W., Jevremovic, I., & Gulbrandsen-Dahl, S. (2023). Laser beam remelting of stainless steel plate for cladding and comparison with conventional CMT process. *International Journal of Advanced Manufacturing Technology*, *127*(1), 911–934. DOI: 10.1007/s00170-023-11567-y

Butt, M. A. (2022). Thin-film coating methods: A successful marriage of high-quality and cost-effectiveness—a brief exploration. *Coatings*, *12*(8), 1115. DOI: 10.3390/coatings12081115

Cao, J., Sun, J., Li, H., Wu, Z., & An, X. (2026). Mastering complexity toward high-performance multi-principal element alloy-based films and coatings: A review on microstructural regulation and property optimization. *Advancement of Science*, *13*(24), e14258. DOI: 10.1002/advs.202514258 PMID: 41870026

Cao, Z., Yang, X., Yang, W., Chen, F., Jiang, W., Zhan, S., Jiang, F., Li, J., Ye, C., Lang, L., Zhang, S., Feng, Z., Lai, X., Liu, Y., Mao, L., Cai, H., Teng, Y., & Xie, J. (2024). Modulation of dendritic cell function via nanoparticle-induced cytosolic calcium changes. *ACS Nano*, *18*(10), 7618–7632. DOI: 10.1021/acsnano.4c00550 PMID: 38422984

Chen, L., Xu, P., Xu, N., Li, L., & Pang, C. (2026). Femtosecond laser-induced gradient architecture of micro-nano structures on Ti-Nb-Zr biomedical claddings: Improving the cytocompatibility of bone implants. *Materials Characterization*, *237*, 116494. DOI: 10.1016/j.matchar.2026.116494

Chen, M., Zhang, J., Qu, Q., Zhang, S., & Liu, D. (2025). Cavity-induced bubbles and pore formation of laser direct energy deposited titanium alloy: Influence of powder melting degree. *Additive Manufacturing*, *105*, 104792. DOI: 10.1016/j.addma.2025.104792

Cheng, J., Xing, Y., Dong, E., Zhao, L., Liu, H., Chang, T., Chen, M., Wang, J., Lu, J., & Wan, J. (2022). An overview of laser metal deposition for cladding: Defect formation mechanisms, defect suppression methods and performance improvements of laser-cladded layers. *Materials (Basel), 15*(16), 5522. DOI: 10.3390/ma15165522 PMID: 36013658

Cherqaoui, A., Paternoster, C., & Mantovani, D. (2026). Biodegradable metals and corrosion control: Challenges, limits and new opportunities for innovating in orthopedic fixations. *Materials (Basel), 19*(9), 1789. DOI: 10.3390/ma19091789 PMID: 42124196

Cui, B., Zhou, P., & Lv, Y. (2025). Research progress in and defect improvement measures for laser cladding. *Materials (Basel), 18*(13), 3206. DOI: 10.3390/ma18133206 PMID: 40649694

Daniyan, A. A., Akpomejero, T. L., Ige, O. O., & Olubambi, P. A. (2023). Corrosion prevention of biomedical implants: Surface coating techniques perspective. *ChemistrySelect, 8*(40), e202300223. DOI: 10.1002/slct.202300223

Das, P., Kumar, T. S., Sahu, K. K., & Gollapudi, S. (2022). Corrosion, stress corrosion cracking and corrosion fatigue behavior of magnesium alloy bioimplants. *Corrosion Reviews, 40*(4), 289–333. DOI: 10.1515/corrrev-2021-0088

De Micco, S., Bellucci, D., & Cannillo, V. (2025). The assessment of bioactivity and biological responsiveness in bioactive glasses and ceramics: A review of available techniques. *Materials (Basel), 18*(18), 4393. DOI: 10.3390/ma18184393 PMID: 41010235

De Riccardis, M. F. (2023). Green coatings: Materials, deposition processes, and applications. In *Handbook of green and sustainable nanotechnology: Fundamentals, developments and applications* (pp. 1627–1654). Springer International Publishing. DOI: 10.1007/978-3-031-16101-8_41

Degli Esposti, L., Zheng, K., Piancastelli, A., Ionescu, A. C., Adamiano, A., Boccaccini, A. R., & Iafisco, M. (2024). Composite materials of amorphous calcium phosphate and bioactive glass nanoparticles for preventive dentistry. *Ceramics International, 50*(1), 593–602. DOI: 10.1016/j.ceramint.2023.10.137

Diwakar, V., Kumar Yadav, S., Yusufzai, M. Z. K., & Vashista, M. (2024). Advances in laser cladding: Material trends, applications, and future perspectives. In *Hybrid materials and advanced manufacturing* (pp. 121–142). CRC Press.

Dorozhkin, S. V. (2022). Calcium orthophosphate (CaPO4)-based bioceramics: Preparation, properties, and applications. *Coatings, 12*(10), 1380. DOI: 10.3390/coatings12101380

Dorozhkin, S. V. (2023). There are over 60 ways to produce biocompatible calcium orthophosphate (CaPO4) deposits on various substrates. *Journal of Composites Science*, *7*(7), 273. DOI: 10.3390/jcs7070273

Drevet, R., Fauré, J., & Benhayoune, H. (2023). Bioactive calcium phosphate coatings for bone implant applications: A review. *Coatings, 13*(6), 1091. DOI: 10.3390/coatings13061091

Du, T., Liu, J., Dong, J., Xie, H., Wang, X., Yang, X., & Yang, Y. (2024). Multifunctional coatings of nickel-titanium implant toward promote osseointegration after operation of bone tumor and clinical application: A review. *Frontiers in Bioengineering and Biotechnology*, *12*, 1325707. DOI: 10.3389/fbioe.2024.1325707 PMID: 38444648

Duan, X., Yang, Y., Zhang, T., Zhu, B., Wei, G., & Li, H. (2024). Research progress of metal biomaterials with potential applications as cardiovascular stents and their surface treatment methods to improve biocompatibility. *Heliyon, 10*(4), e25515. DOI: 10.1016/j.heliyon.2024.e25515 PMID: 38375258

Espiritu, J. W. (2025). Imaging magnesium-based implants and their biological surroundings.

Ferguson, J. I. (2023). Residual stress and microstructural evolution of composites and coatings for extreme environments [Doctoral dissertation, Purdue University].

Fernandes, C., & Taurino, I. (2022). Biodegradable molybdenum (Mo) and tungsten (W) devices: One step closer towards fully-transient biomedical implants. *Sensors (Basel)*, *22*(8), 3062. DOI: 10.3390/s22083062 PMID: 35459047

Fu, W., Yan, X., Gurumukhi, Y., Garimella, V. S., King, W. P., & Miljkovic, N. (2022). High power and energy density dynamic phase change materials using pressure-enhanced close contact melting. *Nature Energy*, *7*(3), 270–280. DOI: 10.1038/s41560-022-00986-y

Gao, B., Sun, Z., Tong, Y., Yu, H., & Wang, F. (2025). Research progress on personalized bone implants based on additive manufacturing. *Micromachines*, *16*(12), 1339. DOI: 10.3390/mi16121339 PMID: 41470504

Gao, B., Zhao, H., Peng, L., & Sun, Z. (2022). A review of research progress in selective laser melting (SLM). *Micromachines, 14*(1), 57. DOI: 10.3390/mi14010057 PMID: 36677118

Garcia-Ortega, D. Y. (2024). Comprehensive treatment strategy for improving surgical resection rate of retroperitoneal sarcomas: A histology-specific approach narrative review. *Frontiers in Oncology*, *14*, 1432900. DOI: 10.3389/fonc.2024.1432900 PMID: 39435281

Gatto, M. L. (2022). Biomaterials for tissue engineering by additive manufacturing.

Gholamzadeh, J., Fatemi, S. M., Mollaei, N., & Abedi, A. (2024). Processing of ultrafine/nano-grained microstructures through additive manufacturing techniques: A critical review. *Journal of Ultrafine Grained and Nanostructured Materials*, *57*(2), 203–221.

Gomah, M. E., Li, G., Jiahui, X., Omar, A. A., Haoran, H., & Zaki, M. M. (2024). Micro to macro-cracking mechanism in thermally treated granodiorite followed by different cooling techniques. *International Journal of Fracture*, *246*(2), 161–180. DOI: 10.1007/s10704-023-00740-9

Gracheva, A., Polozov, I., & Popovich, A. (2025). Additive manufacturing of biodegradable metallic implants by selective laser melting: Current research status and application perspectives. *Metals*, *15*(7), 754. DOI: 10.3390/met15070754

Graeser, H. (2023). *Still on the way: A father's journey to hope*. Wipf and Stock Publishers.

Griffis, J. C., Shahed, K., Meinert, K., Yilmaz, B., Lear, M., & Manogharan, G. (2025). Multi-material laser powder bed fusion: Effects of build orientation on defects, material structure and mechanical properties. npj. *Advances in Manufacturing*, *2*(1), 5.

Grohol, C. M. (2024). Laser cladding of aluminum alloys and high-fidelity modeling of the molten pool dynamics in laser melting of metals [Doctoral dissertation, Purdue University Graduate School].

Grumezescu, V., & Duta, L. (2026). Collagen Type I as a biological barrier interface in biomimetic microfluidic devices: Properties, applications, and challenges. *Biomimetics*, *11*(1), 66. DOI: 10.3390/biomimetics11010066 PMID: 41589983

Gu, Y., Yuan, J., & Chen, L. (2023). Switching of control mechanisms during the rapid solidification of a melt pool. *Physical Review Materials*, *7*(10), 103401. DOI: 10.1103/PhysRevMaterials.7.103401

Gupta, N. K., Ganesan, G., Siddhartha, S., Karade, S. R., Paul, A. K., Dubey, S., Ely, R. H., & Karunakaran, K. P. (2025). In situ pre-heating in wire arc additive manufacturing: Design, development, and experimental investigation on residual stresses and metallurgical and mechanical properties. *Journal of Materials Engineering and Performance*, *34*(12), 12256–12267. DOI: 10.1007/s11665-024-10011-w

Haider, A. J., Alawsi, T., Haider, M. J., Taha, B. A., & Marhoon, H. A. (2022). A comprehensive review on pulsed laser deposition technique to effective nanostructure production: Trends and challenges. *Optical and Quantum Electronics*, *54*(8), 488. DOI: 10.1007/s11082-022-03786-6

Hajjar, S., & Zhou, X. (2023). pH sensing at the intersection of tissue homeostasis and inflammation. *Trends in Immunology*, *44*(10), 807–825. DOI: 10.1016/j.it.2023.08.008 PMID: 37714775

Han, Y., Chen, H., Sun, Y., Liu, J., Wei, S., Xie, B., Zhang, Z., Zhu, Y., Li, M., Yang, J., Chen, W., Cao, P., & Yang, Y. (2024). Ubiquitous short-range order in multi-principal element alloys. *Nature Communications*, *15*(1), 6486. DOI: 10.1038/s41467-024-49606-1 PMID: 39090088

Heimann, R. B. (2024). Plasma-sprayed osseoconductive hydroxylapatite coatings for endoprosthetic hip implants: Phase composition, microstructure, properties, and biomedical functions. *Coatings*, *14*(7), 787. DOI: 10.3390/coatings14070787

Hofmann, S. (2025). Segregation at grain boundaries. In *Surface segregation phenomena* (pp. 107–134). CRC Press.

Homa, K., Zakrzewski, W., Dobrzyński, W., Piszko, P. J., Piszko, A., Matys, J., Wiglusz, R. J., & Dobrzyński, M. (2024). Surface functionalization of titanium-based implants with a nanohydroxyapatite layer and its impact on osteoblasts: A systematic review. *Journal of Functional Biomaterials*, *15*(2), 45. DOI: 10.3390/jfb15020045 PMID: 38391898

Hong, M. H., Lee, J. H., Jung, H. S., Shin, H., & Shin, H. (2022). Biomineralization of bone tissue: Calcium phosphate-based inorganics in collagen fibrillar organic matrices. *Biomaterials Research*, *26*(1), 42. DOI: 10.1186/s40824-022-00288-0 PMID: 36068587

Hsu, K., Yang, L., Medina, F., Mireles, J., Menon, M., Meyer, M., & Godfrey, D. (2026). Laser powder-bed fusion for metals. In *Additive manufacturing of metals* (pp. 139–199). Springer Nature. DOI: 10.1007/978-3-032-04961-2_4

Ikeuba, A. I., Njoku, C. N., Ekerenam, O. O., Njoku, D. I., Udoh, I. I., Daniel, E. F., Uzoma, P. C., Etim, I.-I. N., & Okonkwo, B. O. (2024). A review of the electrochemical and galvanic corrosion behavior of important intermetallic compounds in the context of aluminum alloys. *RSC Advances*, *14*(43), 31921–31953. DOI: 10.1039/D4RA06070A PMID: 39385762

Imani, A. (2024). Corrosion behaviour of a biodegradable rare-earth magnesium alloy in modified simulated body fluid [Doctoral dissertation, University of British Columbia].

Jadav, H., Badheka, V., Upadhyay, G., & Mehta, K. (2022). Dissimilar welding of magnesium alloy to aluminium alloy: A review. *Advances in Materials and Processing Technologies*, *8*(4), 3794–3821. DOI: 10.1080/2374068X.2021.2016564

Jeyaprakash, N., Prabu, G., & Yang, C. H. (Eds.). (2025). *Tailoring of engineering material properties through laser cladding: Metallurgical, mechanical and tribological properties*. CRC Press. DOI: 10.1201/9781032693576

John, M., Kuruveri, U. B., & Menezes, P. L. (2022). Laser cladding-based surface modification of carbon steel and high-alloy steel for extreme condition applications. *Coatings*, *12*(10), 1444. DOI: 10.3390/coatings12101444

Kanaginahal, G. M., Kiran, M. C., Shahapurkar, K., Mahale, R. S., & Kakkamari, P. (2024). Automobile applications of mechanically alloyed magnesium and titanium material. In *Mechanically alloyed novel materials: Processing, applications, and properties* (pp. 379–406). Springer Nature. DOI: 10.1007/978-981-97-6504-1_16

Kang, C., & Sakaguchi, M. (2025). Interfacial bonding and residual stress of single splats on solid substrates: A literature review. *Coatings*, *15*(11), 1259. DOI: 10.3390/coatings15111259

Khan, M. S., Ali, S., Westerbaan, D., Duley, W., Biro, E., & Zhou, Y. N. (2022). The effect of laser impingement angle on the optimization of melt pool geometry to improve process stability during high-speed laser welding of thin-gauge automotive steels. *Journal of Manufacturing Processes*, *78*, 242–253. DOI: 10.1016/j.jmapro.2022.04.022

Khudair, N. Y., Al-Amir, Q. R., & Rashid, H. K. (n.d.). Simulation of the thermal Marangoni effect in a free surface cavity: Heatline and entropy visualizations.

Kim, M. S., Chang, H., Zheng, L., Yan, Q., Pfleger, B. F., Klier, J., Nelson, K., Majumder, E. L.-W., & Huber, G. W. (2023). A review of biodegradable plastics: Chemistry, applications, properties, and future research needs. *Chemical Reviews*, *123*(16), 9915–9939. DOI: 10.1021/acs.chemrev.2c00876 PMID: 37470246

Kong, L., Heydari, Z., Lami, G. H., Saberi, A., Baltatu, M. S., & Vizureanu, P. (2023). A comprehensive review of the current research status of biodegradable zinc alloys and composites for biomedical applications. *Materials (Basel)*, *16*(13), 4797. DOI: 10.3390/ma16134797 PMID: 37445111

Kosińska, E., Sadlik, J., Tomala, A., & Sobczak-Kupiec, A. (2024). Thermal stability of titanium/hydroxyapatite systems. In *Surface modifications of titanium and its alloys for biomedical applications* (pp. 179–196). CRC Press.

Kumar, D., & Ansari, N. (Eds.). (2024). Magnesium alloys for biomedical applications: Advances and challenges.

Kumar, P., Rajak, D. K., & Hussain, M. (Eds.). (2026). *Advanced coating and cladding technologies*. CRC Press. DOI: 10.1201/9781003653226

Kumar, S., Nasna, P., & Ghosh, G. (2025). Recent advancement in biocompatible materials, hybrid bioactive coating, surface modification and post-processing techniques for the fabrication of biomedical implant: Critical review and future prospects. *Proceedings of the Institution of Mechanical Engineers. Part C, Journal of Mechanical Engineering Science*, *239*(9), 3247–3286. DOI: 10.1177/09544062241307455

Kumar Yadav, Y., Kumar Singh, A., & Siddhartha. (2023). Insights to improve the tribo-performance of materials used under slurry erosion applications: A review. Proceedings of the Institution of Mechanical Engineers, Part L: Journal of Materials: Design and Applications, 237(1), 3-32.

Kwiecień, K., Pina, S., Oliveira, J. M., Reis, R. L., & Pamuła, E. (2026). Bioceramics for osteochondral tissue engineering and regeneration. In *Osteochondral tissue engineering* (pp. 1–23). Springer Nature. DOI: 10.1007/978-3-031-80976-7_21-1

Lam, T. N., Lee, A., Chiu, Y. R., Kuo, H. F., Kawasaki, T., Harjo, S., Jain, J., Lee, S. Y., & Huang, E. W. (2022). Estimating fine melt pool, coarse melt pool, and heat affected zone effects on the strengths of additive manufactured AlSi10Mg alloys. *Materials Science and Engineering A*, *856*, 143961. DOI: 10.1016/j.msea.2022.143961

Lata, S., Mondal, S., Mishra, R., Priya, S., Deshwal, R. K., Thapliyal, S., Talukdar, N., Dhasmana, A., Rustagi, S., Bora, J., & Malik, S. (2025). Advancements in biodegradable implants: Evaluation in large animal surgical models. *Annals of Medicine and Surgery (London)*, *87*(10), 6480–6495. DOI: 10.1097/MS9.0000000000003578 PMID: 41181418

Leśniak, D., Zasadziński, J., Libura, W., Gronostajski, Z., Śliwa, R., Leszczyńska-Madej, B., Kaszuba, M., Woźnicki, A., Płonka, B., Widomski, P., & Madura, J. (2024). Latest advances in extrusion processes of light metals. *Archives of Civil and Mechanical Engineering*, *24*(3), 184. DOI: 10.1007/s43452-024-00988-5

Li, C. J., Luo, X. T., Dong, X. Y., Zhang, L., & Li, C. X. (2022). Recent research advances in plasma spraying of bulk-like dense metal coatings with metallurgically bonded lamellae. *Journal of Thermal Spray Technology*, *31*(1), 5–27. DOI: 10.1007/s11666-022-01327-x PMID: 37520913

Li, C. J., Luo, X. T., Yao, S. W., Li, G. R., Li, C. X., & Yang, G. J. (2022). The bonding formation during thermal spraying of ceramic coatings: A review. *Journal of Thermal Spray Technology*, *31*(4), 780–817. DOI: 10.1007/s11666-022-01379-z

Li, D., Dai, D., Xiong, G., Lan, S., & Zhang, C. (2023). Composite nanocoatings of biomedical magnesium alloy implants: Advantages, mechanisms, and design strategies. *Advancement of Science*, *10*(18), 2300658. DOI: 10.1002/advs.202300658 PMID: 37097626

Li, F. M., Huang, L., Zaman, S., Guo, W., Liu, H., Guo, X., & Xia, B. Y. (2022). Corrosion chemistry of electrocatalysts. *Advanced Materials*, *34*(52), 2200840. DOI: 10.1002/adma.202200840 PMID: 35334145

Li, J., Wang, C., Wu, Y., Fan, E., Zhu, G., Zhang, X., Zou, Y., & Shi, S. (2025). Advances in inside-laser material feeding additive manufacturing technology. *International Journal of Advanced Manufacturing Technology*, *140*(11), 5713–5736. DOI: 10.1007/s00170-025-16549-w

Li, K., Zhang, M., Zhang, Z., Jin, P., Wang, Y., Yan, W., Zhu, L., Zhang, D. Z., & Murr, L. E. (2025). High performance realization of functionally graded materials based on integrated optimal design and additive manufacturing: A review. *International Materials Reviews*, *70*(6), 497–547. DOI: 10.1177/09506608251354889

Li, M., Huang, K., & Yi, X. (2023). Crack formation mechanisms and control methods of laser cladding coatings: A review. *Coatings*, *13*(6), 1117. DOI: 10.3390/coatings13061117

Li, M., Zhao, Z., Yang, Q., Wei, Y., & Li, J. (2023). Investigation of the effect of laser energy density on selective laser sintering of PA12 using a multi-physics field simulation method. *International Journal of Advanced Manufacturing Technology*, *129*(5), 1987–1998. DOI: 10.1007/s00170-023-12434-6

Li, T., Zhang, L., Qu, X., & Lei, B. (2025). Advanced thermoactive nanomaterials for thermomedical tissue regeneration: Opportunities and challenges. *Small Methods*, *9*(1), 2400510. DOI: 10.1002/smtd.202400510 PMID: 39588862

Li, T., Zhi, D. D., Guo, Z. H., Li, J. Z., Chen, Y., & Meng, F. B. (2022). 3D porous biomass-derived carbon materials: Biomass sources, controllable transformation and microwave absorption application. *Green Chemistry*, *24*(2), 647–674. DOI: 10.1039/D1GC02566J

Liang, J., Cai, Z., He, X., Luo, Y., Zheng, D., Sun, S., Liu, Q., Li, L., Chu, W., Alfaifi, S., Luo, F., Yao, Y., Tang, B., & Sun, X. (2024). Electroreduction of alkaline/natural seawater: Self-cleaning Pt/carbon cathode and on-site co-synthesis of H2 and Mg hydroxide nanoflakes. *Chem*, *10*(10), 3067–3087. DOI: 10.1016/j.chempr.2024.05.018

Liu, H., Sun, Z., Chen, Y., Zhang, W., Chen, X., & Wong, C. P. (2022). Laser processing of flexible in-plane micro-supercapacitors: Progresses in advanced manufacturing of nanostructured electrodes. *ACS Nano*, *16*(7), 10088–10129. DOI: 10.1021/acsnano.2c02812 PMID: 35786945

Liu, X. (2025). Review of protective coatings for corrosion mitigation in chemical machinery: Performance and mechanical aspects. *Journal of Adhesion Science and Technology*, •••, 1–50.

Liu, Y., Wang, S., Huo, J., Zhang, X., Wen, H., Zhang, D., Zhao, Y., Kang, D., Guo, W., & Ngo, H. H. (2024). Adsorption recovery of phosphorus in contaminated water by calcium modified biochar derived from spent coffee grounds. *The Science of the Total Environment*, *909*, 168426. DOI: 10.1016/j.scitotenv.2023.168426 PMID: 37944608

Lopera, H. A. C., & Godoy, M. A. M. (2025). Orthophosphates synthetized from calcium and calcium magnesium phosphates, types of resources and applications: A review. *ES Materials and Manufacturing*, *31*, 1997.

Lu, G., Ni, E., Jiang, Y., Wu, W., & Li, H. (2024). Room-temperature liquid metals for flexible electronic devices. *Small*, *20*(9), 2304147. DOI: 10.1002/smll.202304147 PMID: 37875665

Lu, J., Wu, X., Wang, Z., Zhang, C., Zeng, Q., Wang, C., & Wen, W. (2025). Study on cooling category and heat transfer mechanisms of piston oscillating cooling in a dual gallery structure. *Applied Thermal Engineering*, *276*, 126862. DOI: 10.1016/j.applthermaleng.2025.126862

Lu, K., Xu, Y., Gao, S., Chen, B., Zhu, D., Zhang, Y., Chen, W., Liu, D., Dong, L., Wang, J., Chen, S., & Li, B. (2025). Biomimetic design of advanced ceramics for hard tissue repair. *Journal of the American Ceramic Society*, *108*(11), e20642. DOI: 10.1111/jace.20642

Lu, K., Zhu, J., Guo, D., Yang, M., Sun, H., Wang, Z., Hui, X., & Wu, Y. (2022). Microstructures, corrosion resistance and wear resistance of high-entropy alloys coatings with various compositions prepared by laser cladding: A review. *Coatings*, *12*(7), 1023. DOI: 10.3390/coatings12071023

Lu, X., Zhao, Y., Peng, X., Lu, C., Wu, Z., Xu, H., Qin, Y., Xu, Y., Wang, Q., Hao, Y., & Geng, D. (2025). Comprehensive overview of interface strategies in implant osseointegration. *Advanced Functional Materials*, *35*(16), 2418849. DOI: 10.1002/adfm.202418849

Luu, C. H., Nguyen, N. T., & Ta, H. T. (2024). Unravelling surface modification strategies for preventing medical device-induced thrombosis. *Advanced Healthcare Materials*, *13*(1), 2301039. DOI: 10.1002/adhm.202301039 PMID: 37725037

Ma, K., Cheng, Y., Jeyaprakash, N., Zhou, J., Wan, Y., & Yang, W. (2023). Temperature gradient and solidification rate simulation model of the microstructure of laser-cladded 27SiMn. *Metals*, *13*(10), 1682. DOI: 10.3390/met13101682

Madhuri, A., Parthasarathy, S., & Swain, B. P. (2025). Mechanism of the wettability of coating materials. In *Advances in mechanical coating: Processing, properties and applications* (pp. 137–163). Springer Nature. DOI: 10.1007/978-981-96-7484-8_5

Maheshwaran, S., Kannan, A. R., & Shanmugam, N. S. (2026). Microstructure and mechanical performance of Inconel 617 thin wall fabricated via wire arc additive manufacturing. *Welding in the World*, *70*(2), 503–519. DOI: 10.1007/s40194-025-02242-9

Mao, J., Kim, J. H., & Seo, S. (2024). Current status and outlook of low-melting-point metals in biomedical applications. *Advanced Functional Materials*, *34*(31), 2307708. DOI: 10.1002/adfm.202307708

Marin, E., & Lanzutti, A. (2025). History of metallic orthopedic materials. *Metals*, *15*(4), 378. DOI: 10.3390/met15040378

Matsui, H., Takao, S., Higashi, K., Kaneko, T., Samjeské, G., Uruga, T., Tada, M., & Iwasawa, Y. (2022). Operando imaging of Ce radical scavengers in a practical polymer electrolyte fuel cell by 3D fluorescence CT–XAFS and depth-profiling nano-XAFS–SEM/EDS techniques. *ACS Applied Materials & Interfaces*, *14*(5), 6762–6776. DOI: 10.1021/acsami.1c22336 PMID: 35077130

Mellon, S. J., & Tanner, K. E. (2012). Bone and its adaptation to mechanical loading: A review. *International Materials Reviews, 57*(5), 235–255. DOI: 10.1179/1743280412Y.0000000008

Meng, D., Shi, X., Jiang, J., Liu, F., Li, H., & Liang, X. (2026). Experimental study on thermal runaway propagation characteristics and combustion behaviors of LFP and NCM lithium-ion batteries induced by dual heat sources. *Applied Thermal Engineering, 294*, 130589. DOI: 10.1016/j.applthermaleng.2026.130589

Meng, F., & Du, Y. (2024). Research progress on laser powder bed fusion additive manufacturing of zinc alloys. *Materials (Basel), 17*(17), 4309. DOI: 10.3390/ma17174309 PMID: 39274701

Miu, E. V., McKone, J. R., & Mpourmpakis, G. (2022). The sensitivity of metal oxide electrocatalysis to bulk hydrogen intercalation: Hydrogen evolution on tungsten oxide. *Journal of the American Chemical Society, 144*(14), 6420–6433. DOI: 10.1021/jacs.2c00825 PMID: 35289172

Mocanu, A. C., Miculescu, F., Stan, G. E., Pasuk, I., Tite, T., Pascu, A., Butte, T. M., & Ciocan, L. T. (2022). Modulated laser cladding of implant-type coatings by bovine-bone-derived hydroxyapatite powder injection on Ti6Al4V substrates—Part I: Fabrication and physico-chemical characterization. *Materials (Basel), 15*(22), 7971. DOI: 10.3390/ma15227971 PMID: 36431457

Mohammed, A., Kannan, S., Pervaiz, S., Ali, S., Thomas, K. K., & Karthikeyan, R. (2022). Sustainable machining of Mg-9Al-1.4 Zn foam used for temporary biomedical implant using cryogenic cooling. *Materials (Basel), 15*(19), 6678. DOI: 10.3390/ma15196678 PMID: 36234019

Mohammed, A. O. A., Hamdan, Y. M., & Ali, H. O. (2026). Additive manufacturing of ceramic components: A review of 3D printing technologies for industrial applications. *Saudi Journal of Engineering and Technology, 11*(1), 1–22. DOI: 10.36348/sjet.2026.v11i01.001

Mokhtari, F., Symes, R. W., Simon, Ž., Dharmasiri, B., Henderson, L. C., Joosten, M. W., & Varley, R. J. (2025). An energy efficient and sustainable approach to structural health monitoring in carbon fiber composites: Harnessing sound-induced vibration with Ti3C2Tx MXene/AgNPs modified P(VDF-TrFE) sensors. *Journal of Materials Chemistry. A, Materials for Energy and Sustainability, 13*(9), 6482–6492. DOI: 10.1039/D4TA07797K

Montesissa, M., Tommasini, V., Rubini, K., Boi, M., Baldini, N., & Boanini, E. (2025). State of art and perspective of calcium phosphate-based coatings coupled with bioactive compounds for orthopedic applications. *Nanomaterials (Basel, Switzerland)*, *15*(15), 1199. DOI: 10.3390/nano15151199 PMID: 40801737

Moriarty, T. F., Metsemakers, W. J., Morgenstern, M., Hofstee, M. I., Vallejo Diaz, A., Cassat, J. E., Wildemann, B., Depypere, M., Schwarz, E. M., & Richards, R. G. (2022). Fracture-related infection. *Nature Reviews. Disease Primers*, *8*(1), 67. DOI: 10.1038/s41572-022-00396-0 PMID: 36266296

Mostafa, A. M. (2025). The influence of various parameters on the ablation and deposition mechanisms in pulsed laser deposition. *Plasmonics*, *20*(7), 5627–5645. DOI: 10.1007/s11468-024-02706-6

Mousavian, S. M. H., & Bautin, V. A. (2026). Microbiologically influenced corrosion in magnesium implants: Mechanisms, challenges, and opportunities for biomedical innovation. *Journal of Bio- and Tribo-Corrosion*, *12*(1), 5. DOI: 10.1007/s40735-025-01065-x

Mubeen, M., Khalid, S., Tabish, M., Mahmood, M., Jawad, M., Malik, M. U., & Yasin, G. (2025). Advancements in magnesium applications, biocompatible implants to nanoparticles in medicine and biomedical innovations. *MetalMat*, *2*(2), e70003. DOI: 10.1002/metm.70003

Muresan, L. M. (2023). Nanocomposite coatings for anti-corrosion properties of metallic substrates. *Materials (Basel)*, *16*(14), 5092. DOI: 10.3390/ma16145092 PMID: 37512373

Murzin, S. P. (2024). Artificial intelligence-driven innovations in laser processing of metallic materials. *Metals*, *14*(12), 1458. DOI: 10.3390/met14121458

Murzin, S. P. (2024). Digital engineering in photonics: Optimizing laser processing. *Photonics*, *11*(10), 935. DOI: 10.3390/photonics11100935

Nascimento, M. D., Brito, T. O., Lima, A. M., & Elias, C. N. (2023). Protein interactions with osseointegrable titanium implants. *Brazilian Journal of Oral Sciences*, *22*, e238749. DOI: 10.20396/bjos.v22i00.8668749

Ni, X., Feng, J., Liang, M., Zhou, F., Xia, Y., Dong, Z., Xue, Q., Li, Z., Pu, F., & Xia, P. (2025). Enhancing bone repair with β-TCP-based composite scaffolds: A review of design strategies and biological mechanisms. *Orthopedic Research and Reviews*, *17*, 313–340. DOI: 10.2147/ORR.S525959 PMID: 40688750

Nikam, S., Sawant, M., & McFadden, S. (Eds.). (2026). *Additive manufacturing of thin-walled structures*. CRC Press. DOI: 10.1201/9781003571209

Nikolova, M. P., & Apostolova, M. D. (2022). Advances in multifunctional bioactive coatings for metallic bone implants. *Materials (Basel), 16*(1), 183. DOI: 10.3390/ma16010183 PMID: 36614523

Oladele, I. O., Adekola, S. A., Agbeboh, N. I., Isola-Makinde, B. A., & Adewuyi, B. O. (2024). Synthesis and application of sustainable tricalcium phosphate based biomaterials from agro-based materials: A review. *Biomedical Engineering and Computational Biology, 15*, 11795972241293525. DOI: 10.1177/11795972241293525 PMID: 39524096

Omigbodun, F. T., Oladapo, B. I., & Osa-uwagboe, N. (2026). Exploring the frontier of polylactic acid/hydroxyapatite composites in bone regeneration and their revolutionary biomedical applications: A review. *Journal of Reinforced Plastics and Composites, 45*(5-6), 1230–1250. DOI: 10.1177/07316844241278045

Pagar, R. R., Musale, S. R., Pawar, G., Kulkarni, D., & Giram, P. S. (2022). Comprehensive review on the degradation chemistry and toxicity studies of functional materials. *ACS Biomaterials Science & Engineering, 8*(6), 2161–2195. DOI: 10.1021/acsbiomaterials.1c01304 PMID: 35522605

Pandey, C., Rokaya, D., & Bhattarai, B. P. (2022). Contemporary concepts in osseointegration of dental implants: A review. *BioMed Research International, 2022*(1), 6170452. DOI: 10.1155/2022/6170452 PMID: 35747499

Park, J., Um, S. H., Seo, Y., Lee, J., Kim, Y. C., Ok, M. R., Hwang, S.-W., Sun, J.-Y., Han, H.-S., & Jeon, H. (2023). Improving hydroxyapatite coating ability on biodegradable metal through laser-induced hydrothermal coating in liquid precursor: Application in orthopedic implants. *Bioactive Materials, 25*, 796–806. DOI: 10.1016/j.bioactmat.2022.06.020 PMID: 37056265

Pastrafidou, M., Binas, V., & Kartsonakis, I. A. (2025). Designing the next generation: A physical chemistry approach to surface coating materials. *Applied Sciences (Basel, Switzerland), 15*(19), 10817. DOI: 10.3390/app151910817

Pearson, I. (2025). Persistent proton corridors as a hidden driver of cervical cancer: A signal-based framework for prevention and intervention.

Pesode, P., & Barve, S. (2023). Additive manufacturing of metallic biomaterials: Sustainability aspect, opportunity, and challenges. *Journal of Industrial and Production Engineering, 40*(6), 464–505. DOI: 10.1080/21681015.2023.2229341

Pethő, Á. G., Fülöp, T., Orosz, P., & Tapolyai, M. (2024). Magnesium is a vital ion in the body—It is time to consider its supplementation on a routine basis. *Clinics and Practice, 14*(2), 521–535. DOI: 10.3390/clinpract14020040 PMID: 38525719

Pramesti, D., Winarto, W., Suharno, B., & Ismadi, S. (2025). With two types of buffer layer on Q55. In Proceedings of the International Conference on Smart and Advanced Manufacturing: ICSAM 2024 (p. 41). Springer Nature.

Prasadh, S., Raguraman, S., Wong, R., & Gupta, M. (2022). Current status and outlook of temporary implants (magnesium/zinc) in cardiovascular applications. *Metals*, *12*(6), 999. DOI: 10.3390/met12060999

Pratiwi, L., Elisa, E., & Sutanto, H. (2024). Probing the protrusions: Lamellipodia and filopodia in cancer invasion and beyond. *Mechanobiology in Medicine*, *2*(2), 100064. DOI: 10.1016/j.mbm.2024.100064 PMID: 40395858

Qazi, M. A. (2025). Zr-induced phase-strengthened high entropy alloy (HEA) coatings for hydrogen embrittlement resistance developed via thermal spraying.

Qin, S. (2026). An advanced modeling framework with well-designed experiments to unveil molten pool dynamics and particle spattering behavior [Doctoral dissertation, University of Dublin].

Raabe, D. (2023). The materials science behind sustainable metals and alloys. *Chemical Reviews*, *123*(5), 2436–2608. DOI: 10.1021/acs.chemrev.2c00799 PMID: 36848879

Raghavendra, G., Prasad, P. S., & Gara, D. K. (Eds.). (2025). *Bioactive glass materials for biological applications*. CRC Press. DOI: 10.1201/9781003660781

Rahman, M. M., Ahammad, R., Neon, A. K., Rhaman, M., Ansari, M. J., Uddin, M. N., Islam, M. M., & Nazir, M. A. (2026). Progresses and challenges in additive manufacturing of bulk metallic glasses. *Journal of Manufacturing and Materials Processing*, *10*(4), 121. DOI: 10.3390/jmmp10040121

Ruff, C., Holt, B., & Trinkaus, E. (2006). Who's afraid of the big bad Wolff?: "Wolff's law" and bone functional adaptation. *American Journal of Physical Anthropology*, *129*(4), 484–498. DOI: 10.1002/ajpa.20371 PMID: 16425178

Saba, T., Saad, K. S. K., & Rashid, A. B. (2024). Precise surface engineering: Leveraging chemical vapor deposition for enhanced biocompatibility and durability in biomedical implants. *Heliyon*, *10*(18), e37976. DOI: 10.1016/j.heliyon.2024.e37976 PMID: 39328539

Saha, S. (2023). Evaluating the potential of wrought and additively manufactured Co-Cr alloy for prosthodontic applications [Master's thesis, The University of North Dakota].

Saha, S., & Roy, S. (2022). Metallic dental implants wear mechanisms, materials, and manufacturing processes: A literature review. *Materials (Basel)*, *16*(1), 161. DOI: 10.3390/ma16010161 PMID: 36614500

Sai Charan, M., Naik, A. K., Kota, N., Laha, T., & Roy, S. (2022). Review on developments of bulk functionally graded composite materials. *International Materials Reviews*, *67*(8), 797–863. DOI: 10.1080/09506608.2022.2026863

Saikia, N. (2024). Inorganic-based nanoparticles and biomaterials as biocompatible scaffolds for regenerative medicine and tissue engineering: Current advances and trends of development. *Inorganics*, *12*(11), 292. DOI: 10.3390/inorganics12110292

Salama, M., Vaz, M. F., Colaço, R., Santos, C., & Carmezim, M. (2022). Biodegradable iron and porous iron: Mechanical properties, degradation behaviour, manufacturing routes and biomedical applications. *Journal of Functional Biomaterials*, *13*(2), 72. DOI: 10.3390/jfb13020072 PMID: 35735927

Sarkar, S. K. (2022). Lasers in materials processing and synthesis. *Handbook on Synthesis Strategies for Advanced Materials*, *II*, 791–831. DOI: 10.1007/978-981-16-1803-1_19

Saxena, P., & Shukla, P. (2023). A review on recent developments and advances in environmental gas sensors to monitor toxic gas pollutants. *Environmental Progress & Sustainable Energy*, *42*(5), e14126. DOI: 10.1002/ep.14126

Seetharaman, S., Sankaranarayanan, D., Wong, E., & Gupta, M. (2025). Magnesium-based syntactic, metastable and nano-composites.

Shabir, S., Khan, N. Z., & Siddiquee, A. N. (2025). Laser-induced surface modifications in biomedical materials: A comprehensive review. *Surface Topography : Metrology and Properties*, *13*(2), 023002. DOI: 10.1088/2051-672X/addc85

Shapagina, N. A., & Dushik, V. V. (2025). Bilayer coating systems: Functional interlayers and top layers for enhanced performance. *Materials (Basel)*, *18*(22), 5217. DOI: 10.3390/ma18225217 PMID: 41304061

Sheriff, P. (2026). A proposed archaeological pedestrian survey on Mt. Rainier–Attempting recovery of R5C Curtis Commando and her crew. Sciences, 1880, 146th.

Shi, G., Zhang, R., Cao, Y., & Yang, G. (2024). A review of the vaporization behavior of some metal elements in the LPBF process. *Micromachines*, *15*(7), 846. DOI: 10.3390/mi15070846 PMID: 39064359

Shrivastava, A., Kumar, A., Aggarwal, L. M., Pradhan, S., Choudhary, S., Ashish, A., Kashyap, K., & Mishra, S. (2024). Evolution of bioelectric membrane potentials: Implications in cancer pathogenesis and therapeutic strategies. *The Journal of Membrane Biology*, *257*(5/6), 281–305. DOI: 10.1007/s00232-024-00323-2 PMID: 39183198

Shuai, C., Li, D., Xie, H., Yao, X., Peng, S., & Gao, C. (2025). Programmable lamellar eutectic Zn-2Al-Mg biodegradable implants manufactured by laser powder bed fusion for synergistic strength-ductility and osteogenesis. *Advanced Healthcare Materials*, *14*(32), e01917. DOI: 10.1002/adhm.202501917 PMID: 40817637

Singh, A. V., Chandrasekar, V., Prabhu, V. M., Bhadra, J., Laux, P., Bhardwaj, P., Al-Ansari, A. A., Aboumarzouk, O. M., Luch, A., & Dakua, S. P. (2024). Sustainable bioinspired materials for regenerative medicine: Balancing toxicology, environmental impact, and ethical considerations. *Biomedical Materials (Bristol, England)*, *19*(6), 060501. DOI: 10.1088/1748-605X/ad85bb PMID: 39389102

Smith, D. S., Lefeuvre, P., Renaux, M., Nait-Ali, B., & Leriche, A. (2023). Study of neck formation and densification in porous hydroxyapatite ceramics using thermal conductivity measurements. *Open Ceramics*, *13*, 100329. DOI: 10.1016/j.oceram.2023.100329

Song, Z., & Li, H. (2022). Plasma spraying with wire feeding: A facile route to enhance the coating/substrate interfacial metallurgical bonding. *Coatings*, *12*(5), 615. DOI: 10.3390/coatings12050615

Subraveti, V., Richter, B., Yeratapally, S. R., & Oskay, C. (2024). Three-dimensional prediction of lack-of-fusion porosity volume fraction and morphology for powder bed fusion additively manufactured Ti–6Al–4V. *Integrating Materials and Manufacturing Innovation*, *13*(2), 511–525. DOI: 10.1007/s40192-024-00347-5

Szczęsny, G., Kopec, M., & Kowalewski, Z. L. (2025). Toxicity, irritation, and allergy of metal implants: Historical perspective and modern solutions. *Coatings*, *15*(3), 361. DOI: 10.3390/coatings15030361

Tabatabaeian, A., Ghasemi, A. R., Shokrieh, M. M., Marzbanrad, B., Baraheni, M., & Fotouhi, M. (2022). Residual stress in engineering materials: A review. *Advanced Engineering Materials*, *24*(3), 2100786. DOI: 10.1002/adem.202100786

Tajabadi, M., Barghamadi, H., Lahiji, Z. P., & Momeninasab, E. (2026). Scaffolds, grafts, and coatings. In Musculoskeletal repair and regeneration: New tools, technology and techniques (pp. 171-294). DOI: 10.1142/9789819819645_0004

Tanaka, Y., Xiao, L., & Miwa, N. (2022). Hydrogen-rich bath with nano-sized bubbles improves antioxidant capacity based on oxygen radical absorbing and inflammation levels in human serum. *Medical Gas Research, 12*(3), 91–99. DOI: 10.4103/2045-9912.330692 PMID: 34854419

Tang, W., Fischer, N. G., Kong, X., Sang, T., & Ye, Z. (2024). Hybrid coatings on dental and orthopedic titanium implants: Current advances and challenges. *BMEmat, 2*(4), e12105. DOI: 10.1002/bmm2.12105

Thakur, A. J. (2022). *The elements of fracture fixation (E-book).* Elsevier Health Sciences.

Thiévenaz, V., Meijer, J. G., Lohse, D., & Sauret, A. (2025). On the shape of air bubbles trapped in ice. *Proceedings of the National Academy of Sciences of the United States of America, 122*(10), e2415027122. DOI: 10.1073/pnas.2415027122 PMID: 40035763

Tzagiollari, A. (2024). Development of a biomimetic phosphoserine modified calcium phosphate adhesive for the regeneration of bone defects [Doctoral dissertation].

Van den Borre, C. E., Zigterman, B. G., Mommaerts, M. Y., & Braem, A. (2022). How surface coatings on titanium implants affect keratinized tissue: A systematic review. *Journal of Biomedical Materials Research. Part B, Applied Biomaterials, 110*(7), 1713–1723. DOI: 10.1002/jbm.b.35025 PMID: 35103386

van Ede, M. C., & Angst, U. (2024). Tafel slopes and exchange current densities of oxygen reduction and hydrogen evolution on steel. *Corrosion Engineering, Science and Technology, 59*(1), 39–55. DOI: 10.1177/1478422X241227829 PMID: 42007378

van Gaalen, K. (2023). An experimental and computational investigation of the corrosion behaviour and mechanical performance of a bioabsorbable medical grade magnesium alloy.

Varut, R. M., Trasca, D. M., Stoica, G. A., Sirbulet, C., Arsenie, C. C., & Popescu, C. (2025). Animal models as foundational tools in preclinical orthopedic implant research. *Biomedicines, 13*(10), 2468. DOI: 10.3390/biomedicines13102468 PMID: 41153751

Vedani, M. (2025). *Metal science in modern manufacturing technologies.* Springer Nature. DOI: 10.1007/978-3-031-97125-9

Verma, P. K., Vasudev, H., & Kumar, V. (2025). Effect of electro-corrosion behavior of different grades of stainless steel used in bio-implants: A review. Advanced Manufacturing Processes, 269-283.

Verma, P. K., Vasudev, H., & Singh, B. (2025). Environmental and economic considerations in the fabrication of coatings and claddings. In *Microwave processing of metallic materials* (pp. 216–236). CRC Press. DOI: 10.1201/9781003614081-12

Vivier, V., & Orazem, M. E. (2022). Impedance analysis of electrochemical systems. *Chemical Reviews, 122*(12), 11131–11168. DOI: 10.1021/acs.chemrev.1c00876 PMID: 35687869

Wang, B., Zhang, Z., & Pan, H. (2023). Bone apatite nanocrystal: Crystalline structure, chemical composition, and architecture. *Biomimetics, 8*(1), 90. DOI: 10.3390/biomimetics8010090 PMID: 36975320

Wang, H., Dong, B. S., Chen, Z. B., Liu, J. Q., Haghdadi, N., Lu, R. Q., Primig, S., Wang, Z. Y., Pan, Z. X., Li, H. J., Ringer, S. P., & Liao, X. Z. (2024). Effect of compositional heterogeneity on the mechanical properties of a single-phase Cu-9Al alloy with different grain sizes. *Acta Materialia, 263*, 119531. DOI: 10.1016/j.actamat.2023.119531

Wang, J. (2023). *Influence of solution temperature on chemical deposited magnesium and corrosion behavior of bare and coated magnesium in simulated body fluid.* Washington State University.

Wang, K., Liu, W., Hong, Y., Sohan, H. S., Tong, Y., Hu, Y., Zhang, M., Zhang, J., Xiang, D., Fu, H., & Ju, J. (2023). An overview of technological parameter optimization in the case of laser cladding. *Coatings, 13*(3), 496. DOI: 10.3390/coatings13030496

Wang, L., Feng, S., Wang, Y., Zhao, X., Ge, J., Gao, T., & Di, F. (2026). Porosity defects in additively manufactured metal materials: Formation mechanisms, impact on performance and regulation. *International Materials Reviews, 71*(2), 97–128. DOI: 10.1177/09506608251371459

Wang, L., Liu, K., Li, J., & Geng, S. (2025). A review on corrosion behavior and surface modification technology of nickel aluminum bronze alloys: Current research and prospects. *Advanced Engineering Materials, 27*(1), 2401779. DOI: 10.1002/adem.202401779

Wang, M. F., Yan, T., Gao, M. C., Han, C. W., Yan, Z. Q., Gao, Y. Z., Zhang, W., & Yi, Z. (2025). A review of the advances in implant technology: Accomplishments and challenges for the design of functionalized surface structures. *Biomedical Materials (Bristol, England), 20*(3), 032003. DOI: 10.1088/1748-605X/adca7c PMID: 40199334

Wang, Y., Wang, H., Peng, S., Xia, B., & Zhu, H. (2024). The dominant role of recrystallization and grain growth behaviors in the simulated welding heat-affected zone of high-Mn steel. *Materials (Basel)*, *17*(10), 2218. DOI: 10.3390/ma17102218 PMID: 38793285

Wang, Z., Qi, F., Luo, H., Xu, G., & Wang, D. (2022). Inflammatory microenvironment of skin wounds. *Frontiers in Immunology*, *13*, 789274. DOI: 10.3389/fimmu.2022.789274 PMID: 35300324

Wang, Z., Tong, S., Niu, J., Cao, C., Gao, A., Jiao, Y., Fu, Y., Li, D., Pan, X., Cui, D., Sheng, N., Yan, L., Cui, S., Lin, S., & Liu, Y. (2025). Microneedles: Multifunctional devices for drug delivery, body fluid extraction, and bio-sensing. *Nanoscale*, *17*(2), 740–773. DOI: 10.1039/D4NR03538K PMID: 39606819

Watson, J. L., Seinkmane, E., Styles, C. T., Mihut, A., Krüger, L. K., McNally, K. E., Planelles-Herrero, V. J., Dudek, M., McCall, P. M., Barbiero, S., Vanden Oever, M., Peak-Chew, S. Y., Porebski, B. T., Zeng, A., Rzechorzek, N. M., Wong, D. C. S., Beale, A. D., Stangherlin, A., Riggi, M., & Derivery, E. (2023). Macromolecular condensation buffers intracellular water potential. *Nature*, *623*(7988), 842–852. DOI: 10.1038/s41586-023-06626-z PMID: 37853127

Wei, F., Wei, Y., Yao, X., Li, X., Wei, Z., Zhang, S., Luo, X., Guo, Z., Liu, Q., & Zhu, Q. (2025). Enhancing bond strength of heterogeneous metal-polymer components the perspective of surface micro-nano morphology construction. *Journal of Materials Science*, *60*(14), 6023–6058. DOI: 10.1007/s10853-025-10795-9

Wen, D., Yuan, R., Cao, K., Yang, F., & Chen, R. (2024). Advancements in atomic-scale interface engineering for flexible electronics: Enhancing flexibility and durability. *Nanotechnology*, *35*(41), 412501. DOI: 10.1088/1361-6528/ad64db PMID: 39025081

Wu, C. H., Chang, F. Y., Lin, C. J., & Teng, P. T. (2026). Development of magnesium alloy stents with layered double hydroxide coating for improved corrosion resistance and biochemical stability in AVF applications. *Journal of Functional Biomaterials*, *17*(2), 76. DOI: 10.3390/jfb17020076 PMID: 41745538

Wu, J., Huang, Y., Tan, Y., Zhou, J., Yao, J., Tan, L., & Liu, F. (2025). Particle size-dependent solidification heterogeneity: Simulation and experimental decoupling of gas-atomized Ni-based superalloy powders. *Powder Technology*, •••, 121658.

Xia, B., Liu, Y., Xing, Y., Shi, Z., & Pan, X. (2025). Biodegradable medical implants: Reshaping future medical practice. *Advancement of Science*, *12*(35), e08014. DOI: 10.1002/advs.202508014 PMID: 40782070

Xia, J., Zhang, J., Jiao, H., Niu, X., Ji, X., He, T., & Wang, Z. (2026). Research progress and prospects of laser coating technology. Light. *Advances in Manufacturing*, *6*(4), 870–895.

Xie, W., Wei, X., Kang, H., Jiang, H., Chu, Z., Lin, Y., Hou, Y., & Wei, Q. (2023). Static and dynamic: Evolving biomaterial mechanical properties to control cellular mechanotransduction. *Advancement of Science*, *10*(9), 2204594. DOI: 10.1002/advs.202204594 PMID: 36658771

Xie, Y., Liu, F., Zhang, X., Jin, Y., Li, Q., Shen, H., Fu, H., & Mao, J. (2022). Benefits and risks of essential trace elements in chronic kidney disease: A narrative review. *Annals of Translational Medicine*, *10*(24), 1400. DOI: 10.21037/atm-22-5969 PMID: 36660676

Yadav, B., Mohan Muralikrishna, G., Vaidya, M., Wilde, G., & Divinski, S. V. (2026). Grain boundary diffusion in compositionally complex alloys: A comprehensive review. *International Materials Reviews*, *71*(2), 129–177. DOI: 10.1177/09506608251390882

Yang, Z., Wu, C., Shi, H., Luo, X., Sun, H., Wang, Q., & Zhang, D. (2022). Advances in barrier membranes for guided bone regeneration techniques. *Frontiers in Bioengineering and Biotechnology*, *10*, 921576. DOI: 10.3389/fbioe.2022.921576 PMID: 35814003

Zhang, H., Ijaz, M., & Blaikie, R. J. (2023). Recent review of surface plasmons and plasmonic hot electron effects in metallic nanostructures. *Frontiers of Physics*, *18*(6), 63602. DOI: 10.1007/s11467-023-1328-9

Zhang, J., Yin, C., Farbiz, F., Jafary-Zadeh, M., & Sing, S. L. (2025). Advancing machine learning applications for in-situ monitoring and control in laser-based metal additive manufacturing: A state-of-the-art review. *Virtual and Physical Prototyping*, *20*(1), e2592732. DOI: 10.1080/17452759.2025.2592732

Zhang, X. D. (2024). Flow dynamics of liquid metal. In *Handbook of liquid metals* (pp. 89–134). Springer Nature. DOI: 10.1007/978-981-97-1614-2_7

Zhao, C., Shi, B., Chen, S., Du, D., Sun, T., Simonds, B. J., Fezzaa, K., & Rollett, A. D. (2022). Laser melting modes in metal powder bed fusion additive manufacturing. *Reviews of Modern Physics*, *94*(4), 045002. DOI: 10.1103/RevModPhys.94.045002

Zhao, C., Yu, Z., Du, W., Zhang, Z., Ma, L., Zhang, T., Bai, Y., & Chen, S. (2025). Laser cladding of high-temperature alloy: Research status and development. *Advanced Engineering Materials*, *27*(10), 2402956. DOI: 10.1002/adem.202402956

Zhou, Y., Wang, J., Yang, Y., Yang, M., Zheng, H., Xie, D., Wang, D., & Shen, L. (2022). Laser additive manufacturing of zinc targeting for biomedical application. *International Journal of Bioprinting*, *8*(1), 501. DOI: 10.18063/ijb.v8i1.501 PMID: 35187283

Zhou, Z., Sun, W., Lu, W., Zhang, S., & Wang, J. (2025). Advancements in understanding the characteristics of gas atomized powders: A review. *Advanced Engineering Materials*, *27*(16), 2500270. DOI: 10.1002/adem.202500270

Zhu, K., & Fuh, Y. H. J. (2026). Modeling and simulation of laser-based additive manufacturing process. In *Monitoring and control of laser additive manufacturing process: Theory and applications* (pp. 93–145). Springer Nature. DOI: 10.1007/978-981-95-5226-9_3

Chapter 5
Laser-Assisted Strategies for Wound Healing and Tissue Engineering Scaffolds

Dileep Pathote
http://orcid.org/0000-0001-8198-175X
Indian Institute of Science, Bangalore, India

Rupesh Kumar
Dr. D.Y. Patil Institute of Technology, Pimpri, India

Ashwani Ranjan
Indian Institute of Technology, Banaras Hindu University, India

Sudipto Datta
http://orcid.org/0000-0003-2161-6878
All Indian Institute of Medical Science, New Delhi, India

ABSTRACT

This chapter reviews laser-assisted technologies as emerging strategies for wound healing and tissue engineering, offering advantages over conventional methods that lack precision and bio-functionalization. Approaches such as laser ablation, laser-induced forward transfer, and two-photon polymerisation enable fabrication of scaffolds with controlled porosity, topography, and biochemical gradients that influence cell adhesion, proliferation, and angiogenesis. Laser-assisted bioprinting allows spatial delivery of cells and growth factors to form biomimetic constructs, while low-level laser therapy accelerates repair by reducing inflammation and promoting

DOI: 10.4018/979-8-3373-4104-0.ch005

vascularisation. Integration of these methods with bioactive materials and photobiomodulation shows synergistic effects, though issues of scalability, reproducibility, and long-term safety remain. Collectively, laser-assisted approaches represent a frontier for next-generation wound-healing systems in regenerative medicine.

1. INTRODUCTION

Regulating the spatial organisation of biomaterials and living cells is fundamental to constructing intricate biological systems. Achieving such fine spatial resolution (below 10 μm) remains challenging with conventional cell-handling approaches, as they often lack the precision, reproducibility, automation, and throughput necessary for the efficient fabrication of engineered tissues. To overcome these limitations, laser-assisted bio-fabrication technologies have emerged, leveraging computer-aided design to pattern and assemble both biological and non-biological components into predefined 2D and 3D architectures. This review highlights various laser-based fabrication strategies, including optical tweezers, multiphoton polymerisation, laser-induced forward transfer (LIFT), matrix-assisted pulsed laser evaporation (MAPLE), and laser ablation, along with their applications in biomedical research and engineering. These innovative methods provide unparalleled control over cellular microenvironments in vitro and enable the creation of complex, heterogeneous, and functional tissue constructs, offering transformative potential in regenerative medicine, drug discovery, and fundamental cell biology. Chronic wounds, burns, and tissue defects pose a persistent challenge in clinical medicine. Conventional treatments often fail due to infection, poor vascularisation, or delayed cellular response. Lasers, through photo-biomodulation (PBM), laser-assisted fabrication, and laser-based imaging/diagnostics, provide unique opportunities to accelerate healing and engineer biomimetic scaffolds with controlled architecture. Recent advances integrate laser fabrication of 3D scaffolds, laser-modified bioactive surfaces, and light-responsive hydrogels that combine structural support with controlled drug delivery.

In this chapter, we explore the latest laser-assisted Biofabrication methods and their ability to mimic natural tissue structures. The major focus on four key strategies: (i) laser tweezers, which use light to trap and arrange individual cells with remarkable precision; (ii) laser-initiated polymerization, where light activates chemical reactions to build intricate 3D structures at the submicron scale; (iii) laser-induced forward transfer (LIFT) and matrix-assisted pulsed laser evaporation (MAPLE), which allow delicate biomaterials to be transferred and deposited without damage; and (iv) laser ablation, where laser beams remove specific parts of biomaterials to create micro-patterns like pores or grids. Together, these methods give researchers the ability to design scaffolds with fine structural detail (below 10 μm), helping us better understand how cells interact with each other and with their environment. Ultimately, such precision is paving the way for breakthroughs in regenerative medicine and tissue engineering. Table 1 shows the Laser-Assisted Biofabrication Strategies: Characteristics, Applications, and Limitations.

Table 1. Laser-Assisted Biofabrication Strategies: Characteristics, Applications, and Limitations

Fabrication Method	Laser Type / Wavelength [nm]	Resolution	Processing Speed	Key Features and Applications	Compatible Materials	Limitations
Multi-photon polymerization	Femtosecond pulsed laser, near-infrared (800 nm)	< 100 nm	Low	• Enables fabrication of complex 3D micro/nanostructures with high spatial resolution. • Supports incorporation of biomaterials within photopolymerizable matrices. Applications: Precision fabrication of scaffolds for tissue engineering, controlled drug delivery systems, and biomaterial–cell constructs.	Photosensitive polymers, ceramics, composites, organic/inorganic hybrids	• Restricted to photo-crosslinkable materials. • Challenging scalability. • Low throughput.
Laser-Induced Forward Transfer (LIFT)	Pulsed laser, ultraviolet–infrared range	~100 nm to ~10 μm	Low	• Facilitates deposition of 2D patterns with high spatial fidelity. • Enables controlled positioning of biomaterials and single cells at high density. Applications: Seeding of cells and biomolecules into pre-designed scaffolds for regenerative medicine.	Polymers, biomolecules, living cells	• Limited scalability to large-area patterns. • Process optimisation is complex.
Matrix-Assisted Pulsed Laser Evaporation (MAPLE)	Excimer pulsed UV laser (ArF: 193; KrF: 248; XeCl: 308 nm)	< 1 μm	High	• Allows deposition of nanoparticulated and multilayered films. • Facilitates coating of organic/inorganic hybrids, polymers, metals, ceramics, and biomolecules. Applications: Development of biosensors, microelectronics, multilayer coatings, and scaffolds for tissue engineering.	Polymers, metals, ceramics, biomolecules	• Process requires multiple parameters. • Limited coating area. • Layer-by-layer fabrication complexity.

continued on following page

Table 1. Continued

Fabrication Method	Laser Type / Wavelength [nm]	Resolution	Processing Speed	Key Features and Applications	Compatible Materials	Limitations
Laser Ablation	Pulsed UV or near-infrared laser (e.g., 400, 532, 800 nm)	~100 nm	High	• Enables direct writing with high fabrication speed. • Produces controllable surface geometries and micro/nano-scale roughness. • Broad material applicability across biological and non-biological systems. Applications: Direct micro-structuring of scaffolds, fabrication of 3D microfluidic channels, and controlled topography for cell guidance.	Polymers, metals, ceramics, biomolecules, cells	• Limited fine control over nanoscale features. • Predominantly top-down fabrication approach.

2. LASER-ASSISTED BIOFABRICATION STRATEGIES

Laser-assisted biofabrication has emerged as a highly versatile and precise approach in tissue engineering and regenerative medicine. Unlike conventional fabrication techniques, laser-based methods enable contact-free processing of both biological and non-biological materials with minimal thermal or mechanical damage, preserving cellular viability and material functionality. These strategies facilitate the creation of scaffolds with highly controlled micro- and nano-scale architectures, which critically influence cell adhesion, proliferation, migration, and tissue integration, processes that are essential for effective wound healing and tissue regeneration. Among the most widely employed laser-assisted biofabrication techniques are multiphoton polymerisation (2PP), laser-induced forward transfer (LIFT), matrix-assisted pulsed laser evaporation (MAPLE), and laser ablation. Each of these methods offers unique advantages, limitations, and potential applications in the fabrication of functional scaffolds for regenerative therapies, as detailed in the following subsections.

Figure 1. Laser-assisted bio-fabrication techniques

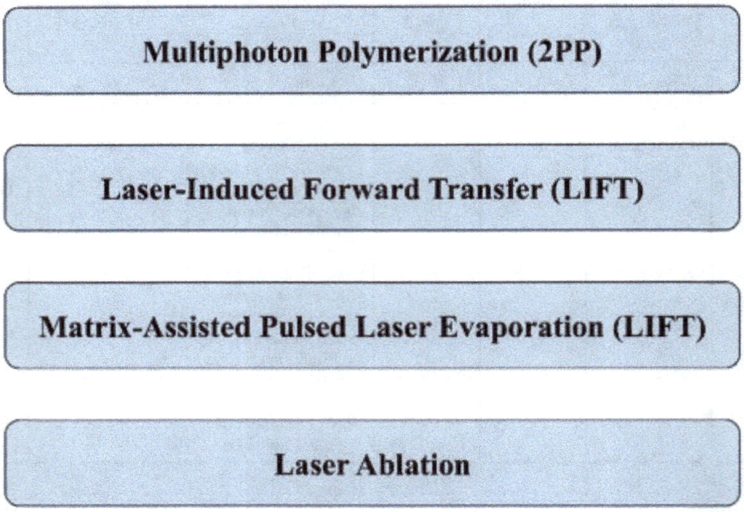

2.1 Multiphoton Polymerization (2PP)

Multiphoton polymerization (2PP) is a nonlinear photopolymerization technique that employs femtosecond laser pulses, typically in the near-infrared region (~800 nm), to induce highly localised polymerisation within photosensitive materials (Maruo et al., 1997). The process relies on the simultaneous absorption of two photons by a photoinitiator, which triggers polymerization only at the laser focal point where photon density is sufficient. This enables the fabrication of intricate three-dimensional micro- and nano-architectures with sub-micron precision, without affecting the surrounding material. 2PP is widely used to produce scaffolds with controlled porosity, microchannels for nutrient transport, and encapsulation of bioactive molecules or growth factors, all of which are crucial for enhanced tissue regeneration and wound healing. Its high spatial resolution and compatibility with computer-aided design (CAD) models enable patient-specific scaffold designs. However, 2PP is limited to photo-cross-linkable materials, exhibits low fabrication throughput, and faces challenges in scaling up for larger tissue constructs (Malinauskas et al., 2013). Figure 1 represents the laser-assisted bio-fabrication techniques.

2.2 Laser-Induced Forward Transfer (LIFT)

Laser-induced forward transfer (LIFT) is a pulsed laser-based deposition technique that allows the precise transfer of biomaterials or living cells from a donor substrate to a receiving substrate without direct contact (Piqué et al., 2003). In LIFT, a pulsed laser irradiates a thin film of donor material, generating a high-pressure bubble that propels a portion of the material onto the target substrate. This method enables high-resolution deposition of cells, proteins, and biomolecules, preserving their viability and functionality. LIFT is particularly effective for seeding cells into predefined scaffold geometries, depositing bioactive molecules in specific patterns, and constructing multilayered tissue constructs. While LIFT provides excellent spatial fidelity and gentle processing of biological materials, it has limited scalability for large-area patterning and requires careful optimisation of laser parameters.

2.3 Matrix-Assisted Pulsed Laser Evaporation (MAPLE)

Matrix-assisted pulsed laser evaporation (MAPLE) is a laser-based thin-film deposition method designed to transfer sensitive biomaterials, polymers, or hybrid compounds without compromising their bioactivity (Chrisey et al., 1998). In MAPLE, a pulsed excimer laser (UV range, e.g., ArF 193 nm) irradiates a frozen target solution containing the material of interest. The laser energy causes sublimation of the solvent, transferring solute molecules to a substrate to form a uniform thin film. MAPLE has been applied in coating scaffolds with bioactive molecules, developing biosensors, and constructing multilayer tissue scaffolds. This method is advantageous for maintaining biomolecule functionality and enabling multilayer deposition, but it requires careful control of process parameters, and the fabrication area is often limited, making large-scale production challenging.

2.4 Laser Ablation

Table 2 represents the comparative summary of different methods, highlighting their advantages and limitations for wound healing and tissue engineering applications. Laser ablation utilises pulsed UV or near-infrared lasers to selectively remove material from a substrate, creating controlled surface micro- and nano-patterns (Bonse et al., 2012). Focused laser pulses vaporise material at precise locations, producing features such as microgrooves, pores, or periodic structures. These patterns influence cell behavior, including adhesion, alignment, proliferation, and migration, which are essential for wound healing and tissue integration. Laser ablation is commonly applied for micro- and nano-patterning of scaffolds, fabrication of 3D microfluidic channels, and surface functionalization to impart antibacterial or bioactive properties. While it provides high fabrication speed and broad material compatibility, laser ablation offers limited nanoscale precision compared to 2PP and predominantly employs a top-down approach, making it less suitable for complex volumetric structures. Table 2 represents the comparison of different wound healing methods and their applications.

Table 2. Comparative summary of different methods, highlighting their advantages and limitations for wound healing and tissue engineering applications

Technique	Resolution	Processing Speed	Key Applications	Advantages	Limitations
Multiphoton Polymerization (2PP)	<100 nm	Low	3D scaffolds, microchannels, and drug delivery	Sub-micron precision; CAD-compatible; supports cell encapsulation	Limited to photo-crosslinkable materials; low throughput; scalability challenges
Laser-Induced Forward Transfer (LIFT)	100 nm–10 μm	Low	Cell deposition, biomolecule patterning, multilayer constructs	High spatial accuracy; gentle on biological materials	Limited large-area scalability; complex parameter optimization; low throughput
Matrix-Assisted Pulsed Laser Evaporation (MAPLE)	<1 μm	Moderate	Thin-film coating, multi-material scaffolds, biosensors	Maintains bioactivity of sensitive biomolecules; supports multilayer deposition	Limited coating area; complex layer-by-layer fabrication
Laser Ablation	~100 nm	High	Micro/nano surface patterning, 3D channels, surface functionalization	High speed; broad material compatibility	Limited nanoscale precision; primarily a top-down approach

3. PHOTO-BIOMODULATION (PBM) AND WOUND HEALING

Photobiomodulation (PBM), also referred to as low-level laser therapy (LLLT), is a non-invasive technique that uses red and near-infrared (NIR) light to stimulate cellular processes and enhance tissue repair (AlGhamdi et al., 2012; Anders et al., 2015). Unlike high-power ablative lasers, PBM does not induce thermal damage; instead, it promotes photochemical and photophysical changes within cells, which accelerate wound healing and tissue regeneration. PBM has been widely explored in regenerative medicine for its ability to modulate inflammation, promote angiogenesis, and improve extracellular matrix (ECM) remodelling, all of which are crucial for effective wound closure and the integration of tissue-engineered scaffolds.

3.1 Mechanisms of PBM

The primary mechanism of PBM involves the absorption of light by cellular chromophores, most notably cytochrome c oxidase in the mitochondrial respiratory chain (Karu, 2010). Upon photon absorption, mitochondrial activity increases, leading to elevated ATP production and controlled modulation of reactive oxygen species (ROS). These changes trigger intracellular signalling cascades, including MAPK, NF-κB, and PI3K/Akt pathways, which regulate cell proliferation, migration, and differentiation (Hamblin, 2017). By stimulating these pathways, PBM enhances fibroblast activity, keratinocyte migration, endothelial cell proliferation, and mesenchymal stem cell (MSC) differentiation, thereby promoting angiogenesis, ECM deposition, and re-epithelialization, key processes in wound healing and scaffold integration.

Figure 2. Mechanism of Photo-biomodulation (PBM) in wound healing

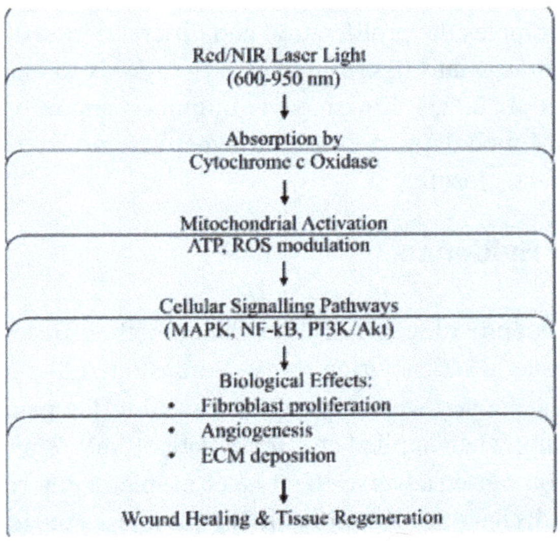

3.2 PBM Parameters and Dosimetry

The therapeutic outcomes of PBM are highly dependent on parameters such as wavelength, power density, energy dose, and treatment duration (AlGhamdi et al., 2012). Red light, in the 600–700 nm range, primarily targets superficial tissues such as skin and mucosa, while NIR light, between 780–950 nm, penetrates deeper structures, including muscle and bone. Optimal energy densities generally range from 1 to 10 J/cm^2, promoting fibroblast proliferation and angiogenesis, whereas excessive irradiation may lead to photoinhibition or tissue damage (Statha et al., 2025). The biphasic dose response, also known as the Arndt-Schulz effect, highlights the importance of precise dosimetry to maximise therapeutic efficacy. Figure 2 represents the mechanism of Photo-biomodulation (PBM) in wound healing.

3.3 Cellular and Molecular Effects

PBM exerts multifaceted effects at the cellular and molecular levels. Fibroblast proliferation and collagen synthesis are significantly enhanced following PBM, resulting in improved tensile strength and ECM remodelling (Eldessoky et al., 2023). Angiogenesis is stimulated through the upregulation of vascular endothelial growth factor (VEGF) and nitric oxide, increasing perfusion and nutrient delivery to the wound site. PBM also modulates inflammation by reducing pro-inflammatory cytokines such as TNF-α and IL-1β while enhancing anti-inflammatory mediators, thereby creating a balanced wound microenvironment (Johns et al., 2014). Furthermore, PBM promotes the proliferation and differentiation of MSCs, which are commonly incorporated into tissue-engineered scaffolds to enhance regenerative outcomes (Yong et al., 2023). Keratinocyte migration and re-epithelialization are accelerated as PBM modulates cytoskeletal dynamics and intercellular signalling, enabling rapid wound closure.

3.4 Preclinical Evidence

Preclinical studies provide compelling evidence for the efficacy of PBM in wound healing. For example, a 2025 murine study demonstrated that 661 nm low-power laser irradiation accelerated wound closure, reduced inflammation, and improved collagen remodelling when applied at 2 J/cm^2 with 50 mW/cm^2 intensity, whereas excessive irradiation caused adverse effects such as microhemorrhage (Statha et al., 2025). Similar results have been reported in full-thickness excisional wound models in rats and rabbits, showing enhanced fibroblast density, capillary formation, and ECM deposition. Notably, PBM has been applied in combination with hydrogel scaffolds and bioactive matrices, which synergistically enhance cell proliferation, scaffold integration, and tissue maturation. These findings highlight the potential of PBM as a complementary therapy to laser-fabricated scaffolds.

3.5 Clinical Applications

PBM has been successfully translated to clinical settings, particularly for chronic wounds, diabetic ulcers, burns, and postoperative healing. Randomised clinical trials have reported faster wound closure, reduced pain, and improved tissue quality in patients receiving PBM compared to conventional care (AlGhamdi et al., 2012; Eldessoky et al., 2023). In tissue engineering, PBM enhances the integration of scaffolds seeded with stem cells or bioactive molecules by stimulating the local wound microenvironment, promoting angiogenesis, and accelerating ECM deposition. When combined with laser-fabricated scaffolds, PBM provides a multifaceted regenerative strategy that integrates structural support with biological stimulation.

3.6 Integration with Tissue-Engineered Scaffolds

Combining PBM with laser-fabricated scaffolds represents a promising approach for regenerative medicine. Scaffolds with controlled porosity, surface topography, and bioactive cues can be further enhanced by PBM, which activates resident and transplanted cells to accelerate tissue regeneration. Recent studies have demonstrated that photobiomodulated hydrogels releasing growth factors or antioxidants on demand significantly improve wound closure, reduce oxidative stress, and promote angiogenesis (Zhang et al., 2025; Xu et al., 2025). This integration ensures that scaffold architecture and biological stimulation work synergistically, creating a microenvironment conducive to effective tissue repair.

3.7 Challenges and Future Directions

Despite the growing body of evidence, PBM faces several challenges in clinical translation. Optimal wavelengths, energy doses, and treatment durations are not yet standardised, leading to variability in outcomes. Patient-specific factors, such as tissue type, wound depth, and comorbidities, further complicate treatment planning. Moreover, scaffold designs must be optimised for PBM absorption and penetration to maximise therapeutic benefits. Future research aims to develop smart laser-responsive scaffolds that integrate PBM with controlled drug release, growth factor delivery, and stem cell activation, enabling personalised regenerative therapies. Advances in fibre-optic, wearable, and portable laser devices are expected to provide precise, minimally invasive delivery of PBM to target tissues, expanding its clinical utility.

4. LASER-BASED SCAFFOLDS FOR TISSUE ENGINEERING

Laser-based fabrication strategies have revolutionised tissue engineering by enabling the precise creation of scaffolds with controlled architecture, surface topography, and functional properties. Unlike conventional fabrication techniques such as solvent casting or particulate leaching, laser-assisted methods provide contact-free, high-resolution processing, minimising thermal or mechanical damage to delicate biomaterials. These techniques allow for the production of scaffolds with highly controlled micro- and nano-scale features, tailored porosity, and spatially defined biochemical gradients, which are crucial for guiding cell adhesion, proliferation, migration, and differentiation (Ranjan et al., 2022; Zhang et al., 2025). The major laser-based scaffold fabrication strategies include additive manufacturing and laser sintering, laser micro/nano-patterning, and smart laser-responsive biomaterials, as described in the following subsections.

4.1 Additive Manufacturing and Laser Sintering

Additive manufacturing techniques, such as selective laser sintering (SLS) and direct metal laser sintering (DMLS), utilise high-energy laser beams to selectively fuse powders in a layer-by-layer fashion, producing scaffolds with high geometric precision and controlled porosity (Zhang et al., 2025). By adjusting laser parameters such as power, scanning speed, and hatch spacing, it is possible to tailor mechanical properties to match those of native tissues, including bone or cartilage. Scaffold porosity, typically ranging from 100–500 μm, facilitates vascular ingrowth, nutrient transport, and cell migration, all critical for successful tissue regeneration. Metals (e.g., Ti, Mg alloys), polymers (e.g., polycaprolactone, PLA), and composite biomaterials have been successfully processed using laser sintering to generate load-bearing and biodegradable scaffolds suitable for orthopaedic and dental applications (Tuikampee et al., 2024). Despite these advantages, laser sintering requires careful control of thermal exposure to avoid residual stresses, microcracks, or phase transformations that may compromise scaffold integrity.

4.2 Laser Micro- and Nano-Patterning

Femtosecond and picosecond lasers enable the generation of micro- and nano-scale surface patterns on scaffolds, which profoundly influence cell behaviour (Ranjan et al., 2022; Xu et al., 2025). Techniques such as laser-induced periodic surface structures (LIPSS) create well-defined topographies that guide cell alignment, adhesion, and migration, thereby promoting functional tissue organisation. Surface modifications can also improve hydrophilicity, protein adsorption, and antibacterial properties, enhancing the integration of scaffolds with host tissue. Polymers like collagen, gelatin, and PCL, as well as metals such as titanium, have been patterned using femtosecond laser ablation to generate hierarchical structures that mimic native extracellular matrix (ECM) architecture. These micro- and nano-scale features not only regulate cellular behaviour but also enhance mechanical interlocking, supporting scaffold stability and long-term functionality in vivo.

4.3 Smart Laser-Responsive Biomaterials

Recent advances in laser-fabricated scaffolds involve smart biomaterials that respond to light, enabling on-demand drug or growth factor release, antibacterial activity, and photothermal stimulation (Zhang et al., 2025; Xu et al., 2025). For instance, near-infrared (NIR)-responsive hydrogels can release antioxidants or bioactive molecules when exposed to specific wavelengths, promoting accelerated wound closure while mitigating oxidative stress. L-arginine-based photothermal hydrogels have been shown to provide both angiogenic stimulation and antibacterial activity, creating a multifunctional scaffold capable of structural support, controlled drug release, and biological modulation. Such laser-responsive systems integrate structural design with therapeutic functionality, bridging scaffold fabrication and active tissue regeneration in a single platform.

4.4 Challenges and Future Perspectives

Despite their transformative potential, laser-fabricated scaffolds face challenges in scalability, reproducibility, and clinical translation. High-precision techniques such as 2PP and LIFT are limited to small constructs due to low throughput, while additive laser sintering faces constraints related to residual stress and thermal management. Integration with vascular networks, especially for large tissues or organs, remains a critical challenge, as current laser-fabricated microchannels cannot yet replicate complex bifurcating vasculature. Future developments are likely to focus on hybrid fabrication strategies, combining laser-assisted micro/nano-patterning with extrusion- or inkjet-based bioprinting, to achieve large-scale, hierarchically organised scaffolds with localised biochemical cues. The development of novel photo-curable biomaterials and light-responsive smart hydrogels will further expand the applicability of laser-based scaffolds in regenerative medicine, enabling personalised tissue constructs with integrated therapeutic functionality.

5. DIAGNOSTIC AND MONITORING APPLICATIONS

Laser Doppler Imaging (LDI) is widely regarded as the gold standard for non-invasive assessment of tissue perfusion. LDI operates by detecting frequency shifts in laser light scattered by moving red blood cells within the microvasculature. This technique allows spatial mapping of blood flow across the wound area, providing insights into tissue viability, angiogenesis, and healing potential (De Mey et al., 2025). Studies have demonstrated that LDI can predict burn wound healing with over 95% accuracy, enabling early intervention and guiding the application of laser-fabricated scaffolds or PBM therapy. LDI is particularly valuable in evaluating the effectiveness of laser-modified biomaterial scaffolds, as it can monitor scaffold-induced neovascularisation and integration with host tissue over time.

5.1 Laser Doppler Imaging (LDI)

Laser Doppler Imaging (LDI) is widely regarded as the gold standard for non-invasive assessment of tissue perfusion. LDI operates by detecting frequency shifts in laser light scattered by moving red blood cells within the microvasculature. This technique allows spatial mapping of blood flow across the wound area, providing insights into tissue viability, angiogenesis, and healing potential (De Mey et al., 2025). Studies have demonstrated that LDI can predict burn wound healing with over 95% accuracy, enabling early intervention and guiding the application of laser-fabricated scaffolds or PBM therapy. LDI is particularly valuable in evaluating the effectiveness of laser-modified biomaterial scaffolds, as it can monitor scaffold-induced neovascularisation and integration with host tissue over time.

5.2 Laser Speckle Contrast Imaging (LSCI)

Laser Speckle Contrast Imaging (LSCI) is a rapid, full-field imaging technique that assesses tissue perfusion by analysing speckle patterns generated by coherent laser light scattering from moving blood cells. LSCI offers real-time monitoring, making it suitable for dynamic wound assessment during surgical procedures or therapeutic interventions (De Mey et al., 2025). While LSCI is advantageous for continuous monitoring, its predictive accuracy in deep dermal wounds is somewhat lower than that of LDI. Nevertheless, LSCI can be integrated with laser-assisted scaffolds and photobiomodulation protocols to evaluate spatial variations in microcirculation, providing critical feedback for optimising treatment parameters.

5.3 Integration With Laser-Fabricated Scaffolds

Combining laser-based imaging modalities with scaffold implantation offers a comprehensive platform for theranostic applications, where diagnostics and therapy are integrated. Laser-fabricated scaffolds with controlled micro- and nano-architecture can be monitored non-invasively for vascularisation, cellular infiltration, and tissue remodelling, while PBM can be applied simultaneously to accelerate healing. Such integration allows for personalised wound management, where scaffold properties and laser treatment parameters are dynamically adjusted based on real-time feedback from imaging modalities. For instance, areas of poor perfusion identified by LDI or LSCI can be selectively targeted with PBM or smart laser-responsive hydrogels, ensuring localised delivery of growth factors or antioxidants.

5.4 Future Perspectives in Diagnostics

Future developments in laser-assisted diagnostics are likely to focus on multimodal imaging, combining LDI, LSCI, optical coherence tomography (OCT), and photoacoustic imaging. These integrated systems will enable high-resolution, depth-resolved assessment of scaffold integration and tissue regeneration, bridging the gap between preclinical studies and clinical translation. Additionally, machine learning algorithms can analyse imaging data to predict wound healing trajectories, optimise PBM protocols, and personalize scaffold designs for individual patients. The convergence of laser-assisted fabrication, photobiomodulation, and real-time diagnostics is poised to establish a closed-loop regenerative medicine platform, enhancing both efficacy and safety in tissue engineering applications.

6. CHALLENGES, LIMITATIONS, AND FUTURE DIRECTIONS

Despite significant advances in laser-assisted strategies for wound healing and tissue engineering, several challenges hinder their widespread clinical translation. The integration of laser-based fabrication, photobiomodulation, and smart scaffolds presents both technical and biological complexities that must be addressed to realize the full potential of these technologies.

6.1 Technical Challenges

One of the primary technical challenges is scalability. Techniques such as multiphoton polymerization (2PP) and laser-induced forward transfer (LIFT) offer sub-micron precision but are limited to small-area constructs due to low processing speeds (Ranjan et al., 2022). Achieving large, clinically relevant tissue constructs with hierarchical vascular networks remains difficult. Additionally, residual thermal stresses during laser sintering of metals or composites can compromise scaffold mechanical integrity and induce microcracks, which may affect both in vivo performance and long-term stability (Tuikampee et al., 2024).

Reproducibility is another key concern. Laser-assisted fabrication often involves multiple parameters—laser power, pulse duration, scan speed, and material properties—all of which must be precisely controlled. Minor variations can result in inconsistent scaffold morphology or cellular responses, complicating standardization for clinical applications. Furthermore, integration with multimodal imaging systems requires sophisticated software and hardware alignment to ensure accurate diagnostics and monitoring.

6.2 Biological and Clinical Limitations

From a biological perspective, stem cell–derived constructs and laser-fabricated scaffolds frequently exhibit immature tissue phenotypes, resembling embryonic or neonatal states rather than fully functional adult tissues. This limits physiological relevance and mechanical performance for load-bearing applications. Incorporating vascularization and innervation remains a major bottleneck, particularly for thick tissues and organs. While PBM can enhance angiogenesis and cellular proliferation, translating these effects from preclinical models to humans requires further validation.

Clinical implementation also faces challenges regarding safety, regulatory approval, and cost-effectiveness. Standardisation of laser parameters for PBM and scaffold fabrication is lacking, resulting in variable outcomes across clinical studies. The cost and complexity of ultrafast laser systems, along with training requirements for operators, pose additional barriers to adoption in routine healthcare settings (Zhang et al., 2025).

6.3 Future Directions

Emerging trends in laser-assisted regenerative medicine aim to address these limitations through technological and material innovations. Hybrid fabrication strategies, combining laser micro/nano-patterning with extrusion or inkjet bioprinting, allow the creation of large-scale scaffolds with precise structural features and spatially controlled biochemical cues. Advances in spatiotemporal beam shaping and high-repetition-rate ultrafast lasers are expected to enhance throughput and scalability while maintaining sub-micron precision. The development of novel photo-curable and laser-responsive biomaterials will enable multifunctional scaffolds capable of on-demand drug delivery, antibacterial action, and localised tissue stimulation. Integration with patient-derived iPSCs and closed-loop PBM systems could facilitate personalised tissue engineering, where scaffold architecture and therapeutic interventions are optimised in real time based on diagnostic feedback.

By analysing imaging data from LDI, LSCI, and other modalities, can predict healing outcomes, optimise laser parameters, and guide scaffold design, reducing trial-and-error approaches and accelerating clinical translation. By integrating fabrication precision, cellular modulation, real-time diagnostics, and smart biomaterials, laser-assisted strategies have the potential to establish a new paradigm in regenerative medicine, offering scalable, personalised, and functionally superior tissue-engineered constructs.

7. CONCLUSIONS

Laser-assisted strategies have emerged as a transformative approach in the fields of wound healing and tissue engineering, offering unprecedented precision in both scaffold fabrication and cellular modulation. Through techniques such as multiphoton polymerisation, laser-induced forward transfer, MAPLE, and laser ablation, researchers can construct scaffolds with controlled micro- and nano-scale architectures that directly influence cell adhesion, proliferation, migration, and angiogenesis. When combined with photobiomodulation therapies, these scaffolds facilitate accelerated wound closure, improved tissue regeneration, and enhanced integration with host tissue. The integration of diagnostic modalities, including laser Doppler imaging and laser speckle contrast imaging, provides real-time assessment of perfusion and scaffold functionality, enabling closed-loop approaches that optimise both scaffold design and therapeutic interventions. Moreover, the development of smart, laser-responsive biomaterials has expanded the functionality of tissue-engineered constructs, allowing for on-demand delivery of growth factors, antioxidants, and antibacterial agents, thereby addressing multiple challenges in complex wound environments. Despite these advancements, the clinical translation of laser-assisted strategies faces technical, biological, and regulatory challenges. Limitations in scaffold scalability, tissue vascularisation, reproducibility of fabrication, and standardisation of photobiomodulation protocols must be addressed. Future directions point toward hybrid fabrication methods, AI-assisted diagnostics, patient-specific scaffold design, and personalised regenerative therapies, which collectively have the potential to overcome current bottlenecks.

In summary, laser-assisted strategies bridge the gap between advanced biomaterial design, precise fabrication, therapeutic modulation, and real-time diagnostics, establishing a new paradigm in regenerative medicine. Continued interdisciplinary research, integrating materials science, photonics, biology, and clinical practice, will be essential to translate these promising technologies into effective, scalable, and personalised solutions for tissue repair and wound healing.

REFERENCES

AlGhamdi, K. M., Kumar, A., & Moussa, N. A. (2012). Low-level laser therapy: A useful technique for enhancing the proliferation of various cultured cells. *Lasers in Medical Science*, *27*(2), 237–249. DOI: 10.1007/s10103-011-0885-2 PMID: 21274733

Anders, J. J., Lanzafame, R. J., & Arany, P. R. (2015). Low-level light/laser therapy versus photobiomodulation therapy. *Photomedicine and Laser Surgery*, *33*(4), 183–184. DOI: 10.1089/pho.2015.9848 PMID: 25844681

Bonse, J., Höhm, S., Kirner, S. V., Rosenfeld, A., & Krüger, J. (2012). Laser-induced periodic surface structures—A scientific evergreen. *IEEE Journal of Selected Topics in Quantum Electronics*, *23*(3), 9000615. DOI: 10.1109/JSTQE.2016.2579013

Chrisey, D. B., Hubler, G. K., Perry, S. S., & Calvert, J. M. (1998). Matrix-assisted pulsed laser evaporation: A new method for thin film deposition. *Applied Physics. A, Materials Science & Processing*, *66*(2), 159–166. DOI: 10.1007/s003390050603

De Mey, K., De Decker, I., & Hoeksema, H.. (2025). Validity of laser speckle contrast imaging for predicting wound healing potential in burns. *Burns*, *51*(4), 107449. Advance online publication. DOI: 10.1016/j.burns.2025.107449 PMID: 40112657

Eldessoky, A. E., Khalefa, M. M., & Abu-Seida, A. M. (2023). Regenerative endodontic therapy in mature teeth with necrotic pulp and apical periodontitis using two disinfection protocols. *BMC Oral Health*, *23*(1), 386. Advance online publication. DOI: 10.1186/s12903-023-02863-w PMID: 36949460

Hamblin, M. R. (2017). Mechanisms and applications of the anti-inflammatory effects of photobiomodulation. *AIMS Biophysics*, *4*(3), 337–361. DOI: 10.3934/biophy.2017.3.337 PMID: 28748217

Johns, D. A., Shivashankar, V. Y., Krishnamma, S., & Johns, M. (2014). Use of photoactivated disinfection and platelet-rich fibrin in regenerative endodontics. *Journal of Conservative Dentistry : JCD*, *17*(5), 487–490. DOI: 10.4103/0972-0707.139850 PMID: 25298655

Karu, T. I. (2010). Multiple roles of cytochrome c oxidase in mammalian cells under red and near-IR radiation. *Photomedicine and Laser Surgery*, *28*(2), 173–180. DOI: 10.1089/pho.2009.2541 PMID: 19743963

Malinauskas, M., Farsari, M., Piskarskas, A., & Juodkazis, S. (2013). Ultrafast laser nanostructuring of photopolymers: A decade of advances. *Physics Reports*, *533*(1), 1–31. DOI: 10.1016/j.physrep.2013.07.005

Maruo, S., Nakamura, O., & Kawata, S. (1997). Three-dimensional microfabrication with two-photon-absorbed photopolymerization. *Optics Letters*, *22*(2), 132–134. DOI: 10.1364/OL.22.000132 PMID: 18183126

Piqué, A., Chrisey, D. B., & Fitz-Gerald, J. M. (2003). Laser-induced forward transfer: Fundamentals and applications. *Applied Physics. A, Materials Science & Processing*, *77*(2), 167–173. DOI: 10.1007/s00339-003-2143-5

Ranjan, A., Jindal, V., & Tyagi, R. (2022). Effect of load on tribological properties of Ti–TiB–Fe composites processed via spark plasma sintering (SPS). *Transactions of the Indian Institute of Metals*, *75*(11), 2847–2856. DOI: 10.1007/s12666-022-02651-0

Ranjan, A., Tyagi, R., & Jindal, V. (2022). Reciprocating wear of Ti–TiB in situ composites synthesised via vacuum arc melting. *Journal of Materials Engineering and Performance*, *31*(12), 9985–9996. DOI: 10.1007/s11665-022-07002-0

Statha, D., Sfiniadakis, I., Rallis, M., Anastassopoulou, J., & Alexandratou, E. (2025). Investigating the wound healing potential of low-power 661 nm laser light in a pigmented hairless murine model. *Free Radical Biology & Medicine*, *233*, S1–S102. DOI: 10.1016/j.freeradbiomed.2025.05.062 PMID: 40338500

Tuikampee, T., Chaijareenont, P., Rungsiyakull, P., & Yavirach, A. (2024). Titanium surface modification techniques to enhance osteoblasts and bone formation. *Metals*, *14*(5), 515. Advance online publication. DOI: 10.3390/met14050515

Xu, L.. (2025). L-arginine-induced versatile hydrogel with skin adaptiveness for wound healing. *Chemical Engineering Journal*. Advance online publication. DOI: 10.1016/j.cej.2025.152376

Yong, J., Gröger, S., Wu, Z., Ruf, S., Ye, Y., & Chen, X. (2023). Photobiomodulation therapy and pulp-regenerative endodontics: A narrative review. *Bioengineering (Basel, Switzerland)*, *10*(3), 371. Advance online publication. DOI: 10.3390/bioengineering10030371 PMID: 36978762

Zhang, Y., & Chen, X.. (2025). Near-infrared light-responsive on-demand puerarin releasing intelligent hydrogel for chronic wound healing. *Materials Letters*. Advance online publication. DOI: 10.1016/j.matlet.2025.135678

Chapter 6
Laser-Based Coatings for Antibacterial and Antifouling Applications

Vyomesh Raseshkumar Buch
http://orcid.org/0000-0002-1101-9597
Parul University, India

ABSTRACT

All healthcare-related infections, biomaterial-associated infections, represent 5070 percent of the total healthcare-related infections, which are in need of advanced solutions in surface engineering. This is a review of laser-based antibacterial and antifouling coatings that are pulsed laser deposition, laser surface texturing, LIFT, LAL, and MAPLE. These methods allow accurate nanoscale control over the surface chemistry and topography, providing an opportunity for contact killing, ROS generation, and controlled ion release. Surface orthopedic, dental, cardiovascular, and catheter implants Laser-fabricated surfaces have proven to be associated with greater than 99.9 percent bacterial reduction and also aid in biocompatibility, but scalability and clinical validation remain problematic.

1.0 INTRODUCTION

The high rate of development of implantable medical technologies has enhanced the survival and quality of life of the patients considerably. Orthopedic prostheses regain movement, dental implants regain mastication and appearance, cardiovascular stents keep the veins open, urinary catheters allow the use of a critical care unit, and ophthalmic implants recover sight. Nevertheless, they are clinically successful but create unnatural surfaces in biological conditions that are naturally prone to

DOI: 10.4018/979-8-3373-4104-0.ch006

microbial settlement. The problem of biomaterial-related infections still is one of the most tenacious and expensive issues in contemporary healthcare.

Figure 1. Healthcare Impact of Biomaterial Infections

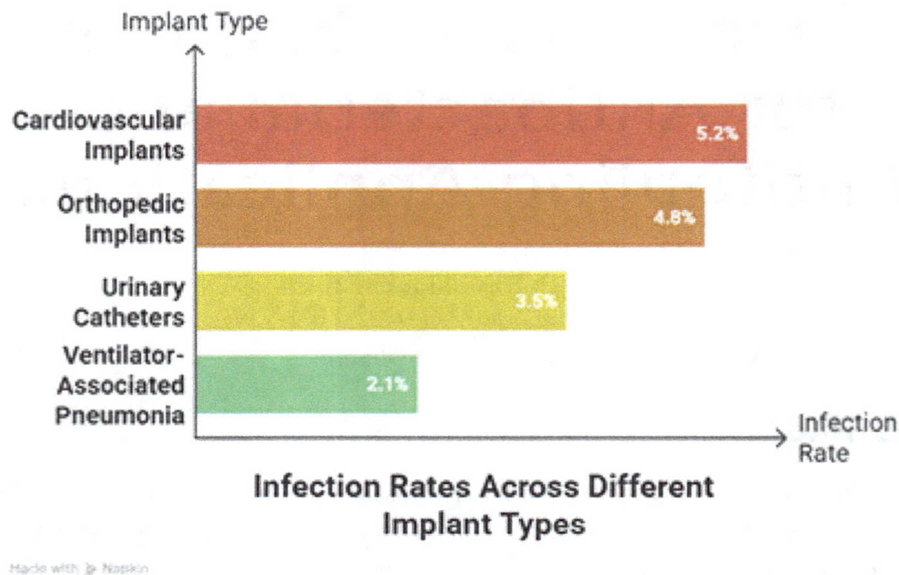

Healthcare-associated infections (HAIs) are considered by the World Health Organization as one of the key health burdens in the world. (World Health Organization, 2011) Among them, the share of device-related infections is quite high, at an almost 50-70 percent rate (Butler et al., 2023). These infections do not just occur as occasional complications; they also tend to cause long lasting inflammations, loosening of implants, systemic infection, and occasionally death. The economic cost is incredible in terms of finances, revision surgeries, prolonged hospital stay, and long-term antibiotic treatment significantly adds to spending on healthcare.

Bacterial adhesion and biofilm formation is at the centre of the problem. When bacteria settle on a biomaterial surface they form an extracellular polymeric structure (EPS) made of polysaccharides, proteins, lipids and extracellular DNA. The microenvironment formed by this matrix offers a lot of shield to protect bacteria against antibiotics and host immune cells. The bacteria in biofilm have the capacity to endure 10-1,000 times more antibiotic concentration that the planktonic ones. (Stewart, 2015; Høiby et al., 2010) As such, when biofilms are formed, the conventional systemic antibiotic approaches often fail.

Figure 2. Biofilm Formation Timeline

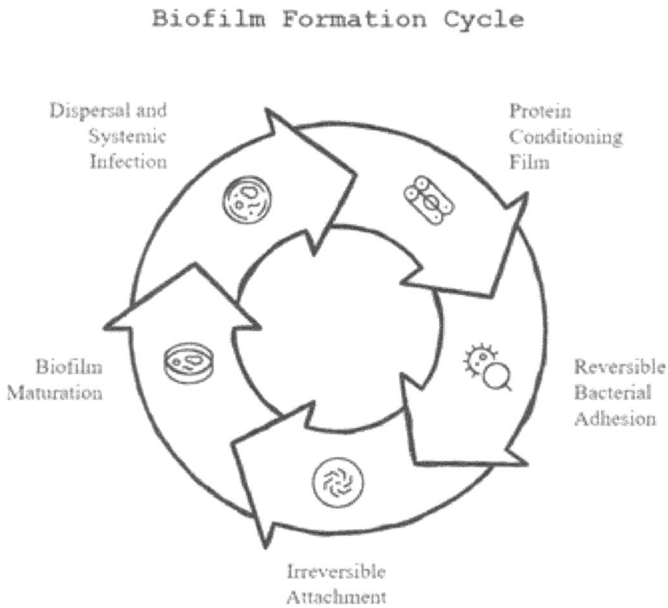

(Image generated by Naplin.ai, [21/02/26]).

That being the case, surface engineering has taken a new course in favour of proactive, as opposed to reactive, approaches. Rather than trying to eliminate existing infections, scientists are trying to fabricate biomaterial surfaces that are naturally resistant to bacterial colonization or to actively eliminate microorganisms when they come into contact with them. Here, laser based surface modification technologies have come out as mighty instruments, which can be used to modify the surface topography, chemistry, crystallinity and nanostructure with an astonishing level of precision.

Laser technologies provide some of the fundamental revolutionizing capabilities:

- Patterning on a nanoscale with no damage of bulk material.
- Transfer Stoichiometric transfer of complex antimicrobial materials.
- Accurate regulation of ion set free dynamics.
- Capacity to develop multiple purpose hierarchical structures.
- Biomolecule compatibility with temperatures.

These characteristics cause the laser-based methods to stand at a unique place at the border of the fields of materials science, microbiology, and clinical medicine.

1.1 Clinical Significance of Biomaterial-Associated Infections

The pathogenesis of the infection with implants is usually a clearly defined process. Following implantation, there is adsorption of host proteins, including fibrinogen, fibronectin, albumin and vitronectin on the material surface that occurs in seconds. (Horbett, 2018) This conditioning film is an adsorption layer through which the cells and bacteria interrelate.

Bacteria have surface adhesins, microbial surface components that identify adhesive matrix molecules (MSCRAMMs) that attach to host proteins. A case in point is Staphylococcus aureus, which colonizes metallic and polymeric implants rapidly by binding fibrinogen and fibronectin using special adhesins. (Hudson et al., 1999; Bhattacharya & Horswill, 2024)

The stages of infection relating to the device include:

- Early reversible adhesion of bacteria.
- Irreversible attachment
- Microcolony formation
- Biofilm maturation
- Dispersal and systemic contagion.

In orthopedic prostheses, the rates of infection are small (1–4%), but the outcomes are dire. (Pulido et al., 2008; Kapadia et al., 2016) Prosthetic joint infections usually necessitate two-step revision surgery; removal of implants, placement of antibiotic spacers, long term antibiotic treatment, and re-implantation. The process exposes patients to physical trauma, psychological distress, and huge economic cost.

1.2 Limitations of the Traditional Coating Processes.

Common antimicrobial coatings are based on many methods of fabrication such as:

Physical Vapor Deposition (PVD), Chemical Vapor Deposition (CVD), Sol–gel processing, Electrochemical deposition, Plasma spraying, Dip or spin coating are: shown in fig.3.

These approaches have been extensively used, but they have intrinsic shortcomings.

To begin with, the adhesion strength may not be adequate when under mechanical loading conditions. Such delamination is not only compromising the antibacterial performance of the product but can also cause particulate debris to be released to the surrounding tissues.

Second, traditional thermal technologies can destroy temperature lithic antimicrobial agents, peptides, or polymer.

Third, it is difficult to uniformly coated complex 3D geometries (porous titanium scaffolds).

Fourth, there are numerous coatings that rely on uncontrolled discharging of antimicrobial agents. Over release of silver ions such as that one may cause cytotoxicity of osteoblasts and fibroblasts. On the other hand, colonization may not be prevented with the release of insufficiency.

Figure 3. Comparison of Laser vs. Traditional Coating Methods

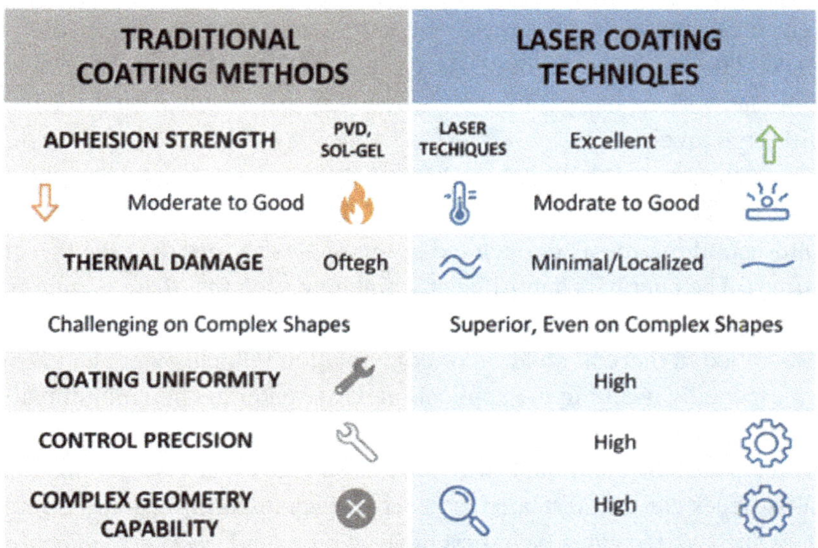

(Image generated by Gemini AI, [21/02/26]).

Lastly, antibiotic eluting surfaces are also a cause of concern regarding development of resistance. The exposure to sub-therapeutic antibiotics encourages the emergence of resistance strains, which can increase the world epidemics of antimicrobial resistance. (Høiby et al., 2010)

These deficiencies drive the development of laser-based procedures that can do precision engineering without necessarily using bulk chemical deposition.

A similar predicament to dental implants is peri-implantitis. Constant colonization by bacteria on the implant soft tissue interface results in inflammatory bone loss and ultimate implant failure. Likewise, urinary catheters are commonly encrusted and develop biofilms, causing catheter-associated urinary tract infections (CAUTIs), that is, one of the most prevalent hospital-acquired infections. (Pelling et al., 2019; Cortese et al., 2018)

Cardiovascular implants are endangered by dual threat, microbial colonization and thrombosis. Platelet activation that is induced by surfaces has the potential to induce synergistic effects in terms of risk of infection.

These clinical facts underscore the need of superficial interventions that inhibit the colonization of bacteria prior to biofilm formation.

1.3 Advantages of Laser-Based Processing

The benefits of Laser-based Processing have been listed as follows:

Laser processing provides a level of control over the energy deposition that is unmatched. Through the control of laser wavelength, pulse duration, fluence and repetition rate, surface morphology and chemistry can be tuned on the sub-micron level with high precision.

Laser At timescales shorter than thermal diffusion times ULTR a short pulse lasers (picosecond and femtosecond regimes) ULTR a short pulse lasers (picosecond and femtosecond regimes) are operated at timescales shorter than the thermal diffusion times. The energy is deposited at a high rate such that there is minimal heat diffusion into the surrounding material and this makes it possible to be referred to as, the so-called, a thermal ablation or cold ablation (Chichkov et al., 1996). This has been especially useful in the application of biomaterials like titanium alloys or polymers in which thermal degradation may destroy mechanical integrity.

Hierarchical micro- and nanostructures of surfaces that resemble natural bactericidal surfaces can be fabricated by laser surface texturing. Nature offers a few examples, such as cicada wings, dragonfly wings, and gecko skin incorporates topographies of nanoscale and mechanically disrupt bacterial membranes (Ivanova et al., 2012). These structures can be replicated on metallic implants using laser fabrication.

Pulsed laser deposition (PLD) provides stoichiometric transfer of multi-component materials of complexity. This is essential in the deposition of composite antibacterial films like Ag-doped TiO_2 or ZnO-hydroxyapatite.

Laser Ablation in Liquid (LAL): Ligand-Free nanoparticles are made using chemical reducing agents. The resultant nanoparticles have smooth surfaces and increased antimicrobial effect.

Taken together, these benefits make the laser-based technologies highly versatile next-generation antibacterial and antifouling surfaces.

2. FUNDAMENTALS OF LASER–MATERIAL INTERACTIONS

Laser–material interactions involve rapid conversion of electromagnetic energy into thermal, mechanical, and chemical effects. The outcome depends on both laser parameters and intrinsic material properties.

2.1 Laser Processing Mechanisms

Figure 4. Laser-Material Interaction Mechanisms

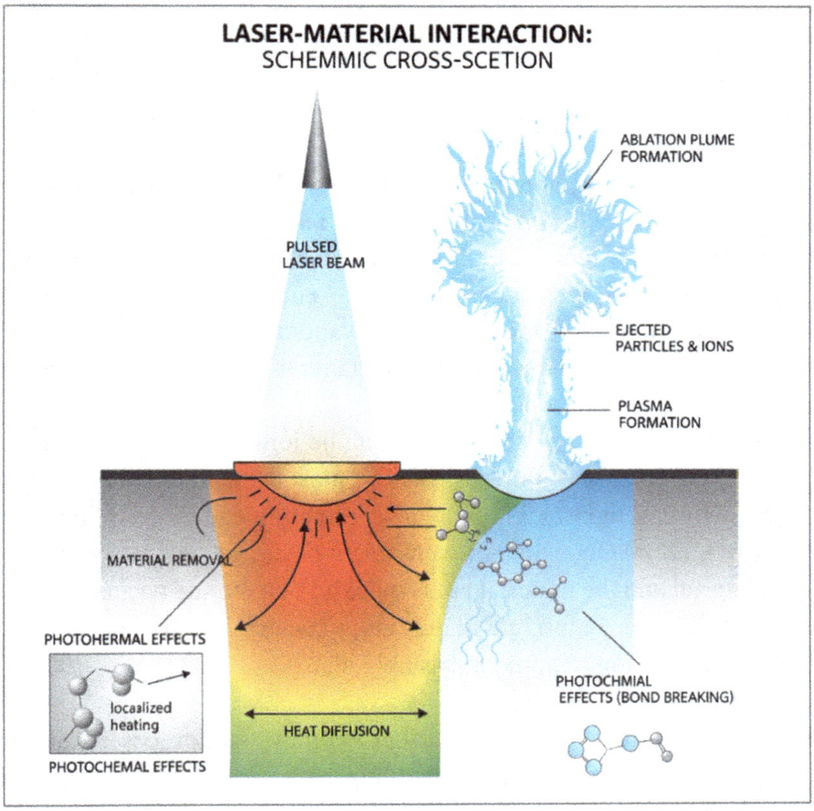

(Image generated by Gemini AI, [21/02/26]).

2.1.1 Photothermal Effects

Photothermal effects are the result of incident laser photons being absorbed by some material and converting their electromagnetic energy into thermal energy. The main conversion of this energy is by electron excitation and electron phonon interaction, in which excited electrons channel their energy to the lattice causing a sudden rise in temperature. Efficiency of this process is determined by the material properties, i.e. absorptivity, reflectivity, thermal conductivity and heat capacity, and the laser parameters i.e. wavelength, pulse duration, fluence and repetition rate.

At nanosecond pulse regime, the pulse duration is much longer than the electron relaxation time. (Phillips et al., 2015; Rethfeld et al., 2017) Due to this, thermal diffusion starts in the course of the pulse. The energy deposited is then propagated into the material surrounding by conduction of heat to form a relatively large heat-affected zone (HAZ). This is in contrast to nanosecond laser processing (compared to ultrafast (femtosecond or picosecond) regimes) where energy deposition is very local and the diffusion of heat during the pulse is negligible.

The common mechanisms of photothermal processes involve a series of thermally activated phase changes and material reactions and these include:

- Melting: Local melt pools are created when the temperature of the surface is above the melting point. The dynamics and morphology of melt pool are affected by surface tension, Marangoni convection and recoil pressure.
- Vaporization: With increased densities of energy, the temperature can rise above the boiling point leading to the evaporation of material. This creates recoil pressure which have the ability to either improve material removal or alter geometry of a melt pool.
- Plasma Formation: The vapor plume can be ionized by intense vaporization to form a plasma above the surface. The plasma has the ability to capture or scatter incoming light radiation by laser affecting energy coupling efficiency.
- Re-solidification: Once the pulse has taken place, the heat is conducted to the bulk material and the material cools quickly. The molten part is solidified and frequently at a very high rate of cooling, which may refine microstructure or cause the formation of some metastable phases.

Photothermal in processes Photothermal Photothermal processing has numerous applications in alloying lasers, cladding, welding, additive manufacturing (e.g. WAAM post-processing), coating deposition and surface hardening. Directed thermal gradients have the capacity to facilitate desirable microstructural changes, such as refinement of the grain, homogenization and enhanced inter-layer bonding.

Nevertheless, over photothermal input can cause unwanted effects as:

- Grains coarsen with extended time in the presence of heat.
- Development of residual tensile stress at sharp thermal gradients.
- Due to thermal shock, microcracking occurs.
- On changes in elevated temperatures during oxidation or composition.

- Deformation or distortion in thin sections

It is thus necessary to optimize laser parameters carefully to ensure that adequate energy is provided to a material to effect the desired changes but also to reduce any undesirable impacts of thermal energy. Numical simulations and analytical heat transfer models are frequently utilized in order to estimate the temperature field, cooling rates, and the size of HAZ to control the process.

2.1.3 Ablation Dynamics

Laser ablation commences when the energy contained in a laser pulse, usually in terms of fluence, is above a critical value depending on the material. At temperatures below this limit, the material can heat, melt or slightly change the surface. Material removal on the other hand takes place in a very dynamic and rapid fashion once the threshold has been exceeded.

Pulse duration is a very important characteristic of ablation. In regimes of very high speed, especially when pulses are of the order of the femtosecond scale, the time over which the laser and the material interact is shorter than the characteristic time of the material in which heat diffusion takes a significant part. This causes the energy to be stored in the electronic system near instantly. The lattice lacks enough time to thermally respond to the pulse itself, which essentially changes the removal process, as compared to nanosecond processing.

At such rapid conditions, the influence of nonlinear optical processes prevails. The multiphoton ionization process is the process in which two or more photons are taken in simultaneously (Harilal et al., 2014), which causes the electrons to be promoted to the conduction band when the photon energy alone would not be sufficient. This is then followed by avalanche ionization, in which the free electrons acquire energy due to the laser field and hit the bound electrons to produce more charge carriers. This is because the accumulated number of free electrons in a matter of no time generates very high electron densities.

In cases where the repulsive effect between positively charged ions is high enough, it may overcome forces of atomic bonding. This effect is commonly known as Coulomb explosion and causes direct ejection of the material without much thermal melting. (Costache & Reif, 2004) Femtosecond ablation is therefore commonly referred to as cold ablation as it reduces areas of heat influence and thermal peripheral damage.

In picoseconds following the pulse, a dense plasma is formed on the surface. This plasma is composed of electrons, ions, neutral atoms and small clusters. The fast growth of this plasma forms a high pressure plume which extends outward of the target surface. The dynamics of expansion are determined by issues like ambient pressure and surrounding medium (vacuum, gas or liquid) and laser energy density.

The actions of this growing plume are very vital in defining the nature of the released species. The plume composition, velocity distribution and angular spread in processes like Pulsed Laser Deposition (PLD) have a direct relationship with the film thickness, uniformity and stoichiometry. With clever expansion of plumes, it is possible to grow thin-film in a controlled manner with high compositional fidelity.

This is further complicated in Laser Ablation in Liquids (LAL). The plume is trapped within the surrounding liquid resulting in a rapid cooling and very high quenching rates. This restriction facilitates the formation of nanoparticles in the liquid medium. Due to the rapid cooling, the particle will be kept small and relatively homogeneous. The nanoparticles formed have a high surface area, a narrow size distribution, and an increased reactivity.

The quick inactivating property of the LAL inhibits the large-scale growth and aggregation of particles and allows stable colloidal suspensions to be built. It makes LAL especially interesting regarding synthesizing metal and metal oxide and composite nanoparticles without chemical precursors and stabilizers.

In general, the ablation processes are controlled by complicated interaction of the laser parameters (pulse duration, wavelength, fluence), material properties (band gap, thermal conductivity, bonding strength), and the environmental factors. The comprehensive knowledge of these interconnected processes is necessary to manage the material erosion rates, reduce the flaws, and shape the features of nanoparticles to suit the state-of-the-art uses.

2.2 Key Laser Parameters

The material does not dictate the result of any laser–material interaction. It is highly determined by the nature of the beam of laser itself. The most important of the various controllable variables are wavelength, pulse duration, fluence, and repeat rate. These parameters define the absorption of energy, development of energy propagation in the material and the eventual responsiveness of the material.

2.2.1 Wavelength

The wavelength has a central role in the determination of interaction of laser energy with matter. The possibility of materials interacting with light in different ways over the electromagnetic spectrum means that the penetration depth, energy coupling efficiency, and the prevailing interaction mechanism are directly dependent on the wavelength of choice.

Photons have fairly high energy in the ultraviolet (UV) region. A large number of polymers and organic materials are highly absorbing at the UV radiation and hence have low penetration depths- in most cases deep within the surface layers.

This renders UV lasers specifically appropriate in the surface functionalization, fine patterning, and photochemical modification of surfaces where little bulk heating is required. Since the energy is concentrated close to the surface, it is possible to come up with precise changes in chemicals without modifying the structure beneath it.

By contrast, near-infrared (NIR) lasers tend to penetrate deeper into metals and ceramics. Nonlinear absorption mechanisms however, can take place when the ultrafast femtosecond pulses apply to the NIR range. Multiphoton absorptions can occur in even materials which are reflective or weakly absorbent at these wavelengths. This allows efficient processing of high precision metals, semiconductors and ceramics.

Therefore it is not just the question of energy delivery that defines the wavelength selection, but it is really what type of interaction can occur, be it photochemical, photothermal, nonlinear in nature.

2.2.2 Pulse Duration

Figure 5. Pulse Duration Effects Comparison

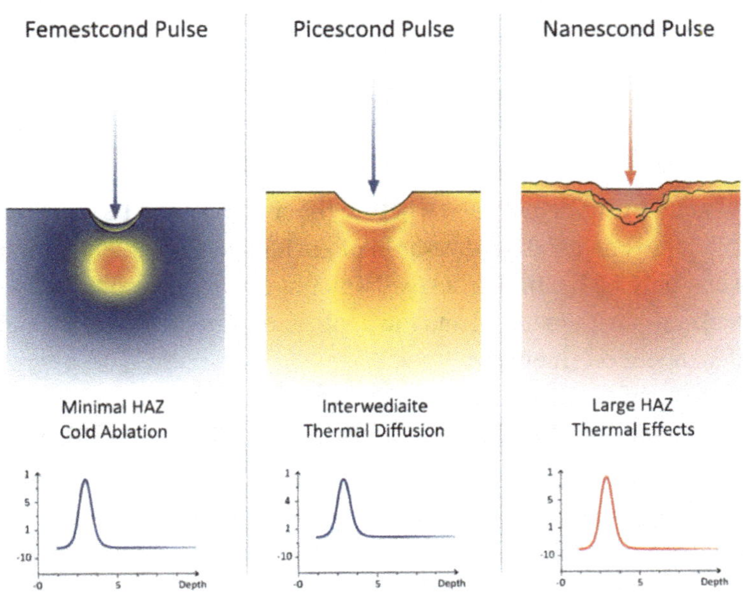

(Image generated by Gemini AI, [21/02/26]).

The pulse duration defines the duration of the material being exposed to the laser energy and has a significant effect on the behaviour of heat diffusion. It is simply a definition of whether the material will have time to feel the heat before the pulse goes off as shown in fig.5.

With femtosecond pulses (approaching 10 -5 seconds), the energy deposits rapidly and thermal diffusion over the pulse is, in fact, insignificant. This is because the lattice does not have adequate time to react thermally leading to a highly localized energy confinement. This results in a small number of heat-affected zones (HAZ), a smaller number of microstructural damage, and very accurate material removal. This is commonly called cold processing because of such ultrafast processing.

An intermediate regime is the picosecond pulses. A certain amount of thermal diffusion can commence during the pulse but it is limited. Balance between precision and processing efficiency is common with this regime.

However, nanosecond pulses enable large amounts of heat to be conducted through the pulse. Consequently, material adjacent to it is thermal loaded, resulting in larger melt and more intense heat-impacted area. Although this may be beneficial when it comes to welding or surface alloying, it can minimize accuracy when it comes to microfabrication activities.

Hence, the duration of the pulse basically limits the trade off between accuracy and thermal effects.

2.2.3 Fluence and Repetition Rate

Fluence is a measure of the amount of energy applied to the material on a given unit area which in turn determines whether the material will be modified, melted or ablated. There is an ablation threshold of each material. At temperatures below this, there can only be heating or a small amount of surface modification. Beyond this, removal starts with material and the depth and morphology of the ablated area is highly dependent on fluence.

Increased fluence tends to enhance the ablation depth per pulse but can additionally cause a plasma shielding effect, surface roughening, or other undesirable structural damage unless controlled. It is therefore necessary to maximize fluence in order to obtain regular morphology and dimensional precision.

Repetition rate, in its turn, determines the frequency of providing pulses. At low rates of repetition, a pulse is in contact with a comparatively cold surface. At increasing rates of repetition residual heat in the place of past pulses may build up resulting in cumulative heating. This effect can make processing more efficient but can also make the area affected by heat larger in case the effect is not addressed properly.

In terms of industry, repetition rate is directly associated with throughput and scalability. Repeat systems are high and they are faster and more productive. Nonetheless, structural accuracy with faster processing speed is only possible when pulse energy, scanning plan, and cooling are optimized.

Practically, the successful industrial laser processing technique is to balance the fluence and repetition rate of the laser so as to gain high productivity along with controlled microstructural results.

Figure 6. Key Laser Parameters and Their Effects

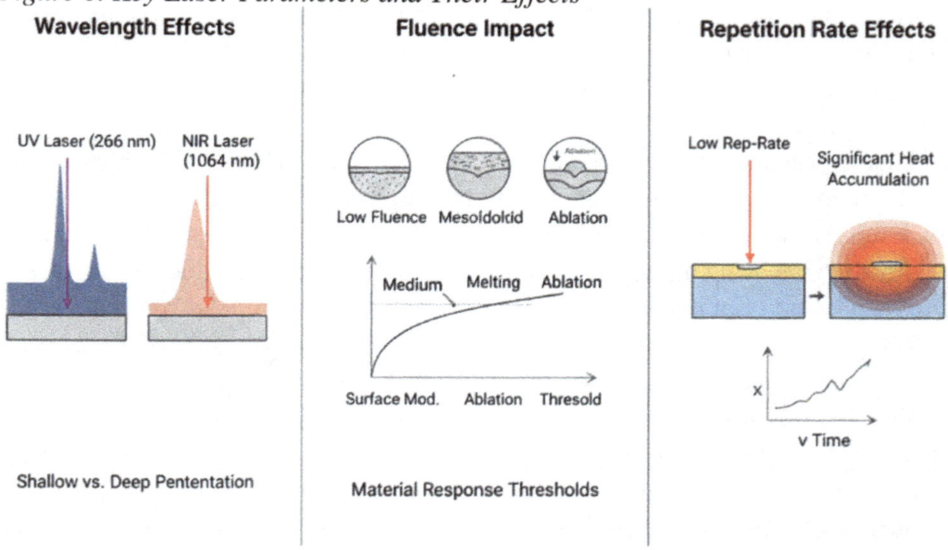

(Image generated by Gemini AI, [21/02/26]).

2.3 Response of the Material to Consider

The way a material behaves when exposed to a laser is not universal; it highly depends on the physical, thermal and optical properties of the material. Reflectivity, absorptivity, thermal conductivity, melting point, electron structure, and so on, are parameters that determine the absorption and redistribution of energy. Thus, the morphologies of surfaces and microstructural changes that various materials possess can be drastically different even with the same laser settings.

Titanium alloys are also the most popular in orthopedic and dental implants and this group of alloys is particularly appealing to laser surface modification. They have moderate thermal conductivity and excellent absorption properties, which ensure controlled energy deposition. Titanium surfaces under the right conditions of the laser may form micro and nano-scale textures that promote osseointegration and cellular-adhesion. Nanostructuring of surfaces can also be enhanced by laser-induced nanostructuring, which is also significant in biological response. Titanium alloys would be the preferred choice in biomedical precision laser operations due to these benefits.

Medical devices that use stainless steel and cobalt chromium alloys are different since they are more reflective and have different thermal characteristics. Such materials can have increased fluence or reduced pulse durations in order to have efficient energy coupling. However, too much thermal input can cause unwanted microstructural changes, e.g. grain growth or the formation of residual stress. As a result, the optimization of parameters is required to trade set surface modification and structural integrity.

There is yet another challenge presented by ceramics. They are brittle in nature and their thermal shock resistance is relatively low which means that must not crack unless the energy deposition is carefully controlled. Ceramic processing is frequently done using ultrafast laser pulses since they reduce the thermal diffusion and the possibility of cracking.

The polymers on the other hand tend to have a lower melting temperature and varying absorption characteristics at varying wavelengths. Polymer surface modification is especially achieved using UV lasers because of high surface absorption, and low bulk heating. Nevertheless, it would lead to degradation or embrittlement on excessive exposure, and thus fine control of fluence and exposure time is important.

Surface oxidation is another element of material response when processing with laser. High temperature of the surface in reactive environments may encourage the development of oxide layers. In case of titanium alloys, the formation of titanium dioxide (TiO_2) is of particular importance. The controlled oxidation has the ability to provide better corrosion resistance, enhance surface wettability and even provide antibacterial performance. Some nanostructured TiO_2 surfaces have photocatalytic properties that can also prevent bacterial adhesion in suitable circumstances.

3. Laser Techniques for Antibacterial Coating Fabrication

Laser technologies encompass a diverse set of processing approaches that differ in physical principles, material transfer mechanisms, scalability, and biomedical suitability. Selection of the appropriate technique depends on desired functionality—whether topographical bactericidal activity, controlled antimicrobial release, photocatalysis, or antifouling chemistry.

3.1 Pulsed Laser Deposition (PLD)

Figure 7. Pulsed Laser Deposition (PLD) Setup and Mechanism.

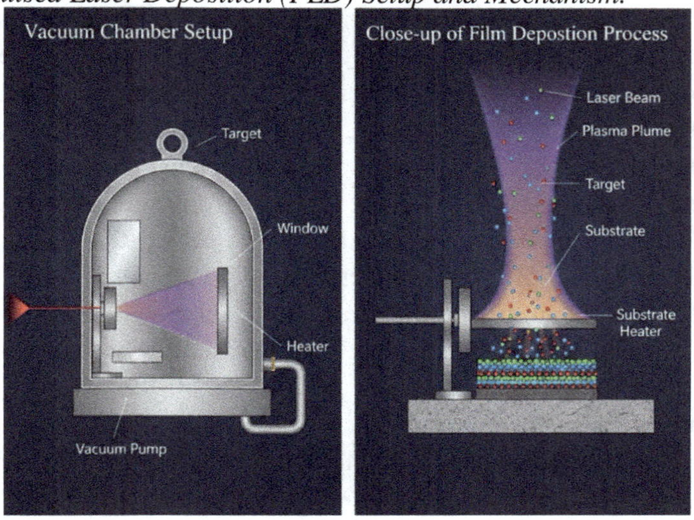

(Image generated by Gemini AI, [21/02/26]).

Pulsed Laser Deposition (PLD) is generally considered to be one of the most versatile laser-based thin film production methods. It is attractive in that it is able to fabricate complex material compositions in a solid target on a substrate with exceptional fidelity. During a standard PLD process, a high-energy pulsed laser beam is directed to a solid target in the form of a dense target located in a vacuum. Above ablation threshold, a beam of laser with high fluence removes a small portion of material on the target surface with a high rate.

This contact creates a very energetic plasma plume that consists of ions, neutral atoms, molecular fragments, aggregates and even nanoparticles. The plume is spread forward in a direction that is away to the target and towards a substrate that is at a controlled distance. These ejected species condense on the substrate surface under strictly controlled vacuum or background gas conditions, creating a film of the species on the surface in layers.

The versatile behavior of plume-growth and the ability to control certain variables, including pressure of the gas and the temperature of the substrate, enables PLD to adjust film density, crystallinity and microstructure with a high degree of accuracy.

3.1.1 Mechanistic Advantages

The achievement of near-stoichiometric transfer is the capability of PLD to transfer electrons. In most standard thin film methods including thermal evaporation or sputtering, the elements with varying vapor pressures or sputtering yields can be deposited at varying rates. This may cause compositional drift especially in multi-element systems or doped systems.

However, PLD is more likely to maintain the chemical properties of the initial target material in the deposited film. The ablation process transfers even complex ceramics and multi-component compounds with low segregation since the material is removed in a very energetic and collective fashion. The latter is critical when it is required to fabricate antibacterial coatings based on specific elemental ratios.

To mention a few, PLD is better suited at manufacturing:

- Coatings of hydroxyapatite dope with silver.
- Silver-titanium dioxide composite films.
- Zinc oxide-calcium phosphate (ZnO-CaP) systems.
- Layered multi-element ceramic composites.

In this type of systems, the dopant concentration can greatly influence antibacterial performance and cytocompatibility with only slight changes in dopant concentration. The fact that PLD has a compositional fidelity is therefore a significant technological advantage.

A variety of parameters are linked to film characteristics in PLD. The quantity and kinetic energy of ablated species depends on the laser fluence. The distance between the target and the substrate influences the density of the plume to the deposition surface. Background gas pressure also affects the collision frequency in the plume and can change the particle size and the film porosity. Nucleation and crystallization behavior is determined by the substrate temperature. These parameters can be optimally controlled to utilize morphology, grain structure, and dispersion of the nanoparticles into the film.

With PLD, antibacterial coatings are created by irradiating the substrate with lasers.

PLD has been shown to have a lot of potential in the process of making antibacterial surfaces on biomedical implants. Ceramic films especially silver-doped ones have demonstrated great antibacterial properties against the most common pathogens like Staphylococcus aureus and Escherichia coli. The bacterial reduction rate has been reported to be more than 99.9 percent, and at the same time, it preserves the osteoblast viability (Jelinek et al., 2013; Ristoscu & Mihailescu, 2014), which is a crucial requirement in orthopedic usage.

One of the main benefits of PLD in those systems is nanoscale dispersion of silver in stable ceramic matrix. Silver can be made to be uniformly distributed at the nanometer scale rather than making large agglomerates. This regulated release allows continued release of the ion with time, giving a longer lasting antibacterial effect without any sudden increases in silver concentration that may lead to cytotoxic action.

In addition to single-layer coatings, it is possible to use PLD to produce multilayer and functional graded structures. To give an example, the common architecture of an implant coating against bacterial infections might be a combination of:

- A titanium dioxide (TiO_2) base layer to improve adhesion and resistance to corrosion.
- A buffer layer with silver nanoparticles as an antibacterial agent.
- It contains a porous top layer or diffusion controlled top layer to control the release of ions.

These multilayer designs allow adjustable antibacterial functionality on a long-term scale, i.e. over weeks, months, but at a time they retain mechanical stability and biocompatibility.

3.1.3 Limitations

PLD is not devoid of limitations, in spite of its accuracy and flexibility. Scalability is one of the major challenges. In the traditional PLD systems, the area of deposition is rather narrow because of the geometry of the plume and the size of the chambers. Also, there is the necessity to install vacuum systems which heightens the cost and complexity of equipment.

High-throughput production is needed in large-scale production of orthopedic implants. System designs may need sophisticated system designs, rotating substrate bearers, more than one laser beam or hybrid processing methods that involve the combination of PLD with other surface engineering techniques to scale PLD to industrial level. However, PLD is still an effective and very dependable method to the use of research-scale fabrication and on-demand biomedical coatings that demand compositional accuracy and control of the structure.

3.2 Laser Surface Texturing (LST)

Laser Surface Texturing (LST) is a completely different method of doing things than the coating-based methods. LST does not include an additional layer of material, but rather changes the surface of the material being used. With a focused beam of laser, one can cut through the surface to form the micro and nano-scaled features by the careful scan of the beam over the material.

Figure 8. Laser Surface Texturing (LST) - Hierarchical Structures

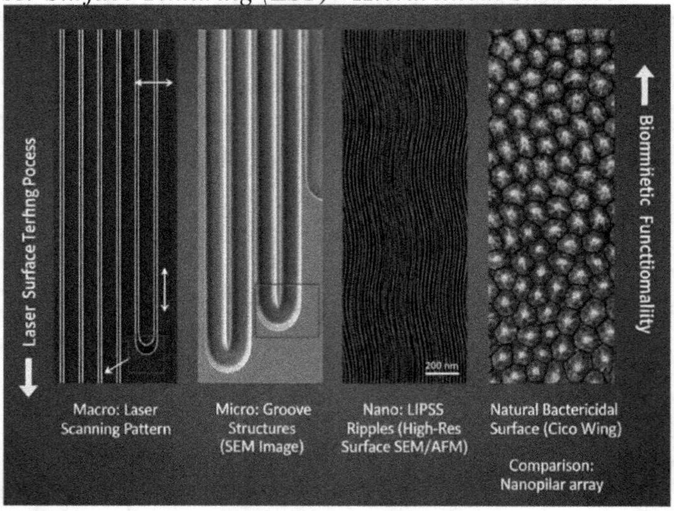

(Image generated by Gemini AI, [21/02/26]).

This method has been especially appealing to biomedical implants since it precludes issues associated with coating delamination, compositional instability, or imbalance ion release with time. The surface can be described as being chemically comparable to the underlying material, and its topography is converted to the form that highly affects biological response.

The resulting structures are dependent on the laser parameters including the pulse duration, wavelength, scanning speed and overlap. Optimally, LST is capable of producing very ordered structures of micrometer-scale grooves to nanoscale ripples and pillars.

3.2.1. Femtosecond Laser-Induced Periodic Surface Structures (LIPSS)

The Laser-Induced Periodic Surface Structures (LIPSS) are considered to be one of the most intriguing results of ultrafast laser processing. When a laser pulse working on a femtosecond scale is applied to the surface of a material a portion of the incident light will interact with the surface-scattered waves. The resulting pattern of periodic distribution of the energy is an interference which is imprinted onto the material in the course of ablation or surface restructuring. The outcome is a succession of closely spaced ripples usually in 200-800 nm. These nanotopography patterns are exceptionally homogenous and self-assembled, which develops without masks or lithography templates. Interestingly enough these nanostructures are very similar to naturally found bactericidal surfaces in biological systems. Wings of cicada and dragonflies, e.g., have arrangements of nanoscale spikes that physically introduce bacteria membranes. Nanostructures induced by lasers are analogous to this protective effect of nature and provide a bio-inspired route to antibacterial surfaces of implants.

3.2.2 Mechanical Bactericidal Mechanism

Figure 9. Mechanical Bactericidal Mechanism

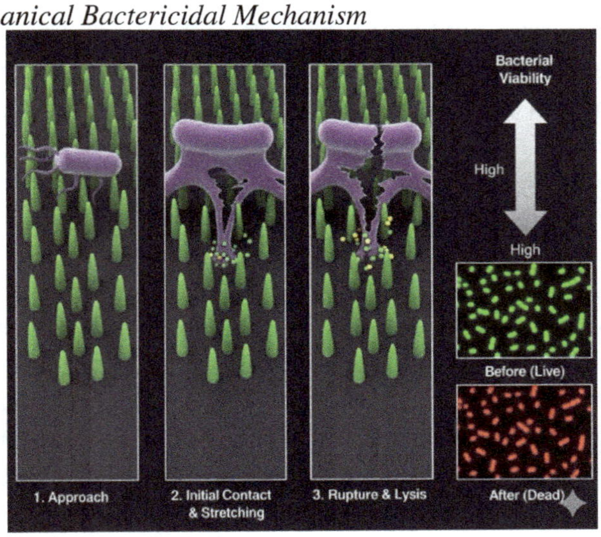

(Image generated by Gemini AI, [21/02/26]).

Unlike chemical deposition of antibacterial surfaces, which are based on the ion release principle, in many cases, nanostructured surfaces prepared using LST operate in a purely mechanical way. Bacteria growing on a nanopillar or sharp nanoscale feature-covered surface have their cell membranes pulled between features, as the bacteria settle down.

When the local mechanical stress is higher than that which the bacterial membrane can withstand tensile stress; which is usually on the order of 1050 Mpa, the membrane breaks. This mechanical interference causes the bacteria to be inactivated without the antibiotics or release of metallic ions.

This mechanism is usually more susceptible to gram-negative bacteria, as they have a thinner peptidoglycan. They are more prone to stretching and rupture due to their outer membrane structure. Gram-positive bacteria with thicker and more rigid cell walls may necessitate a higher aspect ratio or sharpening of the geometries of nanostructures to realize similar antibacterial effects.

On laser-textured surfaces it has been reported that large reductions in bacterial colonization have occurred and include: 85-95 percent decrease in bacterial adhesion. (Pan et al., 2019; Schwibbert et al., 2024)

The reduction of viable bacteria counts to 3 log levels within 24 hours. (Bonse et al., 2020)Notably, since the process is more mechanical than biochemical, chances of bacterial resistance occurrence are low.

3.2.3 Dual Functionality

The dual biological capability of Laser Surface Texturing is one of the strongest benefits of Laser Surface Texturing. Whereas nanoscale features may have a mechanically inhibitive effect to prevent bacterial adsorption, microscale roughness has a complementary role.

The micrometer scale textures augment the surface area and the adsorption of proteins which subsequently augments osteoblast adherence. Cell communication with surface features induces a focal adhesion signaling cascade that facilitates the spreading of cells, proliferation and differentiation. This has been translated into better bone-implant and better osseointegration in orthopedic implants.

LST provides surfaces with both anti-bacterial-scaffolding properties and anti-bacterial-culture properties through the well-designed hierarchical arrangement of micro-scaffold roughness and nanoscale scaffold bactericidal features, which discourages bacterial proliferation and promotes bone formation, respectively.

This trade-off between antibacterial shielding and improved tissue compatibility will make LST a very promising approach on the design of the next generation implants.

3.3 Laser-Induced Forward Transfer (LIFT)

Laser-Induced Forward Transfer (LIFT) is a technique that can be utilized to physically transfer atoms across surfaces using a laser beam. Laser-Induced Forward Transfer (LIFT) Laser-Induced Forward transfer (LIFT) is a process in which atoms can be transported across surfaces through the use of a laser beam.

Laser-Induced Forward Transfer (LIFT) is a laser-based additive manufacturing method that is highly precise and enables the deposition of materials without direct contact. LIFT is in contrast to surface texturing or ablation processes, which lose material in such a way, and LIFT transfers small amounts of material to a receiving material through a controlled method.

A common LIFT system comprises of a transparent substrate (commonly glass) to be deposited with a thin layer of the material to be transferred, which could be a nanoparticle ink, polymer solution, or bioactive compound. A laser pulse is tightly focused and passed through the clear substrate and absorbed at the interface between the substrate and coating layer. This accelerated local energy deposition creates a temporary hype of the sensors in a high pressure that causes the small drop or voxel of material to propel forward onto the substrate receiving it, and that substrate is positioned at the near end of the process.

Since the process is digitally controlled and is highly localized, it allows very fine spatial resolution. Tuning the laser energy, spot size and the thickness of the coating can modify the size of this transferred droplet. Consequently, LIFT has been thought to be one of the most promising methods of precision patterning in biomedical and microfabrication.

3.3.1 Biomedical Applications

Among the strongest benefits of LIFT in biomedical engineering is that it allows the deposition of materials in well-defined patterns and in spatially controlled patterns. Instead of fabricating a uniform coating of the whole surface of an implant, LIFT permits functionalization of certain areas selectively.

In the case of antibacterial uses, this leaves the prospect of creating patterned surfaces that position antimicrobial agents in areas of need only. Examples include:

- Silver nanoparticle microarray deposits in geometries.
- Localization of antimicrobial peptides printing.
- Repetitive polymer-drug composite layers of controlled release.

With this kind of spatial control, it is possible to create selective antibacterial zones. As an example, areas with higher exposure to infection risk can be functionalized with antimicrobial agents, whereas other areas can be left to optimize cell attachment and tissue integration. This selective design method helps to minimize exposure to antibacterial agents and could be used to minimize cytotoxic effects.

The high resolution is also another key advantage of LIFT. The sizes of features in the micrometer (and in some instances, even smaller) are possible. Such precision is most useful in combination with antibacterial patterns with microscale topographies, which facilitate adhesion of osteoblasts. LIFT in other words, facilitates functional patterning without interfering with the biological functionality of the surrounding regions.

In addition, LIFT is a non-contact method and, hence, it can be applied to delicate biomolecules that can include peptides, proteins, or even living cells when optimized. Nozzles are not used which makes them free of clogging problems which are typical of inkjet-based bioprinting systems.

Altogether, Laser-Induced Forward Transfer is an efficient method to create multifunctional surfaces of implants with a high level of spatial resolution of antibacterial and bioactive motifs. It is especially the choice of next-generation biomedical devices fabrication due to its accuracy, versatility, and ability to work with sensitive biomaterials.

3.4 Laser Ablation in Liquid (LAL)

Laser Ablation in Liquid (LAL) is a very simple but strong process of nanoparticles synthesis without using a traditional chemical pathway. Within this technique, a solid target, e.g. silver, titanium, or some other useful material, is located within a liquid, usually water or an organic solvent. An intense beam of a laser is focused on the target surface under water. At laser fluence values above the ablation threshold, ablated material is removed and quickly reacts with the surrounding liquid.

The medium of confined liquid is important in the process. A dense plasma plume is immediately formed on the target surface after ablation. Due to the liquid that surrounds the plume, it cools and compresses at an extremely high rate. This quick quenching facilitates the formation and development of nuclei of nanoparticles in the liquid. It is a colloidal suspension of nanoparticles that is produced without the requirement of chemical precursors and stabilizing agents.

3.4.1 Advantages

The chemical simplicity of LAL is one of the greatest benefits of the compound. LAL does not use reducing agents, surfactants, or stabilizing ligands, unlike conventional wet chemical approaches to the synthesis. This would remove fears of chemical remnants, undesired by-products, or synthesis reagent contamination.

This leads to clean and highly reactive surfaces of the nanoparticles due to the lack of surface bound ligand. In the case of antimicrobial use, it is especially useful. Chemically synthesized nanoparticles that have surface-bound contaminants may eliminate active sites and deactivate ion release. Ligand-free nanoparticles that are formed by LAL, in contrast, tend to be of increased surface activity.

There are major benefits of LAL such as:

- None of such chemical reducing agents or toxic precursors.
- Purities and low levels of contamination.
- Grow nanoparticles upon clean surfaces with increased reactivity.
- Synthesis pathway that is environmentally friendly.

Nanoparticles of silver that have been synthesized through LAL have been demonstrated to exhibit better antimicrobial properties as compared to those that have been synthesized chemically. (Kőrösi et al., 2016; Wu et al., 2024) These enhancements are associated mostly with the fact that there were no organic surface coatings that would hinder the release of the silver ions and direct contact with the bacterial membranes.

Moreover, since particle formation is observed in conditions of rapid quenching, LAL tends to form relatively small and uniform nanoparticles with a large surface area- another factor that leads to increased antibacterial strength.

3.4.2 Coating Integration

Other than synthesis of nanoparticles, LAL-produced colloids can be easily incorporated into coating systems. The flexibility of the technique is not just its ability to produce particles but also in the effectiveness of these particles to be integrated into useful surfaces.

LAL-manufactured nanoparticles can be:

- Placed within polymer substrates in order to form antimicrobial composite coating.
- The substrates of metallic materials are electrophoretically deposited.
- Fixed onto implant surfaces by laser annealing or heating.

Nanoparticles that are dispersed evenly throughout a coating can offer sustained antimicrobial properties and still allow mechanical flexibility in polymer-based coating. Electrophoretic deposition has the benefit of a controlled thickness and uniform distribution on conductive implant surfaces. Laser annealing may also be used to improve the adhesion and stability of the particles by partially embedding the nanoparticles into the substrate without changing the bulk properties in a significant manner.

It is through a combination of the LAL synthesis and advanced deposition methods that one can design coatings that will produce controlled ion release, good substrate bonding and long term antibacterial functionality.

In general, Laser Ablation in Liquid is a highly clean, scalable methodology, and biologically compatible, to synthesize high-purity nanoparticles that can be used in biomedical surface engineering.

3.5.Matrix-Assisted Pulsed Laser Evaporation (MAPLE)

To overcome a major shortcoming of alternative laser deposition methods, Matrix-Assisted Pulsed Laser Evaporation (MAPLE) was created which is the inability to work with fragile organic and biological materials without destroying them. Whilst these methods such as PLD are very useful in ceramics and metals, they may damage delicate biomolecules through direct high-energy interaction. MAPLE eliminates this issue by the introduction of an intermediate matrix which shields the active material during laser irradiation.

The target in the MAPLE process is not a solid. It is rather a frozen dilute solution of the desired biomolecule in a volatile solvent. The solution is cryogenically cooled, thus creating a solid target. The majority of the laser energy is absorbed by the solvent and not the biomolecule when the pulsed laser is irradiated on this frozen matrix. The dissolved molecules move quickly through the solvent as the solvent vaporizes, depositing as a thin film on the substrate.

Since the biomolecules are not in direct contact with the lasers with intense energy, a chemical structure and functional activity of the biomolecule is preserved to a great extent. This type of soft transfer is the strength of MAPLE.

3.3.1 Biomedical Applications

MAPLE has been especially useful in biomedical surface engineering where it is important to preserve the molecular integrity. A number of bioactive compounds become inactive when they are exposed to high temperatures or energetic attacks. MAPLE allows deposition of these types of materials in thin and homogenous layers while preserving their biological functionality.

It is used in: Antimicrobial peptides: These bioactive molecules have the ability to be immobilized to the implant surfaces so as to prevent colonization of the bacteria without the use of metallic ions.

Hydrophilic antifouling polymers Polymers that resist biofilm development and protein adsorption may be deposited in controlled amounts.

Enzyme coating: The enzymes can be deposited without loss of catalytic activity and the coating allows responsive or self-cleaning surfaces.

The capacity of MAPLE to create homogeneous coatings of delicate materials which otherwise disintegrate with standard deposition methods is one of its significant benefits. The laser fluence, type of solvent, concentration of solution and substrate temperature are process parameters which can be fine-tuned to achieve optimal film morphology and thickness.

Since MAPLE maintains molecular activity, it would be particularly applicable in the formation of bioactive antifouling surfaces where biological performance is of high importance over time. The method allows preparing coatings that are antimicrobially active, protein resistant and biocompatible in one engineered surface.

MAPLE, though needing vacuum systems and preparation of cryogenic targets, has the capacity to deposit delicate biomaterials without inactivating them and thus it is an extremely valuable method of preparing the advanced implant coating and biomedical devices.

3.6 Laser Ablation in Liquid (LAL)

Laser Ablation in Liquid (LAL) is a relatively clean and general method of synthesizing nanoparticles directly in liquid. Under this procedure, a dense target, which could be silver or titanium or zinc or any other useful material, is put into a container that has a choice of a suitable liquid, usually deionized water or an organic solvent. A laser beam is focused and then focused on the target surface that is underwater. At a laser energy level that is high compared to the ablation threshold of the material, the atoms and clusters are pushed out of the surface.

It is these high energy species that are immediately introduced into the liquid environment. The expanding plasma plume is contained by the liquid and leads to extremely rapid cooling which can also be termed as quenching. This rapid cooling fosters the nucleation and growth of nanoparticles in the nanosecond. Consequently, a stable suspension of nanoparticles in the liquid is obtained, eliminating the use of complicated chemical synthesis processes. This physical process is simple and this is its greatest strength.

3.6.1 Advantages

Among the most important advantages of LAL is that it is not dependent on any chemical reducing agents or stabilizing ligands. Traditional wet chemical preparation of nanoparticles usually needs precursors, reducing agents, surfactants and capping molecules. Although the chemicals are to be used in regulating the growth of these particles, they can still be left on the surface of the nanoparticle to affect its performance biologically.

Contrarily, LAL synthesizes nanoparticles in a relatively cleaner fashion. The resultant particles usually do not have any surface-bound pollutants as no chemical additives are used. This leads to:

- Dehydrocoupling of toxic reducing agents.
- Crystal-clean and extremely reactive nanoparticle surfaces.
- High chemical purity
- Ecological synthesis.

In the case of antimicrobial applications, the importance of the surface cleanliness is especially relevant. Silver nanoparticles prepared through LAL usually exhibit high antimicrobial efficacy in comparison with chemically prepared analogs. Lack of organic surface coating provides a possibility of more effective interaction of silver and bacterial cell membranes, and more controlled and effective release of ions. This improves antimicrobial efficacy as well as minimizes biological variability.

Also, the high quenching conditions in LAL tend to result in relatively small nanoparticles with high surface area which is an added advantage in terms of enhanced antimicrobial efficacy.

3.6.2 Coating Integration

In addition to the synthesis of nanoparticles, LAL offers versatility in incorporating these particles into working coats to be used in biomedical purposes.

The colloidal nanoparticles which were synthesized using LAL can be:

- Used in polymer matrices to create antimicrobial composite coatings.
- Electrophoretically deposited on conductive substrates.

Precisely attached directly to implant surfaces by laser annealing or controlled thermal treatment.

Nano-particles can also be used when incorporated into polymer coatings as a sustained antimicrobial agent without compromising mechanical flexibility and surface stability. Electrophoretic deposition helps in coating metallic implants with uniform thickness and high adhesion ability. Laser annealing, however, can (in part) implant nanoparticles into the surface increasing mechanical stability but leaving bulk properties largely intact.

LAL-produced nanoparticles can be customized to provide long-term protection against bacteria, controlled ion release, and improved performance of implants with the use of these integration strategies.

In general, Laser Ablation in Liquid provides a biologically compatible, clean and scalable method to achieve high purity nanoparticles that are suitable in the development of sophisticated antimicrobial coating systems.

3.7 Matrix-Assisted Pulsed Laser Evaporation (MAPLE)

To address a severe problem with laser-based thin film deposition, Matrix-Assisted Pulsed Laser Evaporation (MAPLE) was invented: how to deposit fragile organic and biological films without destroying them. Conventional high-energy laser treatments are very successful with metals and ceramics, and delicate biomolecules, including peptides, enzymes, and functional polymers, readily suffer failure during direct laser irradiation. MAPLE solves this problem by coming up with a protective matrix that absorbs a majority of the laser energy.

In MAPLE, the target is not a solid material that is dense. Rather, it is made by dissolving the biomolecule of interest in a low-concentration volatile solvent. This solution is then frozen to become a solid target, which is usually held at cryogenic temperatures. In case the target is frozen and a pulsed laser is applied to it, most of the laser energy is absorbed by the solvent matrix and not the biomolecule. The solvent is then quickly vaporized, and biomolecules move to the substrate intact as they deposit in a thin film.

Due to the indirect nature of energy transfer and the soothing effect of the solvent matrix, the biomolecules suffer little thermal or photochemical damage. The distinguishing feature of MAPLE is this type of soft landing mechanism.

Biomedical and Antifouling Systems. MAPLE has found application especially in biomedical surface engineering where it is important to preserve molecular functionality. A lot of bioactive compounds are not effective when subjected to high temperatures or other harsh conditions of processing. MAPLE can be used to deposit these sensitive materials in the controlled thicknesses without compromising their biological activity.

Common uses are:

- Antimicrobial peptides: These may be attached to surfaces of implants to give specific antibacterial effect without involving release of metal ions.
- Antifouling Polymers hydrophilic-Hydrophilic polymers can be deposited in uniform layers that prevent biofilm and protein adsorption, the performance of implants in the long term.
- Coatings on enzymes: It is possible to transfer functional enzymes and still retain catalytic activity to allow responsive or bioactive surfaces.

The MAPLE is capable of forming homogeneous, thin layers with controlled morphology, an important characteristics of the MAPLE. Through regulation of laser fluence, solvent type, concentration of solution and substrate temperature, researchers are able to control film thickness, roughness, and cover.

The structural integrity of biomolecules is mostly retained, meaning that MAPLE is particularly applicable to the production of bioactive antifouling layers with a combination of antibacterial functionality and biocompatibility. This is why it is a desirable method of next-generation implant coatings in which the balance between infection control and tissue integration has to be very fine.

The fact that vacuum systems and cryogenic preparation are necessitated complicates things, but the fact that MAPLE can deposit delicate biomaterials without damaging its functionality is a useful instrument in high-tech biomedical production.

3.8 Femtosecond Laser Nanostructuring

Femtosecondary (fs) lasers have provided a new opportunity in precision surface engineering. Due to the very short pulse period, 10 -5 seconds, the energy is deposited into the material and substantial heat diffusion is not allowed to take place. This enables a well localized editing with limited thermal damage and femtosecond lasers are especially appropriate to biomedical surface design.

The level of control that is provided by the use of a fs laser processing is one of the most potent. With a smart choice of parameters, including the laser polarization, scanning speed, overlap of the pulse, and the fluence, one can obtain clear hierarchical surface structures. Such structures have the potential to integrate microscale roughness with nanoscale structures in one processing step allowing multifunctional behavior at the surface.

Instead of merely changing chemistry or coating the surface, in femtosecond nanostructuring, the surface is reformed itself, usually forming long-lasting and stable topographies that are independent of any external materials.

3.8.1 Superhydrophobic Surfaces

Low surface energy chemistry, in combination with hierarchical micro–nanostructures, can result in superhydrophobic surfaces. When this occurs, the water droplets sit on top of the structured surface with little solid-liquid contact thus resulting in angles of contact higher than 150 degrees.

This effect is commonly attributed to CassieBaxter wetting model, in which air bubbles within the textures of surfaces decrease the area of contact. Consequently, the water droplets readily roll off and dust particles and contaminants are washed away. Such a self-cleaning effect can be desirable especially in biomedical implants where fluid retention can be minimized to minimize the adhesion of bacteria and biofilms. Practically, microgrooves, microcones, nanopillars are fabricated through femtosecond laser texturing and this leads collectively to this water-repellent property. Long term performance of the superhydrophobic state can be achieved when combined with proper surface chemistry, either inherent or post-treated.

3.8.2 Multifunctionality

The greatest benefit of femtosecond laser nanostructuring is, perhaps, the ability to produce multifunctional surfaces in a single process without using a mask.

To begin with, nanoscale dimensions, including sharp protrusions or ripples, are capable of mechanically damaging bacterial membranes, which results in a bactericidal effect, but it is not dependent on the use of antibiotics or ion release. This eliminates the chance of antimicrobial resistance.

Secondly, antifouling behavior can be offered using superhydrophobicity to reduce bacterial adhesion, as well as, avoiding the deposition of biological fluids.

Third, microscale roughness is capable of facilitating cell attachment and differentiation concomitantly. An example is the osteoblasts which react favorably to the controlled microscale topography by using focal adhesion signaling pathways, which augments the osseointegration of orthopedic implants.

Femtosecond laser processing can offer a compromise between antibacterial activity and improved tissue compatibility by fine-tuning the ratio between the micro- and nanoscale features. This combined capability makes fs nanostructuring a rather promising solution to next generation implant surfaces, in which infection prevention and biological integration are to co-exist.

Both MAPLE and LIFT represent advanced laser-based "soft" deposition techniques that excel where traditional methods fail by protecting the structural integrity of delicate biological matter. Matrix-Assisted Laser Evaporation (MAPLE) achieves this by using a frozen solvent matrix to absorb laser energy and gently evaporate sensitive molecules like proteins or polymers into thin films, while Laser-Induced Forward Transfer (LIFT) utilizes high-precision laser pulses to "click-and-drag" tiny droplets of material—including living cells and hydrogels—directly onto a substrate. Together, these techniques enable the fabrication of sophisticated biomedical devices, ranging from functionalized biosensors to complex, cell-laden scaffolds for tissue engineering shown in Table 1.

Table 1. Comparison of Laser-Based Surface Modification Techniques for Biomedical Applications

Technique	*Primary Advantages*	*Coating Thickness / Resolution*	*Suitable Materials*	*Primary Biomedical Applications*
Pulsed Laser Deposition (PLD)	Stoichiometric transfer, high-density films, excellent adhesion.	nm to μm range; high precision.	Bioceramics (HAp), metals, metal oxides.	Orthopedic/dental implant coatings, bioactive thin films.
Laser Surface Texturing (LST)	Creates hierarchical structures (macro to nano); improves osseointegration.	Surface modification only (depth: nm to μm).	Metals (Titanium), polymers, ceramics.	Antibacterial surfaces, cell-guiding topography, wetting control.
Laser-Induced Forward Transfer (LIFT)	Non-contact, high resolution; capable of transferring sensitive biological matter.	Resolution down to 0 μm.	Hydrogels, living cells, biomolecules, metals.	Tissue engineering, bioprinting, biosensor fabrication.
Laser Ablation in Liquids (LAL)	High-purity nanoparticle synthesis; ligand-free surfaces.	Nanoparticle size: 10–100 nm.	Gold, Silver, Metal oxides, alloys.	Bioimaging probes, drug delivery carriers, antibacterial agents.
Matrix-Assisted Laser Evaporation (MAPLE)	Gentle deposition of fragile organic and biological molecules without degradation.	nm to μm range.	Polymers, proteins, enzymes, drug-loaded layers.	Drug delivery systems, protein arrays, functionalized biosensors.

4. ANTIBACTERIAL MECHANISMS

The operation of laser-fabricated antibacterial coatings takes place in more than one way, which can work either alone or through a complementary relationship. In contrast to traditional antibiotic coating, laser-engineered surfaces can be based on physical, chemical, or photo-reactive effects, which decrease the chances of microbial resistance. Knowing these mechanisms is critical in coming up with effective and biocompatible surfaces.

4.1 Contact-Killing Mechanisms

The contact-based antibacterial approaches kill bacteria directly at the interface with the material, and active agents are not required to be released constantly. The mechanism of altered vascular permeability resulting in vasodilation and increased permeability is clear with a disruption of physical membranes by injury or subsequent infection.

4.1.1 Disruption of Physical Membranes

Purely mechanical is one of the most interesting mechanisms. Nanostructured surfaces (e.g. nanopillars, nanocones, laser-induced ripples) produce localized stress concentrations during the attachment of bacteria cells to them. The bacterial membrane is strained between neighboring features as the bacterial membrane tries to adapt to the topography of the surface. When the mechanical stress induced is in excess of the elastic limit of the membrane, the membrane ruptures and cells die.

Geometry parameters are critical to the effectiveness of this process. According to modeling and experimental tests, it has been shown that nanopillar height (usually 200500 nm), spacing, density, and tip sharpness affect bactericidal performance. Excessively spaced structures can be colonized by bacteria between structures; excessively dense features can lead to lower effective stress concentration. Hence, the ability to minimize antibacterial activity is crucially dependent on the accurate control of nanoscale structure, which is possible with the help of laser processing.

4.1.2 Reactive Oxygen Species (ROS) Generation

The production of reactive oxygen species (ROS) is another significant antibacterial pathway. Photocatalytic nanomaterials like titanium dioxide (TiO_2) can be stimulated to generate ROS by exposure to ultraviolet or visible light. Upon photon absorption:

2 -) that destroy key components of the cell, such as lipid membranes, proteins and DNA. As Butler et al. (2023) state, oxidative stress overloads bacteria with defense, causing permanent cell damage. Laser processing has the potential to dramatically improve photocatalytic performance by: (Korshed et al., 2018; Zimbone et al., 2015)

- Increasing the effective surface area.
- The addition of oxygen vacancies which enhance the separation of charges.
- Favouring the growth or stabilization of anatase crystalline phase.

With these modifications, laser-treated TiO_2 surfaces are able to demonstrate more robust and prolonged antibacterial actions at such presence of light.

4.1.3 Metal Ion Release

Antibacterial systems based on metals especially those containing silver, are based on controlled ion release. Thiol (–SH) groups of bacterial enzymes react with silver ions (Ag +), disrupt electron transport chains, and DNA replication. Such multi-target interactions decrease the chances of development of resistance.

Nevertheless, the concentration and release rate of Ag + needs to be controlled attentively. Over release can be cytotoxic to mammalian cells and under release weaken antimicrobial activity. Fabrication can be carried out using laser methods which enable fine control over the sizes and distributions of nanoparticles and embedding depth to fine tune the release kinetics of antimicrobials to a balance between antimicrobial functionality and cystocompatibility.

4.2 Release-Based Mechanisms

Unlike contact-only systems, release-based coats offer long-term antimicrobial protection by releasing active agents progressively into the immediate environment.

Various material parameters affect the release kinetics and these include:

- Size and morphology of nanoparticles.
- Matrix porosity
- Crystallinity
- Surface area

These features of the structure are exceptionally controlled by laser methods. As an illustration, by adjusting surface porosity by laser texturing, diffusion pathways can be controlled, and nanoparticle synthesis techniques, including laser ablation, can produce clean, ligand-free nanoparticles with predictable release behaviour.

Such tunability is what makes laser-engineered coatings especially favorable in long-term implant applications.

4.3 Bactericidal and Bacteriostatic Effects

Antibacterial coating is usually categorized as bactericidal or bacteriostatic. Bactericidal surfaces obtain a viable count of bacteria (often a 3-log reduction) that is ≥99.9. Bacteriostatic surfaces prevent the growth and multiplication of bacteria but do not entirely destroy the cells.

In the case of medical implantable devices, bacteria-killing capabilities are usually sought. Biofilm formation may happen very quickly following implantation and in patchy position, biofilms will be very stubborn to antibiotics and host defense mechanisms. Consequently, effective and quick surface elimination of bacteria is essential.

Nanostructures processed with laser provide intrinsic bactericidal efficacy without the need to continuously release chemicals, an important solution in reduction of resistance and providing a long-lasting solution.

The antibacterial activity of the product was determined by testing its spectral characteristics in the presence of Ca. difficult bacteriophage (Staphylococcus aureus) and Ca. easy bacteriophage (Escherichia coli).

4.4 Spectrum of Antibacterial Activity

The spectral characteristics of the product were investigated using Ca. difficult bacteriophage (Staphylococcus aureus) and Ca. easy bacteriophage (Escherichia coli).Gram-negative and Gram-positive bacteria are not susceptible to laser-modified surfaces due to difference in their structure.

Gram-negative bacteria have thin peptidoglycan layers and outer membrane, which makes them more susceptible to mechanical rupture by nanoscale features, in general. Gram positive bacteria on the contrary have thicker walls of peptidoglycan that offers more mechanical resistance. Consequently, steeper features or greater aspect ratio nanostructures are usually needed to realize similar bactericidal effects on Gram-positives. Notably, numerous laser-modified surfaces possess broad-spectrum anti-bacterial activity. Their mechanism is predominantly physical or multi-target chemical in nature (e.g. ROS generation or ion interaction), therefore they are not dependent on conventional antibiotics. This lowers the selective pressure towards resistance formation, and laser-based antibacterial surfaces are now predicted to be promising long-term sources of infection-resistant biomaterials.

5. ANTIFOULING STRATEGIES

Antifouling solution is fundamentally different to antibacterial solution. Although both antibacterial and antifouling surfaces are intended to inhibit microorganisms, they do it in different ways: Antibacterial surfaces are formed to kill or inactivate the microorganisms, whereas antifouling surfaces are formed to ensure that the initial attachment is prevented. This difference is especially significant in the biomedical field in which biofilm formation can be inhibited before it can initiate by preventing the adsorption of proteins and early bacterial adhesion.

Instead of using toxicity or chemical means of antifouling in solitude, antifouling systems tend to regulate surface chemistry, wettability, and topography to establish energetically unfavourable conditions in which microbial settlement is prevented.

5.1 Passive Antifouling

Passive antifouling methods do not actively release antimicrobial agents. They instead alter the surface properties to deter adhesion at the initial stages.

5.1.1 Hydration Layer Formation

The hydrophilic surfaces have the ability to bind the water molecules strongly creating a stable and well-arranged hydration layer at the interface. It is a kind of physical and energetic barrier that this layer of water provides so that the direct interaction of proteins and bacteria with the substrate does not occur.

When a bacterium is near such a surface, the layer of structured water has to be pushed aside, which costs a lot of energy. Consequently, protein adsorption and eventual adherence to microbes is significantly lowered. Hydrophilic polymers and zwitterionic coatings are usually employed to do this.

Laser processing can be used to improve the stability of hydration layers by making the surfaces rougher at the nanoscale and therefore improving the ability to hold water and to distribute energy at the interface.

5.1.2 Low Surface Energy

The other passive approach is the lowering of surface energy so that there is less effective contact between microorganisms and the substrate. Superhydrophobic surfaces have more than 150° contact angle with water (Wang et al., 2022) and entrap the air in their micro-nanostructure.

Since the bacteria are suspended in aqueous medium, air pockets reduce the solid-liquid contact area actually. This decreases the adhesion force and enables the contaminants to be washed away with the fluid flow or even low mechanical force.

Hierarchical structuring Laser-induced hierarchical structuring offers an effective alternative in the development of such superhydrophobic surfaces without using complicated chemical surfaces.

5.1.3 Steric Repulsion

A new form of barrier- steric hindrance is presented by polymer brush coatings. In this method, long polymer chains are densely grafted on a surface in a flexible nature. As bacteria approach, such chains squeeze together producing entropic repulsive forces, which keep them apart.

It is not in this mechanism founded on toxicity but rather on using thermodynamic principles. The polymer chains are compressed which makes them natural resistant to deformation and forms a repulsive zone that prevents the settlement of microbes.

Such antifouling layers can be grafted with increased density and controlled spatial density using laser-assisted deposition methods (including precision polymer patterning or controlled surface activation).

5.2 Active Antifouling

Active antifouling technologies are antimicrobial antidepressant anti-adhesive measures. These systems do not just prevent attachment but have an additional defense mechanism.

As one such example, a surface can have micro-nanostructures that decrease the adhesion, and, at the same time, antimicrobial nanoparticles that discharge bactericidal ions. Even in these hybrid designs whereby some of the bacteria get to attach, they are neutralized quite fast.

The laser fabrication techniques are especially useful in the production of these multifunctional systems since it enables the concurrent regulation of topography, distributed nanoparticles, and surface chemistry on a single processing platform.

5.3 Biomimetic Surface Designs

There are many examples of surfaces that are not contaminable, and are not susceptible to microbial colonization, provided by nature. The field of biomimetic engineering aims to imitate such strategies of nature, with the help of high-tech fabrication techniques.

It is the hierarchical micro nanostructure of the lotus leaf (in conjunction with low surface energy chemistry) which can be attributed to the exceptional self-cleaning capabilities (Dai et al., 2013) of the lotus leaf.

Shark skin has riblets which minimizes drag and deters biofouling in the sea. (Pu et al., 2016; Bixler, 2013)

The wings of cicada have nanoscale pillar arrays that can open bacterial membranes with mechanical force.

These biologically inspired architectures can be accurately replicated using laser fabrication technology. Through the control of laser fluence, scanning patterns and polarisation, controlled microgrooves, nanopillars and hierarchal textures can be obtained that are reminiscent of natural antifouling systems.

In comparison to the conventional lithographic techniques, laser processing provides coverage of mask-free, scalable, and material-independent fabrication, and biomimetic antifouling surfaces are now more available to clinical translation.

6. MATERIAL SYSTEMS AND APPLICATIONS

The efficacy of antibacterial as well as antifouling finishes relies not just on the arrangement but also on inherent characteristics of the material system that is utilized. Laser processing allows refined adjustment and incorporation of varied materials like metals, ceramics, carbon based structures and hybrid polymer materials. The different material classes have their own unique mechanisms and benefits and selection of materials is a critical design factor when it comes to biomedical applications.

6.1 Metallic Nanoparticles

The use of nanoparticles of metals is one of the most popular and clinically significant antimicrobial agents because of its broad-spectrum activity and a well-understood mechanism of action.

Silver-Based Systems

Silver has been considered as the gold standard among antimicrobial materials. It is effective because it releases Ag+ ions in a controlled manner and this disrupts bacterial enzymes, breaks DNA replication and disrupts membrane integrity.

In biomedical coating, the optimum levels of silver are usually between 0.5 and 5wts. (Massa et al., 2014; Fielding et al., 2012) Adequate antimicrobial effect can also be attained within this window at minimum cytotoxic effects on host tissues. Nanoparticles Laser-based methods allow the nanoparticle size, dispersion and embedding depth to be controlled finely, which directly determines ion release kinetics and long-term performance.

Copper and Zinc Oxide (ZnO)

Silver can be replaced with copper and zinc oxide which are cost-effective but with good antimicrobial activity.

Copper ions have a bactericidal effect with oxidative stress and membrane damage. Besides the release of Zn^{2+} ions, Zinc oxide (ZnO) shows photocatalytic activity in the UV light or visible light. This catalytic process produces reactive oxygen species (ROS) which adds to the increased antibacterial activity.

Laser treatment has the potential to enhance the activity of ZnO by enhancing surface defects and oxygen vacuity that facilitate the separation of electrons and the formation of ROS.

Composite Coatings

Composite systems are antimicrobial systems containing more than one antimicrobial mechanism in a coating. As an illustration, $AgTiO_2$ hybrid systems have synergistic behaviour: the silver acts as a regulated source of ions and TiO_2 as a source of photocatalytic ROS generation.

These dual-action systems have the potential to improve the overall antibacterial efficacy and potentially improve the necessary concentration of each constituent component. The laser fabrication technique is especially useful in making structures of composite materials whose contents are uniformly distributed and have a good interfacial bonding strength.

6.2 Titanium Dioxide (TiO_2)

Titanium dioxide is widely applied in biomedical implants because it is biocompatible and non-reactive. TiO_2, in addition to its structural application, has photocatalytic and self-cleaning properties which can be used to treat the microbes.

The phase transformation caused by laser is important in optimization of TiO_2. The phase of anatase usually portrays a better photocatalytic prowess than that of rutile. Crystallinity, surface roughness and oxygen vacancy concentration can be adjusted by carefully changing the laser parameters, like fluence and pulse duration.

These alterations promote the ROS generation, better the wettability control, and lead to long-term antifouling behavior.

6.3 Carbon-Based Materials

Another approach to antimicrobial and antifouling design is based on carbon-based materials, whose design frequently focuses on mechanical and tribological performance, as well as on biological functionality.

Graphene oxide (GO) has an antibacterial effect partly because sharp edges of its nanoscale can physically break the bacterial membranes. (Liu et al., 2011; Lu et al., 2017) Also, the bactericidal effect can be due to oxidative stress mechanisms.

Diamond-like carbon (DLC), in its turn, is appreciated because of its low friction coefficient, chemical inertness, and good hemocompatibility. Cards involving contact with blood and cardiovascular devices are especially appealing to the DLC coating, as the reduction of thrombosis and wear is crucial.

6.4 Hybrids of Polymers and Nanoparticles

Polymer-nanoparticle hybrid systems are based on the ability of polymers to be flexible and processable and the antimicrobial performance of encapsulated nanoparticles.

These systems enable the regulated and long-term delivery of antimicrobial agents and also ensure that mechanical compliance- which is a significant demand of the flexible implants, catheters, and soft biomedical devices.

Type of laser surface activation as well as patterning can be used to improve the dispersion of nanoparticles in polymer matrix as well as increase the adhesion between surfaces to provide the resilience of the coating in physiological conditions.

6.5 Diamond-Like Carbon (DLC) having Antimicrobial Doping

The inclusion of antimicrobial agents (especially silver) in DLC coating is a promising multifunctional platform. Such systems have their mechanical strength, wear resistance and hemocompatibility offered by the DLC matrix, and antimicrobial action controlled by silver nanoparticles.

Significantly, silver-doped DLC coatings have been found to be less thrombogenic in blood-contacting devices as well as being able to prevent bacteria colonization. The laser-assisted deposition and surface modification methods will provide a precise regulation of the dopant concentration and distribution, which will allow balancing between antimicrobial activity and biocompatibility.

7. BIOMEDICAL APPLICATIONS

One of the most promising developments in the field of biomedical engineering today is the translation of laser-engineered antibacterial and antifouling surfaces that have been developed at the laboratory level into the clinical setting. Since implant-associated infections and biofilm formation continue to be the leading causes of device failure, surface modification approaches have gained the focus of enhancing long-term clinical outcomes. The use of laser based technologies is especially desirable because of its accuracy, repeatability and its ability to interact with the traditional implant materials.

Figure 10. Clinical Applications Across Implant Types (Image generated by Gemini AI, [21/02/26]).

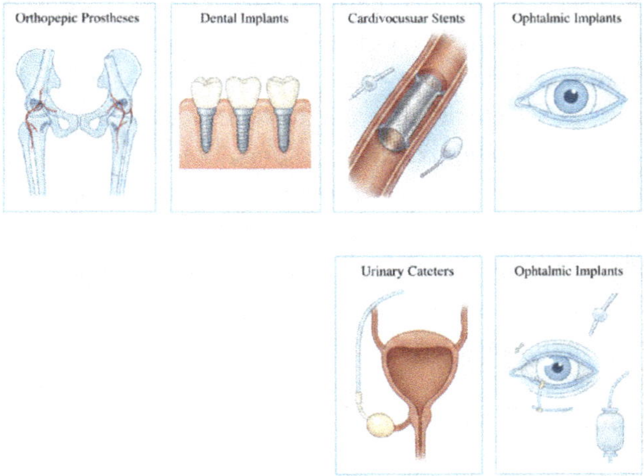

7.1 Orthopedic Implants

Orthopedic implants especially those that are utilized in joint replacement and fracture repair are very prone to infection because of a great deal of surgical exposure and the fact that these implants introduce foreign material in bone tissue. The situation is that even the rather low rates of infections can also cause serious problems, such as the necessity to replace the implants or develop systemic infection.

The laser-textured titanium (Ti) implants have proven to have considerable progress in the initial clinical appraisals. Infection rates have been reported to be lower than 0.5% due to the incorporation of hierarchical micro and nanostructures which combine increased osseointegration and antimicrobial properties in the earlier research, which indicated reduced infection rates at 2-4% to less than 0.5%.

The effectiveness of the systems is that they have a dual mode of functionality, where microscale roughness encourages osteoblast settlement and bone incorporation, whereas nanoscale dimensions are not tolerant to bacterial settlement or mechanical disturbances. Notably, such alterations are accomplished without taking away the mechanical integrity of the implant.

7.2 Dental Applications

Abusive serum is exposed to a complex oral microbiome on a continuous basis of dental implants and abutments. Formation of biofilm in this environment may cause peri-implantitis, which is one of the principal causes of implant failure.

Abutments that are nanostructured and have been treated by laser have demonstrated potential in minimizing bacterial colonization without being incompatible with the gingival and bone tissues. These modified surfaces minimize the deposition of plaque and enhance the incorporation of soft tissues because careful manipulation of the topography and wettability of the surface are controlled.

Due to the highly dynamic and salivated environment dental implants have to operate in, a stable antifouling behavior is especially important in this setting.

7.3 Cardiovascular Devices

Specific devices like stents, heart valves or vascular grafts are special because they are in direct physical contact with blood. The surfaces in such systems should reduce the bacterial adhesion as well as platelet activation to prevent infection and thrombosis.

A promising solution is diamond-like carbon (DLC) which is doped with silver (Ag). The DLC matrix has low friction, chemical stability and high hemocompatibility with built-in silver nanoparticles offering antimicrobial security. These types of coating have been shown to reduce platelet adhesion and prevent the colonization of bacteria, covering two serious risks at once.

Depositing and surface modification can be performed through the laser technique allowing the control of the thickness of the coating and the distribution of dopants, which are the key to the balance between antimicrobial activity and blood compatibility.

7.4 Urinary Catheters

One of the most widespread causes of hospital-acquired infections is urinary catheters. The presence of catheters in the long run is followed by biofilm formation, encrustment, and blockage.

Hydrophilic coatings produced using lasers have demonstrated the possibility of decreasing the mineral encrustation as well as the growth of bacterial biofilms. These coatings provide an environment that prevents the attachment of microbes by increasing the surface hydration and reducing protein adsorption.

Due to the flexibility of catheters, only mechanical compliance of alterations on these surfaces can be done. Treatments using lasers provide topographical refinements with a localized approach with minor changes to bulk material characteristics.

7.5 Ophthalmic Implants

Special smooth and biocompatible surfaces are demanded of the ophthalmic implants such as intraocular lenses and corneal devices. Small surface defects can also cause inflammation or protein deposition to affect visual performance.

The creation of ultra-smooth antifouling surfaces that minimise inflammatory response without optical impairment can be achieved by laser-based precision polishing and nanoscale surface engineering. In these applications, aggressive bactericidal activity can be less important than antifouling performance since it is the survival of delicate ocular tissue compatibility that is important.

8. PERFORMANCE EVALUATION AND STANDARDIZATION

Innovative design is not the only key to the successful application of laser-engineered antibacterial and antifouling surfaces into clinical practice: intense performance testing and compliance with regulations are also necessary. Normalized testing procedures are necessary to guarantee predictability, security and protracted viability. The next step of evaluation is usually controlled in vitro tests followed by in vivo validation and long-term durability tests.

8.1 In Vitro Testing

In vitro testing gives the initial stage of quantitative analysis in the controlled laboratory condition. Such experiments can be used to evaluate systematically antibacterial efficacy, biofilm resistance and cytocompatibility and then go on to animal studies or clinical studies.

Common methods include:

Colony-forming unit (CFU) counting: This is a quantitative test that determines viable bacteria that have been exposed to the modified surface. Reduction of $\geq$3-log (99.9) is typically taken to be bactericidal.

Live/dead staining experiments The Fluorescent labeling of viable and membrane-compromised bacteria can be used to visualize the bactericidal process at a microlevel.

Assays to quantify biofilm: Biofilm biomass and maturity are determined by methods like crystal violet staining or metabolic activity measurements.

Protein adsorption tests and cell adhesion experiments are frequently done as well as the antimicrobial performance to test the antifouling behavior and compatibility with host cells.

Although in vitro experiments are considered to be very useful in understanding the mechanisms, these studies do not completely capture the complexity of the physiological environment. Thus, additional validation is needed.

8.2 In Vivo Assessment

The next important step to be evaluated is animal implantation studies. The models evaluate the resistance to infection and tissue incorporation in living systems.

In vivo studies are generally investigations of:

The growth of bacteria at the location of implant.

Local inflammation.

Bone or tissue integration (orthopedic and dental devices)

Systemic toxicity or reactions.

These tests give a better picture of the performance of laser-modified surfaces under dynamic biological tests, such as immune responses, fluid flow, and mechanical loading.

Before regulatory approval and human clinical trials can occur, one needs to have successful in vivo results.

Biocompatibility Considerations To be taken into account.

All implantable medical devices must meet the conditions of biocompatibility. Biological evaluation is thoroughly covered by international standards, and especially ISO 10993.

Testing under ISO 10993 includes:

Cytotoxicity assays

Sensitization testing

Irritation evaluation and inflammatory response evaluation.

Blood contacting devices Hemocompatibility testing.

In the case of coating that involves the use of antimicrobial agents like silver or copper, great care must be taken to ensure that antimicrobial activity is balanced with minimum cytotoxic effects on host cells. Laser processing may help to obtain this balance by providing a strict control over chemical composition of surfaces and distribution of nanoparticles.

8.4 Long-Term Stability

Long-term stability is one of the most crucial but difficult things to design an implant. A number of orthopedic and cardiovascular implants will have a lifetime of 10-20 years and above.

The long term validation should consider:

- Cyclic mechanical durability.
- Wear resistance and coating adhesion.
- Nanostructures Stability of nanostructures
- Long-term antimicrobial efficacy.

In the case of release-based systems, the ion release kinetics should be maintained within the interval of the implant lifetime without any toxic concentration. In the case of topography-based antibacterial surfaces, nanostructures should not be subjected to mechanical degradation which may diminish functionality.

Tests to predict the long-term behavior are usually accelerated aging tests, corrosion tests and fatigue tests. Finally, strong durability information is needed in order to get regulatory approval and clinical confidence.

9. DIFFICULTIES AND FUTURE OUTLOOKS

Despite the promise of laser-engineered antibacterial and antifouling surfaces in laboratory and clinical trials, there are several technical, regulatory, economic issues that have to be resolved before it becomes possible to adopt them widely. Meanwhile, the high pace of development in materials science, artificial intelligence and intelligent surface engineering suggests opportunities to introduce new ways.

9.1 Current Limitations

Laser based surface modification technologies have limitations to practicality in spite of their accuracy and flexibility.

The main constraint is that the cost of femtosecond laser systems is too high. These systems need involved optical configurations, environmental regulation and daily upkeep, that may restrict availability in a large scale production setting.

Scalability is also another issue. Laser processing is not usually parallel, although it is very accurate. When large areas of implants are being treated, or large volumes of production, processing time can consume a lot of time unless parallelization or high-speed scanning measures are adopted.

In the case of ion-releasing systems, what is most important is long-term control. To obtain durable antimicrobial action over years without surpassing cytotoxic levels, it is important to carefully engineer nanoparticle size, dispersion and embedding depth. Unpredictability of the physiological conditions also adds to the predictability of release.

To overcome these constraints, technological refinement will not be enough but it should be combined with scalable manufacturing processes.

9.2 Regulatory Considerations

Clinical approval is also an important milestone in the translation process. Before approval of modified surface of implants, agencies like the U. S food and drug administration (FDA) and the European Medicines Agency (EMA) demand thorough biological and toxicological assessment.

The usual assessments required are:

- Cytotoxicity testing
- Genotoxicity evaluation
- Sensitization and irritation experiments.
- Systemic toxicity analysis 5-10 years of studies on implants.

In case of coating containing nanoparticles release, regulators can insist on further characterization of degradation products, the kinetics of ion release, and the possible organ accumulation.

Since, in most instances, laser modification does not involve modifying bulk composition, but an alteration in surface structure, regulatory classification can change according to whether the technology is viewed as a surface treatment or a novel combination device. The most critical requirements of successful regulatory submission are clear documentation, reproducibility, and standardized testing protocols.

9.3 Emerging Technologies

New developments are bound to improve performance and manufacturability.

The use of artificial intelligence (AI) has been done to optimize the parameters of laser. Machine learning software can be used to process the correlation between fluence, pulse length, scan speed, and the morphology obtained to make predictive design decisions instead of trial-and-error design methods. This data optimization would help to develop a lot faster.

Another promising direction is set by smart responsive surfaces. These systems can change their characteristics based on environmental stimulus like PH, presence of bacteria or mechanical stress. An example includes the use of coats that are inactive in normal conditions but release antimicrobials when the formation of biofilms starts.

Hybrid nanostructure-chemical coatings also are receiving interest. The integration of mechanical bactericidal topographies with regulated chemical functionality is intended to generate robust, durable and multi-mechanistic protection by these systems. C.T. Pathways Clinical translation represents the fourth key concept in transcriptional research.

9.4 Clinical Translation Pathways

Clinical Translation Pathways Clinical translation The fourth important concept in transcriptional research is clinical translation.

Technological innovation should be followed by stringent clinical testing and economic rationalization to be adopted widely.

To establish the safety and efficacy of the technique over a wide range of patients and surgery types, multicenter clinical trials are necessary to prove uniform safety and effectiveness. The results of such trials are statistically powerful indicators that surface modifications have a significant impact on reducing the rate of infection without causing harmful effects. Concurrently, cost-benefit analysis is very essential. Though the higher cost of laser processing now and in the future might barely cover the higher cost of manufacture, the overall healthcare cost might save so much through the cutback of revision surgeries, the number of hours in the hospital, and also through the complications that arise due to infections. A coordinated strategy of engineers, clinicians, regulatory experts, and industry partners will therefore be the key to successful clinical translation.

10. CONCLUSION

Antibacterial and antifouling surfaces made of lasers are a revolutionary development in biomaterial surface engineering. The use of laser technologies can be contrasted with traditional methods that use only antibiotics or passive coating to modify the architecture and chemistry of surfaces directly and with precision at the micro- and nanoscale. This has enabled the combination of various functional principles - mechanical bactericidal effects, controlled ion release, photocatalytic activity and antifouling behavior into one, long lasting platform.

Laser processing provides a very high level of control of surface morphology and material phase by adjusting parameters, including pulse length, fluence, polarization and pattern of scan. The outcome is new generation multifunctional implant surfaces that have the ability to prevent bacteria colonization, prevent biofilm formation and facilitate tissue integration.

Notably, these changes can be typically attained without significantly changing the bulk mechanical characteristics of the implant, maintaining structural integrity and improving biological performance. The lack of antibiotics in research systems that have been engineered using lasers also diminishes the concerns about antimicrobial resistance.

Gaps still exist in laser technology, automation, and materials design, but with continued developments in laser technology, automation, and material design, such gaps continue to decrease. With the enhanced accessibility of systems, coupled with higher manufacturing throughput, laser-based surface engineering is bound to move out of the specialized research use to regular use in industry.

These technologies hold high promise of becoming standard components of next generation medical implants in the next decade helping to generate significant improvements in the number of device related infections, reduced numbers of revision surgery as well as better long term patient outcomes.

REFERENCES

Bhattacharya, M., & Horswill, A. R. (2024). The role of human extracellular matrix proteins in defining Staphylococcus aureus biofilm infections. *FEMS Microbiology Reviews*, *48*(2), fuae002. Advance online publication. DOI: 10.1093/femsre/fuae002 PMID: 38337187

Bixler, G. D. (2013). Bioinspired surface for low drag, self-cleaning, and antifouling: Shark skin, butterfly and rice leaf effects [Doctoral dissertation, The Ohio State University]. OhioLINK Electronic Theses and Dissertations Center.

Bonse, J., Kirner, S. V., & Krüger, J. (2020). Laser-induced periodic surface structures (LIPSS). In Handbook of Laser Micro- and Nano-Engineering (pp. 1-59). Springer. DOI: 10.1007/978-3-319-69537-2_17-2

Butler, M. S., Henderson, I. R., Capon, R. J., & Blaskovich, M. A. T. (2023). Antibiotics in the clinical pipeline as of December 2022. *The Journal of Antibiotics*, *76*(8), 431–473. DOI: 10.1038/s41429-023-00629-8 PMID: 37291465

Chichkov, B. N., Momma, C., Nolte, S., von Alvensleben, F., & Tünnermann, A. (1996). Femtosecond, picosecond and nanosecond laser ablation of solids. *Applied Physics. A, Materials Science & Processing*, *63*(2), 109–115. DOI: 10.1007/BF01567637

Cortese, Y. J., Wagner, V. E., Tierney, M., Devine, D., & Fogarty, A. (2018). Review of catheter-associated urinary tract infections and in vitro urinary tract models. *Journal of Healthcare Engineering*, *2986742*, 1–16. Advance online publication. DOI: 10.1155/2018/2986742 PMID: 30405898

Costache, F., & Reif, J. (2004). Femtosecond laser induced Coulomb explosion from calcium fluoride. *Thin Solid Films*, *453-454*, 334–339. DOI: 10.1016/j.tsf.2003.11.096

Dai, S., Zhang, D., Shi, Q., Han, X., Wang, S., & Du, Z. (2013). Biomimetic fabrication and tunable wetting properties of three-dimensional hierarchical ZnO structures by combining soft lithography templated with lotus leaf and hydrothermal growth. *CrystEngComm*, *15*(25), 5040–5047. DOI: 10.1039/c3ce40238j

Fielding, G. A., Roy, M., Bandyopadhyay, A., & Bose, S. (2012). Antibacterial and biological characteristics of plasma sprayed silver and strontium doped hydroxyapatite coatings. *Acta Biomaterialia*, *8*(8), 3144–3152. DOI: 10.1016/j.actbio.2012.04.004 PMID: 22487928

Harilal, S. S., Freeman, J. R., Diwakar, P. K., & Hassanein, A. (2014). Femtosecond laser ablation: Fundamentals and applications. In *Laser-Induced Breakdown Spectroscopy* (pp. 143–166). Springer., DOI: 10.1007/978-3-642-45085-3_6

Høiby, N., Bjarnsholt, T., Givskov, M., Molin, S., & Ciofu, O. (2010). Antibiotic resistance of bacterial biofilms. *International Journal of Antimicrobial Agents, 35*(4), 322–332. DOI: 10.1016/j.ijantimicag.2009.12.011 PMID: 20149602

Horbett, T. A. (2018). Fibrinogen adsorption to biomaterials. *Journal of Biomedical Materials Research. Part A, 106*(10), 2777–2788. DOI: 10.1002/jbm.a.36460 PMID: 29896846

Hudson, M. C., Ramp, W. K., Nicholson, N. C., Williams, A. S., & Nousiainen, M. T. (1999). Staphylococcus aureus adhesion to bone matrix and bone-associated biomaterials. *FEMS Microbiology Letters, 173*(2), 279–284. DOI: 10.1111/j.1574-6968.1999.tb13514.x PMID: 10227156

Ivanova, E. P., Hasan, J., Webb, H. K., Truong, V. K., Watson, G. S., Watson, J. A., Baulin, V. A., Pogodin, S., Wang, J. Y., Tobin, M. J., Löbbe, C., & Crawford, R. J. (2012). Natural bactericidal surfaces: Mechanical rupture of Pseudomonas aeruginosa cells by cicada wings. *Small, 8*(16), 2489–2494. DOI: 10.1002/smll.201200528 PMID: 22674670

Jelinek, M., Kocourek, T., Remsa, J., Weiserová, M., Jurek, K., Miksovsky, J., Strnad, J., Galandáková, A., & Ulrichová, J. (2013). Antibacterial, cytotoxicity and physical properties of laser—Silver doped hydroxyapatite layers. *Materials Science and Engineering C, 33*(3), 1242–1246. DOI: 10.1016/j.msec.2012.12.018 PMID: 23827567

Kapadia, B. H., Berg, R. A., Daley, J. A., Fritz, J., Bhave, A., & Mont, M. A. (2016). Periprosthetic joint infection. *Lancet, 387*(10016), 386–394. DOI: 10.1016/S0140-6736(14)61798-0 PMID: 26135702

Kőrösi, L., Rodio, M., Dömötör, D., Kovács, T., Papp, S., Diaspro, A., Intartaglia, R., & Dékány, I. (2016). Ultrasmall, ligand-free Ag nanoparticles with high antibacterial activity prepared by pulsed laser ablation in liquid. *Journal of Chemistry, 4143560*, 1–8. Advance online publication. DOI: 10.1155/2016/4143560

Korshed, P., Li, L., Liu, Z., Mironov, A., & Wang, T. (2018). Antibacterial mechanisms of a novel type picosecond laser-generated silver-titanium nanoparticles and their toxicity to human cells. *International Journal of Nanomedicine, 13*, 1007–1019. DOI: 10.2147/IJN.S140222 PMID: 29317818

Liu, S., Zeng, T. H., Hofmann, M., Burcombe, E., Wei, J., Jiang, R., Kong, J., & Chen, Y. (2011). Antibacterial activity of graphite, graphite oxide, graphene oxide, and reduced graphene oxide: Membrane and oxidative stress. *ACS Nano*, *5*(9), 6971–6980. DOI: 10.1021/nn202451x PMID: 21851105

Lu, X., Feng, X., Werber, J. R., Chu, C., Zucker, I., Kim, J. H., Osuji, C. O., & Elimelech, M. (2017). Enhanced antibacterial activity through the controlled alignment of graphene oxide nanosheets. *Proceedings of the National Academy of Sciences of the United States of America*, *114*(46), E9793–E9801. DOI: 10.1073/pnas.1710996114 PMID: 29078354

Massa, M. A., Covarrubias, C., Bittner, M., Fuentevilla, I. A., Capetillo, P., Von Marttens, A., & Carvajal, J. C. (2014). Synthesis of new antibacterial composite coating for titanium based on highly ordered nanoporous silica and silver nanoparticles. *Materials Science and Engineering C*, *45*, 146–153. DOI: 10.1016/j.msec.2014.08.057 PMID: 25491813

Pan, Q., Cao, Y., Xue, W., Zhu, D., & Liu, W. (2019). Picosecond laser-textured stainless steel superhydrophobic surface with an antibacterial adhesion property. *Langmuir*, *35*(33), 11414–11421. DOI: 10.1021/acs.langmuir.9b01333 PMID: 31397568

Pelling, H., Nzakizwanayo, J., Milo, S., Denissen, J., Smits, W. P., Thet, N. T., & Jenkins, A. T. A. (2019). Bacterial biofilm formation on indwelling urethral catheters. *Letters in Applied Microbiology*, *68*(4), 277–293. DOI: 10.1111/lam.13144 PMID: 30811615

Phillips, K. C., Gandhi, H. H., Mazur, E., & Sundaram, S. K. (2015). Ultrafast laser processing of materials: A review. *Advances in Optics and Photonics*, *7*(4), 684–712. DOI: 10.1364/AOP.7.000684

Pu, X., Li, G., & Huang, H. (2016). Preparation, anti-biofouling and drag-reduction properties of a biomimetic shark skin surface. *Biology Open*, *5*(4), 389–396. DOI: 10.1242/bio.016899 PMID: 26941105

Pulido, L., Ghanem, E., Joshi, A., Purtill, J. J., & Parvizi, J. (2008). Periprosthetic joint infection: The incidence, timing, and predisposing factors. *Clinical Orthopaedics and Related Research*, *466*(7), 1710–1715. DOI: 10.1007/s11999-008-0209-4 PMID: 18421542

Rethfeld, B., Ivanov, D. S., Garcia, M. E., & Anisimov, S. I. (2017). Modelling ultrafast laser ablation. *Journal of Physics. D, Applied Physics*, *50*(19), 193001. DOI: 10.1088/1361-6463/50/19/193001

Ristoscu, C., & Mihailescu, I. N. (2014). Biomimetic coatings by pulsed laser deposition. In Laser Technology in Biomimetics: Basics and Applications (pp. 179-208). Springer. DOI: 10.1007/978-3-642-41341-4_7

Schwibbert, K., Richter, A. M., Krüger, J., & Bonse, J. (2024). Laser-textured surfaces: A way to control biofilm formation? *Laser & Photonics Reviews*, *18*(5), 2300753. DOI: 10.1002/lpor.202300753

Stewart, P. S. (2015). Antimicrobial tolerance in biofilms. *Microbiology Spectrum*, *3*(3), MB-0010–MB-2014. DOI: 10.1128/microbiolspec.MB-0010-2014 PMID: 26185072

Wang, Q., Xu, S., Xing, X., & Wang, N. (2022). Progress in fabrication and applications of micro/nanostructured superhydrophobic surfaces. *Surface Innovations*, *10*(2), 89–109. DOI: 10.1680/jsuin.21.00031

World Health Organization. (2011). Report on the burden of endemic health care-associated infection worldwide. https://iris.who.int/bitstream/handle/10665/80135/

Wu, L., Wei, S., Cheng, X., He, N., Kang, X., Zhou, H., Yang, S., & Pan, X. (2024). Release of ions enhanced the antibacterial performance of laser-generated, uncoated Ag nanoparticles. *Colloids and Surfaces. B, Biointerfaces*, *238*, 113894. DOI: 10.1016/j.colsurfb.2024.114131 PMID: 39094211

Zimbone, M., Buccheri, M. A., Cacciato, G., Sanz, R., Rappazzo, G., Boninelli, S., Reitano, R., Romano, L., Privitera, V., & Grimaldi, M. G. (2015). Photocatalytical and antibacterial activity of TiO2 nanoparticles obtained by laser ablation in water. *Applied Catalysis B: Environmental*, *165*, 487–494. DOI: 10.1016/j.apcatb.2014.10.031

Chapter 7
Laser-Coated Orthopedic and Dental Implants Advances in Surface Engineering for Bioactivity and Longevity

Dileep Pathote
http://orcid.org/0000-0001-8198-175X
Indian Institute of Science, Bangalore, India

Rupesh Kumar
Dr. D.Y. Patil Institute of Technology, Pimpri, India

Ashwani Ranjan
Indian Institute of Technology, Banaras Hindu University, India

Sudipto Datta
http://orcid.org/0000-0003-2161-6878
All India Institute of Medical Science, New Delhi, India

ABSTRACT

This chapter reviews laser-based surface engineering for orthopedic and dental implants, highlighting its advantages over conventional methods such as grit blasting, acid etching, and plasma spraying, which often suffer from weak adhesion and poor nanoscale control. Laser-assisted techniques enable precise, reproducible modification and deposition of bioactive coatings with tailored micro–nano structures that enhance protein adsorption, cell adhesion, and bone integration. Coatings

DOI: 10.4018/979-8-3373-4104-0.ch007

including hydroxyapatite, calcium phosphates, and bioactive glasses provide added antibacterial, angiogenic, and wear-resistant properties. Integration with additive manufacturing offers pathways to patient-specific multifunctional implants, though challenges remain in coating uniformity, long-term stability, and scalability.

1. INTRODUCTION

The use of orthopedic and dental implants has transformed clinical practice by offering reliable solutions for restoring mobility, functionality, and aesthetics to patients suffering from degenerative diseases, traumatic injuries, or tooth loss. Over the past three decades, a steady increase in life expectancy and the growing prevalence of musculoskeletal disorders have accelerated the global demand for implants. According to recent market analyses, the orthopedic implant industry alone is valued at more than USD 50 billion, while dental implant technologies are projected to surpass USD 8 billion by the end of this decade. This expanding demand has been matched by continuous research into biomaterials and surface modification strategies that can ensure mechanical durability, resistance to infection, and long-term integration with host tissues (Geetha et al., 2009; Goriainov et al., 2014; Liu et al., 2021).

Titanium and its alloys, cobalt–chromium-based alloys, and zirconia ceramics have emerged as the dominant materials of choice for implant fabrication due to their exceptional mechanical strength, fatigue resistance, corrosion resistance, and biocompatibility. Among these, commercially pure titanium (cp-Ti) and Ti–6Al–4V are widely used in both orthopedic and dental contexts. Cobalt–chromium alloys are particularly suitable for high-load bearing prostheses, while zirconia has become popular in dental applications because of its excellent aesthetics and bioinert nature. Nevertheless, despite these intrinsic advantages, these materials share a fundamental limitation: they are generally bioinert. In clinical settings, bioinertness often results in suboptimal osseointegration, leading to poor bonding with the surrounding bone tissue. This deficiency manifests as delayed healing, increased susceptibility to aseptic loosening, and, in some cases, long-term implant failure (Hanawa, 2011). Reports indicate that aseptic loosening accounts for more than half of long-term orthopedic implant failures, while peri-implantitis remains one of the leading causes of dental implant loss (Albrektsson et al., 2017).

The quest to address these challenges has traditionally focused on modifying the implant surface to optimise biological interactions at the tissue–implant interface. Conventional surface treatments such as grit blasting, acid etching, plasma spraying, sol–gel coating, anodization, and physical vapor deposition (PVD) have all been widely employed in both research and commercial practice (Rupp et al.,

2014; Surmenev et al., 2014). While these approaches have delivered incremental improvements, they also present inherent drawbacks. For instance, grit blasting and acid etching create micro-rough surfaces that can enhance osteoblast attachment, but they also risk introducing contaminants and lack reproducibility at the nanoscale, which is increasingly recognised as critical for protein adsorption and cellular differentiation (Bigerelle et al., 2002). Plasma spraying of hydroxyapatite (HA), one of the most widely adopted coating strategies, produces relatively thick bioactive layers but often suffers from poor adhesion strength, high porosity, and phase instability at physiological temperatures. Over time, these coatings are prone to delamination, particularly under cyclic loading in weight-bearing applications (Sun et al., 2001; Goriainov et al., 2014). Sol–gel techniques and PVD methods, on the other hand, are capable of producing thin, uniform films with controlled chemical composition. However, such films are often brittle and exhibit poor adhesion when applied to metallic substrates subjected to dynamic mechanical stresses, making them unsuitable for long-term clinical translation (Surmenev et al., 2014).

In recent years, laser-based surface modification technologies have emerged as promising alternatives to overcome the inherent limitations of conventional methods. The interaction of a high-energy laser beam with the surface of an implant material provides a unique route to tailor its physicochemical and topographical properties across multiple length scales. Unlike conventional thermal or chemical processes, laser treatments are non-contact, contamination-free, and highly controllable, enabling the fabrication of surfaces with hierarchical micro- and nano-features that closely mimic the extracellular matrix (ECM) of bone tissue (Vorobyev & Guo, 2013). This biomimetic structuring is particularly relevant because it directly influences protein adsorption, cell adhesion, and subsequent bone remodelling, thereby promoting more robust and rapid osseointegration (Zhao et al., 2021).

Laser surface texturing (LST) represents one of the most well-studied approaches in this regard. Through precise modulation of laser fluence, pulse duration, and scanning strategies, LST can generate defined microgrooves, pits, ridges, or cross-linked patterns on implant surfaces. Such modifications provide both mechanical anchorage and biological cues, guiding cell migration, alignment, and proliferation. Studies on zirconia implants, for example, have shown that microgrooves with widths between 30 and 100 μm can promote mesenchymal stem cell alignment and significantly enhance osteoblast proliferation compared to untreated controls (Pereira et al., 2025). When combined with biphasic calcium phosphate (CaP) coatings, the synergistic effects of topographical guidance and bioactivity lead to markedly improved osseointegration. Similarly, ultrafast lasers operating in the femtosecond or picosecond regimes have been employed to create nanoscale textures with minimal thermal damage, thereby avoiding microcracking or undesired phase transformations. These nano-patterned surfaces have been demonstrated to modulate protein adsorption profiles and influence the integrin-mediated signaling pathways of osteoblasts, accelerating the early stages of bone healing (Vorobyev & Guo, 2013; Zhao et al., 2021).

Beyond texturing, lasers have also been utilised for the deposition and sintering of bioactive coatings. Pulsed Nd: YAG and excimer lasers, among others, can facilitate the direct incorporation of calcium phosphates such as hydroxyapatite (HA), tricalcium phosphate (TCP), and biphasic calcium phosphate onto metallic and ceramic substrates. These coatings typically exhibit superior adhesion strength and phase stability compared to plasma-sprayed counterparts, as the localised and rapid heating induced by lasers minimises undesirable phase decomposition (Hussein et al., 2019). Moreover, laser-assisted deposition offers the advantage of fabricating dense, crack-free coatings with elastic moduli closer to those of cortical bone, thereby reducing the mismatch in mechanical properties and mitigating stress-shielding effects. Recent innovations have further extended this approach to multifunctional coatings incorporating antibacterial nanoparticles such as silver, copper, and zinc, or carbon nanomaterials such as graphene oxide and multiwalled carbon nanotubes (MWCNTs). These additions endow implants with combined antibacterial, angiogenic, and osteogenic properties, addressing multiple clinical challenges simultaneously (Logesh et al., 2024). Another critical advantage of laser-based processing lies in its compatibility with additive manufacturing.

Figure 1. Comparison of conventional coating vs laser-assisted coating

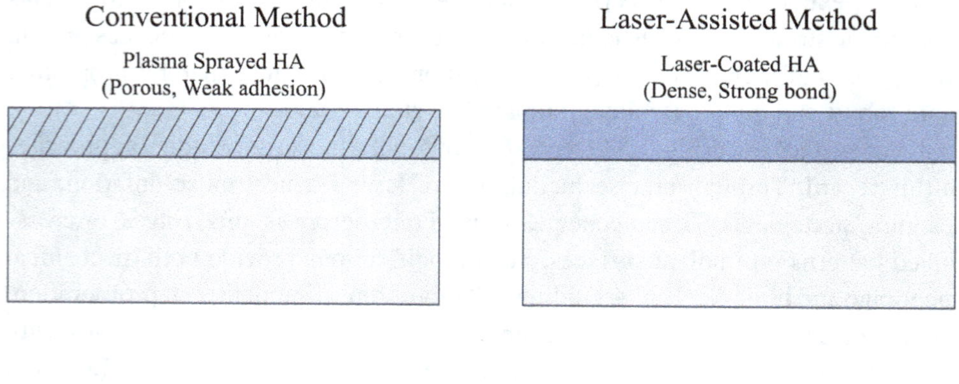

With the increasing adoption of selective laser melting (SLM) and electron beam melting (EBM) for patient-specific implant fabrication, the integration of laser texturing and coating processes allows for in situ functionalization of complex geometries. This approach not only reduces manufacturing steps but also ensures uniform coating quality on intricate structures such as lattice-porous implants designed to reduce stiffness mismatch and promote vascularized bone ingrowth (Robau-Porrua et al., 2025). As such, the convergence of additive manufacturing with laser-based surface engineering is poised to redefine the future of orthopaedic and dental implants, paving the way for personalised, multifunctional devices with superior clinical performance. Figure 1 represents the difference between lased assisted and conventional coating.

The scope of this chapter is to provide a comprehensive and critical assessment of the advances in laser-based coating and surface modification technologies for orthopaedic and dental implants. Particular emphasis is placed on the processing–structure–property correlations that govern biological performance, the comparative evaluation of laser-assisted versus conventional techniques, the development of advanced nanocomposite coatings, and the failure mechanisms at the coating–substrate interface. In addition, we highlight clinical and translational considerations, including regulatory challenges and the need for long-term in vivo validation. Finally, we present future directions in adaptive, patient-specific coatings and integrated additive manufacturing strategies, which may collectively contribute to the emergence of next-generation implants with enhanced bioactivity, mechanical durability, and longevity.

2. LASER SURFACE MODIFICATION AND TEXTURING

Laser surface modification and texturing have emerged as transformative approaches in implant surface engineering, offering unparalleled precision, reproducibility, and versatility compared to traditional methods such as grit blasting, acid etching, and plasma spraying. By directing focused, high-energy laser beams onto the implant surface, controlled ablation, localised melting, or microstructural reorganisation can be achieved. This enables the fabrication of well-defined micro- and nano-topographies such as grooves, ridges, pits, and hierarchical textures that directly influence cellular responses. Importantly, such laser-based methods avoid contamination from abrasive particles or chemical residues, which are common drawbacks of conventional surface treatments. Moreover, laser texturing has been successfully applied to a wide range of biomaterials, including titanium alloys, zirconia ceramics, and hybrid composites, making it highly adaptable for both orthopaedic and dental implants (Martínez-Calderon et al., 2020; Brazete et al., 2021). The capacity of these engineered surfaces to modulate protein adsorption, osteoblast adhesion, stem cell differentiation, and bacterial interactions underlines their clinical relevance. Laser-based surface modification, therefore, stands as a pivotal technique for enhancing the bioactivity and longevity of orthopaedic and dental implants. Table 1 represents the comparison between conventional and laser assisted implants surface modification methods.

Table 1. Comparison of Conventional vs. Laser-Assisted Surface Modification for Implants

Feature	Conventional Methods (Plasma Spray, Acid Etching, Grit Blasting)	Laser-Assisted Methods (LST, Laser Coating, Laser Cladding)
Surface Roughness Control	Limited, often irregular	Highly precise, tunable from nano to micro scale
Contamination Risk	High (residual grit, acid traces)	Very low, clean and contact-free
Coating Adhesion	Weak to moderate	Strong metallurgical bonding
Thermal Damage	Often significant (e.g., phase decomposition)	Minimal (esp. with ultrafast lasers)
Bioactivity	Moderate, may need secondary treatment	High can combine texturing and bioactive coating

2.1 Principles of Laser Surface Texturing

Laser Surface Texturing (LST) is a precision surface engineering technique that utilises high-energy laser beams to modify the morphology and chemistry of biomaterial surfaces at micro- and nanoscale levels. Unlike conventional methods such as grit blasting or acid etching, which rely on mechanical abrasion or chemical corrosion, LST provides a non-contact, highly controllable, and contamination-free means of tailoring implant topographies. By adjusting laser parameters such as wavelength, pulse duration, repetition rate, fluence, and scanning strategy, it is possible to engineer a wide variety of surface features, ranging from microgrooves and dimples to hierarchical micro–nano structures. These engineered features have been shown to significantly influence protein adsorption, osteoblast adhesion, mesenchymal stem cell (MSC) differentiation, and bacterial interactions, making LST particularly relevant for orthopaedic and dental implant applications (Sobolev et al., 2021; Daskalova et al., 2022). At the core of LST lies the phenomenon of laser–material interaction, which depends on the material's optical absorption properties and the energy density delivered. When laser energy exceeds the ablation threshold of a material, localised melting, vaporisation, or plasma formation occurs, leading to the formation of microcavities, grooves, or pits. In ceramics such as zirconia, short-pulsed lasers (e.g., femtosecond and nanosecond) are often preferred to minimise thermal damage and microcrack formation, while in metals such as titanium and its alloys, nanosecond or picosecond pulsed Nd: YAG and fibre lasers are widely employed to generate reproducible patterns (Kunz et al., 2018). The ability to control these microstructures without compromising the bulk mechanical integrity of the substrate provides a decisive advantage for biomedical implants, where mechanical reliability is critical.

One of the distinctive strengths of LST is the ability to create functional surface hierarchies. For instance, microgrooves ranging from 20–100 μm can guide osteoblast alignment along their axis, mimicking the natural anisotropy of bone tissue, while superimposed nanostructures enhance protein adsorption and integrin-mediated focal adhesion formation (Zhang et al., 2021). Moreover, the increased surface roughness and wettability generated by laser texturing enhance initial blood protein adsorption (fibrinogen, vitronectin, and fibronectin), which is crucial for clot formation and subsequent osseointegration. Recent reports demonstrate that LST-modified titanium implants exhibit up to 40–70% higher cell viability and more rapid new bone formation compared to conventionally etched implants (da Rocha Silva et al., 2023).

Beyond osteointegration, LST has also shown promise in imparting antibacterial properties by designing topographies that mechanically disrupt bacterial membranes or reduce bacterial adhesion without the need for antibiotic coatings. For example, femtosecond laser-induced periodic surface structures (LIPSS) on titanium have been reported to selectively reduce *Staphylococcus aureus* colonisation while supporting osteoblast attachment (Kisala J. et al., 2021). This dual effect—promoting eukaryotic cell growth while resisting bacterial colonisation—is particularly desirable for implants, where infection remains one of the leading causes of failure.

Figure 2. Schematic diagram of Laser surface texturing

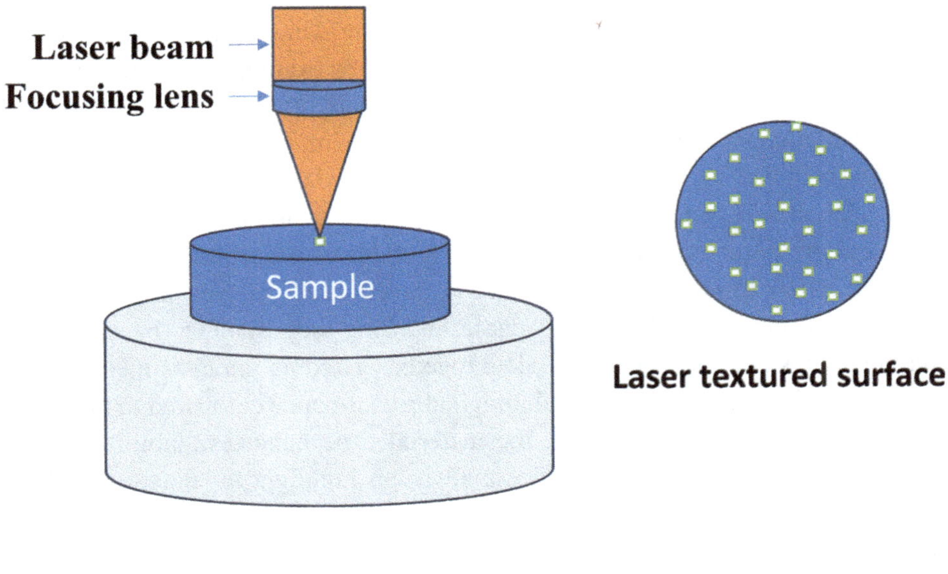

2.2 Zirconia Implants With Laser Surface Texturing

Zirconia (ZrO_2) has emerged as a promising alternative to titanium in orthopedic and dental implants due to its superior aesthetics, corrosion resistance, and favorable biomechanical compatibility. However, its inherent bioinertness limits spontaneous osseointegration, necessitating surface modifications to promote biological interactions. Conventional roughening approaches such as grit-blasting and acid-etching are associated with residual contamination and limited reproducibility. In contrast, Laser Surface Texturing (LST) offers a contamination-free, highly precise, and reproducible method for engineering implant topographies that guide cellular behavior and enhance biointegration (Jin et al., 2024).

LST enables the fabrication of microgrooves, micropits, and cross-linked ridge patterns that can be tailored to specific biological outcomes. For instance, femtosecond and nanosecond laser systems can generate microgrooves in the range of 30–100 μm, which have been shown to align mesenchymal stem cells and significantly promote osteoblast adhesion and proliferation (Cao et al., 2020; Xing et al., 2021). Similarly, cross-linked ridges and micropits not only increase the available surface area but also enhance early cell viability, with studies reporting improvements of up to 40–70% compared to smooth zirconia surfaces (Duta et al., 2021). These engineered textures influence the initial protein adsorption layer, which in turn regulates downstream cell adhesion, cytoskeletal organization, and osteogenic differentiation (Anselme et al., 2010; Hanawa, 2011). Figure 2 represents the schematic diagram of the laser surface texturing method.

A particularly promising strategy involves hybrid modification approaches, where LST is combined with bioactive coatings. For example, applying biphasic calcium phosphate (BCP) — typically composed of 70% hydroxyapatite (HA) and 30% β-tricalcium phosphate (β-TCP) — onto zirconia implants with laser-generated microtopographies has shown synergistic effects. While the LST patterns provide mechanical and topographic cues for cellular alignment, the BCP coating delivers osteoconductive and bioresorbable phases that accelerate bone–implant integration (Narayan, 2010). This combination strategy offers dual advantages: improved bone anchorage during the early healing phase and long-term stability by mitigating interfacial gaps and micromotion. Taken together, zirconia implants modified via LST exhibit significantly enhanced biological performance compared to untreated controls, bridging the gap between the material's mechanical reliability and the biological demand for bioactivity. These advances highlight the potential of LST to not only match but, in certain contexts, surpass the performance of traditional titanium implants, making zirconia a competitive candidate for next-generation orthopedic and dental applications.

2.3 Titanium Implants With Laser Surface Texturing

Titanium and its alloys (e.g., Ti-6Al-4V, Ti-15Nb, and Ti-10Zr) remain the most widely used materials in orthopaedic and dental implants due to their favourable mechanical strength, corrosion resistance, and biocompatibility. Nevertheless, challenges such as stress shielding, bacterial colonisation, and insufficient long-term osseointegration have motivated extensive research on surface modifications. Among these, Laser Surface Texturing (LST) has emerged as a powerful, contamination-free, and controllable method to optimise titanium implant interfaces for enhanced biological performance (Bhaduri et al., 2021; Gnilitskyi et al., 2022). LST enables the precise fabrication of micro- and nano-scale features that significantly influence protein adsorption, osteoblast adhesion, and early bone ingrowth. Microgrooves in the range of 20–80 μm have been shown to align osteoblasts along the laser-induced ridges, thereby facilitating anisotropic bone tissue organization and faster osseointegration (Yang et al., 2020). In parallel, nanopatterned pits and ripples produced via femtosecond lasers mimic the extracellular matrix (ECM), supporting mesenchymal stem cell adhesion and differentiation into osteogenic lineages (Liu et al., 2019). Importantly, these engineered surfaces enhance interfacial bonding strength, reducing micromotion and the risk of implant loosening.

Beyond osseointegration, LST has demonstrated promise in imparting antibacterial properties. Laser-induced periodic surface structures (LIPSS) on titanium can rupture bacterial membranes mechanically while maintaining mammalian cell compatibility, thereby reducing the risk of peri-implantitis (Zhang et al., 2021). Compared to silver or antibiotic coatings, LST provides a long-lasting, chemical-free antibacterial strategy that minimizes concerns of cytotoxicity or antimicrobial resistance (Costa et al., 2020).

Hybrid strategies combining LST with bioactive coatings further expand the potential of titanium implants. For example, LST coupled with hydroxyapatite or calcium phosphate deposition promotes synergistic effects: laser-created microtopographies serve as mechanical anchors for coating adhesion, while the coatings provide osteoconductive cues for bone regeneration (Zhou et al., 2020). Additionally, innovative approaches such as LST combined with graphene oxide or titanium dioxide nanostructures are being explored to simultaneously improve osseointegration, corrosion resistance, and antibacterial efficacy (Drevet, R et. al., 2023). Taken together, titanium implants modified by LST offer a multi-functional platform that not only accelerates bone healing but also mitigates post-surgical complications. As femtosecond and picosecond laser systems advance, the reproducibility and scalability of LST are expected to further strengthen the clinical translation of these technologies, cementing their role in the design of next-generation orthopedic and dental implants.

While laser surface texturing of zirconia and titanium implants has demonstrated significant improvements in early cell attachment, bone–implant contact, and antibacterial response, these benefits are often limited to the initial stages of osseointegration. Long-term implant performance requires not only optimized topography but also bioactivity derived from surface chemistry. Here, coatings enriched with calcium phosphates, bioactive ceramics, or nanocomposites provide sustained osteoconductivity and tailored resorption kinetics. Importantly, the textured surfaces produced by laser modification act as ideal anchoring platforms, ensuring superior adhesion and stability of these coatings under physiological loading. Thus, the natural progression from laser-induced texturing is the integration of laser-assisted bioactive coatings, which combine topographic cues with biochemical functionality to achieve durable osseointegration and implant longevity.

3. LASER-ASSISTED BIOACTIVE COATINGS

Laser-assisted bioactive coatings represent a significant advancement in surface engineering of orthopedic and dental implants, combining the topographic precision of laser modification with the biochemical functionality of osteoconductive and osteoinductive materials. Unlike conventional plasma spraying or sol–gel deposition, laser-based methods provide localized melting, rapid solidification, and strong metallurgical bonding between the coating and substrate, thereby reducing the risk of delamination and coating failure under cyclic loading conditions (Balu et al., 2021; Singh et al., 2023). Moreover, laser processing minimizes thermal degradation of bioactive phases such as hydroxyapatite (HA) or tricalcium phosphate (TCP), enabling the fabrication of coatings with crystallinity and stoichiometry closely resembling natural bone (Paital & Dahotre, 2009).

Figure 3. Schematic of the laser deposition technique

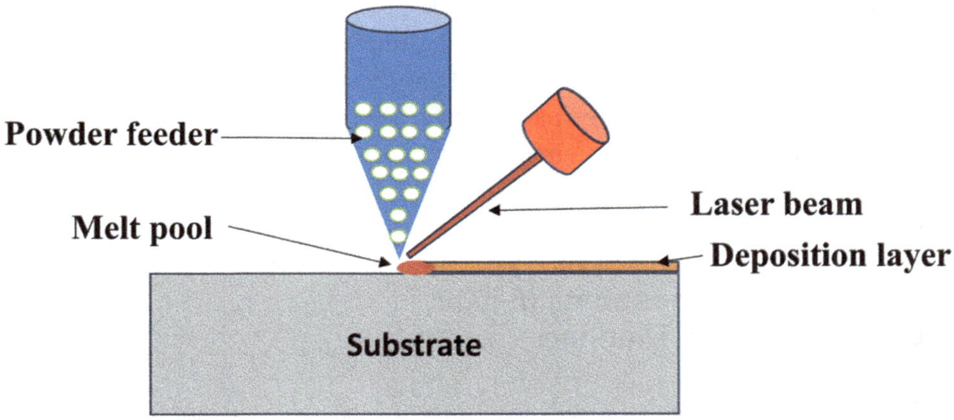

Table 2. Types of Laser-Assisted Bioactive Coatings and Their Key Functions

Coating Type	Substrate	Key Functions	Advantages
β-TCP	Titanium	Oste conduction + controlled dissolution	Enhances early bone bonding
BCP (HA + β-TCP)	Zirconia	Long-term stability + gradual ion release	Balanced bioactivity and resorption
Hydroxyapatite (HA)	Ti & Zr	Strong bone bonding	High similarity to natural bone mineral
Bioactive Glass	Ti & Zr	Ion release (Ca, Si, P), angiogenesis, antibacterial	Multifunctional properties
Hybrid Nanocomposites	Ti & Zr	Osseointegration + antibacterial + mechanical strength	Tailored multifunctionality

The rationale for such coatings lies in addressing the dual challenge of implant integration: while micro- and nano-texturing enhance early osteoblast adhesion, chemical bioactivity is crucial for long-term bone remodeling and mineralisation. Calcium phosphate–based coatings, especially β-tricalcium phosphate (β-TCP) and biphasic calcium phosphate (BCP; HA + TCP), are widely explored due to their compositional similarity to bone mineral and their balance between mechanical stability and controlled resorption (Surmenev et al., 2021). Recent advances also extend towards laser-assisted nanocomposite coatings incorporating carbon nanotubes, graphene, and metallic nanoparticles, offering multifunctional properties such as antibacterial activity, corrosion resistance, and improved mechanical resilience (Eliaz, 2019). Figure 3 represents the schematic of the laser deposition technique.

By integrating topographic cues from laser structuring with chemical cues from bioactive coatings, laser-assisted approaches provide a synergistic strategy for achieving superior osseointegration, reducing implant failure, and extending the functional lifespan of orthopedic and dental prostheses. The following subsections discuss specific coating systems, including β-TCP coatings on titanium, biphasic calcium phosphate coatings on zirconia, and advanced nanocomposite coatings developed through laser-assisted hybrid techniques. Table 2 presents various laser-assistive bioactive coatings.

3.1 β-Tricalcium Phosphate (β-TCP) Laser Coatings on Titanium

Titanium and its alloys are widely recognized as the gold standard for orthopedic and dental implants due to their favorable mechanical properties, biocompatibility, and corrosion resistance. However, their bioinert nature often results in insufficient osseointegration, particularly in patients with compromised bone quality. To address this limitation, β-tricalcium phosphate (β-TCP) coatings deposited via laser-assisted techniques have emerged as a promising strategy to enhance implant bioactivity while maintaining structural integrity.

β-TCP is particularly attractive for biomedical applications due to its moderate solubility, osteoconductivity, and bioresorbability. Unlike hydroxyapatite (HA), which provides long-term stability but slower remodeling, β-TCP undergoes controlled dissolution, releasing calcium and phosphate ions that stimulate osteogenic cell proliferation and new bone formation (Dorozhkin, 2017). The degradation rate of β-TCP aligns with the bone healing process, allowing for dynamic bone–implant interaction and gradual replacement of the coating by natural bone tissue (Ibrahim et al., 2020). Laser-assisted deposition techniques -such as laser cladding, laser ablation-assisted coating, and pulsed laser deposition (PLD)-enable precise incorporation of β-TCP on titanium substrates. These methods generate a strong metallurgical bond at the interface due to localized melting and rapid solidification, which minimizes the risk of delamination that is often observed in plasma-sprayed coatings (Paital & Dahotre, 2009). Furthermore, laser control over parameters such as power density, pulse duration, and scanning speed allows tuning of coating crystallinity and microstructure, thereby tailoring dissolution kinetics and cellular responses (Singh et al., 2023).

In vitro and in vivo studies have demonstrated that titanium implants coated with β-TCP via laser methods exhibit enhanced osteoblast adhesion, proliferation, and mineralization, compared to uncoated or conventionally coated implants. For instance, laser-deposited β-TCP coatings with a thickness of 20–50 μm have been shown to increase new bone-to-implant contact by over 40% within the first 4-6 weeks post-implantation in rabbit femoral models (Duta, L. (2021). Additionally, the presence of micro-nano hierarchical features introduced during laser processing synergistically promotes stem cell differentiation into osteogenic lineages, accelerating osseointegration (Surmenev et al., 2021).

Another critical advantage of laser-assisted β-TCP coatings lies in their ability to integrate antibacterial functionalities by co-depositing trace elements such as silver, zinc, or strontium into the coating matrix. Such composite coatings not only maintain osteoconductivity but also inhibit bacterial colonization, thereby reducing the risk of peri-implantitis, one of the major causes of late implant failure (Kumar et al., 2025). Overall, β-TCP coatings applied through laser-assisted methods represent a clinically relevant approach to bridge the gap between mechanical stability and biological activity in titanium-based implants. Their controlled resorption, osteoinductive ion release, and strong adhesion to the metallic substrate make them highly suitable for next-generation load-bearing orthopedic and dental devices.

Using pulsed Nd: YAG lasers, β-TCP can be sintered onto titanium implant surfaces, forming crystalline coatings with Ca/P ratios close to bone. Optimised laser parameters (2 kW, 0.2 mm/s) yielded adherent, crack-free coatings with elastic modulus (≈19 GPa) similar to cortical bone, improving biomechanical compatibility.

3.2 Biphasic Calcium Phosphate Coatings on Zirconia

Zirconia has emerged as a reliable alternative to titanium for dental implants, largely due to its excellent biocompatibility, favorable aesthetics, and absence of metallic ion release. However, as with titanium, its bioinert surface chemistry limits direct bone bonding, often requiring extended healing times before achieving functional osseointegration. To overcome this drawback, biphasic calcium phosphate (BCP) coatings, composed of hydroxyapatite (HA) and β-tricalcium phosphate (β-TCP) in tailored ratios, have been extensively investigated for zirconia implants. BCP combines the stability of HA with the bio-resorbability of β-TCP, providing a balanced material that supports both immediate implant stability and long-term bone remodeling (Dorozhkin, 2017). The HA phase ensures sustained bioactivity, while β-TCP gradually dissolves, releasing Ca^{2+} and PO_4^{3-} ions that stimulate osteoblast proliferation and extracellular matrix mineralization.

Laser-assisted deposition techniques, such as pulsed laser deposition (PLD), laser cladding, and laser ablation-assisted coating, offer significant advantages in applying BCP coatings onto zirconia substrates. These methods ensure contamination-free processing, dense and uniform coating formation, and strong interfacial adhesion due to localized surface melting of zirconia and rapid solidification of the BCP layer (Surmenev et al., 2021).

Experimental evidence supports the clinical relevance of such coatings. In vitro studies report that zirconia implants coated with 70% HA–30% β-TCP via PLD show up to 65% higher osteoblast viability compared to uncoated zirconia (Liu et al., 2021). Animal studies further demonstrate significantly greater bone-to-implant contact (BIC) for BCP-coated zirconia compared to bare surfaces, with early bone formation observed within 3-4 weeks of implantation (Zhang et al., 2019). Additionally, the gradual dissolution of β-TCP ensures a continuous supply of osteogenic ions during the early healing phase, while HA provides long-term anchorage of the implant within the bone matrix. Another important aspect of laser-assisted BCP coatings is their ability to incorporate biofunctional dopants, such as strontium (Sr), magnesium (Mg), or silicon (Si), during deposition. These dopants enhance angiogenesis, modulate osteoclast activity, and improve the mechanical resilience of the coating (Afonso et al., 2017). Furthermore, combining BCP coatings with laser-induced surface texturing has been shown to create synergistic effects, where topographic guidance and bioactive ion release collectively accelerate osseointegration and reduce healing time.

In summary, laser-assisted BCP coatings on zirconia implants represent a transformative approach for dental and orthopedic applications. By integrating the stability of HA with the resorbability of β-TCP, and leveraging the precision of laser deposition, these coatings offer a clinically viable pathway to enhance bone bonding, reduce healing time, and improve long-term implant success rates. Laser-assisted deposition of biphasic CaP on zirconia yielded smooth, adherent coatings without conventional furnace treatments. This dual-phase (HA + TCP) system balanced osteoconduction with controlled resorption, promoting long-term stability.

3.3 Hydroxyapatite (HA) Coatings on Titanium and Zirconia via Laser-Assisted Deposition

Hydroxyapatite (HA), a crystalline calcium phosphate with chemical similarity to the mineral phase of natural bone, is one of the most widely investigated coatings for orthopedic and dental implants. Its biocompatibility, osteoconductivity, and ability to form direct chemical bonds with bone make it highly attractive for improving osseointegration of otherwise bioinert implant substrates such as titanium and zirconia (Narayanan et al., 2025). However, conventional plasma spraying of HA, though widely employed, often results in coatings with variable crystallinity, poor interfacial adhesion, and phase decomposition, which compromise long-term clinical outcomes.

Laser-assisted deposition methods-including pulsed laser deposition (PLD), laser cladding, laser ablation, and laser melt injection-offer several advantages over traditional coating techniques. These include precise control of coating thickness (typically 1–10 μm), strong adhesion due to localised surface melting of the substrate, high phase purity of HA, and the ability to generate dense, uniform, and crack-free coatings (Surmenev et al., 2021).

On titanium implants, laser-deposited HA coatings have demonstrated significant biological benefits. In vitro studies show that HA-coated Ti implants enhance osteoblast adhesion and alkaline phosphatase (ALP) activity by 40–60% compared to uncoated controls (Wei et al., 2020). In vivo animal models report up to 70% higher bone-to-implant contact (BIC) within 4 weeks, underscoring the rapid osseointegration potential (Huang et al., 2020). Furthermore, combining HA coatings with laser surface texturing (LST) has been shown to create a synergistic effect: the micro–nano topographies guide cell alignment, while HA provides bioactive cues for osteogenic differentiation.

On zirconia implants, HA coatings deposited via PLD and laser ablation have overcome some of the limitations of zirconia's bioinertness. Liu et al. (2021) demonstrated that zirconia implants coated with crystalline HA films exhibited superior protein adsorption, early osteoblast attachment, and faster bone mineral apposition rates compared to bare zirconia. Additionally, HA coatings improved soft tissue integration, a crucial factor for preventing peri-implantitis in dental applications. The ability to deposit HA with nanoscale precision enables tailoring of surface energy and wettability, further enhancing cell–material interactions (Cheng et al., 2018).

Clinically, HA coatings deposited by laser methods offer the potential for longer-lasting and mechanically stable integration compared to plasma-sprayed coatings. Strong interfacial bonding reduces the risk of delamination under cyclic loading, a critical limitation in conventional HA-coated implants. Moreover, the high reproducibility and contamination-free nature of laser processes align well with medical-grade manufacturing requirements, supporting their translation into orthopedic and dental implant production.

In summary, laser-assisted HA coatings on titanium and zirconia implants combine the osteoconductive benefits of HA with the precision, purity, and durability of laser deposition techniques. These coatings provide enhanced early-stage osseointegration, long-term stability, and multifunctional bioactivity, representing a major advance in surface engineering strategies for next-generation implants.

3.4 Bioactive Glass and Composite Coatings via Laser Deposition

Bioactive glasses (BGs), first introduced by Hench in 1969, have emerged as a promising class of implant coatings due to their ability to form a hydroxycarbonate apatite (HCA) layer in physiological fluids, which establishes a direct bond with bone tissue (Hench, 2006). Unlike crystalline hydroxyapatite coatings, BGs also release biologically active ions such as Si, Ca, P, and Na, which stimulate osteogenic differentiation, angiogenesis, and antibacterial responses (Rahaman et al., 2014). When combined with laser-assisted deposition techniques, bioactive glass coatings achieve enhanced interfacial adhesion, structural stability, and compositional control, offering a multifunctional surface modification strategy for orthopedic and dental implants.

Laser-based approaches—including pulsed laser deposition (PLD), laser cladding, and laser ablation—enable the deposition of dense, uniform BG films on metallic and ceramic substrates. Compared to conventional sol–gel or thermal spraying methods, laser deposition offers high purity, controlled stoichiometry, and tailored micro–nanostructures, which are crucial for reliable bioactivity (Baino & Vitale-Brovarone, 2018).

On zirconia implants, BG coatings applied via PLD and laser ablation enhance the otherwise bioinert ceramic surface. Studies have shown that BG-coated zirconia promotes early protein adsorption, mesenchymal stem cell differentiation, and angiogenic marker expression compared to bare zirconia (Cheng et al., 2019). Such coatings not only improve osseointegration but also contribute to soft tissue sealing, a critical factor in dental implantology to prevent peri-implantitis. Moreover, by adjusting the SiO_2, CaO, and P_2O_5 ratios during deposition, laser processing allows tailoring of the dissolution kinetics and mechanical strength of the coatings, thereby customising bioactivity to clinical needs (Baino & Kargozar, 2022).

Beyond single-phase BG coatings, laser techniques also enable composite and doped bioactive glass coatings for multifunctional implants. For instance, Ag- or Cu-doped BG coatings impart potent antibacterial properties while maintaining osteoconductivity, effectively reducing the risk of implant-associated infections (Cacciotti, 2017). Similarly, Sr- or Mg-doped BG coatings enhance osteogenesis and vascularization, making them particularly beneficial in osteoporotic or compromised bone environments (Zhang et al., 2021). Composite coatings that integrate BG with polymers (e.g., chitosan, PEEK) have also been successfully deposited via laser cladding, creating bioactive yet flexible interfaces suitable for load-bearing orthopedic applications. From a clinical perspective, laser-assisted BG and composite coatings combine the bioactivity, ion release functionality, and antibacterial potential of bioactive glasses with the precision, durability, and reproducibility of laser deposition. These coatings provide a versatile surface modification strategy that not only enhances osseointegration but also offers long-term protection against bacterial colonization and implant loosening, thus addressing two major causes of implant failure.

In summary, laser-deposited bioactive glass and composite coatings represent a significant advancement in implant surface engineering. Their ability to activate cellular pathways, release therapeutic ions, and maintain strong mechanical bonding to both titanium and zirconia substrates makes them highly promising for next-generation orthopaedic and dental implants.

3.5 Multifunctional Hybrid Coatings

While single-component coatings such as hydroxyapatite (HA) or bioactive glass (BG) offer substantial improvements in osseointegration, their long-term clinical performance can be limited by issues such as mechanical fragility, dissolution kinetics, or lack of antibacterial protection. To overcome these drawbacks, recent research has focused on the development of multifunctional hybrid coatings that combine the benefits of different bioactive materials with advanced laser-assisted deposition techniques. Such coatings are designed to deliver synergistic effects, including enhanced osteoconductivity, controlled ion release, antibacterial activity, and mechanical durability.

Laser processing plays a pivotal role in hybrid coating fabrication, as it enables the precise layering, doping, and structural integration of multiple materials onto implant surfaces. Unlike conventional thermal spraying or plasma deposition, laser-assisted methods reduce contamination, preserve amorphous bioactive phases, and allow site-specific patterning for tailored functionality (Yue et al., 2024).

HA–BG Hybrid Coatings: The combination of HA and BG in laser-deposited coatings addresses the limitations of each material. HA provides structural stability and bone-mimicking crystallinity, whereas BG contributes rapid ion release and angiogenic stimulation. In vivo studies further demonstrated enhanced bone-to-implant contact and reduced fibrous tissue formation (Liu et al., 2021).

Ion-Doped Hybrid Coatings: Incorporation of therapeutic ions (Sr, Mg, Zn, Cu, Ag) into HA–BG laser coatings provide multifunctionality. Sr- and Mg-doped hybrids accelerate osteoblast differentiation and enhance angiogenesis, making them suitable for osteoporotic bone (Zhang et al., 2021). Functionally Graded Hybrid Coatings: Laser deposition also allows fabrication of functionally graded coatings (FGCs), where the composition gradually transitions from the implant substrate (Ti or ZrO_2) to a bioactive surface. For example, a gradient from Ti–BG composite at the base to pure HA or HA–BG hybrid at the surface reduces the risk of delamination, improves stress distribution, and mimics the natural transition between bone and implant (Yue et al., 2024). Such graded structures are particularly valuable for load-bearing orthopaedic implants.

Multifunctional hybrid coatings provide a comprehensive solution for orthopaedic and dental implants by combining osseointegration, antibacterial protection, angiogenesis, and mechanical reliability. Compared to conventional single-phase coatings, these advanced hybrids reduce implant failure caused by aseptic loosening, infection, or insufficient bone integration. With increasing success in preclinical trials, laser-deposited multifunctional hybrids are expected to move closer to clinical application, representing a significant step toward the next generation of long-lasting bioactive implants.

4. ADVANCED NANOCOMPOSITE LASER COATINGS

4.1 Laser-Fabricated Polymer–Ceramic Hybrid Coatings

Polymer–ceramic hybrid coatings produced via laser-assisted techniques are emerging as a powerful route to combine the osteoconductive properties of ceramic phases (hydroxyapatite, bioactive glass) with the toughness, flexibility, and drug-delivery capability of biodegradable polymers (chitosan, PLGA, PEEK derivatives). Laser methods—particularly pulsed laser deposition (PLD), matrix-assisted pulsed laser evaporation (MAPLE) and laser cladding—enable deposition of thin, conformal ceramic layers with high phase purity and strong adhesion to the substrate; subsequent or co-deposition of polymeric phases can be achieved either by laser-assisted co-ablation or by low-temperature laser patterning that fixes polymer layers into porous architectures. The resulting hybrids combine mechanical compliance with sustained release kinetics: the ceramic phase provides a bioactive scaffold for bone bonding while the polymeric matrix controls porosity and local elution of antibiotics, growth factors or osteogenic drugs. Recent studies highlight that laser processing allows spatial control of polymer content and microstructure, producing coatings with tunable drug release profiles and improved mechanical fatigue resistance attributes especially valuable in dental implants and localised orthopaedic reconstructions where on-demand, local therapy reduces systemic side effects. Key technical challenges remain in avoiding thermal degradation of polymers during laser steps and in ensuring long-term polymer stability in vivo, but process innovations (low-fluence PLD, MAPLE) and hybrid thermal management strategies are rapidly advancing this field.

4.2 Laser-Induced Metal–Ceramic Nanocomposites

Laser cladding, selective laser melting (SLM) and laser remelting have enabled the fabrication of metal–ceramic nanocomposite surfaces and surface layers in which hard, bioactive ceramic phases (e.g., HA, TiC, TiB, or bioactive glass particulates) are uniformly embedded within a ductile metallic matrix (Ti, Ti–Zr, Ti–Ta). These laser routes produce metallurgically bonded coatings with minimal interfacial porosity and graded microstructures, overcoming the adhesion and delamination problems classically associated with thermally sprayed ceramics. The in-situ formation of ceramic reinforcements during rapid solidification (for example TiC/TiB formation during SLM or laser cladding) yields significant gains in surface hardness, wear resistance and fatigue life while retaining adequate toughness under cyclic loads — critical for load-bearing orthopedic applications. Recent reviews and experimental reports document improved tribological behavior and corrosion resistance of such laser-fabricated titanium matrix composites, and show that careful control of powder mixes, laser energy density and scan strategy can limit cracking and porosity formation that otherwise degrade performance. Translationally, laser-induced metal–ceramic nanocomposites are attractive because they can be produced directly on complex geometries and integrate with SLM workflows for patient-specific implants, although standardization of feedstock and qualification under physiologic fatigue/corrosion regimes remains necessary.

4.3 Laser-Surface Alloyed Antibacterial Coatings

Laser surface alloying and laser cladding permit precise incorporation of antibacterial metallic ions (Ag, Cu, Zn) or formation of ion-doped oxide/ceramic matrices (Ag-HA, Zn-doped CaP, Cu-doped TiO_2) within the topmost microns of an implant surface. The rapid melting and solidification inherent to laser processing enables trapping of dopants in metastable or finely dispersed phases, which moderates ion release and extends antibacterial efficacy without the high burst release and cytotoxicity seen in some conventional coatings. Several recent studies demonstrate that laser-alloyed surfaces can achieve potent reductions in *Staphylococcus aureus* and *Escherichia coli* colonization while preserving or even enhancing osteoblast viability and differentiation — a dual osteogenic/antibacterial function essential for clinical longevity. The laser toolbox additionally supports graded dopant profiles (higher antibacterial ion content at the outermost surface, lower concentration toward the substrate) to balance early antimicrobial protection with later tissue compatibility. Remaining challenges include rigorous in vivo quantification of ion release kinetics under dynamic loading and body fluid flow, and ensuring regulatory-acceptable safety margins for released metal ions; nonetheless, laser alloying provides one of the most controllable routes for creating durable, dual-function implant surfaces.

4.4 Future Perspectives: Smart and Multifunctional Coatings Designed by Laser Processing

Looking forward, laser processing is uniquely positioned to create smart and stimuli-responsive coatings that combine structural bioactivity with on-demand therapeutic functions. Three convergent trends are noteworthy. First, photothermal/photodynamic strategies integrate plasmonic or carbonaceous photothermal agents (gold nanostructures, graphene derivatives, polypyrrole) into laser-deposited matrices; under near-infrared illumination these agents produce localized heating sufficient to eradicate biofilms while preserving adjacent tissue when carefully controlled — an attractive adjunct for in situ treatment of peri-implant infections. Second, laser-crafted micro-architectures can incorporate reservoirs, channels or graded porosity designed for controlled, stimulus-triggered drug release (e.g., heat-triggered polymer phase transitions or light-triggered release), enabling on-demand therapies without systemic exposure. Third, coupling laser fabrication with advances in smart feedstock chemistry (ion-release morphologies, biodegradable carriers, and bioresponsive polymers) will permit coatings that dynamically adapt to the healing environment — for example, releasing pro-angiogenic factors during early repair and then switching to osteoinductive release profiles. While these directions are scientifically promising, clinical translation will hinge on standardizing laser process descriptors, comprehensive long-term in vivo studies that combine mechanical fatigue, corrosion and biological response, and regulatory pathways for multifunctional implants. The coming decade is likely to see laser-engineered smart coatings move from proof-of-concept to translational demonstration as additive manufacturing and laser-processing pipelines become more tightly integrated in medical device manufacturing.

5. FAILURE MECHANISMS IN LASER-COATED IMPLANTS

Despite the clear advantages of laser-assisted surface coatings in enhancing osseointegration, corrosion resistance, and mechanical performance, failure of the coating–substrate system remains a critical challenge in both orthopedic and dental applications. Understanding these mechanisms is essential to optimize laser parameters, coating design, and clinical durability.

5.1 Mechanical Failure Modes

Laser-coated implants are subject to mechanical stresses from cyclic loading, mastication (in dental implants), or body movement (orthopedic implants). Failure typically occurs at the coating–substrate interface or within the coating itself, manifesting as adhesive or cohesive failure. Selective laser melting (SLM) Co–Cr dental alloys with laser-deposited ceramic coatings often fail cohesively due to strong interfacial bonding, whereas cast alloys with lower metallurgical compatibility show adhesive failures, where the coating delaminates under tensile or shear loads (Dikova et al., 2019). Laser parameters—energy density, scanning speed, and beam overlap—directly influence coating adhesion, residual stress, and microstructure, thereby affecting mechanical reliability.

5.2 Stress Concentration and Crack Initiation

Rapid melting and solidification inherent to laser processing can induce thermal gradients, generating residual stresses that concentrate at sharp edges, corners, or coating discontinuities. Microcracks often initiate at these high-stress regions and can propagate under cyclic physiological loads, ultimately leading to premature coating delamination or fracture. Finite element analysis (FEA) studies reveal that graded or hybrid coatings (dense core with porous surface) can mitigate stress concentrations, enhance fatigue resistance while maintain bioactive surface features (Robau-Porrua et al., 2025).

5.3 Corrosion and Wear-Induced Degradation

In vivo, laser-coated implants encounter corrosive bodily fluids, ionic exchange, and wear from articulating surfaces. Even coatings with strong adhesion can undergo localized degradation if microdefects or porosity exist. Studies on β-TCP and HA coatings on titanium show that laser parameters optimized for crystallinity and densification reduce dissolution and maintain bioactivity over longer periods (Hussein et al., 2021). Incorporating laser-induced graded layers or nanocomposite reinforcements further improves wear resistance, reducing the risk of particle release and subsequent inflammatory responses.

6. CONCLUSION

Laser coating technologies represent a transformative advancement in the surface engineering of orthopaedic and dental implants. By enabling precise control over surface chemistry, topography, and bioactivity, laser-based processes overcome the limitations of conventional coating methods and address key clinical challenges, including poor osseointegration, stress shielding, and infection. The strategic incorporation of biphasic calcium phosphate (BCP) and β-tricalcium phosphate (β-TCP) provides osteoconductive and bioresorbable functionalities that align with the natural bone healing process, while nanocarbon reinforcements impart mechanical robustness, wear resistance, and antimicrobial activity. This multifunctional synergy positions laser-engineered coatings as a next-generation solution for long-lasting and biologically active implants. Moving forward, translational studies and clinical validation will be critical to establishing standardised protocols and fully realising the potential of these eco-friendly, adaptive, and highly effective technologies in improving patient outcomes.

REFERENCES

Afonso, C. R. M., Vidilli, A. L., Spinelli, J. E., Riva, R., Amigó, V., & Kiminami, C. S. (2017). An assessment of microstructure and properties of laser clad coatings of ultrafine eutectic β Ti-Fe-Nb-Sn composite for implants. *Surface and Coatings Technology*, *328*, 161–171. DOI: 10.1016/j.surfcoat.2017.08.035

Albrektsson, T., Chrcanovic, B., Mölne, J., & Wennerberg, A. (2017). Foreign body reactions, marginal bone loss, and allergies in relation to titanium implants. *European Journal of Oral Implantology*, *10*(1), 7–19. PMID: 30109298

Anselme, K., Davidson, P., Popa, A. M., Giazzon, M., Liley, M., & Ploux, L. (2010). The interaction of cells and bacteria with surfaces structured at the nanometre scale. *Acta Biomaterialia*, *6*(10), 3824–3846. DOI: 10.1016/j.actbio.2010.04.001 PMID: 20371386

Baino, F., & Kargozar, S. (Eds.). (2022). *Bioactive glasses and glass-ceramics: Fundamentals and applications*. John Wiley & Sons. DOI: 10.1002/9781119724193

Bhaduri, S. B., Narayan, R., & Balani, K. (2021). *Engineering Biomaterials for Regenerative Medicine: Novel Technologies for Clinical Applications*. Springer.

Bigerelle, M., Anselme, K., Noel, B., Ruderman, I., Hardouin, P., & Iost, A. (2002). Improvement in the morphology of Ti-based surfaces: A new process to increase in vitro human osteoblast response. *Biomaterials*, *23*(7), 1563–1577. DOI: 10.1016/S0142-9612(01)00271-X PMID: 11922462

Cacciotti, I. (2017). Bivalent cationic ions doped bioactive glasses: The influence of magnesium, zinc, strontium and copper on the physical and biological properties. *Journal of Materials Science*, *52*(15), 8812–8831. DOI: 10.1007/s10853-017-1010-0

Cao, X., Meng, X., Zhang, L., Li, X., & Wang, C. (2020). Femtosecond laser micro/nanostructuring of zirconia ceramics for improved osteoblast responses. *Applied Surface Science*, *518*, 146178. DOI: 10.1016/j.apsusc.2020.146178

Cheng, L., Wu, H., Li, J., Zhao, H., & Wang, L. (2021). Polydopamine modified ultrathin hydroxyapatite nanosheets for anti-corrosion reinforcement in polymeric coatings. *Corrosion Science*, *178*, 109064. DOI: 10.1016/j.corsci.2020.109064

Costa, B. C., Ponciano Gomes, J. A., & Soares, P. (2020). Surface texturing of titanium for biomedical applications: From mechanical to biological functionality. *Applied Surface Science*, *532*, 147464.

da Rocha Silva, I., da Silva Barreto, A. T., Seixas, R. S., Paes, P. N. G., do Nascimento Lunz, J., Thiré, R. M. D. S. M., & Jardim, P. M. (2023). Novel Strategy for Surface Modification of Titanium Implants towards the Improvement of Osseointegration Property and Antibiotic Local Delivery.

Daskalova, A., Bliznakova, I., & Kochev, V. (2022). Surface engineering of biomedical implants using femtosecond lasers: Advances and perspectives. *Journal of Biomedical Materials Research. Part B, Applied Biomaterials*, *110*(3), 451–468. DOI: 10.1002/jbm.b.35027

Dikova, T., Vasilev, T., & Dolgov, N. (2019). Failure of ceramic coatings on cast and selective laser melted Co-Cr dental alloys under tensile test: Experiment and finite element analysis. *Engineering Failure Analysis*, *105*, 1045–1054. DOI: 10.1016/j.engfailanal.2019.07.018

Drevet, R., Fauré, J., & Benhayoune, H. (2023). Bioactive calcium phosphate coatings for bone implant applications: A review. *Coatings*, *13*(6), 1091. DOI: 10.3390/coatings13061091

Duta, L. (2021). In Vivo Assessment of Synthetic and Biological-Derived Calcium Phosphate-Based Coatings Fabricated by Pulsed Laser Deposition: A Review. *Coatings*, *11*(1), 99. DOI: 10.3390/coatings11010099

Geetha, M., Singh, A. K., Asokamani, R., & Gogia, A. K. (2009). Ti based biomaterials, the ultimate choice for orthopaedic implants – A review. *Progress in Materials Science*, *54*(3), 397–425. DOI: 10.1016/j.pmatsci.2008.06.004

Gnilitskyi, I.. (2022). Femtosecond laser-induced surface modifications of Ti-6Al-4V alloys: Toward improved bioactivity and antibacterial performance. *Applied Surface Science*, *573*, 151511.

Goriainov, V., Cook, R., Latham, J. M., Dunlop, D. G., & Oreffo, R. O. C. (2014). Bone and metal: An orthopaedic perspective on osseointegration of metals. *Acta Biomaterialia*, *10*(10), 4043–4057. DOI: 10.1016/j.actbio.2014.06.004 PMID: 24932769

Hanawa, T. (2011). A comprehensive review of techniques for biofunctionalization of titanium. *Journal of Periodontal & Implant Science*, *41*(6), 263. DOI: 10.5051/jpis.2011.41.6.263 PMID: 22324003

Hussein, B. M. A., Safi, I. N., Al Shammari, A. M., & Tawfiq, T. A. (2019). Implementation and characterization of coating pure titanium dental implant with sintered β-TCP using Nd:YAG laser. *The Saudi Dental Journal*, *31*(2), 242–250. DOI: 10.1016/j.sdentj.2018.12.004 PMID: 30983835

Hussein, M. A., Kumar, A. M., Ankah, N., & Azeem, M. A. (2021). Thermal treatment effect on the surface and in vitro corrosion characteristics of arc deposited TiN coating on Ti alloy for orthopedic applications. *Ceramics International*, *47*(16), 23203–23213. DOI: 10.1016/j.ceramint.2021.05.032

Jin, H. W., Noumbissi, S., & Wiedemann, T. G. (2024). Comparison of zirconia implant surface modifications for optimal osseointegration. *Journal of Functional Biomaterials*, *15*(4), 91. DOI: 10.3390/jfb15040091 PMID: 38667548

Kisała, J., Gnilitskyi, I., Cieniek, B., Krzemiński, P., Marchewka, M., Barylyak, A., & Bobitski, Y. (2021). Synthesis of micro-spikes and herringbones structures by femtosecond laser pulses on a titanium plate—A new material for water organic pollutants degradation. *Materials (Basel)*, *14*(19), 5556. DOI: 10.3390/ma14195556 PMID: 34639953

Kunz, C., Müller, F. A., & Gräf, S. (2018). Multifunctional hierarchical surface structures by femtosecond laser processing. *Materials (Basel)*, *11*(5), 789. DOI: 10.3390/ma11050789 PMID: 29757240

Liu, X., Chu, P. K., & Ding, C. (2019). Surface modification of titanium, titanium alloys, and related materials for biomedical applications. *Materials Science and Engineering R Reports*, *47*(3-4), 49–121. DOI: 10.1016/j.mser.2004.11.001

Liu, Y., Rui, Z., Cheng, W., Song, L., Xu, Y., Li, R., & Zhang, X. (2021). Characterization and evaluation of a femtosecond laser-induced osseointegration and an anti-inflammatory structure generated on a titanium alloy. *Regenerative Biomaterials*, *8*(2), rbab006. DOI: 10.1093/rb/rbab006 PMID: 33738120

Logesh, M., Toan, N. K., Ahn, S.-G., & Choe, H.-C. (2024). Engineering bioactive MWCNT-reinforced hydroxyapatite coatings on Ti29Nb5Zr alloy by plasma electrolytic oxidation method. *Journal of Alloys and Compounds*, *1006*, 176339. DOI: 10.1016/j.jallcom.2024.176339

Martínez-Calderon, M., Martín-Palma, R. J., Rodríguez, A., Gómez-Aranzadi, M., García-Ruiz, J. P., Olaizola, S. M., & Manso-Silván, M. (2020). Biomimetic hierarchical micro/nano texturing of TiAlV alloys by femtosecond laser processing for the control of cell adhesion and migration. *Physical Review Materials*, *4*(5), 056008. DOI: 10.1103/PhysRevMaterials.4.056008

Narayan, R. J. (2010). The next generation of biomaterial development. Philosophical Transactions of the Royal Society A: Mathematical, Physical and Engineering Sciences, 368(1917), 1831-1837.

Narayanan, S., Singh, M., & Kandasubramanian, B. (2025). Innovative Approaches in Regenerative Medicine for the Development of Sustainable and Bioactive Implants. *Biomedical Materials & Devices*, 1–25.

Pereira, L. M., Nawaz, Q., Ferreira, N. M., Pereira, A. B., Detsch, R., Boccaccini, A. R., & Mesquita-Guimarães, J. (2025). Effects of S53P4 bioactive glass coated on double-size laser-textured zirconia substrates: Cell adhesion and mineralisation behaviour. *Ceramics International*.

Rahaman, M. N., Xiao, W., & Bal, B. S. (2014). Development of Implants Composed of Hollow Hydroxyapatite Microspheres for Bone Regeneration. *Biomaterials Science: Processing. Properties and Applications IV: Ceramic Transactions*, *251*, 45–56.

Robau-Porrua, A., González, J. E., Arancibia-Castillo, R., Picardo, A., Araneda-Hernández, E., & Torres, Y. (2025). Design, fabrication, and characterization of novel dental implants with porosity gradient obtained by selective laser melting. *Materials & Design*, *251*, 113660. DOI: 10.1016/j.matdes.2025.113660

Rupp, F., Gittens, R. A., Scheideler, L., Marmur, A., Boyan, B. D., Schwartz, Z., & Geis-Gerstorfer, J. (2014). A review on the wettability of dental implant surfaces I: Theoretical and experimental aspects. *Acta Biomaterialia*, *10*(7), 2894–2906. DOI: 10.1016/j.actbio.2014.02.040 PMID: 24590162

Sobolev, A., Sviridov, A., & Romanov, D. (2021). Surface functionalization of zirconia and titanium implants by laser microstructuring: A review. *Optics & Laser Technology*, *140*, 106992. DOI: 10.1016/j.optlastec.2021.106992

Sun, L., Berndt, C. C., Gross, K. A., & Kucuk, A. (2001). Material fundamentals and clinical performance of plasma-sprayed hydroxyapatite coatings: A review. *Journal of Biomedical Materials Research*, *58*(5), 570–592. DOI: 10.1002/jbm.1056 PMID: 11505433

Surmenev, R. A., Surmeneva, M. A., & Ivanova, A. A. (2014). Significance of calcium phosphate coatings for the enhancement of new bone osteogenesis – A review. *Acta Biomaterialia*, *10*(2), 557–579. DOI: 10.1016/j.actbio.2013.10.036 PMID: 24211734

Surmenev, M., Grubova, I., Glukhova, N., Khrapov, D., Koptyug, A., Volkova, A., Ivanov, Y., Cotrut, C. M., Vladescu, A., Teresov, A., Koval, N., Tyurin, A., & Surmenev, R. (2021). New Ti–35Nb–7Zr–5Ta alloy manufacturing by electron beam melting for medical application followed by high current pulsed electron beam treatment. *Metals*, *11*(7), 1066. DOI: 10.3390/met11071066

Vorobyev, A. Y., & Guo, C. (2013). Direct femtosecond laser surface nano/micro-structuring and its applications. *Laser & Photonics Reviews*, *7*(3), 385–407. DOI: 10.1002/lpor.201200017

Xing, Y., Xiong, W., Li, X., & Zhang, J. (2021). Biofunctionalization of zirconia implant surfaces via femtosecond laser-induced microstructures. *Journal of Biomedical Materials Research. Part A*, *109*(10), 1875–1886.

Yang, Y., Wen, C., & Hodgson, P. D. (2020). Laser surface modification of titanium alloys for biomedical applications. *Materials & Design*, *191*, 108626.

Yue, W., Zhang, Y., Zheng, Z., & Lai, Y. (2024). Hybrid laser additive manufacturing of metals: A review. *Coatings*, *14*(3), 315. DOI: 10.3390/coatings14030315

Zhang, L.. (2021). Laser-induced periodic surface structures (LIPSS) on titanium with simultaneous antibacterial and biocompatible effects. *Journal of Biomedical Materials Research. Part A*, *109*(3), 362–373. PMID: 32506802

Zhang, L., Xiong, N., Liu, Y., & Gan, L. (2021). Biomimetic cell-adhesive ligand-functionalized peptide composite hydrogels maintain stemness of human amniotic mesenchymal stem cells. *Regenerative Biomaterials*, *8*(2), rbaa057. DOI: 10.1093/rb/rbaa057 PMID: 33738111

Zhang, T., Zhang, X., Mao, M., Li, J., Wei, T., & Sun, H. (2020). Chitosan/hydroxyapatite composite coatings on porous Ti6Al4V titanium implants: In vitro and in vivo studies. *Journal of Periodontal & Implant Science*, *50*(6), 392. DOI: 10.5051/jpis.1905680284 PMID: 33350179

Zhang, Y., Wu, C., & Ding, J. (2021). Laser-induced hierarchical surface structures for guiding cell behavior and tissue integration. *Bioactive Materials*, *6*(7), 2091–2105. DOI: 10.1016/j.bioactmat.2021.01.033

Zhou, X., Yu, W., & Huang, Q. (2020). Synergistic effect of laser microtexturing and hydroxyapatite coating on titanium implants: Enhancing osseointegration. *Ceramics International*, *46*(15), 23677–23685.

Chapter 8
Surface Functionalization of Biomedical Textiles Through Laser-Induced Coatings

Sushila Sahani
http://orcid.org/0000-0002-5739-144X
Lovely Professional University, Punjab, India

ABSTRACT

The incorporation of advanced surface engineering techniques in biomedical textiles has opened new pathways for developing high performance, multifunctional textiles which can be tailored for healthcare applications. This chapter explores the role of laser-induced coatings and treatments in the surface functionalization of biomedical textiles. These technologies offer contactless, solvent-free, and localized surface modification. Such textile treatments enable the deposition of biocompatible and antimicrobial coatings on diverse textile substrates. These processes fine-tune surface properties like roughness, wettability, cell adhesion, and composition. These modifications are vital for wound dressings and coatings, offering antibacterial, biosensing, and tissue compatible functions. Laser-induced surface functionalization not only extends the functional limits of biomedical fabrics but also aligns with eco-friendly manufacturing goals. This chapter highlights research, challenges, and future directions, positioning laser-induced coating as key to next-generation medical textiles.

DOI: 10.4018/979-8-3373-4104-0.ch008

INTRODUCTION

Biomedical textiles are composed of fibrous materials which are specifically designed for use in medical surroundings. The medical and health sector textiles have seen significantly growing in recent years and this development is happening due to advancement of technology in the textile as well as medical field. Biomedical textiles have many applications in medical settings. They can be used as simple cleaning wipes, sophisticated barrier textiles of Operation Theater, thread suture as well as intricate composite constructs for bone replacement (Akter et al., 2014). To fulfill such diverse and highly specialized medical requirements, it is essential to enhance the performance of textile substrates beyond their intrinsic properties. One major and effective approach is surface functionalization. Surface functionalization of textile materials is a technique that involves modifying the outermost layer of fibers or fabrics. The goal is to create enhanced or entirely new properties without compromising the bulk characteristics of the material (Nadi et al, 2018). The primary aim is to improve qualities such as biocompatibility, antimicrobial activity, hydrophobicity, and cell adhesion, which are crucial for a broad range of high-performance medical textiles (Zhang et al., 2023). To create various such medical textiles, various techniques are employed. These techniques include plasma treatment, polymerization, nanotechnology applications, and chemical or photo-induced grafting. These techniques allow targeted addition of functional groups, nanoparticles, or bioactive molecules on textile surfaces making them suitable for various medical applications (Nasef et al., 2021). For instance, plasma treatment can increase surface energy which enhances the adhesion of coatings or functional layers (Hegemann et al., 2003), while chemical grafting and nanocoatings can introduce properties such as bacterial resistance, water repellency, or improved interaction with biological tissues (Elashnikov et al., 2021). Methods like photo-grafting and sol-gel processes can covalently bond functional agents to the surface which ensures durability and continued performance even after repeated washing or exposure to environmental stress (Ghosh et al., 2025). This surface specific approach is predominantly valuable in contemporary textile engineering where adding multi-functionality (eg: self-cleaning, flame retardancy, or responsive sensing) is appreciated and must be achieved sustainably and safely. Given current environmental concerns, the goal is to avoid use of environmentally hazardous chemicals and preserving the underlying structure and comfort of the fabric.

Laser induced treatment of biomedical textiles can prove to be sustainable in current landscape. Unlike traditional methods such as mechanical abrasion, chemical etching, or dip-coating, laser-induced techniques do not require physical contact with the substrate, thereby eradicating the issues of tool wear, physical damage, and contamination (Al Javed & Bin Rashid, 2024). In a typical laser induced treatment process, a high-energy laser beam is directed and precisely focused onto the material's surface and the energy input can rapidly melt or modify thin surface layer. This process can sometimes incorporate additional materials supplied as powders, gases, or pre-coatings to create functional coatings or micro/nanopatterns (Park et al., 2024). It is a non-invasive approach which enables excellent control over the area being modified which allow for the selective treatment of designated regions. The scale of these changes can be brought down to micrometer scales without affecting the surrounding material. This aspect is particularly advantageous for delicate or high-value materials and complex geometries. Key advantages over traditional techniques include speed and automation suitability, high repeatability, and minimized environmental impact by avoiding toxic chemicals or by-products. It also has potential for integration with digital manufacturing workflows for rapid customization or small-batch, high-value production. Since laser-induced coatings do not compromise the bulk properties or hand-feel of textiles and polymers, they are increasingly favored for high-performance textile applications. These applications range from smart textiles which are capable of hosting electronics to medical devices. Such advanced textiles require strict contamination control, and advanced garments where aesthetic, comfort, and durability must coexist.

This chapter aims to explore the potential of precise, non-contact laser-based techniques for enhancing the surface properties of biomedical textiles applications. Its scope covers fundamental principles, recent advancements, and applications of laser-induced coatings in biomedical textiles. The chapter provides an overview of how laser processing enables the integration of various functional agents such as graphene, metallic nanoparticles, and bioactive molecules onto textile substrates with micrometer precision. Furthermore, it discusses evolving challenges and future directions, such as improving adhesion, wash resistance, and multi-functionality for advanced medical and healthcare needs, ultimately positioning laser-induced coatings as a cornerstone technology in next-generation biomedical textile engineering.

FIBERS USED IN BIOMEDICAL TEXTILES AND THEIR RELEVENT PROPERTIES

Fibers used in biomedical textiles can be broadly classified into natural and synthetic fibers. Each of these serves distinct roles in medical applications due to their inherent structural and functional characteristics. Natural fibers, such as cotton, wool and silk, are derived from plant or animal sources and are widely valued for their biocompatibility, comfort, and high moisture absorbency. Synthetic fibers, on the other hand, are produced from petrochemical or bioengineered raw materials through controlled chemical processes. Some examples include polyester, polypropylene, polyamide (nylon), and polyethylene. The careful selection between natural and synthetic fibers or their blends is determined by application requirements. These requirements include biocompatibility, absorbency, durability, mechanical properties, and the need for use in medical devices and healthcare products. Below mentioned Table 1 provides an overview of biomedical textile fibers, highlighting their types, specific fibers, key characteristics, and detailed roles in medical applications. The table distinguishes between natural fibers (such as cotton, silk, and wool) and synthetic fibers (including polyester, polypropylene, polyamide, polyethylene, polylactic acid, and polyglycolic acid). Each fiber type demonstrates unique physicochemical properties which directly influence their suitability for different medical applications.

Table 1. Biomedical textile fiber types, properties, and detailed roles in medical applications: natural fibers

Specific Fibers	Characteristics	Roles in Medical Applications	References
Cotton	• Biocompatible, Biodegradable, • Highly Absorbent, • Soft, • Hydrophilic, • Air Permeable, • Hygroscopic, • Good Mechanical Strength • Susceptible To Microbial Contamination	• Primarily in non-implantable textiles such as wound dressings, bandages, gauzes, absorbent pads, and hospital bedding/linens; • Absorbs exudates, protects against infection, promotes healing in minor wounds or as protective wipes; • Can be functionalized with antioxidants or antibacterial for enhanced bioactivity, antioxidant activity and moderate antibacterial efficacy in bioactive variants.	(Shahriari Khalaji & Lugoloobi, 2020), (Abou Elmaaty, Elsisi, Elsayad, Elhadad, & Plutino, 2022), (Oliveira et al., 2025)
Silk	• Biocompatible, • Biodegradable, • Low inflammatory response, • Excellent mechanical strength, • High oxygen/vapor permeability, • Promotes cell adhesion • Can trigger immunogenic reactions if not properly processed.	• Used in implantable and non-implantable applications including surgical sutures (biodegradable, braided/ monofilament for wound closure) • Wound dressings (promotes epithelialization), nerve guidance conduits • Artificial tendons/ ligaments (RGD-modified for collagen production), and tissue engineering (e.g., artificial skin with keratinocytes/fibroblasts); • Also in drug delivery for controlled release and bioactive variants antibacterial efficacy • Ideal for post-surgical dressings and anti-aging textiles.	(Li et al., 2015), (Yazawa et al., 2016), (Majumder et al., 2024), (Oduor et al., 2021), (Magaz et al., 2018), (Jao et al., 2016), (Yergöz, 2017), (Bernardes et al., 2024), (Ivanovska et al., 2025)

continued on following page

Table 1. Continued

Specific Fibers	Characteristics	Roles in Medical Applications	References
Wool	• Biocompatible, • Biodegradable, • Air permeable, • Moisture-managing, • High porosity, • Mimics extracellular matrix (ECM) with cellular recognition sites, • Resilient to compression	• Applied in wound dressings, bandages, diabetic socks, antibacterial hospital linens, and scaffolds for tissue engineering (e.g., bone repair via fibril sponges supporting SAOS-2 cell growth); • Accelerates wound closure (97.8% by day 14 in animal models), enhances collagen synthesis, boosts antioxidant defenses, and reduces oxidative stress.	(Singha, Pandit, & Maity, 2022), (Patrucco et al., 2016), (Morris & Murray, 2021), (Patrucco, 2017), (Ivanovska et al., 2025)
Polyester	• Durable, • Biocompatible, • Resistant to biological degradation, • Good mechanical and thermal stability, • Customizable (e.g., strength, degradation), • Low porosity, • Infection risk, • May lack cell adhesion sites requiring modifications like collagen impregnation.	• In implantable devices such as vascular grafts/ prostheses (e.g., aortofemoral bypass) • Non-implantable uses include hospital apparel, patient gowns, compression garments, masks, and diapers	(Mogosanu et al., 2014), (de Sá Guimarães, 2021), (Wu, 2015), (Morris & Murray, 2020)

Table 2. Biomedical textile fiber types, properties, and detailed roles in medical applications: synthetic fibers

Specific Fibers	Characteristics	Roles in Medical Applications	References
Poly-propylene	• Durable • Good mechanical strength/ flexibility • Lightweight, • Chemically inert, • Resistant to degradation, • Biocompatible but often needs enhancements for endothelialization/ anticalcification.	• Used in bioprosthetic heart valves (e.g., hydrogel hybrids for long-term functionality), orthopedic implants (bone/ joint replacement), surgical meshes for hernia repair, sutures, and non-implantable PPE (masks, respirators); • Provides structural support, prevents infection, and aids in extracorporeal devices like artificial lungs.	(Polshettiwar, 2025), (Ghanbari et al., 2009), (Patel et al., 2012), (Patel et al., 2012), (Molla et al., 2024), (Morris & Murray, 2020)
Polyamide (Nylon)	• High tensile strength, • Durable, flexible, • Good mechanical/ thermal stability, • Require modifications for cell adhesion.	• Employed in non-biodegradable sutures (for tissue securing), artificial tendons, vascular grafts, compression bandages/ garments (for wound support), and orthopedic supports; • Inhibits pathogens (e.g., MRSA), controls infection in dressings, and offers durability in surgical applications.	(Deng et al., 2021), (Singh et al., 2015), (Sikka et al., 2024), (Shakiba et al., 2021), (Liu et al., 2014), (Morris & Murray, 2020)
Poly-ethylene	• High tensile strength/durability, • biocompatible, • hydrophilic, • wear-resistant, • good mechanical properties, • Customizable density (e.g., low for cartilage mimicry), but may need hybrids for better integration.	• In orthopedic implants (artificial bones/joints, cartilage), joint replacements, hybrid ligament constructs (e.g., with silk for ACL reconstruction, improving collagen regeneration), operating room drapes, and drug delivery (e.g., insulin carriers via micelles); • Provides long-term wear resistance and structural support in load-bearing applications.	(Suhag et al., 2024), (Moro et al., 2011), (Adamovic et al., 2017), (Al Ali, 2024), (Jiang et al., 2014), (Hahn et al., 2023), (Kim et al., 2016), (Ashjari et al., 2012)

continued on following page

Table 2. Continued

Specific Fibers	Characteristics	Roles in Medical Applications	References
Poly-lactic Acid (PLA)	• Biodegradable, • Biocompatible, • Non-toxic, • Good mechanical strength/ flexibility, • Electrospinnable for nanofibers, but brittle and may lack ideal surface chemistry.	• Used in bioresorbable stents (e.g., for coronary stenosis), biodegradable sutures (monofilament/braided for tissue securing), scaffolds for bone/cartilage regeneration (supports cell proliferation), bioink for 3D-printed hydrogels (organ printing), drug delivery systems (controlled release), and tissue engineering; • Degrades into non-toxic byproducts, providing temporary support.	(Elmowafy et al., 2019), (Maleki et al., 2022), (Verdoia et al., 2020), (Xu et al., 2022), (Fattahi, 2020), (Wu et al., 2019), (Li et al., 2017), (Ranakoti et al., 2023)
Poly-glycolic Acid (PGA)	• Biodegradable, • Biocompatible, • High initial mechanical strength/tensile modulus, • Hydrophilic, • Rapid degradation, but sensitive to hydrolysis and challenging for porous scaffolds.	• Integrated into warp-knitted meshes for dermal scaffolds (e.g., with collagen-chitosan for wound healing, reduced contraction), biodegradable sutures (absorbable multifilament for temporary support), bioink for 3D printing (cell growth/ differentiation), and tissue engineering (bone/cartilage regeneration); • Offers porosity for nutrient flow and degrades without toxicity.	(Sharma et al., 2021), (Low et al., 2020), (Baylón et al., 2017), (Singhal et al.,1988), (Somasekhar, 2021), (Yeo et al., 2021), (General et al., 2017), (Farjaminejad et al., 2024).

KEY PROPERTIES REQUIRED FOR BIOMEDICAL TEXTILES

Biomedical textiles are designed to interact directly with biological tissues, requiring properties like biocompatibility, mechanical durability, and controlled interactions with bodily fluids (Morris & Murray, 2020). Unmodified natural fibers, derived from plant or animal sources, offer inherent renewability and similarity to extracellular matrices, which can promote initial cell adhesion (Abdul Khalil et al., 2014). Balancing these properties is essential, as excessive porosity might compromise mechanical strength, while poor degradation control could lead to inflammation (Shamim et al., 2025). There is a need of tailoring these traits to specific applications. Natural fibers often excel in biocompatibility but synthetics provide superior mechanical durability, though controversies exist around long-term tissue responses and environmental impacts. For instance, cotton's hygroscopic nature aids in absorbent applications, while silk's tensile strength (610-690 MPa) suits sutures (Altman et al., 2003). Synthetic fibers, engineered from polymers, provide customizable strength and resistance to degradation, making them ideal for implants like vascular grafts (Zia et al., 2022). However, without modifications such as antimicrobial impregnation or surface functionalization, both types fall short in demanding biomedical environments (Thakur et al., 2022). Unmodified textiles often require enhancements to meet ISO standards for biocompatibility (e.g., ISO 10993), as their native properties can lead to suboptimal performance (Al-Zyoud et al., 2023). The challenges stem from fiber chemistry: natural fibers' porous, hydrophilic structures causes microbial growth and variable degradation, while synthetics' hydrophobic, inert surfaces resist biological integration and may accumulate toxins. Environmental impacts further complicate usage, with synthetic fibers contributing to microplastic pollution and natural fibers generating wastewater during processing. These issues highlight the need for a balanced approach, where modifications address deficiencies without increasing ecological concerns. Table 2 outlines the fundamental properties that govern the performance of biomedical textiles, including biocompatibility, mechanical strength, porosity, and degradation behavior.

PRINCIPLES OF SURFACE FUNCTIONALIZATION

Fundamentally, surface functionalization operates on principles of surface energy manipulation, chemical reactivity, and topographic structuring. For instance, altering surface energy via functional groups can shift a material from hydrophobic to hydrophilic, facilitating better wetting or cell attachment (Sun et al., 2021). This is achieved through thermodynamic considerations, where minimization of interfacial free energy guides the selection of modification techniques (Klee & Höcker, 1999). General approaches to surface modification are classified into physical, chemical, and biological categories. Each method offers distinct advantages and limitations in terms of precision, durability, scalability, and environmental impact. A comparative overview is provided in Table 3.

Table 3. Key properties of biomedical textiles

Property	Definition	Importance in Biomedical Use	Evaluation Methods	Factors Affecting	Examples in Biomedical Textiles	References
Bio-compatibility	The ability of a material to work safely in the body without causing harmful reactions like toxicity or inflammation.	Prevents rejection and supports integration; mandatory for implants to ensure safety and efficacy in treating various medical conditions	In vitro (MTT assay, viability ≥70%), in vivo (implantation tests, histological scoring), hemocompatibility (hemolysis).	Surface chemistry, hydrophobicity, sterilization; hydrophobic surfaces increase activation.	Silk scaffolds for skin regeneration (low thrombogenicity); chitosan dressings (non-toxic hemostasis).	(Kyriakides et al., 2021), (Scarlat & Mitu, 2020), (Dalton & Cook, 1995), ASTM. Standard Test Method,
Mechanical Strength	The ability of a material to handle tensile, compressive, and shear forces, including elasticity and toughness.	Ensures durability under physiological loads; mismatches between the material and the biological environment can lead to failure in grafts or scaffolds.	Tensile/fatigue testing, rheometry	Crystallinity, weaving density, humidity	Polyester for vascular prostheses (high tensile); PLA/PCL weaves for bone scaffolds (customizable elasticity).	(Da Silva, 2006), (Zia et al., 2022), (Chowdhury et al., 2024), (Dorishetty et al., 2020), (Li et al., 2017)
Porosity	The proportion of void spaces within a material, determined by the size, interconnectivity, and overall distribution of these voids.	Facilitates cell growth, nutrient flow, vascularization; reduces fibrosis but risks infection if excessive.	SEM for morphology, porosimetry for volume	Fabrication method, material type.	Electrospun silk for nerve guides (cell infiltration); low-porosity Dacron grafts (vascular stability).	(Lawrence & Jiang, 2017), (Moore et al., 2004), (Lawrence & Madihally, 2008), (Zhao et al., 2022), (Oliviero et al., 2012), (Zhang et al., 2021), (Jonas et al., 1986)

continued on following page

Table 3. Continued

Property	Definition	Importance in Biomedical Use	Evaluation Methods	Factors Affecting	Examples in Biomedical Textiles	References
Degradation Behavior	The process by which a material breaks down through hydrolysis or enzymatic action, producing non-toxic byproducts at a rate compatible with tissue healing.	Enables temporary support without residue; mismatches between the material and the biological environment can cause inflammation or premature failure.	Gravimetry, nuclear magnetic resonance (NMR) Scanning electron microscope (SEM) ; ASTM F1635-11 guidelines.	Crosslinking, pH, polymer geometry.	PLA/PGA meshes for dermal scaffolds (controlled hydrolysis); silk sutures	(Gering, 2023), (Mndlovu et al., 2024), (Chłopek et al., 2009), (Mndlovu et al., 2024), (Chasse et al., 2019), (Kirchhof et al., 2015), (Sanjari et al., 2025), (Devine, 2017)
Wettability Control	The ability of a textile surface to regulate its interaction with fluids; inadequate control can prevent optimal absorption, spreading, or repelling of liquids.	Governs protein adsorption, cell adhesion, proliferation, and fluid absorption critical for wound healing, tissue scaffolds, and drug delivery systems.	Contact angle measurement, Surface energy analysis, Fluid uptake/absorption tests, In vitro cell adhesion assays.	Fiber chemistry, Surface roughness and porosity, Presence of coatings/finis, Environmental factors (humidity, pH).	Poly(dimethylsiloxane) (PDMS) substrate, Natural cotton is hydrophilic, but uncontrolled wettability) resulting in inconsistent wound dressing performance, Plasma-treated PET nanofibers for enhanced wettability for scaffolds.	(Li et al., 2024), (Sun et al., 2021), (Odugbemi, 2018), (Lampin et al., 1997), (Ahmad Khalili & Ahmad, 2015), (Samipour et al., 2020)

Table 4. Surface Functionalization Methods for Biomedical Textiles

Method	Principle / Mechanism	Advantages	Limitations	References
Physical	Methods that alter the surface topography and energy of a material without changing its bulk properties. Examples include plasma treatment, where ionized gases (e.g., Ar, O_2) introduce hydrophilic groups; etching, which removes surface layers to enhance roughness; and laser processing, which enables precise micro- or nano-scale texturing such as grooves and pillars.	• Environmental friendly (no chemicals) • Short reaction tismes • Improves wettability, • Cell adhesion, and cell proliferation • Laser is highly precise, reagent-free. Laser-textured stents improve endothelialization	Laser systems can be costly, scalability debated.	(Nemani et al., 2018), (Sahani, 2025), (Obilor et al., 2022), (Cheng et al., 2013), (Ibrahim et al., 2025)
Chemical	Involves covalent or non-covalent bonding to introduce new functional groups onto a material's surface. Common approaches include grafting, which initiates or attaches polymer chains; sol-gel processing, where hydrolysis and condensation of precursors form hybrid coatings; and surface coatings, which deposit functional or structural layers such as hydroxyapatite.	• Highly specific, • durable modifications • Enables customizable porosity, wettability, and bioactivity. • Controlled drug release and antibacterial coatings possible. • Allows grafted peptides for targeted drug release, Sol-gel coatings for antibacterial implants, Hydroxyapatite coatings to promote bone integration and Polymer coatings on stents to reduce thrombosis.	• May require solvents causing environmental risks. • Potential contamination. • Risk of coating delamination under stress.	(Azadmanjiri et al., 2020), **(Sherazi et al., 2015),** (Catauro & Vecchio Ciprioti, 2019), (Xie et al.,, 2023)

continued on following page

Table 4. Continued

Method	Principle / Mechanism	Advantages	Limitations	References
Biological	Utilizes biomolecules such as enzymes, peptides, or proteins to create bio-mimetic interfaces. These biomolecules are immobilized on the material surface through covalent or non-covalent bonding to enhance biological interactions.	• High specificity and biocompatibility. • Promotes cell-specific adhesion, osteogenesis, and reduced protein fouling. • Enhances implant-tissue integration. • Allows enzyme-functionalized wound dressings, Integrating binding peptides to enhance scaffold adhesion, bone scaffolds with immobilized biomolecules	• Complex and costly. • Stability and scalability issues. • Sensitive to sterilization and storage.	(Veerapandian & Yun, 2011), **(Ramakrishna et al., 2018), (Mohammed Sadhakathullah et al., 2023), (Shabanloo et al., 2025), (Bösiger et al., 2018), (Karimi et al., 2018), (Lee et al., 2016)**

LASER TECHNOLOGIES FOR SURFACE FUNCTIONALIZATION

Laser-Material Interactions

Laser-material interactions involve the transfer of photon energy to a material, triggering physical and chemical changes at the surface or subsurface level (Brown & Arnold, 2010). The primary mechanisms are governed by the relationship of laser wavelength (e.g., UV, IR, visible), pulse duration (continuous wave to femtoseconds), and energy density (fluence, typically in J/cm^2) (Bulgakova, 2013). These parameters influencing outcomes like surface topography, chemistry, or structural integrity (Buchberger et al., 2023). In biomedical textiles, the goal is to achieve precise modifications, such as micro-patterns for cell adhesion or hydrophilic surfaces for drug delivery while preserving the material's biocompatibility and mechanical properties (Sun et al., 2021). Depending on parameters such as wavelength, pulse duration, and energy density, lasers can induce distinct physical effects on textile fibers, ranging from surface activation to material ablation (Dutta Majumdar & Manna, 2011). These mechanisms have significant implications for biocompatibility, wettability, and structural performance. **Table 4** provides a comparative overview of these mechanisms, outlining their principles, influencing parameters, biomedical impacts, and associated challenges.

Laser Systems for Biomedical Textile Processing

Biomedical textiles require tailored laser processing to enhance properties. Laser-based techniques have emerged as powerful tools for modifying textile and biomedical materials, offering precise, controllable, and often reagent-free processing methods. Following are the laser systems which are mainly used for textile materials for biomedical applications (Table 5):

CO_2 Lasers

CO_2 lasers, operating at a wavelength of approximately 10.6 μm, rely on strong absorption in organic fibers to induce photothermal melting or vaporization, making them suitable for etching and marking. They offer high throughput and are cost-effective. However, they have limitations, including a larger heat-affected zone (HAZ), and a risk of charring (Bäuerle, 2013). In biomedical applications, CO_2 lasers have been used to etch cotton for higher porosity in wound dressings and to mark grafts. Challenges include managing thermal damage in biodegradable fibers, such as PLA chain scission, and achieving uniform etching in woven textiles (Ramasubramanian et al., 2022; Chow et al., 2011; Todorov, 2020)

Femtosecond Lasers

Femtosecond lasers, utilize multiphoton absorption and plasma formation. This results in ultrafast ablation with minimal thermal damage, enabling submicron resolution. Their advantages include sub-micron precision, negligible thermal effects, and preservation of biocompatibility (Lin & Hong, 2021; Sidhu & Dhingra, 2022). However, there are some limitations which include high equipment costs, slower processing for large areas, and the need for technical expertise (König, 2023; Omenogor & Adeniran, 2024). These lasers are applied in texturing silk scaffolds for nerve conduits and ablating PLA scaffolds for bone regeneration (Ribba et al., 2014; Angelova et al., 2024). For instance, silk nanopatterning enhanced keratinocyte adhesion (Visan & Popescu-Pelin, 2024) and PLA micropores increased cell proliferation (Daskalova et al., 2018). The main challenges faces are high cost, scalability issues, and optimizing parameters across diverse textile materials.

Table 5. Laser–Material Interaction Mechanisms in Biomedical Textiles

Mechanism	Principle	Biomedical Impact	Examples of biomedical applications	Challenges	References
Absorption	The process in which photons are taken up by a material, leading either to electron excitation that produces localized photothermal heating or to direct photochemical bond breaking.	Enhances wettability, introduces functional groups for protein adsorption.	UV excimer treatment of polyester grafts endothelial cell attachment.	Avoiding polymer degradation (e.g., PLA chain scission); ensuring uniform exposure on textile weaves.	(Bäuerle, 2011), (Peethan et al., 2020), (Doi et al., 1996), (Stępak et al., 2014)
Ablation	A process in which high-energy photons induce multiphoton absorption, leading to plasma formation and subsequent ejection of material from the surface.	Creates micro/ nano pores, promoting angiogenesis and cell infiltration in scaffolds.	Femtosecond laser ablation of PLA scaffolds	Managing debris, preventing inflammatory response from residues.	(Sidhu & Dhingra, 2022), (Zhang et al., 2023), (Lee et al., 2012), (Ramirez, 2023)
Melting	Photothermal heating elevates the material's temperature above its melting point, resulting in surface reshaping without significant material loss.	Produces smoother/ porous textures; enhances wettability and mechanical stability.	CO_2 laser melting of polyester for porous surfaces; reduction in contact angle	Avoiding excessive melting to prevent weakening of fiber structure in biodegradable textiles.	(Yadroitsev et al., 2014), (Manne et al., 2018), (Yan et al., 2016), (Tu & Chung, 2016), (Kobielarz et al., 2019), (Riveiro et al., 2018),
Vaporization	A process in which high photon energy converts material directly into the gaseous phase, enabling deep removal or precise cutting.	Enables deep grooves, microchannels, or cutting for sensor/drug-delivery textiles.	Nd:YAG vaporization of nylon resulting in microchannels enhancing drug release	Thermal damage risk; toxic byproducts from vaporized material.	(Chen et al., 2024), (Li et al., 2020), (Kumar et al., 2023), (Fader & Ratner, 2000)

Diode lasers

Diode lasers induce photothermal heating that causes surface melting or recrystallization which is useful for annealing (controlled heating and cooling to alter material properties) (Wu et al., 2019; Lee et al., 2005). They are compact and cost-effective, and particularly suited for synthetic fibers. However, they can cause thermal stress in natural fibers. Biomedical applications include annealing nylon sutures to improve flexibility ((Shein et al., 2008) and restructuring polypropylene meshes for hernia repair (Lanzafame et al., 2005). Key challenges involve ensuring uniform heating in woven structures and avoiding degradation in heat-sensitive fibers.

SPECIFIC LASER-INDUCED TECHNIQUES FOR BIOMEDICAL APPLICATIONS

Laser-Induced Periodic Surface Structures (LIPSS) for creating biomimetic textures

LIPSS are a potent tool for creating biomimetic textures on biomedical textiles, enhancing their functionality for applications like tissue engineering and implants (Visan & Popescu-Pelin, 2024). LIPSS are nanoscale to microscale periodic patterns (ripples, grooves) formed by laser interference with material surfaces, replicating natural tissue structures like collagen fibers to promote cell alignment and growth (Bonse et al., 2017). In biomedical textiles, LIPSS enhance surface properties of fibers, improving wettability, cell proliferation, and antimicrobial resistance, critical for scaffolds and grafts (Sirdeshmukh & Dongre, 2021). These periodic nanostructures or microstructures, formed through interference between incident laser light and surface-scattered waves, mimic the topographical features of natural biological tissues, such as the extracellular matrix (ECM) or collagen fibril patterns (Matsugaki et al., 2015). The formation mechanism involves a feedback loop where initial surface irregularities enhance interference, leading to self-organized periodic structures (Huang et al., 2022). For biomedical textiles, LIPSS are typically created using femtosecond lasers due to their precision and minimal thermal impact (Florian et al., 2020), preserving the biocompatibility of fibers like silk, PLA, or polyester. Two main types of LIPSS are relevant:

- **Low Spatial Frequency LIPSS (LSFL)**: Periods close to the laser wavelength (e.g., 780 nm for Ti:sapphire lasers) (Schwarz, Rung, & Hellmann, 2017), formed via surface plasmon interference (Perrakis et al., 2024), ideal for mimicking ECM textures (Petrović et al., 2019).

- **High Spatial Frequency LIPSS (HSFL)**: Sub-wavelength periods (~80 to 240 nm) (Pimenov et al., 2022), suitable for nanoscale biointeractions (Mettler, 2024).

Advantages include high reproducibility, chemical-free processing, and the ability to tailor textures to specific biological needs (Fan et al., 2024). Limitations include high equipment costs, complexity in scaling for large textile areas, and potential material degradation if parameters are not optimized. In biomedical textiles, LIPSS are applied to both natural (e.g., silk, cotton) and synthetic (e.g., PLA, polyester) fibers to create surfaces that mimic biological interfaces, improving integration with tissues like skin, bone, or blood vessels. Continued advancements in laser technology and hybrid processing will further their impact, addressing current limitations and expanding their role in regenerative medicine.

Laser-Induced Graphene (LIG) for conductive and antimicrobial coatings on textiles

In biomedical textiles, LIG coatings address critical challenges such as microbial contamination and the need for conductive interfaces, making them ideal for applications like wearable biosensors, smart wound dressings, and infection-resistant implants.

Laser-Induced Graphene (LIG) for Conductive Textiles

Laser-Induced Graphene (LIG) enables the direct conversion of polymer textiles into conductive, antimicrobial coatings via laser irradiation (Cheng et al., 2021). This single-step, mask-free process transforms carbon-rich substrates like polyimide (PI), polyethersulfone (PES), or polyester into three-dimensional, porous graphene structures with high electrical conductivity and tunable surface properties (Ameku et al., 2022). Key principles include:

- **Photothermal Carbonization**: Laser energy heats the polymer surface, causing carbonization and graphitization, forming a porous graphene network (Biswas, 2023).
- **Parameter Control**: Wavelength, pulse duration (continuous wave to femtosecond), and laser power dictate LIG morphology, porosity, and conductivity. For example, CO2 lasers produce thicker graphene layers (Lin et al., 2013), while femtosecond lasers enable finer patterns (Shi et al., 2015).
- **Material Compatibility**: Suitable for polymers with high carbon content (e.g., PI, PES) (Thakur et al., 2021).

Advantages include rapid fabrication (seconds to minutes), and precise patterning. However it has some limitations such as dependency on carbon-rich substrates and potential variability in graphene quality, requiring optimization of laser parameters (Le et al., 2022). Conductive coatings using laser-induced graphene (LIG) result in textiles (eg: polyester and nylon) which are flexible, lightweight, and conductive materials (Lipovka et al., 2023). While LIG offers high flexibility and integration with textiles, challenges such as conductivity degradation under stress or fluids necessitate protective strategies (Zhu, Huang, & Song, 2021).

Laser-Induced Graphene (LIG) for Antimicrobial Coatings in Biomedical Textiles

LIG's antimicrobial properties stem from its sharp-edged graphene structure, which disrupts bacterial membranes (Vázquez-López et al., 2024), and can be significantly enhanced by doping with antimicrobial agents like silver nanoparticles (AgNPs) (Sharma et al., 2024). Un-doped LIG exhibits moderate anti-biofouling due to its porous, hydrophobic surface. Doping with AgNPs enhances toxicity via Ag^{+} ion release and surface contact, achieving significant bacterial reduction (Gupta et al., 2019). This makes LIG-coated textiles ideal for infection-resistant biomedical applications. Key applications include:

- **Wound Dressings**: Ag@LIG on PES textiles reduces biofilm formation, ideal for chronic wound management (Gupta et al., 2019).
- **Water Filtration Membranes**: LIG-coated PES membranes reduce biofouling, extending operational life in medical water systems (Barbhuiya, Kumar, & Singh, 2021).
- **Implantable Devices**: LIG on polymer scaffolds enhances antimicrobial resistance, preventing infections in orthopedic implants (Pandit et al., 2021).

This allows chemical-free synthesis, customizable antimicrobial strength via doping, and compatibility with flexible textiles. However, there are potential of AgNP toxicity and environmental release of these coatings. The long-term efficacy may also decrease due to ion depletion.

Pulsed Laser Deposition (PLD) and Matrix-Assisted Pulsed Laser Evaporation (MAPLE) For Thin-Film Coatings

Pulsed Laser Deposition (PLD) and Matrix-Assisted Pulsed Laser Evaporation (MAPLE) are prominent techniques in the field of thin-film deposition. It uses laser-material interactions to create high-quality coatings for a wide array of applications, from electronics to biomedicine (Papavlu et al., 2017). These methods fall under physical vapor deposition (PVD) processes, where material is transferred from a source to a substrate in a controlled vacuum environment.

PLD involves directing a high-energy pulsed laser at a target material in a vacuum chamber, vaporizing it to form a plasma plume that deposits onto a substrate. This allows precise control over film properties through parameters like laser fluence and substrate temperature.

MAPLE modifies PLD by dissolving the target in a volatile solvent, freezing it, and using laser energy to evaporate the matrix gently, transferring the material intact to the substrate. This protects sensitive compounds from thermal or photochemical damage.

Both techniques face scalability issues, with PLD limited by plume size and MAPLE by cryogenic needs. Environmental considerations favor their chemical-free nature, but energy use and waste (e.g., particulates) require attention (Yang & Zhang, 2017; Mostafa, 2025).

Laser-Induced Forward Transfer (LIFT) For Precise Deposition of Biomolecules Or Nanoparticles

Laser-Induced Forward Transfer (LIFT) has emerged as a pivotal technique in advanced materials deposition, particularly for biomedical applications requiring precise control over biomolecules and nanoparticles. LIFT allows laser-material interactions to enable nozzle-free, high-resolution patterning (Serra & Piqué, 2019). This process addresses limitations in traditional methods like inkjet or extrusion printing, such as clogging and shear stress on sensitive materials, by facilitating non-contact transfer (Das et al., 2024). In biomedical contexts, LIFT supports the creation of complex 3D structures for tissue engineering, drug delivery, and biosensors, with its ability to maintain high cell viability and material functionality (Serra & Piqué, 2019). At its core, LIFT involves directing a pulsed laser through a transparent donor substrate onto an absorbing layer. This action vaporizes and propels the overlying biomaterial toward a receiver substrate. The donor typically consists of a transparent base (e.g., quartz or glass), an energy-absorbing layer (e.g., metals like titanium or gold, or polymers like triazene), and a bioink layer containing cells, biomolecules, or nanoparticles (Nagel & Lippert, 2012). Mechanisms vary by regime such as sub-threshold (insufficient energy for transfer), jetting (stable droplet formation for precise deposition), and plume (unstable spray from excess energy) (Ali, 2014). Bubble dynamics are central, with expansion and collapse generating pressures that eject material as microdroplets (volumes from nanoliters to sub-picoliters) (Kryou & Zergioti, 2023).

Variants enhance specificity, for example Matrix-Assisted Pulsed Laser Evaporation Direct-Write (MAPLE-DW) embeds materials in a frozen solvent matrix for gentle transfer, ideal for delicate biomolecules. Whereas, Biological Laser Printing (BioLP) focuses on cell-laden inks useful for cell-laden matrixes. Another variant is Continuous Wave LIFT, which uses steady lasers for spray-like deposition of nanoparticle inks (Fitz-Gerald et al., 2000; Ghorbani et al., 2021; Pohl, 2015). These adaptations reduce thermal damage, with picosecond pulses showing superior performance in maintaining nanoparticle luminescence or bio-molecule activity.

The efficiency of Laser-Induced Forward Transfer (LIFT) depends on carefully tuned parameters, including laser settings (wavelength, pulse duration, and energy, where optimal ranges avoid plume formation), bioink properties (viscosity, cell density, and stabilizing additives), substrate factors (absorber thickness, donor–receiver distance, and biocompatible coatings to enhance cell viability), and environmental controls (ambient conditions, repetition rates, and real-time monitoring of jet dynamics). Together, these factors determine jet stability, cell survival, and printing precision.

LIFT's precision enables controlled deposition of biomolecules and nanoparticles in biomedical applications. It prints DNA/RNA for microarrays, proteins and enzymes for diagnostics and biosensors, and lipid vesicles or peptides for drug delivery. Nanoparticles such as silver, silicon, carbon nanofibers, and nano-hydroxyapatite are patterned for sensors, antimicrobial coatings, and bone regeneration (Chang & Sun, 2023). Broader uses include tissue engineering of skin, vascular, neural, cardiac, and bone constructs with high cell viability. However, high costs, slower speeds, jet dynamics, and bio-ink variability pose challenges to scalability and in vivo integration.

Direct Laser Interference Patterning (DLIP) and Direct Laser Writing (DLW) For Patterned Functionalization

Direct Laser Interference Patterning (DLIP) and Direct Laser Writing (DLW) represent sophisticated laser-based methodologies for achieving patterned functionalization on surfaces, with significant implications for biomedical textiles and related applications. These techniques enable precise modification of material properties at micro- and nano-scales, enhancing aspects such as wettability, biocompatibility, and antimicrobial resistance (Huerta Murillo, 2019). DLIP operates on the principle of laser beam interference which enables periodic ablation by using multiple beams with controlled angles and wavelengths across the UV-IR range making it suitable for large-area, maskless surface modification (Lasagni et al., 2017). However, coherence management and high energy requirements present notable limitations. Biomedical applications of DLIP include reducing biofilm formation on implants and inducing cell alignment on scaffolds, while in textiles, it has been explored for patterning polyurethanes to create antimicrobial fabrics ((Kligman et al., 2021; Estevam-Alves et al., 2016) .

Direct Laser Writing (DLW), in contrast, relies on tightly focused laser beams for direct writing or polymerization, using pulses in the femtosecond to nanosecond range (Balena et al., 2023). DLW is highly versatile, allowing fabrication of complex 3D patterns with high resolution. Its drawbacks include slower throughput and potential risks of thermal damage (Sedghamiz et al., 2022). In biomedical contexts, DLW has been used to fabricate microneedles and scaffolds (Young et al., 2024).

MATERIALS IN LASER-INDUCED COATINGS FOR BIOMEDICAL TEXTILES

Laser-induced coatings on biomedical textiles commonly utilize carbon-based materials like graphene for conductivity, metallic nanoparticles such as silver for antibacterial properties, and polymers including poly-dopamine for enhanced bio-adhesion. Materials are selected for their compatibility with laser ablation, deposition, or transformation techniques, ensuring enhanced performance without compromising the substrate's integrity.

- **Carbon-based** (Diamond-like carbon (DLC), Graphene (LIG)): These coating materials provide strength, conductivity, and biocompatibility (Claro et al., 2022). DLC offers wear resistance (Lin et al., 2019), while LIG formed by photothermal carbonization of polymers, enables flexible electrodes for biosensing (Wang et al., 2022).
- **Metallic (AgNPs):** Such coating delivers potent antibacterial activity (E. coli, S. aureus), with efficacy linked to nanoparticle size (Gouyau et al., 2021). Laser ablation and pretreatment enhance uniform deposition and stability on cotton or polyester (Raslan et al., 2020).
- **Polymeric (PDA):** These acts as a bio-adhesive, improving cell attachment and mechanical resilience (Pei et al., 2021). Laser-annealed PDA coatings also support hybrid systems (e.g., PDA–AgNP) that combine adhesion with antimicrobial effects (Li et al., 2024).

ADVANTAGES, CHALLENGES, AND LIMITATIONS OF LASER-INDUCED COATINGS FOR BIOMEDICAL TEXTILES

Laser-induced coatings offer unique opportunities for enhancing the performance of biomedical textiles. However, there are certain drawbacks and limitations of this technology which prevents its full utilizations.

Advantages of Laser-Induced Coatings

Laser-induced coatings offer several benefits that make them highly suitable for biomedical textiles. These are as follows:

- **High Precision**: Techniques like LIFT and DLW enable intricate patterns such as LIPSS for biomimetic textures or precise deposition of biomolecules. For example, LIFT deposits cells with significant viability, creating vascular structures with micron-level accuracy (Mei et al., 2025).
- **Eco-Friendly Processing**: Laser methods eliminate the need for harsh chemicals used in traditional coating processes like sol-gel or chemical vapor deposition, reducing environmental impact and waste. LIG and MAPLE, for instance, rely on photothermal or photochemical mechanisms, minimizing reagent use (Cheng et al., 2021; Yang & Zhang, 2017).
- **Versatility across Materials**: These techniques process a wide range of materials, including carbon-based (e.g., LIG, DLC), metallic (e.g., AgNPs), and polymeric (e.g., polydopamine) coatings, compatible with textiles like polyester, silk, and Kevlar. This versatility supports diverse applications; from conductive e-textiles to antibacterial wound dressings (Morris & Murray, 2020).
- **Complex Patterns**: DLIP and DLW enable mask-free patterning, creating intricate designs like hierarchical LIPSS or conductive LIG paths without photolithography, reducing processing steps and costs (Le et al., 2022).

CHALLENGES OF LASER-INDUCED COATINGS

Despite their benefits, several challenges restrict the widespread adoption of laser-induced coatings, particularly in large-scale biomedical textile applications.

- **High Initial Costs**: Laser systems, such as femtosecond lasers for LIPSS or LIFT are very costly and come with maintenance costs which add to expenses. This limits accessibility for small-scale manufacturers, especially compared to traditional coating methods like dip-coating.

- **Potential Thermal Damage**: Heat-sensitive fibers like polyester or PLA risk melting or chain scission during laser processing, particularly with CO_2 lasers. Low-energy pulses mitigate this but require precise control. For example, excessive fluence can reduce polyester tensile strength (Kan, 2008).
- **Scalability for Large-Scale Production**: Techniques like LIFT and MAPLE are slow, making them less viable for mass production compared to DLIP, which achieves higher throughput. Scaling requires advanced scanning systems or multiple laser sources, increasing complexity.
- **Optimization of Parameters**: Achieving uniform coatings demands fine-tuning laser, scanning speed, and fluence, which differ by material and technique. This process is iterative and material-specific, complicating reproducibility.

These challenges necessitate advancements in cost-effective laser systems and automated parameter optimization to enhance practical implementation.

LIMITATIONS OF LASER-INDUCED COATINGS

Inherent limitations in laser-induced coating processes impact their performance and applicability, particularly for biomedical textiles.

- **Dependence on Material Absorption Properties**: Effective coating requires materials to absorb specific laser wavelengths. Non-absorbing textiles like cotton require pre-treatments, adding complexity. For instance, low-carbon fibers yield poor LIG quality without modification.
- **Risks of Uneven Coating in Porous Textiles**: The porous, woven structure of textiles can lead to inconsistent coating deposition, especially with PLD or LIFT, where plume dynamics may miss interstitial spaces. This results in non-uniform properties, such as patchy conductivity in LIG. Techniques like DLIP lessen this by covering larger areas but struggle with complex weaves.

These limitations highlight the need for material-specific process design and advanced deposition strategies to ensure uniformity across diverse textile substrates.

FUTURE PERSPECTIVES AND EMERGING TRENDS

As laser-induced surface functionalization continues to advance, several emerging directions promise to expand its utility in biomedical textiles. These trends highlight the intersection of sustainability, digital optimization, and translational research, all of which can shape the future of this field.

Sustainability and Green Manufacturing

Sustainability is now central to material science and biomedical engineering, and laser-based techniques offer distinctive advantages over conventional wet-chemical processes. Unlike solvent-intensive chemical coatings, laser functionalization is clean, and requires minimal chemical reagents. This aligns with the principles of the circular economy, reducing both toxic effluents and resource consumption in textile manufacturing. If fully exploited, such approaches may serve as benchmarks for developing environmentally conscious biomedical devices.

Emerging Innovations in Processing

Frontier innovations are expected to transform the flexibility and customization of laser-induced coatings. Artificial intelligence (AI) and machine learning algorithms can optimize process parameters such as wavelength, fluence, and scanning speed allowing real-time feedback control to achieve consistent and patient-specific coatings. Similarly, hybrid approaches, such as laser-plasma, or laser-assisted bioprinting, may unlock superior coating adhesion, structural definition, and bioactivity. These methods dovetail well with the growing emphasis on personalized medicine, where wearable or implantable biomedical textiles can be tailored for individual patients' physiological needs, enhancing therapeutic efficacy and comfort.

Research Gaps and Translational Challenges

Despite these opportunities, several critical gaps remain before widespread clinical translation becomes feasible. The long-term in vivo performance of laser-engineered coatings, particularly their biocompatibility, biodegradation, and impact on tissue integration, requires systematic study. Cost reduction and scalability remain barriers, as high-energy laser setups and precise nanomaterial integration may not yet be commercially viable for large-scale textile production. Additionally, standardization of protocols, including reproducible evaluation methods for coating durability, sterilization compatibility, and regulatory compliance is vital to ensure safe and effective biomedical textile products. Addressing these gaps will demand coordinated efforts from material scientists, biomedical engineers, clinicians, and regulatory agencies.

CONCLUSION

Laser-induced coatings have redefined the landscape of biomedical textiles by offering precise, non-contact methods to impart advanced functionalities directly onto textile surfaces without compromising structural integrity or comfort. By the assistance of photon-driven interactions, these techniques enable the integration of antimicrobial, conductive, and bioactive features at micro and nanoscale. These techniques surpass the capabilities of conventional chemical and physical approaches in both selectivity and environmental sustainability. The adoption of technologies such as Laser-Induced Periodic Surface Structures (LIPSS), Laser-Induced Graphene (LIG), Pulsed Laser Deposition (PLD), Matrix-Assisted Pulsed Laser Evaporation (MAPLE), Laser-Induced Forward Transfer (LIFT), and Direct Laser Writing (DLW) has enabled the rapid development of smart, multifunctional textiles for diverse medical applications. These applications include wound dressings, biosensors, tissue engineering scaffolds, and antimicrobial hospital fabrics. These approaches deliver unmatched control over functionalization, reduce the need for hazardous chemicals, and support the growing demand for customization and green manufacturing in healthcare materials. Despite their considerable promise, challenges remain. These include the high cost and complexity of laser systems, scalability for industrial production, thermal management for sensitive substrates, and the need for standardized characterization and regulatory validation to ensure long-term performance in clinical settings. Innovations such as AI-driven process optimization, the integration of sustainable materials, and hybrid laser technologies are actively addressing these limitations, promising broader deployment in the near future. In summary, laser-induced surface functionalization is poised to become a cornerstone technology in next-generation biomedical textiles, driving advances in patient-specific therapy, infection control, wearable health monitoring, and sustainable manufacturing. Continued multidisciplinary collaboration, rigorous evaluation methods, and a clear commitment to eco-friendly practices will be critical to fully realizing the transformative impact of these cutting-edge coatings on the future of healthcare textiles.

REFERENCES

Abdul Khalil, H. P. S., Bhat, A. H., Abu Bakar, A., Tahir, P. M., Zaidul, I. S. M., & Jawaid, M. (2014). *Cellulosic nanocomposites from natural fibers for medical applications: A review. Handbook of Polymer Nanocomposites. Processing, Performance and Application* (Vol. C). Polymer Nanocomposites of Cellulose Nanoparticles.

Adamovic, D., Ristic, B., & Zivic, F. (2017). Review of existing biomaterials—method of material selection for specific applications in orthopedics. Biomaterials in clinical practice: advances in clinical research and medical devices, 47-99.

Ahmad Khalili, A., & Ahmad, M. R. (2015). A review of cell adhesion studies for biomedical and biological applications. *International Journal of Molecular Sciences*, *16*(8), 18149–18184. DOI: 10.3390/ijms160818149 PMID: 26251901

Akter, S., Azim, A. Y. M. A., & Al Faruque, M. A. (2014). Medical textiles: Significance and future prospect in Bangladesh. *European Scientific Journal, 10*(12).

Al Ali, M. (2024). Biomechanics of the Human Body. In *Fundamentals of Orthopedic Design with Non-parametric Optimization* (pp. 91–103). Springer Nature Singapore. DOI: 10.1007/978-981-97-1040-9_4

Al Javed, M. O., & Bin Rashid, A. (2024). Laser-assisted micromachining techniques: An overview of principles, processes, and applications. *Advances in Materials and Processing Technologies*, 1–44.

Al-Zyoud, W., Haddadin, D., Hasan, S. A., Jaradat, H., & Kanoun, O. (2023). Biocompatibility testing for implants: A novel tool for selection and characterization. *Materials (Basel), 16*(21), 6881. DOI: 10.3390/ma16216881 PMID: 37959478

Ali, M. (2014). Study of a novel configuration of laser Assisted Bioprinting (Doctoral dissertation, Université de Bordeaux).

Altman, G. H., Diaz, F., Jakuba, C., Calabro, T., Horan, R. L., Chen, J., ... & Kaplan, D. L. (2003). Silk-based biomaterials. biomaterials, 24(3), 401-416.

Ameku, W. A., Negahdary, M., Lima, I. S., Santos, B. G., Oliveira, T. G., Paixão, T. R., & Angnes, L. (2022). Laser-scribed graphene-based electrochemical sensors: A review. *Chemosensors (Basel, Switzerland), 10*(12), 505. DOI: 10.3390/chemosensors10120505

Angelova, L., Daskalova, A., Mincheva, R., Filipov, E., Dikovska, A., Fernandes, M. H., Vig, S., & Buchvarov, I. (2024). Design of femtosecond microstructured poly lactic acid temporal scaffolds coated with hydroxyapatite by pulse laser deposition method for bone tissue regeneration. *Optical and Quantum Electronics*, *56*(7), 1195. DOI: 10.1007/s11082-024-06500-w

Ashjari, M., Khoee, S., Mahdavian, A. R., & Rahmatolahzadeh, R. (2012). Self-assembled nanomicelles using PLGA–PEG amphiphilic block copolymer for insulin delivery: A physicochemical investigation and determination of CMC values. *Journal of Materials Science. Materials in Medicine*, *23*(4), 943–953. DOI: 10.1007/s10856-012-4562-1 PMID: 22354326

ASTM. (2013). *Standard Test Method for Analysis of Hemolytic Properties of Nanoparticles*. ASTM.

Azadmanjiri, J., Kumar, P., Srivastava, V. K., & Sofer, Z. (2020). Surface functionalization of 2D transition metal oxides and dichalcogenides via covalent and non-covalent bonding for sustainable energy and biomedical applications. *ACS Applied Nano Materials*, *3*(4), 3116–3143. DOI: 10.1021/acsanm.0c00120

Balena, A., Bianco, M., Pisanello, F., & De Vittorio, M. (2023). Recent advances on high-speed and holographic two-photon direct laser writing. *Advanced Functional Materials*, *33*(39), 2211773. DOI: 10.1002/adfm.202211773

Barbhuiya, N. H., Kumar, A., & Singh, S. P. (2021). A journey of laser-induced graphene in water treatment. *Transactions of the Indian National Academy of Engineering : an International Journal of Engineering and Technology*, *6*(2), 159–171. DOI: 10.1007/s41403-021-00205-2

Bäuerle, D. (2011). Thermal, photophysical, and photochemical processes. In *Laser Processing and Chemistry* (pp. 13–38). Springer Berlin Heidelberg. DOI: 10.1007/978-3-642-17613-5_2

Bäuerle, D. (2013). *Laser processing and chemistry*. Springer Science & Business Media.

Baylón, K., Rodríguez-Camarillo, P., Elías-Zúñiga, A., Díaz-Elizondo, J. A., Gilkerson, R., & Lozano, K. (2017). Past, present and future of surgical meshes: A review. *Membranes*, *7*(3), 47. DOI: 10.3390/membranes7030047 PMID: 28829367

Biswas, R. K. (2023). Carbonization of Polyimide using CO2 laser and Femtosecond laser for Sensor Applications (Doctoral dissertation, NUI Galway).

Bonse, J., Kirner, S. V., Höhm, S., Epperlein, N., Spaltmann, D., Rosenfeld, A., & Krüger, J. (2017, February). Applications of laser-induced periodic surface structures (LIPSS). In Laser-based Micro-and Nanoprocessing XI (Vol. 10092, pp. 114-122). SPIE.

Bösiger, P., Tegl, G., Richard, I. M., Le Gat, L., Huber, L., Stagl, V., & Fortunato, G. (2018). Enzyme functionalized electrospun chitosan mats for antimicrobial treatment. *Carbohydrate Polymers*, *181*, 551–559. DOI: 10.1016/j.carbpol.2017.12.002 PMID: 29254006

Brown, M. S., & Arnold, C. B. (2010). Fundamentals of laser-material interaction and application to multiscale surface modification. In *Laser precision microfabrication* (pp. 91–120). Springer Berlin Heidelberg. DOI: 10.1007/978-3-642-10523-4_4

Buchberger, G., Muck, M., Plamadeala, C., & Heitz, J. (2023). Laser Structuring for Biomedical Applications. In *Ultrafast Laser Nanostructuring: The Pursuit of Extreme Scales* (pp. 1105–1165). Springer International Publishing. DOI: 10.1007/978-3-031-14752-4_31

Bulgakova, N. M. (2013). Fundamentals of ultrafast laser processing. Ultrafast laser processing: from micro-to nanoscale, 99-182.

Catauro, M., & Vecchio Ciprioti, S. (2019). Sol-gel synthesis and characterization of hybrid materials for biomedical applications. In *Thermodynamics and Biophysics of Biomedical Nanosystems: Applications and Practical Considerations* (pp. 445–475). Springer Singapore. DOI: 10.1007/978-981-13-0989-2_13

Chang, J., & Sun, X. (2023). Laser-induced forward transfer based laser bioprinting in biomedical applications. *Frontiers in Bioengineering and Biotechnology*, *11*, 1255782. DOI: 10.3389/fbioe.2023.1255782 PMID: 37671193

Chasse, B., Xu, H., & Budhlall, B. M. (2019). In-vitro biodegradation study of poly (ε-caprolactone) films using a 3D printed helical flow prototype to simulate the physiological conditions for cardiovascular implanted devices. *Biomedical Physics & Engineering Express*, *5*(6), 065021. DOI: 10.1088/2057-1976/ab4e2b

Chen, Z., Xing, X., Yang, Y., & Hong, M. (2024). Laser processing for medical devices. In Teoh, S.-h. (Ed.), *Designing materials for medical devices: Fundamentals* (p. 464). World Scientific., DOI: 10.1142/9789811295898_0013

Cheng, L., Guo, W., Cao, X., Dou, Y., Huang, L., Song, Y., Su, J., Zeng, Z., & Ye, R. (2021). Laser-induced graphene for environmental applications: Progress and opportunities. *Materials Chemistry Frontiers*, *5*(13), 4874–4891. DOI: 10.1039/D1QM00437A

Cheng, Q., Lee, B. L. P., Komvopoulos, K., Yan, Z., & Li, S. (2013). Plasma surface chemical treatment of electrospun poly (L-lactide) microfibrous scaffolds for enhanced cell adhesion, growth, and infiltration. *Tissue Engineering. Part A, 19*(9-10), 1188–1198. DOI: 10.1089/ten.tea.2011.0725 PMID: 23281641

Chow, Y. L., Chan, C. K., & Kan, C. W. (2011). Effect of CO2 laser treatment on cotton surface. *Cellulose (London, England), 18*(6), 1635–1641. DOI: 10.1007/s10570-011-9603-2

Chowdhury, N., Faisal, S., Kumar, A., Choudhari, A., Gupta, A. K., Kumar, A., & Kumar, A. (2024). Characterization of Hydrogel Properties in the Advancement of Bio-Tribology. In *Applications of biotribology in biomedical systems* (pp. 27–78). Springer Nature Switzerland. DOI: 10.1007/978-3-031-58327-8_2

Claro, C, P. I., Pinheiro, T., Silvestre, S. L., Marques, A. C., Coelho, J., Marconcini, J. M., ... & Martins, R. (. (2022). Sustainable carbon sources for green laser-induced graphene: A perspective on fundamental principles, applications, and challenges. *Applied Physics Reviews, 9*(4). Advance online publication. DOI: 10.1063/5.0100785

Da Silva, V. D. (2006). *Mechanics and strength of materials*. Springer Berlin Heidelberg. DOI: 10.1007/3-540-30813-X

Dalton, J. E., & Cook, S. D. (1995). In vivo mechanical and histological characteristics of HA-coated implants vary with coating vendor. *Journal of Biomedical Materials Research, 29*(2), 239–245. DOI: 10.1002/jbm.820290215 PMID: 7738072

Das, A., Majumder, S. D., Kozak, D., & Ding, C. F. (2024). Laser-Induced Forward Transfer for Biosensor Application. Mechanical Engineering in Biomedical Applications: Bio-3D Printing, Biofluid Mechanics, Implant Design, Biomaterials, Computational Biomechanics, Tissue Mechanics, 77-117.

Daskalova, A., Ostrowska, B., Zhelyazkova, A., Święszkowski, W., Trifonov, A., Declercq, H., Nathala, C., Szlazak, K., Lojkowski, M., Husinsky, W., & Buchvarov, I. (2018). Improving osteoblasts cells proliferation via femtosecond laser surface modification of 3D-printed poly-ε-caprolactone scaffolds for bone tissue engineering applications. *Applied Physics. A, Materials Science & Processing, 124*(6), 413. DOI: 10.1007/s00339-018-1831-y

de Sá Guimarães, A. M. (2021). Developing an antimicrobial coating to prevent abdominal aortic graft infections (AGIs).

Deng, X., Qasim, M., & Ali, A. (2021). Engineering and polymeric composition of drug-eluting suture: A review. *Journal of Biomedical Materials Research. Part A, 109*(10), 2065–2081. DOI: 10.1002/jbm.a.37194 PMID: 33830631

Devine, D. M. (Ed.). (2017). *Bioresorbable polymers and their biomedical applications*. Smithers Rapra.

Doi, K., Nakayama, Y., & Matsuda, T. (1996). Novel compliant and tissue-permeable microporous polyurethane vascular prosthesis fabricated using an excimer laser ablation technique. *Journal of Biomedical Materials Research, 31*(1), 27–33. DOI: 10.1002/(SICI)1097-4636(199605)31:1<27::AID-JBM4>3.0.CO;2-S PMID: 8731146

Dorishetty, P., Balu, R., Athukoralalage, S. S., Greaves, T. L., Mata, J., De Campo, L., Saha, N., Zannettino, A. C. W., Dutta, N. K., & Choudhury, N. R. (2020). Tunable biomimetic hydrogels from silk fibroin and nanocellulose. *ACS Sustainable Chemistry & Engineering, 8*(6), 2375–2389. DOI: 10.1021/acssuschemeng.9b05317

Dutta Majumdar, J., & Manna, I. (2011). Laser material processing. *International Materials Reviews, 56*(5-6), 341–388. DOI: 10.1179/1743280411Y.0000000003

Elashnikov, R., Ulbrich, P., Vokatá, B., Pavlíčková, V. S., Švorčík, V., Lyutakov, O., & Rimpelová, S. (2021). Physically switchable antimicrobial surfaces and coatings: General concept and recent achievements. *Nanomaterials (Basel, Switzerland), 11*(11), 3083. DOI: 10.3390/nano11113083 PMID: 34835852

Elmowafy, E. M., Tiboni, M., & Soliman, M. E. (2019). Biocompatibility, biodegradation and biomedical applications of poly (lactic acid)/poly (lactic-co-glycolic acid) micro and nanoparticles. *Journal of Pharmaceutical Investigation, 49*(4), 347–380. DOI: 10.1007/s40005-019-00439-x

Estevam-Alves, R., Günther, D., Dani, S., Eckhardt, S., Roch, T., Mendonca, C. R., Cestari, I. N., & Lasagni, A. F. (2016). UV direct laser interference patterning of polyurethane substrates as tool for tuning its surface wettability. *Applied Surface Science, 374*, 222–228. DOI: 10.1016/j.apsusc.2015.11.119

Fader, D. J., & Ratner, D. (2000). Principles of CO2/erbium laser safety. *Dermatologic Surgery, 26*(3), 235–239. DOI: 10.1046/j.1524-4725.2000.09187.x PMID: 10759800

Fan, L., Xue, X., Wu, L., Zhang, S., Zhao, T., Wang, T., Qian, H., Xie, B., Tofil, S., & Yao, J. (2024). Regulation of High-Index Crystal Facets with Laser-Induced Periodic Surface Structures on CoFe2O4 Epitaxial Films for Ethanol Gas Sensing. *Advanced Materials Interfaces, 11*(27), 2400217. DOI: 10.1002/admi.202400217

Farjaminejad, S., Farjaminejad, R., Hasani, M., Garcia-Godoy, F., Abdouss, M., Marya, A., Harsoputranto, A., & Jamilian, A. (2024). Advances and challenges in polymer-based scaffolds for bone tissue engineering: A path towards personalized regenerative medicine. *Polymers*, *16*(23), 3303. DOI: 10.3390/polym16233303 PMID: 39684048

Fattahi, F. S. (2020). Poly (lactic acid) nano-structures for cartilage regeneration and joint repair: Strategies and ideas. *Recent Trends Nanosci. Technol*, *1*, 1–19.

Fitz-Gerald, J. M., Chrisey, D. B., Piqu, A., Auyeung, R. C. Y., Mohdi, R., Young, H. D., Wu, H. D., Lakeou, S., & Chung, R. (2000). Matrix assisted pulsed laser evaporation direct write (MAPLE DW): A new method to rapidly prototype active and passive electronic circuit elements. [OPL]. *Proceedings of the Materials Research Society*, *624*, 143. DOI: 10.1557/PROC-624-143

Florian, C., Kirner, S. V., Krüger, J., & Bonse, J. (2020). Surface functionalization by laser-induced periodic surface structures. *Journal of Laser Applications*, *32*(2), 022063. DOI: 10.2351/7.0000103

GERING. (2023). *C*. Design Strategies for Polysaccharide Hydrogels Used in Soft Tissue Engineering.

Ghanbari, H., Viatge, H., Kidane, A. G., Burriesci, G., Tavakoli, M., & Seifalian, A. M. (2009). Polymeric heart valves: New materials, emerging hopes. *Trends in Biotechnology*, *27*(6), 359–367. DOI: 10.1016/j.tibtech.2009.03.002 PMID: 19406497

Ghorbani, F., Li, D., Zhong, Z., Sahranavard, M., Qian, Z., Ni, S., Zhang, Z., Zamanian, A., & Yu, B. (2021). Bioprinting a cell-laden matrix for bone regeneration: A focused review. *Journal of Applied Polymer Science*, *138*(8), 49888. DOI: 10.1002/app.49888

Ghosh, J., Rupanty, N. S., Khan, F., Noor, T., Jahangir, R., Mirmohammadsadeghi, S., & Islam, T. (2025). Discover Applied Sciences. *Discover*, *7*, 49.

Gouyau, J., Duval, R. E., Boudier, A., & Lamouroux, E. (2021). Investigation of nanoparticle metallic core antibacterial activity: Gold and silver nanoparticles against Escherichia coli and Staphylococcus aureus. *International Journal of Molecular Sciences*, *22*(4), 1905. DOI: 10.3390/ijms22041905 PMID: 33672995

Gupta, A., Holoidovsky, L., Thamaraiselvan, C., Thakur, A. K., Singh, S. P., Meijler, M. M., & Arnusch, C. J. (2019). Silver-doped laser-induced graphene for potent surface antibacterial activity and anti-biofilm action. *Chemical Communications*, *55*(48), 6890–6893. DOI: 10.1039/C9CC02415H PMID: 31134255

Hahn, J., Gögele, C., & Schulze-Tanzil, G. (2023). Could an Anterior Cruciate Ligament Be Tissue-Engineered from Silk? *Cells*, *12*(19), 2350. DOI: 10.3390/cells12192350 PMID: 37830564

Hegemann, D., Brunner, H., & Oehr, C. (2003). Plasma treatment of polymers for surface and adhesion improvement. *Nuclear Instruments & Methods in Physics Research. Section B, Beam Interactions with Materials and Atoms*, *208*, 281–286. DOI: 10.1016/S0168-583X(03)00644-X

Huang, J., Liu, Y., Jin, S., Wang, Z., Qi, Y., Zhang, J., Wang, K., & Qiu, R. (2022). Uniformity control of laser-induced periodic surface structures. *Frontiers in Physics (Lausanne)*, *10*, 932284. DOI: 10.3389/fphy.2022.932284

Huerta Murillo, D. (2019). Development of Laser Induced Hierarchical Surface Structures on Ti-6Al-4V using short and ultra-short pulsed lasers for modified wettability applications (Doctoral dissertation, Industriales).

Ibrahim, M. S., Seresht, H. B., Kum, C. H., Cho, J. H., Jin, G., An, S. H., Ye, S., Kim, S., Wagner, W. R., & Chun, Y. (2025). Novel laser-textured grooves extended to the sidewall edges of CoCr surfaces for rapid and selective endothelialization following coronary artery stenting. *Biomaterials*, *321*, 123299. DOI: 10.1016/j.biomaterials.2025.123299 PMID: 40188719

Ivanovska, A., Petrović, A., Lazarević-Pašti, T., Ilic-Tomic, T., Dimić-Mišić, K., Lađarević, J., & Bradić, J. (2025). Development of Bioactive Cotton, Wool, and Silk Fabrics Functionalized with Origanum vulgare L. for Healthcare and Medical Applications: An In Vivo Study. *Pharmaceutics*, *17*(7), 856. DOI: 10.3390/pharmaceutics17070856 PMID: 40733065

Jiang, J., Wan, F., Yang, J., Hao, W., Wang, Y., Yao, J., Shao, Z., Zhang, P., Chen, J., Zhou, L., & Chen, S. (2014). Enhancement of osseointegration of polyethylene terephthalate artificial ligament by coating of silk fibroin and depositing of hydroxyapatite. *International Journal of Nanomedicine*, •••, 4569–4580. DOI: 10.2147/IJN.S69137 PMID: 25302023

Jonas, R. A., Schoen, F. J., Levy, R. J., & Castaneda, A. R. (1986). Biological sealants and knitted Dacron: Porosity and histological comparisons of vascular graft materials with and without collagen and fibrin glue pretreatments. *The Annals of Thoracic Surgery*, *41*(6), 657–663. DOI: 10.1016/S0003-4975(10)63083-4 PMID: 2940983

Kan, C. W. (2008). Effects of laser irradiation on polyester textile properties. *Journal of Applied Polymer Science*, *107*(3), 1584–1589. DOI: 10.1002/app.26632

Karimi, F., O'Connor, A. J., Qiao, G. G., & Heath, D. E. (2018). Integrin clustering matters: A review of biomaterials functionalized with multivalent integrin-binding ligands to improve cell adhesion, migration, differentiation, angiogenesis, and biomedical device integration. *Advanced Healthcare Materials*, *7*(12), 1701324. DOI: 10.1002/adhm.201701324 PMID: 29577678

Kim, C. S., Jung, K. H., Kim, H., Kim, C. B., & Kang, I. K. (2016). Collagen-grafted porous HDPE/PEAA scaffolds for bone reconstruction. *Biomaterials Research*, *20*(1), 23. DOI: 10.1186/s40824-016-0071-5 PMID: 27468356

Kirchhof, S., Strasser, A., Wittmann, H. J., Messmann, V., Hammer, N., Goepferich, A. M., & Brandl, F. P. (2015). New insights into the cross-linking and degradation mechanism of Diels–Alder hydrogels. *Journal of Materials Chemistry. B*, *3*(3), 449–457. DOI: 10.1039/C4TB01680G PMID: 32262047

Klee, D., & Höcker, H. (1999). Polymers for biomedical applications: improvement of the interface compatibility. In *Biomedical Applications Polymer Blends* (pp. 1–57). Springer Berlin Heidelberg. DOI: 10.1007/3-540-48838-3_1

Kligman, S., Ren, Z., Chung, C. H., Perillo, M. A., Chang, Y. C., Koo, H., Zheng, Z., & Li, C. (2021). The impact of dental implant surface modifications on osseointegration and biofilm formation. *Journal of Clinical Medicine*, *10*(8), 1641. DOI: 10.3390/jcm10081641 PMID: 33921531

Kobielarz, M., Gazińska, M., Tomanik, M., Stępak, B., Szustakiewicz, K., Filipiak, J., Antończak, A., & Pezowicz, C. (2019). Physicochemical and mechanical properties of CO2 laser-modified biodegradable polymers for medical applications. *Polymer Degradation & Stability*, *165*, 182–195. DOI: 10.1016/j.polymdegradstab.2019.05.010

König, K. (2023). Medical femtosecond laser. *Journal of the European Optical Society. Rapid Publications*, *19*(2), 36. DOI: 10.1051/jeos/2023032

Kryou, C., & Zergioti, I. (2023). Laser-induced forward transfer on regenerative medicine applications. *Biomedical Materials & Devices*, *1*(1), 5–20. DOI: 10.1007/s44174-022-00040-1

Kumar, A., Byadwal, M., Kumar, A., & Kumar, A. (2023). Laser micromachining in biomedical industry. In *Laser-based technologies for sustainable manufacturing* (pp. 169–206). CRC Press. DOI: 10.1201/9781003402398-8

Kyriakides, T. R., Raj, A., Tseng, T. H., Xiao, H., Nguyen, R., Mohammed, F. S., Halder, S., Xu, M., Wu, M. J., Bao, S., & Sheu, W. C. (2021). Biocompatibility of nanomaterials and their immunological properties. *Biomedical Materials (Bristol, England)*, *16*(4), 042005. DOI: 10.1088/1748-605X/abe5fa PMID: 33578402

Lampin, M., Warocquier-Clérout, R., Legris, C., Degrange, M., & Sigot-Luizard, M. F. (1997). Correlation between substratum roughness and wettability, cell adhesion, and cell migration. *Journal of Biomedical Materials Research: An Official Journal of The Society for Biomaterials and The Japanese Society for Biomaterials*, *36*(1), 99–108. DOI: 10.1002/(SICI)1097-4636(199707)36:1<99::AID-JBM12>3.0.CO;2-E PMID: 9212394

Lanzafame, R. J., Brondon, P., Stadler, I., DeVore, D. P., Soltz, R., & Soltz, B. A. (2005). Histologic assessment of mesh fixation following laser-assisted tissue soldering in a lapine model. *Lasers in Surgery and Medicine*, *37*(2), 130–137. DOI: 10.1002/lsm.20201 PMID: 16044423

Lasagni, A. F., Gachot, C., Trinh, K. E., Hans, M., Rosenkranz, A., Roch, T., & Mücklich, F. (2017, March). Direct laser interference patterning, 20 years of development: from the basics to industrial applications. In *Laser-based micro-and nanoprocessing XI* (Vol. 10092, pp. 186–196). SPIE.

Lawrence, B. J., & Madihally, S. V. (2008). Cell colonization in degradable 3D porous matrices. *Cell Adhesion & Migration*, *2*(1), 9–16. DOI: 10.4161/cam.2.1.5884 PMID: 19262124

Lawrence, M., & Jiang, Y. (2017). Porosity, pore size distribution, micro-structure. In *Bio-aggregates based building materials: state-of-the-art report of the RILEM technical committee 236-BBM* (pp. 39–71). Springer Netherlands. DOI: 10.1007/978-94-024-1031-0_2

Le, T. S. D., Phan, H. P., Kwon, S., Park, S., Jung, Y., Min, J., Chun, B. J., Yoon, H., Ko, S. H., Kim, S.-W., & Kim, Y. J. (2022). Recent advances in laser-induced graphene: Mechanism, fabrication, properties, and applications in flexible electronics. *Advanced Functional Materials*, *32*(48), 2205158. DOI: 10.1002/adfm.202205158

Lee, B. L. P., Jeon, H., Wang, A., Yan, Z., Yu, J., Grigoropoulos, C., & Li, S. (2012). Femtosecond laser ablation enhances cell infiltration into three-dimensional electrospun scaffolds. *Acta Biomaterialia*, *8*(7), 2648–2658. DOI: 10.1016/j.actbio.2012.04.023 PMID: 22522128

Lee, S. J., Lee, D., Yoon, T. R., Kim, H. K., Jo, H. H., Park, J. S., Lee, J. H., Kim, W. D., Kwon, I. K., & Park, S. A. (2016). Surface modification of 3D-printed porous scaffolds via mussel-inspired polydopamine and effective immobilization of rhBMP-2 to promote osteogenic differentiation for bone tissue engineering. *Acta Biomaterialia*, *40*, 182–191. DOI: 10.1016/j.actbio.2016.02.006 PMID: 26868173

Lee, S. Y., Shim, E. S., Kang, H. S., Pang, S. S., & Kang, J. S. (2005). Fabrication of ZnO thin film diode using laser annealing. *Thin Solid Films*, *473*(1), 31–34. DOI: 10.1016/j.tsf.2004.06.194

Li, C., Wang, F., Douglas, G., Zhang, Z., Guidoin, R., & Wang, L. (2017). Comprehensive mechanical characterization of PLA fabric combined with PCL to form a composite structure vascular graft. *Journal of the Mechanical Behavior of Biomedical Materials*, *69*, 39–49. DOI: 10.1016/j.jmbbm.2016.11.005 PMID: 28038405

Li, J., Ding, J., Liu, T., Liu, J. F., Yan, L., & Chen, X. (2017). Poly (lactic acid) controlled drug delivery. In *Industrial Applications of Poly (lactic acid)* (pp. 109–138). Springer International Publishing. DOI: 10.1007/12_2017_11

Li, J., Zhang, F., Lyu, H., Yin, P., Shi, L., Li, Z., Zhang, L., Di, C., & Tang, P. (2024). Evolution of musculoskeletal electronics. *Advanced Materials*, *36*(28), 2303311. DOI: 10.1002/adma.202303311 PMID: 38561020

Li, W., Yang, L., Huang, J., Zheng, C., Chen, Y., Li, Y., Yang, D., Li, S., Chen, Z., Cai, W., & Lai, Y. (2024). Progress on fiber engineering for fabric innovation in ecological hydrophobic design and multifunctional applications. *Industrial Chemistry & Materials*, *2*(3), 393–423. DOI: 10.1039/D4IM00048J

Li, Y., Fischer, R., Zboray, R., Boillat, P., Camenzind, M., Toncelli, C., & Rossi, R. M. (2020). Laser-engraved textiles for engineering capillary flow and application in microfluidics. *ACS Applied Materials & Interfaces*, *12*(26), 29908–29916. DOI: 10.1021/acsami.0c03988 PMID: 32506905

Lin, Y., Zhou, Z., & Li, K. Y. (2019). Improved wear resistance at high contact stresses of hydrogen-free diamond-like carbon coatings by carbon/carbon multilayer architecture. *Applied Surface Science*, *477*, 137–146. DOI: 10.1016/j.apsusc.2017.11.118

Lin, Z., & Hong, M. (2021). *Femtosecond laser precision engineering: from micron, submicron, to nanoscale.* Ultrafast Science.

Lin, Z., Huang, T., Ye, X., Zhong, M., Li, L., Jiang, J., Zhang, W., Fan, L., & Zhu, H. (2013). Thinning of large-area graphene film from multilayer to bilayer with a low-power CO2 laser. *Nanotechnology*, *24*(27), 275302. DOI: 10.1088/0957-4484/24/27/275302 PMID: 23764487

Lipovka, A., Fatkullin, M., Shchadenko, S., Petrov, I., Chernova, A., Plotnikov, E., Menzelintsev, V., Li, S., Qiu, L., Cheng, C., Rodriguez, R. D., & Sheremet, E. (2023). Textile electronics with laser-induced graphene/polymer hybrid fibers. *ACS Applied Materials & Interfaces*, *15*(32), 38946–38955. DOI: 10.1021/acsami.3c06968 PMID: 37466067

Liu, R., Suárez, J. M., Weisblum, B., Gellman, S. H., & McBride, S. M. (2014). Synthetic polymers active against Clostridium difficile vegetative cell growth and spore outgrowth. *Journal of the American Chemical Society*, *136*(41), 14498–14504. DOI: 10.1021/ja506798e PMID: 25279431

Low, Y. J., Andriyana, A., Ang, B. C., & Zainal Abidin, N. I. (2020). Bioresorbable and degradable behaviors of PGA: Current state and future prospects. *Polymer Engineering and Science*, *60*(11), 2657–2675. DOI: 10.1002/pen.25508

Maleki, H., Azimi, B., Ismaeilimoghadam, S., & Danti, S. (2022). Poly (lactic acid)-based electrospun fibrous structures for biomedical applications. *Applied Sciences (Basel, Switzerland)*, *12*(6), 3192. DOI: 10.3390/app12063192

Manne, B., Thiruvayapati, H., Bontha, S., Rangarasaiah, R. M., Das, M., & Balla, V. K. (2018). Surface design of Mg-Zn alloy temporary orthopaedic implants: Tailoring wettability and biodegradability using laser surface melting. *Surface and Coatings Technology*, *347*, 337–349. DOI: 10.1016/j.surfcoat.2018.05.017

Matsugaki, A., Aramoto, G., Ninomiya, T., Sawada, H., Hata, S., & Nakano, T. (2015). Abnormal arrangement of a collagen/apatite extracellular matrix orthogonal to osteoblast alignment is constructed by a nanoscale periodic surface structure. *Biomaterials*, *37*, 134–143. DOI: 10.1016/j.biomaterials.2014.10.025 PMID: 25453944

Mei, X., Yang, Z., Wang, X., Shi, A., Blanchard, J., Elahi, F., Kang, H., Orive, G., & Zhang, Y. S. (2025). Integrating microfluidic and bioprinting technologies: Advanced strategies for tissue vascularization. *Lab on a Chip*, *25*(5), 764–786. DOI: 10.1039/D4LC00280F PMID: 39775452

Mettler, J. J. H. (2024). Investigation of high and low spatial frequency laser induced periodic surface structures on metals (Doctoral dissertation, University of Illinois at Urbana-Champaign).

Mndlovu, H., Kumar, P., du Toit, L. C., & Choonara, Y. E. (2024). A review of bio material degradation assessment approaches employed in the biomedical field. *NPJ Materials Degradation*, *8*(1), 66. DOI: 10.1038/s41529-024-00487-1

Mogosanu, D. E., Giol, E. D., Vandenhaute, M., Dragusin, D. M., Samal, S. K., & Dubruel, P. (2014). Polyester biomaterials for regenerative medicine. In The Design, Synthetic Strategies and Biocompatibility of Polymer Scaffolds for Biomedical Application (pp. 155-197). Bentham Science Publishers. DOI: 10.2174/9781608058761114010009

Mohammed-Sadhakathullah, A. H., Paulo-Mirasol, S., Torras, J., & Armelin, E. (2023). Advances in functionalization of bioresorbable nanomembranes and nanoparticles for their use in biomedicine. *International Journal of Molecular Sciences*, *24*(12), 10312. DOI: 10.3390/ijms241210312 PMID: 37373461

Molla, S., Abedin, M. M., & Siddique, I. M. (2024). Exploring the versatility of medical textiles: Applications in implantable and non-implantable medical textiles. *World Journal of Advanced Research and Reviews*, *21*(1), 603–615. DOI: 10.30574/wjarr.2024.21.1.0058

Moore, M. J., Jabbari, E., Ritman, E. L., Lu, L., Currier, B. L., Windebank, A. J., & Yaszemski, M. J. (2004). Quantitative analysis of interconnectivity of porous biodegradable scaffolds with micro-computed tomography. *Journal of Biomedical Materials Research. Part A*, *71*(2), 258–267. DOI: 10.1002/jbm.a.30138 PMID: 15376269

Moro, T., Yoshio Takatori, M. D., & Kawaguchi, H. (2011). Cartilage-mimicking, High-density Brush Structure Improves Wear Resistance of Crosslinked Polyethylene. *Clinical Orthopaedics and Related Research*, *469*(8), 2327–2336. DOI: 10.1007/s11999-010-1718-5 PMID: 21132412

Morris, H., & Murray, R. (2020). Medical textiles. Textile progress, 52(1-2), 1-127.

Morris, H., & Murray, R. (2021). Textile Materials for Healthcare and Medical Applications. In Medical Textiles (pp. 35-62). CRC Press. DOI: 10.1201/9781003170570-3

Mostafa, A. M. (2025). The Influence of various parameters on the ablation and deposition mechanisms in pulsed laser deposition. *Plasmonics*, *20*(7), 1–19. DOI: 10.1007/s11468-024-02706-6

Nadi, A., Boukhriss, A., Bentis, A., Jabrane, E., & Gmouh, S. (2018). Evolution in the surface modification of textiles: A review. *Textile Progress*, *50*(2), 67–108. DOI: 10.1080/00405167.2018.1533659

Nagel, M., & Lippert, T. (2012). Laser-induced forward transfer for the fabrication of devices. Nanomaterials: processing and characterization with lasers, 255-306.

Nasef, M. M., Gupta, B., Shameli, K., Verma, C., Ali, R. R., & Ting, T. M. (2021). Engineered bioactive polymeric surfaces by radiation induced graft copolymerization: Strategies and applications. *Polymers*, *13*(18), 3102. DOI: 10.3390/polym13183102 PMID: 34578003

Nemani, S. K., Annavarapu, R. K., Mohammadian, B., Raiyan, A., Heil, J., Haque, M. A., Abdelaal, A., & Sojoudi, H. (2018). Surface modification of polymers: Methods and applications. *Advanced Materials Interfaces*, *5*(24), 1801247. DOI: 10.1002/admi.201801247

Obilor, A. F., Pacella, M., Wilson, A., & Silberschmidt, V. V. (2022). Micro-texturing of polymer surfaces using lasers: A review. *International Journal of Advanced Manufacturing Technology*, *120*(1), 103–135. DOI: 10.1007/s00170-022-08731-1

Odugbemi, M. (2018). Biopolymers for bioartificial pancreas (Doctoral dissertation, University of Westminster).

Oliviero, O., Ventre, M., & Netti, P. A. (2012). Functional porous hydrogels to study angiogenesis under the effect of controlled release of vascular endothelial growth factor. *Acta Biomaterialia*, *8*(9), 3294–3301. DOI: 10.1016/j.actbio.2012.05.019 PMID: 22641106

Omenogor, C. E., & Adeniran, A. A. (2024). Advancing Precision Healthcare: The Integration of Nanotechnology, Millimeter Wave Sensing, Laser Technology, Fibre Bragg Grating, and Deep Learning Models. International Journal of Research Publication and Reviews.

Pandit, S., Gaska, K., Kádár, R., & Mijakovic, I. (2021). Graphene-based antimicrobial biomedical surfaces. *ChemPhysChem*, *22*(3), 250–263. DOI: 10.1002/cphc.202000769 PMID: 33244859

Papavlu, A. P., Dinca, V., Filipescu, M., & Dinescu, M. (2017). Matrix-Assisted Pulsed Laser Evaporation of Organic Thin Films: Applications in Biology and Chemical. Laser Ablation: From Fundamentals to Applications, 171.

Park, H., Park, J. J., Bui, P. D., Yoon, H., Grigoropoulos, C. P., Lee, D., & Ko, S. H. (2024). Laser-based selective material processing for next-generation additive manufacturing. *Advanced Materials*, *36*(34), 2307586. DOI: 10.1002/adma.202307586 PMID: 37740699

Patel, H., Ostergard, D. R., & Sternschuss, G. (2012). Polypropylene mesh and the host response. *International Urogynecology Journal*, *23*(6), 669–679. DOI: 10.1007/s00192-012-1718-y PMID: 22430945

Patrucco, A. (2017). KERATIN-BASED 3D SCAFFOLD DESIGN FOR BONE TISSUE ENGINEERING.

Patrucco, A., Cristofaro, F., Simionati, M., Zoccola, M., Bruni, G., Fassina, L., Visai, L., Magenes, G., Mossotti, R., Montarsolo, A., & Tonin, C. (2016). Wool fibril sponges with perspective biomedical applications. *Materials Science and Engineering C*, *61*, 42–50. DOI: 10.1016/j.msec.2015.11.073 PMID: 26838822

Peethan, A., Unnikrishnan, V. K., Chidangil, S., & George, S. D. (2020). Laser-Assisted Tailoring of Surface Wettability-Fundamentals and Applications: A Critical Review. *Progress in Adhesion and Adhesives*, *5*, 331–365. DOI: 10.1002/9781119749882.ch11

Pei, X., Wang, J., Cong, Y., & Fu, J. (2021). Recent progress in polymer hydrogel bioadhesives. *Journal of Polymer Science*, *59*(13), 1312–1337. DOI: 10.1002/pol.20210249

Perrakis, G., Tsilipakos, O., Tsibidis, G. D., & Stratakis, E. (2024). Impact of hybrid electromagnetic surface modes on the formation of low spatial frequency LIPSS: A universal approach. *Laser & Photonics Reviews*, *18*(7), 2301090. DOI: 10.1002/lpor.202301090

Petrović, S., Peruško, D., Skoulas, E., Kovač, J., Mitrić, M., Potočnik, J., Rakočević, Z., & Stratakis, E. (2019). Laser-assisted surface texturing of Ti/Zr multilayers for mesenchymal stem cell response. *Coatings*, *9*(12), 854. DOI: 10.3390/coatings9120854

Pohl, R. (2015). Laser-induced forward transfer of pure metals//Towards 3D printing.

Polshettiwar, S. (2025). Polymers for Vascular Applications. Materials Research Foundations, 172.

Ramakrishna, T. R., Nalder, T. D., Yang, W., Marshall, S. N., & Barrow, C. J. (2018). Controlling enzyme function through immobilisation on graphene, graphene derivatives and other two dimensional nanomaterials. *Journal of Materials Chemistry. B*, *6*(20), 3200–3218. DOI: 10.1039/C8TB00313K PMID: 32254379

Ramasubramanian, B., Sundarrajan, S., Rao, R. P., Reddy, M. V., Chellappan, V., & Ramakrishna, S. (2022). Novel low-carbon energy solutions for powering emerging wearables, smart textiles, and medical devices. *Energy & Environmental Science*, *15*(12), 4928–4981. DOI: 10.1039/D2EE02695C

Ramirez, G. (2023). *Cytotoxic Effects of Dental Implant Debris After Decontamination Treatment Using Er: YAG Laser*. Temple University.

Ranakoti, L., Gangil, B., Bhandari, P., Singh, T., Sharma, S., Singh, J., & Singh, S. (2023). Promising role of polylactic acid as an ingenious biomaterial in scaffolds, drug delivery, tissue engineering, and medical implants: Research developments, and prospective applications. *Molecules (Basel, Switzerland)*, *28*(2), 485. DOI: 10.3390/molecules28020485 PMID: 36677545

RASLAN, W. M., EL-HALWAGY, A. A., & ELSAYAD, H. (2020). Recent advances in plasma/nanoparticles treatments of textile fibers. Journal of Textiles. *Coloration and Polymer Science*, *17*(2), 87–105.

Ribba, L., Parisi, M., D'accorso, N. B., & Goyanes, S. (2014). Electrospun nanofibrous mats: From vascular repair to osteointegration. *Journal of Biomedical Nanotechnology*, *10*(12), 3508–3535. DOI: 10.1166/jbn.2014.2046 PMID: 26000368

Riveiro, A., Maçon, A. L., del Val, J., Comesaña, R., & Pou, J. (2018). Laser surface texturing of polymers for biomedical applications. *Frontiers in Physics (Lausanne)*, *6*, 16. DOI: 10.3389/fphy.2018.00016

Sahani, S. (2025). Plasma Innovations: Revolutionizing Sustainable Textile Treatments. In: Dutta, S., Bansal, P., Kapoor, A., Reddy, C.C., Venkatraman, P. (eds) Selected Articles from the International Conference on Sustainable Fashion and Technical Textiles. ICSFT 2024. Proceedings in Technology Transfer. Springer, pp 78–94. Singapore. DOI: 10.1007/978-981-96-3169-8_7

Samipour, S., Taghvaei, H., Mohebbi-Kalhori, D., & Rahimpour, M. R. (2020). Plasma treatment and chitosan coating: A combination for improving PET surface properties. *Surface Innovations*, *8*(1-2), 76–88. DOI: 10.1680/jsuin.19.00030

Sanjari, S., Etemadi Haghighi, S., Saraeian, P., & Alinia-ziazi, A. (2025). Modeling Degradation Behavior of (Bio) Degradable Polymers for Medical Devices: A Comparative Review of Physio-chemical Approaches. *Journal of Polymers and the Environment*, *33*(2), 660–697. DOI: 10.1007/s10924-024-03376-5

Scarlat, E. V. R., & Mitu, B. (2020, June). the Biocompatibility of the Textile Medical. In Advances in Human Factors and Ergonomics in Healthcare and Medical Devices: Proceedings of the AHFE 2020 Virtual Conference on Human Factors and Ergonomics in Healthcare and Medical Devices, July 16-20, 2020, USA (Vol. 1205, p. 217). Springer Nature.

Schwarz, S., Rung, S., & Hellmann, R. (2017). Two-dimensional low spatial frequency laser-induced periodic surface structuring of sapphire. *Journal of Laser Micro Nanoengineering*, *12*(2), 67–71. DOI: 10.2961/jlmn.2017.02.0003

Sedghamiz, E., Liu, M., & Wenzel, W. (2022). Challenges and limits of mechanical stability in 3D direct laser writing. *Nature Communications*, *13*(1), 2115. DOI: 10.1038/s41467-022-29749-9 PMID: 35440637

Serra, P., & Piqué, A. (2019). Laser-induced forward transfer: Fundamentals and applications. *Advanced Materials Technologies*, *4*(1), 1800099. DOI: 10.1002/admt.201800099

Shabanloo, R., Montazer, M., Farahani, A., & Karimi, N. (2025). A review on surface modification of nanofibrous textiles for diverse applications: Focus on medical uses. *Heliyon*, *11*(2), e41863. DOI: 10.1016/j.heliyon.2025.e41863 PMID: 39897825

Shakiba, M., Rezvani Ghomi, E., Khosravi, F., Jouybar, S., Bigham, A., Zare, M., Abdouss, M., Moaref, R., & Ramakrishna, S. (2021). Nylon—A material introduction and overview for biomedical applications. *Polymers for Advanced Technologies*, *32*(9), 3368–3383. DOI: 10.1002/pat.5372

Shamim, Z., Awan, M. T., Jamal, A., & Shahid, M. N. (2025). Structural and Mechanical Features of Biomedical Textiles. In Biomedical Textiles (pp. 69-99). Jenny Stanford Publishing. DOI: 10.1201/9781003651697-3

Sharma, C. P., Gupta, A., Khadka, M., Ben-Yoav, H., Ronen, A., & Arnusch, C. J. (2024). Silver nanoparticle doped laser-induced graphene fabrication methodology affects silver nanoparticle size, distribution, biological and electrochemical properties. *Environmental Science. Nano*, *11*(4), 1582–1596. DOI: 10.1039/D3EN00371J

Sharma, S., Sudhakara, P., Singh, J., Ilyas, R. A., Asyraf, M. R. M., & Razman, M. R. (2021). Critical review of biodegradable and bioactive polymer composites for bone tissue engineering and drug delivery applications. *Polymers*, *13*(16), 2623. DOI: 10.3390/polym13162623 PMID: 34451161

Sharma, S., Sudhakara, P., Singh, J., Ilyas, R. A., Asyraf, M. R. M., & Razman, M. R. (2021). Critical review of biodegradable and bioactive polymer composites for bone tissue engineering and drug delivery applications. *Polymers*, *13*(16), 2623. DOI: 10.3390/polym13162623 PMID: 34451161

Shein, P., Cilip, C. M., Quinto, G., Behrens, A., & Fried, N. M. (2008, August). Selective laser suture lysis with a compact, low-cost, red diode laser. In 2008 30th Annual International Conference of the IEEE Engineering in Medicine and Biology Society (pp. 4358-4360). IEEE. DOI: 10.1109/IEMBS.2008.4650175

Sherazi, T. A., Rehman, T., Naqvi, S. A. R., Shaikh, A. J., Shahzad, S. A., Abbas, G., Raza, R., & Waseem, A. (2015). Surface functionalization of solid state ultra-high molecular weight polyethylene through chemical grafting. *Applied Surface Science*, *359*, 593–601. DOI: 10.1016/j.apsusc.2015.10.080

Shi, X., Li, X., Jiang, L., Qu, L., Zhao, Y., Ran, P., Wang, Q., Cao, Q., Ma, T., & Lu, Y. (2015). Femtosecond laser rapid fabrication of large-area rose-like micro-patterns on freestanding flexible graphene films. *Scientific Reports*, *5*(1), 17557. DOI: 10.1038/srep17557 PMID: 26615800

Sidhu, M. S., & Dhingra, N. (2022). Ablation of materials using femtosecond lasers and electron beams. In *Terahertz*. Ultrafast Lasers and Their Medical and Industrial Applications. IntechOpen. DOI: 10.5772/intechopen.106198

Sikka, M., Garg, S., & Kumari, N. (2024). Bandages (compression, retention, and support). In *Textile Materials for Good Health and Wellbeing: Design and Applications* (pp. 155–170). Springer Nature Singapore.

Singh, C., Wong, C. S., & Wang, X. (2015). Medical textiles as vascular implants and their success to mimic natural arteries. *Journal of Functional Biomaterials*, *6*(3), 500–525. DOI: 10.3390/jfb6030500 PMID: 26133386

Singha, K., Pandit, P., & Maity, S. (2022). Wool composites for hygienic/medical applications. In *Wool Fiber Reinforced Polymer Composites* (pp. 387–406). Woodhead Publishing. DOI: 10.1016/B978-0-12-824056-4.00007-8

Singhal, J. P., Singh, H., & Ray, A. R. (1988). Absorbable suture materials: Preparation and properties. *Polymer Reviews (Philadelphia, Pa.)*, *28*(3-4), 475–502.

Sirdeshmukh, N., & Dongre, G. (2021). Laser micro & nano surface texturing for enhancing osseointegration and antimicrobial effect of biomaterials: A review. *Materials Today: Proceedings*, *44*, 2348–2355. DOI: 10.1016/j.matpr.2020.12.433

Somasekhar, L. (2021). Bioink Optimization and Effects of Microgravity on 3D Bioprinted Cell Laden Constructs.

Stępak, B. D., Antończak, A. J., Szustakiewicz, K., Kozioł, P. E., & Abramski, K. M. (2014). Degradation of poly (L-lactide) under KrF excimer laser treatment. *Polymer Degradation & Stability*, *110*, 156–164. DOI: 10.1016/j.polymdegradstab.2014.08.028

Suhag, D., Kaushik, S., & Taxak, V. B. (2024). *Handbook of Biomaterials for Medical Applications* (Vol. 2).

Sun, L., Guo, J., Chen, H., Zhang, D., Shang, L., Zhang, B., & Zhao, Y. (2021). Tailoring materials with specific wettability in biomedical engineering. *Advancement of Science*, *8*(19), 2100126. DOI: 10.1002/advs.202100126 PMID: 34369090

Thakur, A., Kumar, A., Kaya, S., Marzouki, R., Zhang, F., & Guo, L. (2022). Recent advancements in surface modification, characterization and functionalization for enhancing the biocompatibility and corrosion resistance of biomedical implants. *Coatings*, *12*(10), 1459. DOI: 10.3390/coatings12101459

Thakur, A. K., Lin, B., Nowrin, F. H., & Malmali, M. (2021). Comparing structure and sorption characteristics of laser-induced graphene (LIG) from various polymeric substrates. *ACS ES&T Water*, *2*(1), 75–87. DOI: 10.1021/acsestwater.1c00259

Todorov, D. N. (2020, November). Practical research of marking and cutting of textiles with increased resistance, using CO2 laser. [). IOP Publishing.]. *Journal of Physics: Conference Series*, *1681*(1), 012014. DOI: 10.1088/1742-6596/1681/1/012014

Tu, K. T., & Chung, C. K. (2016). Rapid prototyping of biodegradable microneedle arrays by integrating CO2 laser processing and polymer molding. *Journal of Micromechanics and Microengineering*, *26*(6), 065015. DOI: 10.1088/0960-1317/26/6/065015

Vázquez-López, A., de la Vega, J., Collado, I., Carmona, F. J., Pradanos, P., Prolongo, S. G., & Wang, D. Y. (2024). Graphene Oxide/Polylactic Acid-Based Face Mask to Combat H3N2: A Strategy against Influenza. *ACS Applied Nano Materials*, *7*(6), 6460–6470. DOI: 10.1021/acsanm.4c00183

Veerapandian, M., & Yun, K. (2011). Functionalization of biomolecules on nanoparticles: Specialized for antibacterial applications. *Applied Microbiology and Biotechnology*, *90*(5), 1655–1667. DOI: 10.1007/s00253-011-3291-6 PMID: 21523475

Verdoia, M., Kedhi, E., Suryapranata, H., Galasso, G., Dudek, D., & De Luca, G. (2020). Poly (L-lactic acid) bioresorbable scaffolds versus metallic drug-eluting stents for the treatment of coronary artery disease: A meta-analysis of 11 randomized trials. *Catheterization and Cardiovascular Interventions*, *96*(4), 813–824. DOI: 10.1002/ccd.28594 PMID: 31730255

Visan, A. I., & Popescu-Pelin, G. F. (2024). Advanced laser techniques for the development of nature-inspired biomimetic surfaces applied in the medical field. *Coatings*, *14*(10), 1290. DOI: 10.3390/coatings14101290

Wang, H., Zhao, Z., Liu, P., & Guo, X. (2022). Laser-induced graphene based flexible electronic devices. *Biosensors (Basel)*, *12*(2), 55. DOI: 10.3390/bios12020055 PMID: 35200316

Wu, C. S. (2015). Influence of modified polyester on the material properties of collagen-based biocomposites and in vitro evaluation of cytocompatibility. *Materials Science and Engineering C*, *48*, 310–319. DOI: 10.1016/j.msec.2014.12.013 PMID: 25579928

Wu, T., Li, H., Xue, J., Mo, X., & Xia, Y. (2019). Photothermal welding, melting, and patterned expansion of nonwoven mats of polymer nanofibers for biomedical and printing applications. *Angewandte Chemie International Edition*, *58*(46), 16416–16421. DOI: 10.1002/anie.201907876 PMID: 31373102

Wu, Y., Heikal, L., Ferns, G., Ghezzi, P., Nokhodchi, A., & Maniruzzaman, M. (2019). 3D bioprinting of novel biocompatible scaffolds for endothelial cell repair. *Polymers*, *11*(12), 1924. DOI: 10.3390/polym11121924 PMID: 31766610

Xie, H., Hong, G., Zhou, Q., Wu, Y., Chen, Y., You, R., & Han, F. (2023). Nanocomposite surface modification. In *Innovative Development in Micromanufacturing Processes* (pp. 296–371). CRC Press. DOI: 10.1201/9781003364948-15

Xu, L., Liu, Y., Zhou, W., & Yu, D. G. (2022). Electrospun Medical Sutures for Wound Healing: A Review. *Polymers*, *2022*(14), 1637. DOI: 10.3390/polym14091637 PMID: 35566807

Yadroitsev, I., Krakhmalev, P., & Yadroitsava, I. (2014). Selective laser melting of Ti6Al4V alloy for biomedical applications: Temperature monitoring and microstructural evolution. *Journal of Alloys and Compounds*, *583*, 404–409. DOI: 10.1016/j.jallcom.2013.08.183

Yan, L., Yuan, Y., Ouyang, L., Li, H., Mirzasadeghi, A., & Li, L. (2016). Improved mechanical properties of the new Ti-15Ta-xZr alloys fabricated by selective laser melting for biomedical application. *Journal of Alloys and Compounds*, *688*, 156–162. DOI: 10.1016/j.jallcom.2016.07.002

Yang, S., & Zhang, J. (2017). Matrix-Assisted Pulsed Laser Evaporation (MAPLE) technique for deposition of hybrid nanostructures. *Front. Nanosci. Nanotechnol*, *3*(1). Advance online publication. DOI: 10.15761/FNN.1000146

Yeo, T., Ko, Y. G., Kim, E. J., Kwon, O. K., Chung, H. Y., & Kwon, O. H. (2021). Promoting bone regeneration by 3D-printed poly (glycolic acid)/hydroxyapatite composite scaffolds. *Journal of Industrial and Engineering Chemistry*, *94*, 343–351. DOI: 10.1016/j.jiec.2020.11.004

Young, O. M., Xu, X., Sarker, S., & Sochol, R. D. (2024). Direct laser writing-enabled 3D printing strategies for microfluidic applications. *Lab on a Chip*, *24*(9), 2371–2396. DOI: 10.1039/D3LC00743J PMID: 38576361

Zhang, S., Wang, J., Zheng, Z., Yan, J., Zhang, L., Li, Y., Zhang, J., Li, G., Wang, X., & Kaplan, D. (2021). Porous nerve guidance conduits reinforced with braided composite structures of silk/magnesium filaments for peripheral nerve repair. *Acta Biomaterialia*, *134*, 116–130. DOI: 10.1016/j.actbio.2021.07.028 PMID: 34289421

Zhang, Y., Wang, X., Yan, K., Zhu, H., Wang, B., & Zou, B. (2023). Laser micro/nano-structuring pushes forward smart sensing: Opportunities and challenges. *Advanced Functional Materials*, *33*(8), 2211272. DOI: 10.1002/adfm.202211272

Zhang, Y., Xia, X., Ma, K., Xia, G., Wu, M., Cheung, Y. H., Yu, H., Zou, B., Zhang, X., Farha, O. K., & Xin, J. H. (2023). Functional textiles with smart properties: Their fabrications and sustainable applications. *Advanced Functional Materials*, *33*(33), 2301607. DOI: 10.1002/adfm.202301607

Zhao, W., Zhu, J., Hang, J., & Zeng, W. (2022). Biomaterials to promote vascularization in tissue engineering organs and ischemic fibrotic diseases. *MedComm - Biomaterials and Applications*, *1*(1), e16. DOI: 10.1002/mba2.16

Zhu, J., Huang, X., & Song, W. (2021). Physical and chemical sensors on the basis of laser-induced graphene: Mechanisms, applications, and perspectives. *ACS Nano*, *15*(12), 18708–18741. DOI: 10.1021/acsnano.1c05806 PMID: 34881870

Zia, A. W., Liu, R., & Wu, X. (2022). Structural design and mechanical performance of composite vascular grafts. *Bio-Design and Manufacturing*, *5*(4), 757–785. DOI: 10.1007/s42242-022-00201-7

Chapter 9
Ultrafast Laser Surface Processing for Medical Implants Functionalization

Hassane Toiwiya
INSERM, SAINBIOSE U1059, Université Jean Monnet Saint-Étienne, France

Alain Guignandon
http://orcid.org/0000-0002-0745-5417
INSERM, SAINBIOSE U1059, Université Jean Monnet Saint-Étienne, France

Virginie Dumas
CNRS, ENTPE, LTDS, UMR5513, ENISE, Ecole Centrale de Lyon, Université de Lyon, France

Sedao
http://orcid.org/0000-0003-4072-492X
Hubert Curien Laboratory, UMR 5516, Jean Monnet University, France

ABSTRACT

Ultrafast femtosecond laser surface functionalization offers a versatile single-step strategy to engineer medical implants with finely tuned biological responses. Using dental implants as a model system, this chapter illustrates how micro et nanoscale laser patterns enhance osseointegration, improve soft-tissue sealing and reduce bacterial colonization: three key challenges associated with implant failures. A major emerging direction is the integration of extracellular vesicles (EVs) through laser-induced nanostructures that act as passive topographical traps, enabling selective EV capture, protection, and controlled presentation at the tissue-implant interface.

DOI: 10.4018/979-8-3373-4104-0.ch009

This approach opens promising routes toward next-generation bioactive implants capable of modulating the Healing process. Beyond dentistry, femtosecond laser texturing is applicable to various devices, including pacemakers, orthopedic screws, and vascular stents. The chapter closes with industrial, regulatory and technological considérations necessary for successful clinical translation.

INTRODUCTION

The advent of ultrashort-pulse lasers has made a major impact across a wide range of technological and scientific fields. Their development has expanded the possibilities for high-resolution imaging, advanced diagnostic techniques, and precision materials processing. Thanks to their ability to interact with dielectrics, semiconductors, and metals with exceptional precision, ultrafast lasers have become indispensable tools in fields such as micromachining, photonics, and biomedical research. A fundamental aspect of these systems lies in the extremely short duration of the emitted pulses, which correspond to highly concentrated packets of photons, confined both spatially and temporally, often on the femtosecond scale.

Let's consider an ultrafast laser pulse of 100 femtosecond (fs) pulse duration (1 fs = 10^{-15} s), a time duration often used in the ultrafast laser science and technology world. Notwithstanding light propagates at an amazingly great speed (which is 3.10^{8} $m.s^{-1}$, at this pace one would travel from Earth to the Moon in a little bit more than a single second), the light contained in this ultrafast laser pulse does not get to travel a far distance, the distance it travels is around 30 μm, a dimension less than the diameter of a human hair. Such small pockets of light, when shine at the surface of a material, bring about fascinating facts.

In the time domain, the interaction always starts from the encounter between the photons and electrons. Excited electrons then react with the material lattice. The lattice then takes over the lead, reacts, and generates hydrodynamic movements. Figure 1 shows a diagram illustrating the interaction between a femtosecond laser pulse and the material being exposed to the laser irradiation. The laser-induced processes can be roughly divided into three distinct time windows, listed in chronological order and represented by different colors in the figure: first, in the sub-picosecond (ps) regime (1 ps = 10^{-12} s), laser excitation of optically active electronic states in the irradiated material and the initial ultrafast processes occurring in response to the electronic excitation (yellow boxes); Second, in the time window between 1 and 100 ps, rapid out-of-equilibrium phase transitions triggered by the transfer of energy from the excited electrons to atomic vibrations (green boxes); finally, in the nanosecond (ns) time scale (1 ns = 10^{-9} s), the subsequent cooling and solidification of the surface region of the irradiated target (blue boxes).

Viewed in the spatial domain, these aforementioned subatomic-scale reactions take place within a comparatively large volume defined by the laser spot size and the penetration depth of the light into the material. The intertwined time-space dynamics of ultrafast laser-matter interaction ultimately dictate the material's response to irradiation. Ultrafast laser-based processing operates in an athermal regime, producing minimal heat and causing almost no collateral damage near the irradiated region. It enables high-precision, high-quality structuring with micrometre- to nanometre-scale resolution, supports volumetric or bulk modification through nonlinear absorption, and offers many additional advantages.

When the laser pulse width is shorter than the electron–phonon coupling time in laser–matter interactions (typically 1-100 ps), thermal diffusion, the region over which laser energy is transferred as heat, becomes negligible and remains below a fraction of a micrometre. In contrast, with conventional nanosecond laser pulses, thermal diffusion extends well beyond 1 μm. This disparity has a pronounced impact on the quality of the processed material. Figures 1b and 1c show scanning electron microscope (SEM) images of holes made in metal plates using laser ablation with pulses of 200 fs and 6 ns, respectively (Zhao et al., 2020). With a femtosecond laser, the holes obtained have sharp edges, steep walls, and very few melted or resolidified areas. Conversely, ablation performed with nanosecond pulses generates significant melting of the material and is often accompanied by visible burrs around the hole.

Figure 1. (a) Energy transfer and phase transitions following the excitation of a metal surface irradiated by an intense ultrashort laser pulse; (b) and (c) SEM images of cavities ablated in thin foils of steel by ultrafast laser pulses at 200 fs pulse duration, and by nanosecond laser pulses at a pulse duration of 6 ns, respectively

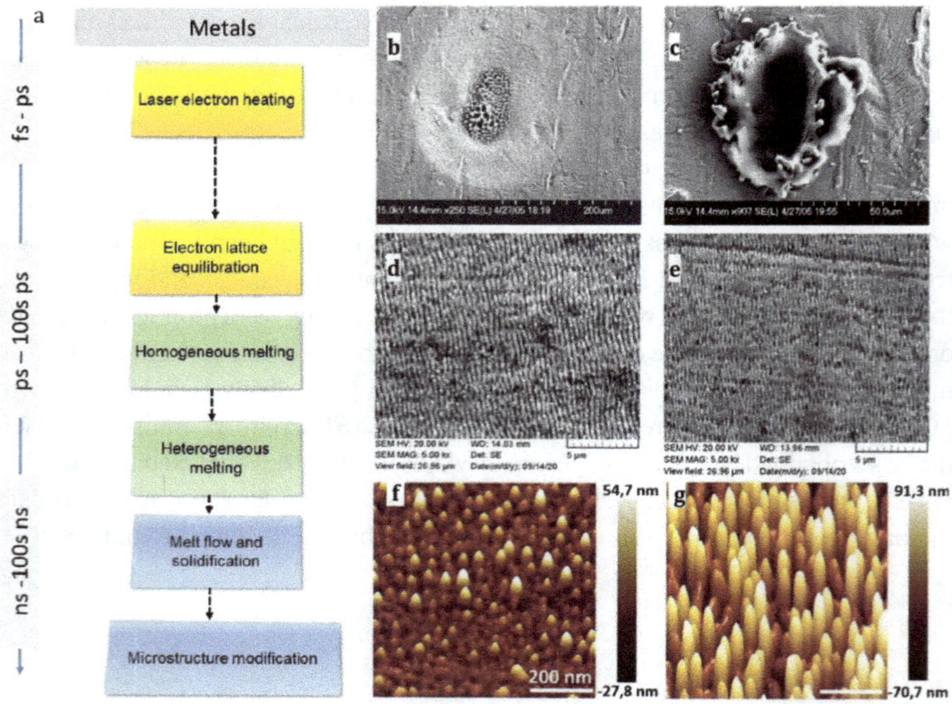

(Adapted and reprinted from, Jiantao Zhao et al., Nanosecond Laser Ablation of Ti–6Al–4V under Different Temperature, Applied sciences (MDPI, Basel, Switzerland), Copyright 2020 under Creative Commons BY 4.0 license. Retrieved from https://doi.org/10.3390/app10134657) (Zhao et al., 2020). The supreme cleanliness and burr-free surface quality of femtosecond laser ablation over longer pulse laser machining are evident; (d) and (e) Laser Induced Periodic Structures (LIPSS) with dimensions below the laser wavelength; (f) and (g) AFM images of nano-spikes with dimensions far below the wavelength of the laser source (Adapted and reprinted from, Anthony Nakhoul et al., Boosted Spontaneous Formation of High-Aspect Ratio Nanopeaks on Ultrafast Laser-Irradiated Ni Surface, Advanced Science (Wiley, Germany), Copyright 2022 under Creative Commons BY 4.0 license. Retrieved from https://doi:10.1002/advs.202200761) (Nakhoul et al., 2022).

The diffraction limit dictates that a laser beam cannot be focused to an arbitrarily small point; in practice, the minimum spot size is typically on the order of the laser wavelength. For visible light at around 500 nm, the focal spot is no smaller than roughly 0.2 μm, which conventionally defines the achievable spatial resolution in laser micromachining. However, by exploiting laser interference effects and the superposition of scattered near fields, the electromagnetic energy distribution within the focal region can be further modulated at dimensions well below the incident wavelength. This enables the fabrication of nanoscale features far smaller than what the diffraction limit alone would suggest. One prominent example is the formation of Laser-Induced Periodic Surface Structures (LIPSS), highly ordered nanopatterns whose periodicity ranges from near the laser wavelength down to below 100 nm. Figures 1d and 1e present SEM images of LIPSS with their periodicities below the laser wavelength. Furthermore, Figure 1f and 1g are atomic force microscopic (AFM) images of 2-dimentional quasi-periodic structures with their periodicities far below the laser wavelength (~10 fold) (Nakhoul et al., 2022). These sub-wavelength structures have been observed on a broad variety of materials, both on surfaces and within bulk volumes. Their precise, reproducible nanoscale features have led to applications across tribology, biology, energy technologies, and more, some of which are discussed in later sections.

Ultrafast laser texturing has emerged as a promising approach for surface biofunctionalization. Laser-based technologies are increasingly relevant in the orthopaedic and dental implant sectors, which together accounted for €41.8 billion in 2017 and are projected to reach €60.6 billion by 2025, corresponding to a compound annual growth rate (CAGR) of 4.7% from 2018 to 2025. Trauma implants represent roughly 17% of this market, while dental implants account for about 8%. In this section, orthopaedic implants (specifically bone screws) and dental implants are selected for focused discussion for the sake of clarity and representativeness. However, many other medical implant categories also stand to benefit from laser-induced surface functionalization, particularly in applications involving controlled cell adhesion. This extends beyond orthopaedic and dental devices to include, for example, cardiovascular implants, which constitute an additional multi-billion-euro market of their own.

Medical implants are being used increasingly across virtually all clinical disciplines. One major application area is dental prosthetics, a rapidly expanding market driven by ageing populations. These implants, typically made of titanium or titanium alloys, must promote strong and rapid osseointegration with the jawbone. In contrast, other implants, such as bone screws and fixation plates, may need to be removed after several months or years. For these devices, excessive or complete overgrowth by surrounding cells is undesirable. Therefore, a single-step surface functionalization method capable of controlling cellular attachment and growth is highly sought after. Laser-based surface processing offers a particularly competitive and versatile solution to this challenge.

Due to their mechanical strength, resistance to degradation in physiological environments, and biocompatibility, titanium-based materials, such as Ti-6Al-4V, have become the standard choice for many applications in orthopedics and dental implantology. Nevertheless, long-term clinical outcomes can be compromised by poor bone integration or the development of infections around the implant. Promoting effective osseointegration, characterized by adequate bone formation and mineralization at the implant-bone interface, remains a key objective of biomaterials research. Since the biological response is triggered immediately after implantation, interactions between the implant surface and surrounding cells are critical factors influencing the implant's performance and durability. Human mesenchymal stem cells (hMSCs) are attracted to the implant site and start interacting with the implant surface. This step is followed by cell proliferation and differentiation toward an osteoblastic lineage. Mature osteoblasts deposit an extracellular matrix, rich in collagen type I and fibronectin, and then produce a new calcified and mineralized matrix in contact with the implant surface. Once osseointegration is achieved, the tissue formed at the bone-implant interface exhibits structural and compositional characteristics similar to those of natural mineralized bone. To enhance this process, various surface engineering techniques have been developed with the aim of enhancing the biological activity of implant materials. Surface roughness is one of the most studied approaches, as it allows for the modulation of cell-material interactions and the stimulation of osteogenic responses. This parameter reflects the degree of surface irregularity and is generally assessed based on the average dimensions of the microscopic protrusions and depressions present on the material's surface.

Surface roughness (Sa, refers to the arithmetical mean height of a surface and is one of the most widely used 3D roughness parameters in surface metrology). The mean surface height (Sa) parameter, widely used to characterize the three-dimensional topography of a surface, is used to measure its roughness. Studies have shown that microscopic-scale roughness, typically ranging from 1 to 2 µm, promotes bone apposition and enhances the osteogenic activity of cells compared to smooth implant surfaces. However, these topographic features can also create conditions conducive to bacterial adhesion and the subsequent development of a biofilm, which can compromise implant success. Conversely, surface modifications at the nanoscale have attracted considerable interest, as they can prevent bacterial adhesion while maintaining desirable biological interactions with host tissues.

In this context, it is worth mentioning that, once placed in the oral cavity, implants are quickly exposed to microbial colonization, which can lead to biofilm formation and the development of peri-implantitis. This condition remains one of the most significant biological complications associated with modern dental implant therapy. Peri-implantitis is characterized by an inflammatory process affecting both the soft tissues and the supporting bone surrounding the implant, ultimately leading to progressive bone loss. According to recent reviews, this condition is implicated in a significant proportion of implant failures, with reported prevalence rates reaching up to 47% among failed dental implantsimplants (Derks & Tomasi, 2015). In transmucosal implant systems, bacterial contamination is primarily caused by the migration of oral microorganisms through the peri-implant soft tissues to the implant surface, where a biofilm can then form. A critical vulnerability of dental implants lies at the interface between the oral cavity and the underlying bone, which can serve as a route for microbial invasion. When soft tissue attachment to the implant surface is insufficient, bacteria can rapidly infiltrate, triggering inflammatory responses, gingival recession, and progressive bone loss that can ultimately compromise implant stability. Therefore, establishing a stable biological seal through effective adhesion of peri-implant soft tissues is essential for preventing peri-implantitis and ensuring the long-term success of the implant.

To address these challenges, there is growing interest in the development of multifunctional implant surfaces capable of both promoting host cell adhesion and limiting bacterial colonization. These "dual-function" surfaces are particularly important in the context of rising antimicrobial resistance, which reduces the effectiveness of conventional infection control strategies. Recent approaches have focused on modifying surface properties to selectively influence the adhesion behavior of mammalian and bacterial cells through precise control of surface topography. Consequently, surface modifications at the nanoscale (Sa < 100 nm) have emerged as promising alternatives to conventional micrometer-scale roughening, offering the potential to accelerate osseointegration while reducing biofilm formation.

Among the available surface engineering techniques, laser-induced nanotexturing of titanium has attracted significant interest. This approach allows for the creation of functional nanostructures directly on the implant surface without the need for additional coatings, thereby offering a versatile and highly controllable strategy for improving both biological integration and antibacterial properties.

A new potential application of laser texturing in implantology, which we will discuss in more detail in this chapter, is its use in the biofunctionalization of surfaces through the capture of EVs, which are nanoscopic particles secreted by cells. It constitutes an innovative approach to improving tissue-material interactions. By exploiting the biological and communicational properties of EVs, it becomes possible to modulate cellular response and optimize tissue regeneration and healing. In this context, laser texturing appears to be a promising technology for modifying the implant surface topography on a precise scale in order to promote selective EV anchoring. This combination paves the way for a new generation of bioactive surfaces capable of stimulating targeted biological responses and improving implant integration.

Other approaches to surface modification exist and combine physical, chemical, and biological treatments to adjust topography, chemistry, and bioactivity, as shown in Table 1.

Table 1. Overview of Surface Modification Strategies for Implants

	Main Techniques	**Goals**	**Biological Effects**
Chemical	**Oxidation, plasma treatment, silanization, CVD/DLC coatings**	**Adjust surface energy, wettability, and protein adsorption**	**Improved osseointegration surfaces**
Bioactive / Mineral	Hydroxyapatite coatings, bioactive glasses, sol-gel layers	Promote bone formation and bioactivity	Dental and bone implants
Biomolecular	Grafting of RGD peptides, collagen, or growth factors	Stimulate specific cell adhesion and signaling	Vascular and bone implant coatings
Antimicrobial / Antifouling	PEGylation, silver/ copper ions, antimicrobial peptides	Prevent infection and biofilm formation	Catheters, joint replacements
Physical / Topographical	Sandblasting, acid etching, anodization	Increase roughness, enhance cell adhesion and differentiation	Titanium and orthopedic implants
Laser Texturing	fs, ps, or ns laser structuring	Create controlled micro/nanopatterns; enhance cell adhesion, protein adsorption, and EV capture; modulate surface chemistry	Smart bioactive implants, dental and orthopedic devices
EV-based (Emerging)	Passive capture on textured surfaces, anchoring via antibodies or binding motifs	Local delivery of regenerative and anti-inflammatory signals	Next-generation biofunctional surfaces

Ultrafast laser texturing has emerged in recent years as an alternative to existing techniques for modifying implants in order to improve their integration. It has the advantage of not altering the surface chemistry of the materials, thus preserving their biocompatibility, and offering a wider choice of textures compared to other topographical modification techniques.

Taking dental implants as an example, we have several medical questions concerning, in particular, improving their osseointegration, promoting healing and tissue regeneration, and also preventing bacterial infections. It is therefore possible to use laser texturing to create multifunctional implants by adapting the texture to the contact area and responding to a specific need (Figure 2).

Figure 2. Functional Zonation of a Laser-Textured Dental Implant Surface.

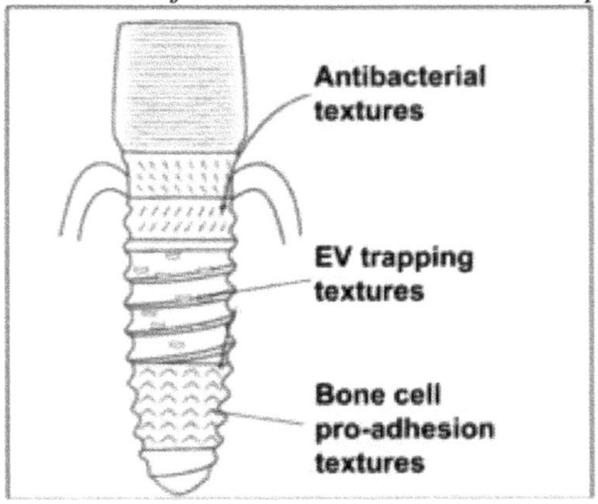

Source: Adaptation, Google images.

After describing the physical foundations of laser-matter interaction and the possibilities offered by femtosecond lasers for structuring matter at scales ranging from microns to nanometers, it is relevant to examine how these capabilities can be exploited in a field where control of surface topography is crucial: implantology. Indeed, the extreme precision of the textures obtained by Fs-L paves the way for surfaces whose biological properties can be controlled with unparalleled finesse. We can now focus on how these periodic patterns, and in particular LIPSS, influence cellular response and contribute to improving the integration of medical implants.

I. FEMTOSECOND LASER TEXTURING AND OPTIMIZATION OF OSSEOINTEGRATION AND IMPROVEMENT OF CELL ADHESION IN GINGIVAL TISSUE

I.1. Enhancement of Osteoblastic Differentiation through Laser Surface Texturing

As we mentioned earlier, Fs-L surface texturing is now considered a promising approach to optimize the osseointegration of metallic implants, particularly titanium implants. By generating periodic patterns at the nanoscale, this technique modifies the physicochemical and mechanical properties of the implant-bone interface. These changes directly influence mechanotransduction mechanisms, promoting improved osteoblast adhesion, enhanced cytoskeletal organization, and increased extracellular matrix secretion.

LIPSS are nanoscale patterns whose geometry depends on the polarization and wavelength of the laser beam. According to the study of Maalouf et al. (2022), the use of a radially polarized Fs-L to generate LIPSS (Figure 3) on titanium alloys significantly enhances osteogenesis.

Figure 3. Dental implant on which two screw threads have been textured with a Radial LIPSS configuration.

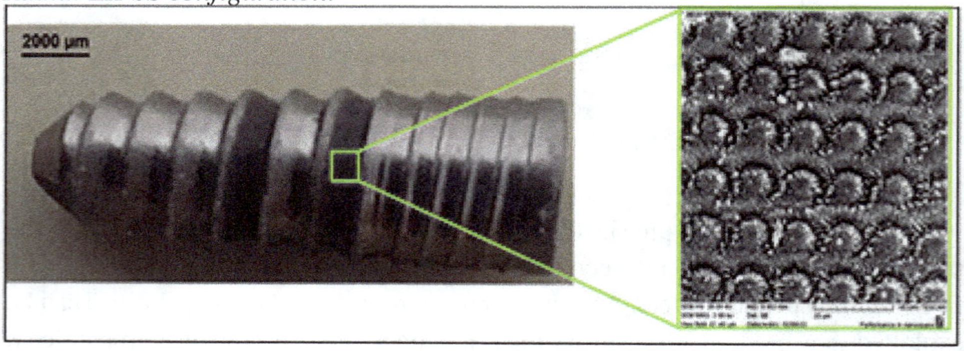

These nanoscale features stimulate the differentiation of hMSCs into osteoblasts even under conditions where biochemical signals are limited, highlighting the importance of topographical cues in activating osteogenic pathways. The results demonstrate improved adhesion, fibrillogenesis, and formation of mineralized nodules on the textured surfaces, confirming that nanostructuring acts as a key biophysical signal in regulating stem cell fate. *In vitro* experiments conducted with human mesenchymal stem cells (hMSCs) have demonstrated a significant increase in mineralization around laser-textured implants, as confirmed by fluorescent calcium labeling and scanning electron microscopy observations.

In vitro experiments with hMSCs on these new surfaces to verify if the osteogenic enhancement effect is still preserved when moving from a flat shape to a three-dimensional and complex topography. We therefore seeded hMSCs on the screws in osteoblastic medium for 4 weeks and carried out mineralization assessment with calcein blue (Figure 4).

Figure 4. Mineralization clusters labeled by calcein blue on a dental implant with two textured screw threads at 4 weeks post seeding.

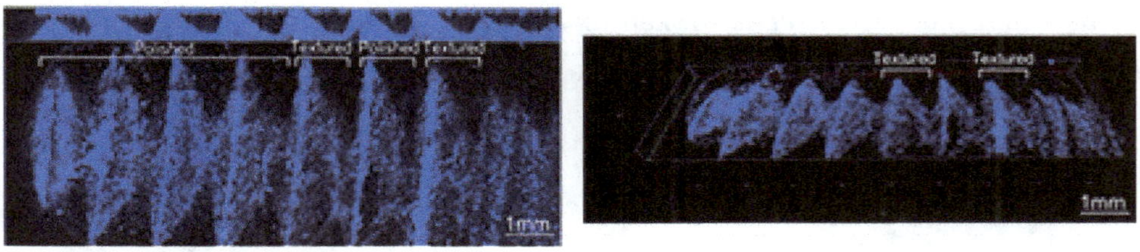

Cell adhesion was imaged at 4 weeks by scanning electron microscopy (SEM) and our observations showed the existence of regions of cell detachment on the polished surfaces that are not found on the surfaces textured by Radial LIPSS (Figure 5). These preliminary results confirm the improvement of long term cell adhesion on screw threads with a radial LIPSS texture.

Figure 5. Representative image of hMSCs grown during 4 weeks on a dental implant displaying Radial LIPSS textured areas and polished areas. White arrows indicate areas where the cell layer is broken or detached. Of interest, these regions are present only in the polished areas.

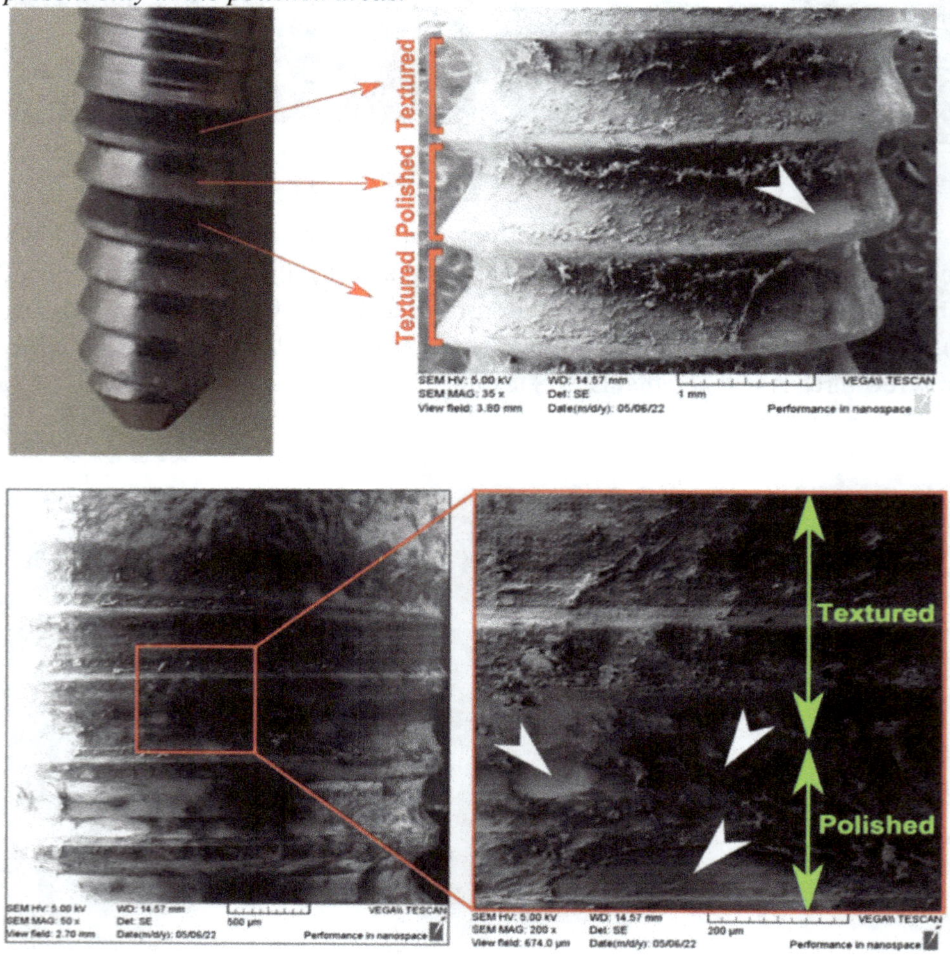

I.2. Impact of Morphology and Dimensions on the Speed of Osteoblast Differentiation

The effectiveness of laser texturing depends not only on the presence of periodic structures but also on their dimensions and morphology. Experiments performed on cylindrical implants have shown that titanium screws featuring a radial LIPSS texture retain their positive effects on cell adhesion and long-term cell layer stability, unlike polished areas where detachment occurs. However, the industrial production of such structures remains challenging, as the small laser spot size (13 µm) limits the depth of focus and slows down the processing speed. Ongoing studies aim to evaluate the impact of larger beam diameters (50-200 µm) on LIPSS formation and their ability to maintain or even enhance osseointegration. Thus, optimizing texturing parameters will help reconcile biological performance with industrial feasibility, paving the way for faster and more robust bone integration of dental and orthopedic implants.

I.3. Improvement of Cell adhesion: Application to Gingival Tissue

Cell adhesion is an essential process in tissue formation. It involves the binding of cells to the extracellular matrix, to other cells, or to a surface via specific surface proteins. This interaction generates forces on the cytoskeleton that propagate through the cell to the nucleus, leading to the activation of signaling pathways. When a cell binds to a surface via adhesion molecules and this binding triggers the transmission of a signal to the cytoskeleton, this process is called "focal adhesion" (Figure 6).

Focal adhesion points are points of intersection where mechanical signals from the environment are converted into intracellular forces and chemical signals, thanks to cytoskeletal connections and signaling proteins. The protein layers formed in the first few minutes of contact with the culture medium consist almost exclusively of albumin. The initial adsorption of proteins, which affects fibronectin adhesion among other things, could then regulate cell adhesion through focal contacts. Once tensile forces are generated on the cytoskeleton, they are transmitted to the various structures of the cell, with a significant impact on the nucleus in particular. Indeed, cellular events such as differentiation or changes in gene expression profiles are indirectly influenced by the forces exerted on the nucleus.

Figure 6. Schematic representation of focal adhesions. Adherent cell. During spreading and migration, the cell adheres to ECM ligands via nascent adhesions that transform into focal complexes in the lamellipodium (LP) and then into focal adhesions in the lamella (LM).

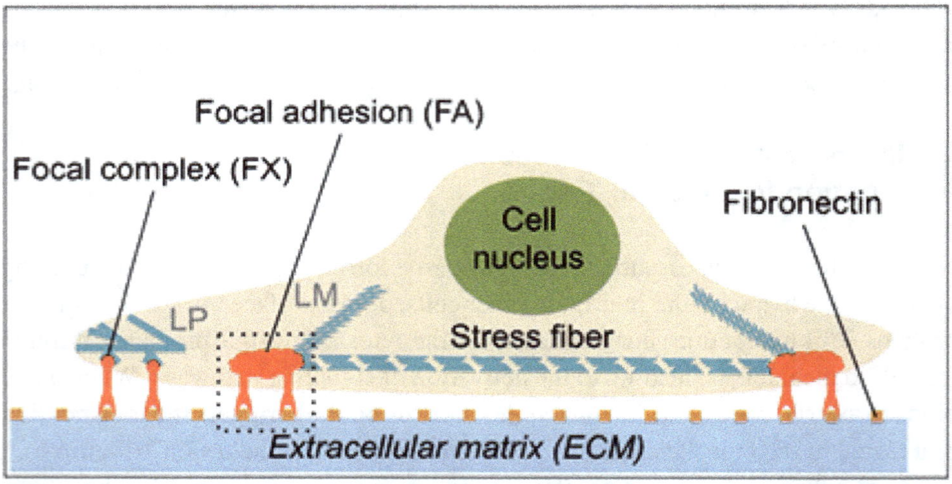

According to Hoffmann et al, A kinetic model for RNA-interference of focal adhesions, BMC Systems Biology, Springer Nature, 2013, license cc BY 2.0, https://doi.org/10.1186/1752-0509-7-2 (Hoffmann & Schwarz, 2013)

I.4. Role of Laser Texturing in Gingival Fibroblast Adhesion: Application to Dental Implants

The long-term success of a dental implant relies not only on osseointegration but also on the quality of its interface with soft tissues, particularly the gingiva. Strong adhesion of gingival tissue to the implant surface serves as a critical biological barrier, preventing bacterial infiltration and contributing to the mechanical stability of the implant. In this context, Fs-L surface texturing offers promising opportunities to enhance soft tissue integration. In particular, LIPSS can be generated using two different wavelengths: 1030 nm (LIPSS IR) and 515 nm (LIPSS GR). These structures increase surface roughness in a controlled manner, potentially enhancing cell adhesion. To mimic the long-term contact between implant and gingival tissue, primary human gingival fibroblasts (HGnF) were cultured for 21 days on both untreated and fs-L treated Ti-alloy surfaces (Papa, 2023). A new adhesion measurement protocol was developed to evaluate the resistance of a mature fibroblast cell layer (rather than isolated cells) to mechanical disruption. A tribometer was used to apply a calibrated local force to the cell layer, producing a controlled indentation. The resulting cell disruption was imaged using fluorescence microscopy, and the area of the detached cell layer was quantitatively measured (Figure 7). This area served as an indicator of the adhesion strength between the fibroblast layer and the surface. Results showed that fibroblast layers adhered significantly better to fs-L textured surfaces (LIPSS IR) compared to polished, untreated surfaces: LIPSS formation increased the surface roughness (parameter Sa), which appears to promote better anchoring of gingival fibroblasts. These findings demonstrate that Fs-L surface modification enhances gingival fibroblast adhesion by altering surface topography in a favorable way. By improving the adhesion of gingival tissue to the implant surface, fs-L textured surfaces help establish a stable soft tissue seal. This seal acts as a first line of defense against microbial infiltration and contributes to the long-term functional success of dental implants.

Figure 7. (a) Scratch tests developed in LTDS lab to evaluate cell layer adhesion on surfaces. (b) Representative fluorescence images of torn off cells layer after tribometer scratch on polished surface and on different percentages of textured area. (preliminary results from LTDS, Sainbiose, Hubert Curien Laboratories)

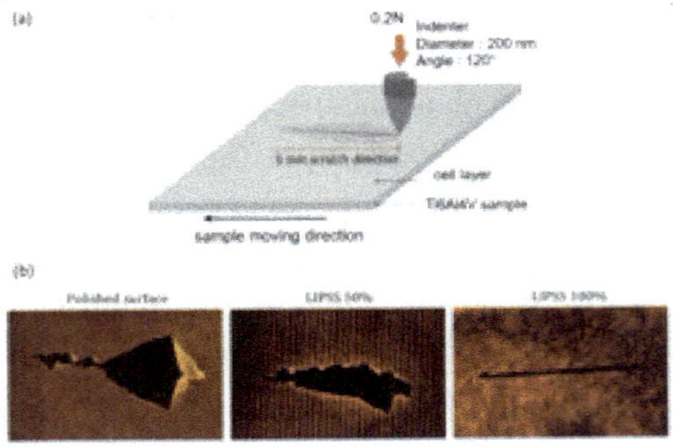

The results presented demonstrate that surfaces textured by femtosecond laser not only improve osseointegration, but also strengthen soft tissue adhesion, particularly gingival fibroblasts. This dual capability, optimizing biological interactions and stabilizing the implant-tissue interface, is a major asset in preventing post-surgical complications. However, beyond improving cellular responses, another major challenge remains: limiting bacterial adhesion and preventing the formation of biofilms, which are responsible for severe infections and implant failures. The following section continues in this vein, focusing on the engineering of high aspect ratio nanostructures generated by ultrafast lasers and their potential for creating mechanically bactericidal surfaces.

II. ULTRAFAST LASER PROCESSING OF HIGH-ASPECT-RATIO NANOSTRUCTURES: A PATHWAY TO MECHANO-BACTERICIDAL IMPLANT SURFACES

Ultrafast photonics technologies enable a rapid, one-step, reliable, flexible and contamination-free surface engineering process. However, it is a relatively recent technology in the field of medical implants, therefore, there is not a significant amount of research on how bacteria interact with these new nano-topographies. There is a limited body of research on how bacterial cells interact with these laser-induced nanostructures, particularly at scales relevant to mechano-bactericidal effects. Understanding these interactions is crucial to harnessing the full potential of laser-structured surfaces in preventing implant-associated infections.

II.1. Mechano-Bactericidal Surfaces: A Promising Alternative to Antibiotics

The fight against antibiotic resistance rises to the ecological challenges of our era ((Tao et al., 2019; Antimicrobial Resistance Collaborators, 2022). Antibiotic resistance is one of the most serious concerns to public health as routine treatments of bacterial infections become increasingly inefficient. Antibiotic resistance leads to longer hospital stays, more medical expenses and greater mortality. *Streptococci* and *Staphylococci* among others, which cause infections that can compromise implant integration and whose antibio-resistance is starting to emerge, are frequently responsible for biomedical device-associated infections that arise from various metallic medical devices (Khatoon et al., 2018) including orthopedic and dental implants, vascular stents, fracture fixation devices (screws, plates). Biofilms can form within hours of implantation, so preventive methods are required. Various techniques (Chug & Brisbois, 2022) have been directed towards preventing bacterial contamination and biofilm formation by modifying the surfaces using chemistry, reactive oxides, nanoparticles or metallic ions. However, these techniques have limitations such as uncontrollable release inducing loss of efficiency over time, even toxicity. The development of physically antibacterial surfaces provides an excellent alternative to eradicate pathogens, especially bacteria.

Indeed, antibacterial surfaces discovered in nature have stimulated great interest in the use of surface topography to engineer biomimetic surfaces that exhibit an antimicrobial contact killing mechanism (Tripathy et al., 2017). Surface nanostructures discovered on insects promote self-cleaning and direct rupture of bacterial cell membranes. For instance, the study conducted by Kelleher et al. (2016) on the nanopillar architecture and antibacterial properties of cicada wings revealed that bactericidal efficacy is closely linked to the geometric characteristics of the surface nanotopography. Parameters such as the height, diameter, spacing, periodicity, and aspect ratio of the pillars were identified as key factors influencing antibacterial activity. Furthermore, it was found that the efficacy of these nanotextured surfaces depended on the bacterial strain in question, with variations observed depending on cell wall type (Gram-positive or Gram-negative), bacterial morphology, and biofilm organization (Lee et al., 2021).

II.2 The Capabilities of Ultrafast Laser Irradiation to Generate High Aspect Ratio Nanostructures

Ultrafast laser irradiation is a well-established method to produce modifications of topography of a wide range of micro- and sub micro-structures (ripples, columns, pits, or spikes among others). These periodic or full self-assembled structures, guided by light coupling on surface roughness, have already presented interest for biological performances and especially for the integration of implants (Müller et al., 2016). More recently, publications (Nakhoul et al., 2022; Nakhoul et al., 2022) from Hubert Curien laboratory demonstrate that ultrafast-laser irradiations can even produce polymorphic nanopatterns (Figure 8).

Figure 8. Different types of randomly and periodically distributed nanostructures with a regular energy dose

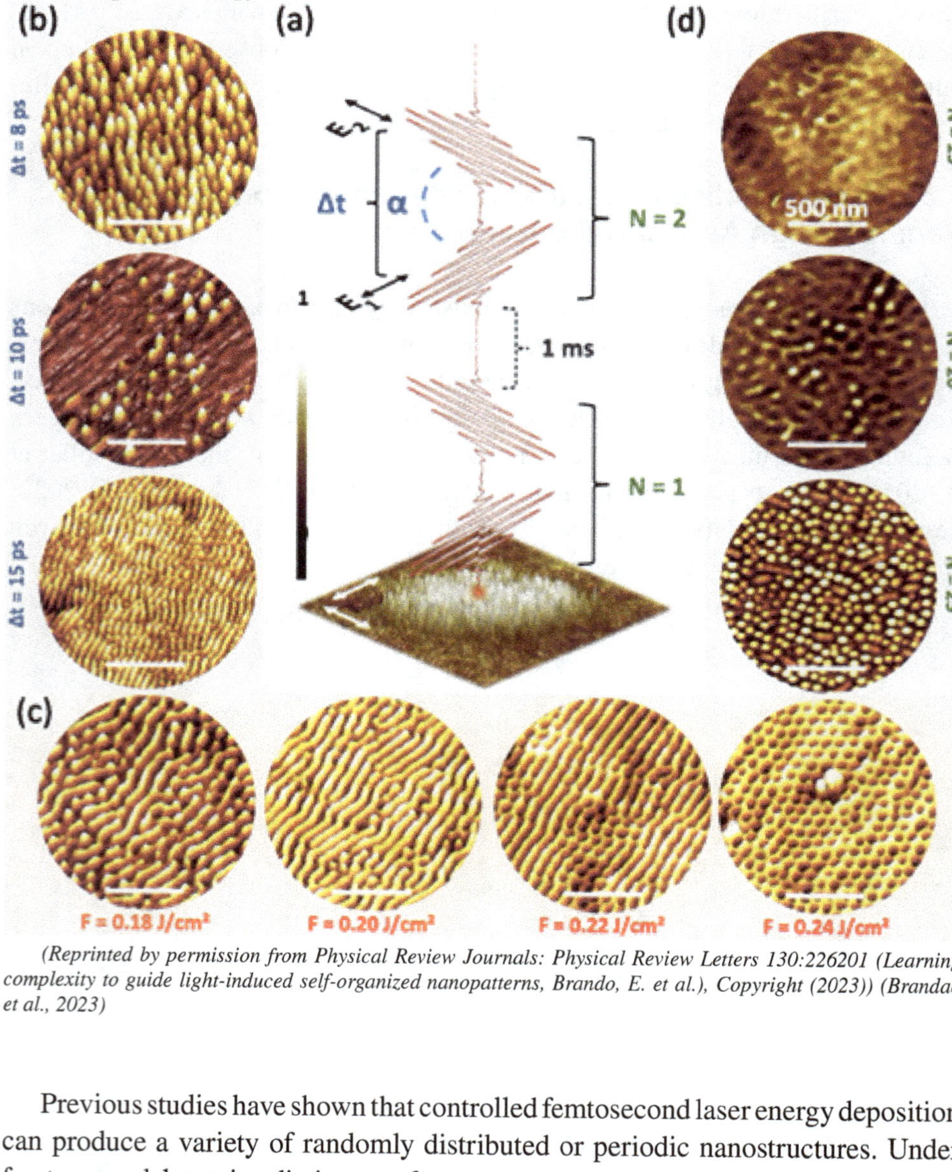

(Reprinted by permission from Physical Review Journals: Physical Review Letters 130:226201 (Learning complexity to guide light-induced self-organized nanopatterns, Brando, E. et al.), Copyright (2023)) (Brandao et al., 2023)

Previous studies have shown that controlled femtosecond laser energy deposition can produce a variety of randomly distributed or periodic nanostructures. Under femtosecond laser irradiation, surface patterns such as nanopeaks, nanobumps, nanohumps, and nanocavities can be formed, with characteristic lateral dimensions ranging from 20 to 80 nm and heights reaching approximately 100 nm. Surface morphologies, tailored by an interplay between optical and hydrodynamical resonances, strongly depend on the laser process parameters such as polarization, time pulse shaping, fluence, wavelength, pulse duration, overlapping or repetition rate.

The ability to tailor such sharp structures on-demand opens up new possibilities for "function-driven" implant design, particularly in the development of mechano-bactericidal surfaces on a range of biomedical metals and alloys.

Despite these advances, further research is needed to fully explore the mechano-bactericidal potential of these nanostructures and optimize their performance in clinical settings

II.3. Interactions Between Bacterial Cells and Laser-Nanostructured Surfaces

The antibacterial properties of laser engineered nanostructures are still poorly reported: there are only few publications (Epperlein et al., 2017; Jalil et al., 2020) yet limited to micro or sub-micrometric textures (Papa et al., 2023). No publications are found at nano-metric scales, below bacteria size (<500nm). The influence of the LIPSS period on the anti-bacterial adhesion was recently carried out on polymers (Richter et al., 2021) but with a limited effect. Moreover, few investigations have been performed on metals.

Papa et al. (2023) (Figure 9) has already identified certain types of topographies that help limit colonization by *Streptococcus mutans*, *Porphyromonas gingivalis or Staphylococcus aureus* with potential applications to prevent infections in dental and bone implants.

Figure 9. Streptococcus mutans staining on two different nano-micro-topographies from a preliminary study by the SAINBIOSE, LTDS, and LabHC laboratories

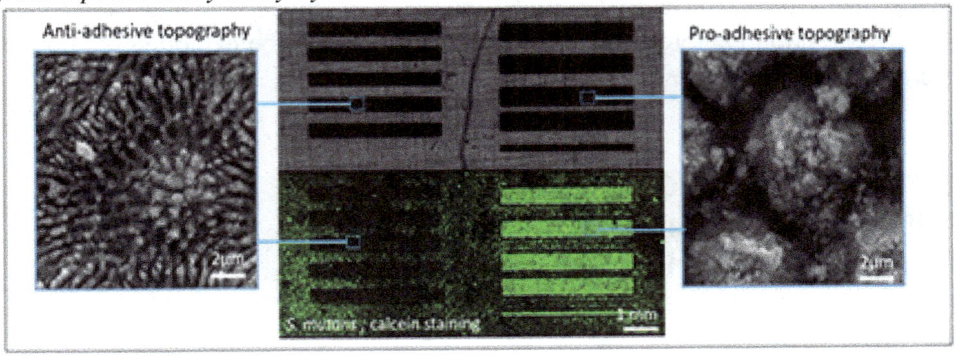

(S. Papa, PhD thesis)

These studies were able to establish correlations between surface parameters and the number of bacteria. SEM images have been taken to observe *Streptococcus mutans* and *Staphylococcus aureus* on nanopatterned surfaces (Figure 10).

Figure 10. (a) SEM images of Streptococcus mutans on 600nm LIPSS period, (b) 300nm LIPSS period and (c) Staphylococcus aureus on polymer nanopeaks

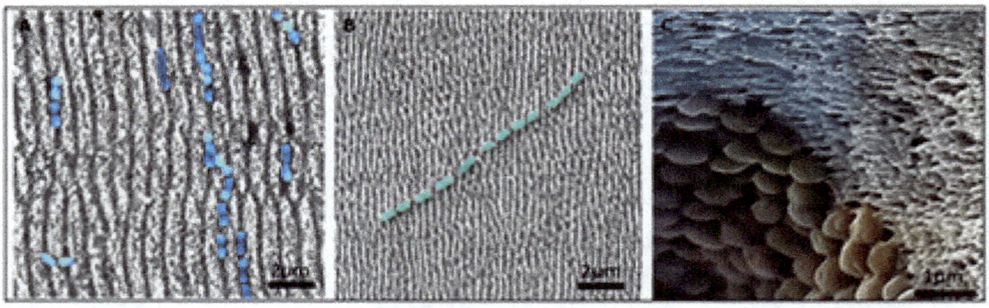

(preliminary results from Sainbiose, LTDS and Hubert Curien Laboratory)

To clearly confirm the differences in bacterial attachment depending on surface architecture, further quantitative analyses are required to identify correlations between topographic parameters and bacterial adhesion. Future studies should systematically investigate the specific roles of pattern size, density, and aspect ratio on bacterial behavior, adopting a comprehensive and multidisciplinary approach. Such investigations will be essential to highlight the critical importance of precise surface engineering in the development of next-generation antibacterial implants.

Previous results show that nanostructures generated by femtosecond lasers can modulate bacterial adhesion and offer a promising mechanobactericidal strategy. However, controlling topography at these scales also opens the door to other biological applications. Among these is interaction with extracellular vesicles (EVs), whose size closely matches the dimensions of Fs-L-induced nanopatterns. The following section explores how these nanostructures can be exploited to capture and functionalize EVs on the surface of implants.

III. FS-L TEXTURING FOR THE CAPTURE AND FUNCTIONALIZATION OF SURFACES BY EVS: A NEW CHALLENGE

III.1. Introduction

EVs are defined as small particles enclosed in a lipid bilayer membrane and naturally secreted by cells. They can be found in biological fluids such as blood, cerebrospinal fluid, saliva, and urine. As described in Figure 11, there are two main biogenesis pathways that allow them to be classified: the endosomal pathway, which leads to the formation of exosomes ranging from 50 to 150 nm, and membrane budding, which leads to the formation of ectosomes that can reach sizes up to 1000 nm. However, current knowledge of EV biogenesis pathways is insufficient to clearly distinguish exosomes from ectosomes. There is still no consensus on specific protein markers capable of differentiating them. Moreover, smaller ectosomes (<50 nm) have also been reported (Nabhan et al., 2012). Consequently, neither size nor exosomal markers alone are sufficient to discriminate between exosomes and ectosomes. For this reason, EVs are typically classified by size: small EVs (sEVs), with a diameter below 200 nm, and medium/large EVs (m/lEVs), above 200 nm.

Figure 11. EV biogenesis. EVs are largely categorized into (a) exosomes, (b) ectosomes, and (c) other types such as apoptotic bodies or exomeres. Exosome biogenesis occurs through both (a-i) ESCRT-dependent and (a-ii) ESCRT-independent pathways. PM: plasma membrane.

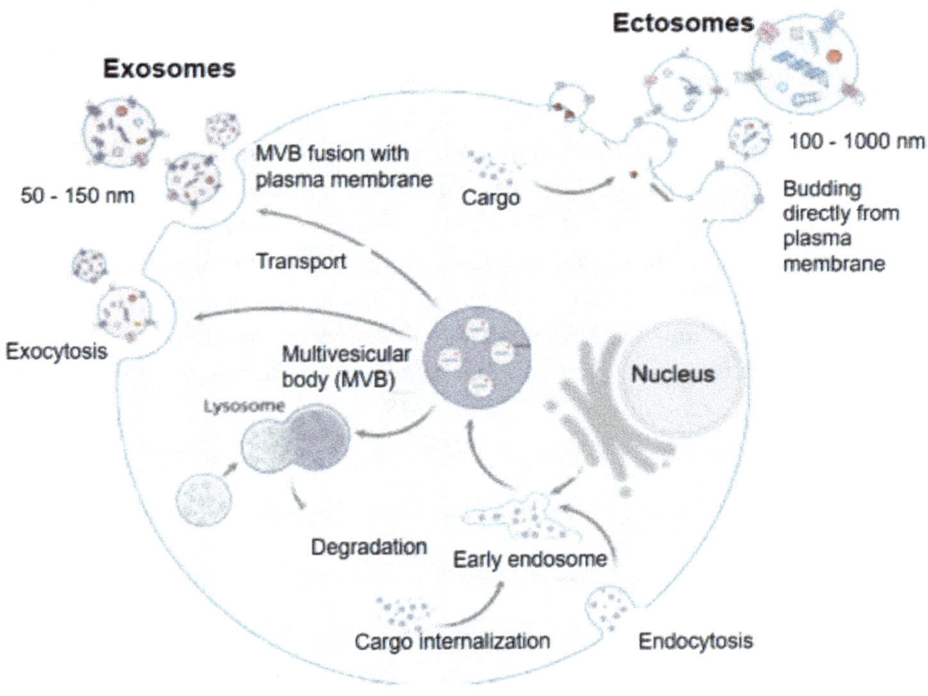

(Adapted from Yu et al., 2024. © 2024 Yu et al. Open Access under CC BY 4.0.) (Yu et al., 2024).

When released by cells, EVs carry biomolecules such as proteins, lipids, and nucleic acids, facilitating the exchange of information between cells. This has opened up new avenues for understanding various physiological and pathological phenomena. EVs circulating in biological fluids may also carry specific markers for several diseases, such as cancers, cardiovascular and neurodegenerative diseases.

This characteristic makes EVs particularly interesting as non-invasive liquid diagnostic tools, enabling early detection and continuous monitoring of diseases (Chen et al., 2024). Lastly, numerous studies have demonstrated the therapeutic potential of EVs in many pathologies. For instance, in 2010, studies showed for the first time in vivo that exosomes purified from mesenchymal stem cell (MSC) culture medium significantly reduced the size of myocardial ischemia in a mouse model via cardiac perfusion (Chen et al., 2024). In the field of bone healing and implantology, a study using h-MSC-derived EVs encapsulated in a hyaluronic acid hydrogel demonstrated effective repair of bone defects via stimulation of angiogenesis and osteogenesis in vivo (Takeda et al., 2023). Indeed, the cargo contained in certain EVs gives them the ability to initiate or accelerate important mechanisms of bone homeostasis, as described in Figure 12.

Figure 12. Extracellular vesicles (EVs) originating from different cell types exhibit diverse molecular compositions and biological activities. Depending on their origin, these vesicles may either interact directly with bone-forming cells or promote bone repair through indirect mechanisms, such as modulation of immune responses, suppression of osteoclast activity, and enhancement of endothelial cell regeneration. Black arrows indicate EV production and their biological actions and red arrows denote inhibitory effects exerted by EVs or the corresponding cells. The black arrows represent activations and the red lines show inhibitions. EVs produced by dendritic cells participate in the inhibition of osteoclasts, whose role is to degrade the bone matrix. Meanwhile, MSCs and osteoblast EVs promote the differentiation of preosteoblasts into osteoblasts, whose role is to synthesize and mineralize the bone matrix, While macrophage EVs act on MSCs.

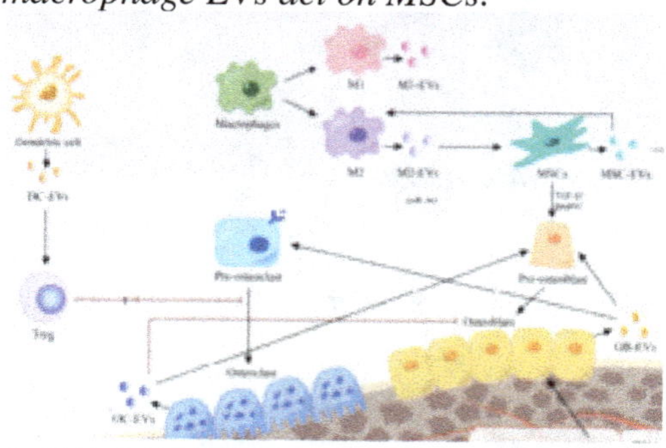

(Adapted from Jiao et al., 2024. © 2024 The Authors, published by MDPI. Open Access under CC BY 4.0.).

Our research aims to use the capabilities of femtosecond lasers to engineer nanoscale surface patterns designed to act as "EV traps". By precisely structuring the surface of materials at the nanometric level, these laser-generated patterns can passively capture and retain EVs without modifying their chemical composition. This controlled trapping process should help to preserve the EVs morphological integrity while they remain on the material's surface. Once the implant is placed in contact with biological tissue, the trapped EVs can be naturally released back to the surrounding cells, thereby promoting biological communication and enhancing tissue integration.

Therefore, it is useful to know the EV sizes in order to adapt the patterns generated on the surfaces so that a given subpopulation of EVs can be trapped. Furthermore, given their different biological origins, exosomes and ectosomes have different biological effects. Thus, assuming that each fraction (small or large) is enriched in an exosomal or ectosomal subpopulation, we could imagine smart surfaces capable of sorting the EVs retained and therefore having a different biological effect.

III.2. Challenges in Capturing and Preserving EVs on Implants

The capture, preservation, and selection of EVs on the surface of implants is a major challenge, given their biological properties and therapeutic potential. The aim is to make these EVs accessible to surrounding cells in order to favorably modulate the tissue response and promote regeneration. However, unlike cells, EVs do not produce an extracellular matrix that would facilitate their anchoring, and behave more like colloidal particles in suspension (Xie et al., 2016). This colloidal nature, combined with their structural fragility, explains the difficulty in immobilizing them stably without altering their function. Indeed, EVs are susceptible to aggregation, membrane fusion, and degradation of their cargo (proteins, RNAs, lipids) over time or under mechanical stress (Théry et al., 2018; Lötvall et al., 2014).

To overcome these limitations, several approaches have been developed to permanently attach EVs to metal surfaces. The first involves using binding intermediates, such as adhesive polymers or antibodies. Polydopamine, for example, has been widely explored for its ability to create a bioadhesive layer and effectively immobilize Evs (Konoshenko et al., 2018). However, some studies have highlighted its potential cytotoxicity and disruption of cell adhesion when used at high densities (Zhao et al., 2019), limiting its direct clinical application. Another strategy relies on the use of antibodies directed against typical EV surface proteins, mainly the tetraspanins CD9, CD63, and CD81 (Antich-Rosselló et al., 2022), which are often used as vesicular markers. Although this approach allows for specific binding, it has several drawbacks: high cost, limited yield, and partial coverage of EV populations. Indeed, tetraspanins are not universally expressed: some EVs lack them, and expression can vary between subpopulations. Combined analyses have shown that only about 50% of EVs strongly express a given marker, and even though a significant proportion are triple-positive for CD9/CD63/CD81, a large fraction escapes capture (Andreu & Yáñez-Mó, 2014).

Furthermore, EVs are heterogeneous. Their biological role differs depending on their origin (exosomes, microvesicles/ectosomes), but there is currently no clear consensus on how to reliably distinguish these subpopulations solely by surface markers (Kowal et al., 2016). Exosomes produced via ESCRT-dependent pathways express certain characteristic proteins, but exosomes derived from ESCRT-independent pathways do not, which complicates their selective isolation. As for ectosomes, no specific marker has yet been validated. In practice, separation is often based on size: fractions smaller than 200 nm are enriched in exosomes and fractions larger than 200 nm are enriched in ectosomes, eliminating apoptotic bodies (>500 nm) (Théry et al., 2018), and current techniques for grafting EVs onto surfaces do not allow sorting according to this size criterion.

These limitations highlight the value of alternative strategies, such as nanometric surface texturing, which could offer a means of preferentially capturing certain populations of EVs based on their size and physicochemical properties, paving the way for a new generation of bioactive implants better suited to the specificities of intercellular communication.

III.3. Role of Femtosecond Nanostructures in EV Capture

A promising avenue for overcoming these obstacles lies in nanometric surface texturing. By creating patterns measuring between 30 and 500 nm, it becomes theoretically possible to mechanically trap EVs based on their size, without resorting to potentially harmful chemical coupling agents. In this regard, the use of femtosecond lasers (Fs-L) offers a considerable advantage due to its ability to generate precise surface structures.

In these research studies conducted in our laboratory, using smooth and textured titanium with a 515 nm wavelength (green, Gr) fs-L, fluorescent EVs were deposited on each material at a rate of 100 million EVs per material and incubated for 2 hours at 4°C. The materials were then gently rinsed to assess their ability to retain EVs on their surface. They were then visualized using confocal fluorescence microscopy. Our results showed that EVs preferentially adhere to textured titanium surfaces compared to smooth surfaces, as shown in Figure 13.

This gives us an initial insight into the role of nanoscale topography in modulating EV-surface interactions. The increased adhesion observed on laser-textured areas suggests that surface roughness and the presence of specific nanostructures may enhance physical anchoring and retention of EVs. These findings support the hypothesis that surface patterning can be strategically designed to control EV attachment, stability, and subsequent release, paving the way for the development of bioactive implant surfaces that promote improved tissue integration.

Figure 13. (a) SEM images of smooth and fs-L textured materials, before EV deposition; (b): Confocal fluorescence microscopy of EVs deposited on smooth and laser-textured materials at a wavelength of 515 nm.

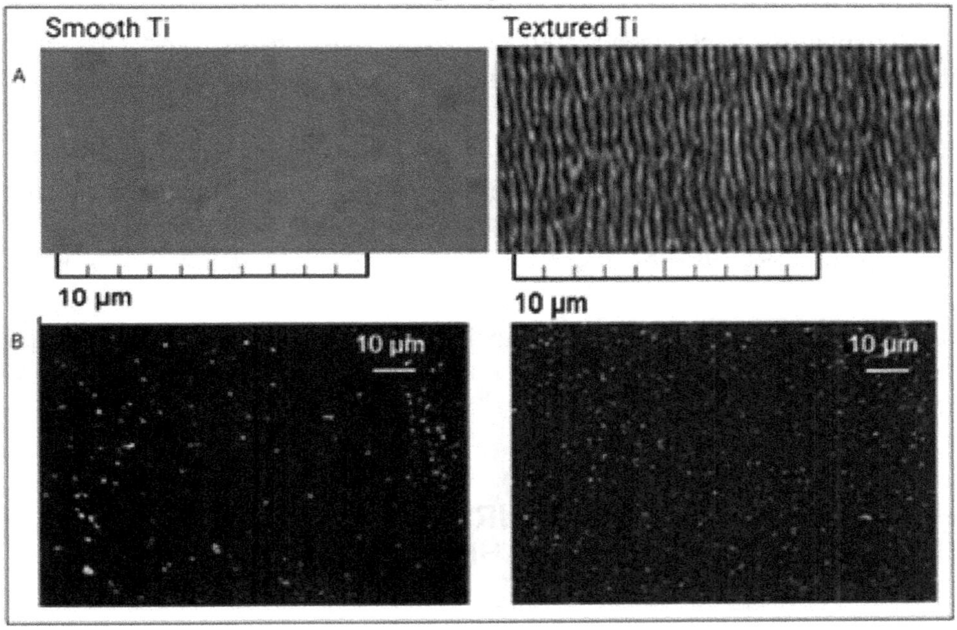

III.4. Interaction Between Laser-Generated Nanostructures and EVs

While no direct studies have yet analyzed the interaction between EVs and surfaces textured by Fs-L, work on other texturing approaches, notably electrochemical anodization, suggests that nanostructuring can influence EV adsorption (Willms et al., 2018). However, these methods suffer from experimental limitations: EVs are often deposited in high density by droplets on the surface, then incubated at 37°C under vacuum for several hours, conditions that may compromise their integrity (membrane fusion, cargo degradation).

From a physicochemical point of view, EVs are colloidal particles with a negative zeta potential (average of -10 to -30 mV), reflecting a surface charge that determines their electrostatic interactions with substrates (Pfeiffer et al., 2010).

Thus, their interaction with a textured surface depends not only on the size of the patterns, but also on the surface chemistry (hydrophilicity/hydrophobicity, polarity) and local electrostatic forces. The Fs-L has a dual advantage here: it not only sculpts the topography, but also modifies the wettability and surface energy of materials, combining mechanical trapping and chemical affinity effects. This synergy could therefore promote both the stable capture of EVs and their gradual release in a biological environment.

The ability of femtosecond lasers to generate high-precision nanostructures, already considered for creating mechanically bactericidal surfaces, also opens up prospects that go far beyond simply fighting infection. Indeed, the same principles of nanoscale structuring can be exploited to interact not with bacteria, but with much smaller biological particles that are essential for cellular communication: extracellular vesicles (EVs). Thus, after considering the potential of Fs-L-induced nanostructures to prevent bacterial adhesion, it seems natural to explore their ability to selectively capture EVs, a major challenge for the development of bioactive implants capable of modulating tissue response. This technological and biological continuity justifies the transition to a new application: the use of Fs-L textured surfaces as platforms for EV trapping and functionalization.

VI. FEMTOSECOND LASER SURFACE TEXTURING AND FUNCTIONALIZATION FOR OTHER MEDICAL DEVICES

In contrast to dental implants, where rapid and robust osseointegration is essential, certain orthopedic implants require *limited* bone integration. Bone screws, for example, are used to secure orthopedic plates to fractured bones, stabilizing the assembly during healing. Since these screws, typically made of titanium or titanium alloys, often need to be removed after several months or years, such as following bone remodeling, they should not become fully overgrown by bone cells or encapsulated in dense scar tissue. Ultrafast laser texturing offers an effective strategy for achieving controlled cell repellency. Hierarchical micro- and nano-scale structures generated on titanium, particularly on anodized Ti surfaces, have been shown to significantly reduce cell adhesion. This makes ultrafast laser–processed surfaces highly suitable for removable medical hardware such as orthopedic bone screws (Muck et al., 2021).

A proof-of-concept experiment is presented in Figure 14, where a ring-shaped region on a titanium bone screw was laser-patterned to create a hierarchical surface topography consisting of conical micrometer-scale spikes and periodic nanometer-scale ripples (Figures 14 a–c). Muck and co-workers demonstrated the effect of this laser treatment on osteoblast adhesion, shown in Figures 14 d–e. On the non-treated portion of the screw (Figure 14d, central area), the surface is covered by a dense, confluent layer of osteoblasts, which even begin to grow three-dimensionally along the edges of the screw threads. In contrast, many regions of the laser-treated area on the same screw (Figures 14c and 14d, upper right) remain largely cell-free, allowing the underlying hierarchical micro- and nanostructures to remain visible (Muck et al., 2021). In some cases, cells that do attach (indicated by a white star in Figure 14e) tend to cluster together and lose substantial contact with the surface. These results clearly show that the combination of micro-cones and nano-ripples obtained by ultrafast laser gives the surface a strong ability to repel osteoblasts. This type of texture is therefore particularly suitable for orthopedic implants that are intended to be removed after healing, where excessive cell adhesion would be problematic.

Figure 14. Proof-of-concept of bone screw texturing for reduced osteoblast cells adhesion. (a) and (b) laser patterning of a narrow band area on a bone screw; (c) the SEM micrograph illustrates the hierarchical microstructures of spikes and LIPSS;(d) overview of cells seeding on the screw between laser patterned and not laser patterned areas; (e) close-up view revealing osteoblast cells detaching from laser treated surface.

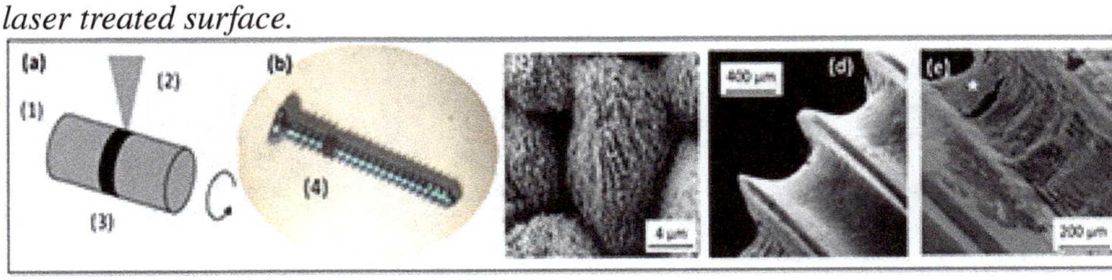

(Adapted and reprinted from, Martina Muck et al., Femtosecond Laser-Processing of Pre-Anodized Ti-Based Bone Implants for Cell-Repellent Functionalization, Nanomaterials (MDPI, Basel, Switzerland), Copyright 2021 under Creative Commons BY 4.0 license. Retrieved from https://doi.org/10.3390/nano11051342) (Muck et al., 2021)

Implantable medical devices are used in many parts of the body and interact with a wide variety of tissues and organs. A prime example is the pacemaker. This device sends electrical impulses to one or more chambers of the heart to cause them to contract, maintain adequate blood flow, and stabilize the heart rhythm. Pacemakers are generally indicated when the heart's natural pacemaker is failing or when a blockage occurs in the electrical conduction system. The latest miniaturized versions can now be implanted directly into the right ventricle via catheterization, reducing the complexity of the procedure and limiting the risks associated with traditional probes. Since these devices, mainly composed of titanium alloys, may need to be removed after several years (e.g., at the end of the battery's lifespan), their surfaces should not be fully overgrown by cells or scar tissue. Heitz *et al.* demonstrated that femtosecond laser irradiation of medical-grade titanium alloys can produce sharp conical spikes overlaid with sub-wavelength LIPSS (Heitz et al., 2017). When combined with controlled electrochemical oxidation to tune the oxide layer thickness, these surfaces exhibit hydrophilic, cell-repellent properties. Such surfaces are highly suitable for removable miniaturized pacemakers (Figure 15) and could be extended to other implants or prostheses that require partial tissue integration or eventual removal.

Figure 15. Laser patterning of hierarchical micro-/nanostructures on miniature pacemaker surface for avoiding scar tissue overgrowth. (a) miniature pacemaker with a ring of the hierarchical patterns placed in scar tissue culture for the proof-of-concept test; (b) after one week of seeding the fibroblast overgrew surface without laser pattern whereas limited coverage observed in laser patterned area; (c) close-up view of the hierarchical structure, the structure surface remains cell free

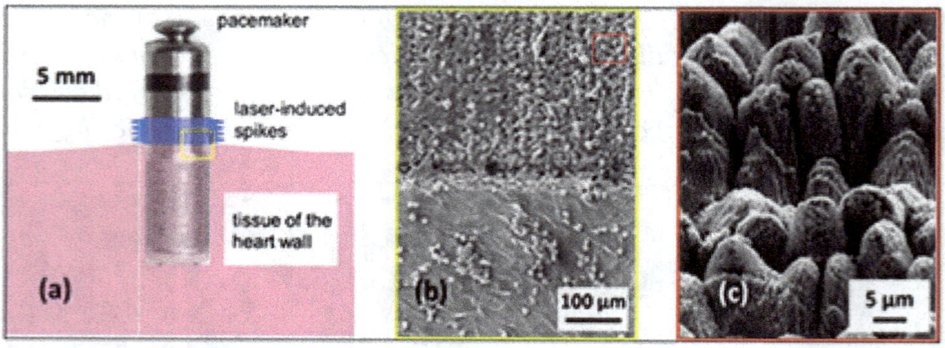

(Adapted and reprinted from, Johannes Heitz et al., Femtosecond laser-induced microstructures on Ti substrates for reduced cell adhesion, Applied Physics A (Springer Nature Group, Heidelberg & Berlin, Germany), Copyright 2017 under Creative Commons BY 4.0 license. Retrieved from https://doi.org/10.1007/s00339-017-1352-0) (Heitz et al., 2017)

Ultrafast laser machining is increasingly employed in the fabrication of implantable medical devices, including stents. Stents are specialized devices designed to provide structural support and maintain the patency of vessels and tubular structures in the body. They are widely used to treat conditions involving narrowing or obstruction of arteries, veins, or other passageways. Medical stents are typically made from metals or polymers. Among self-expandable metal stents, commonly used biomaterials include stainless steel, nickel-titanium (NiTi, nitinol), and niobium–zirconium alloys, which offer mechanical strength, corrosion resistance, and biocompatibility. Unlike conventional nanosecond lasers, which can induce burr formation due to melting during drilling, ultrashort femtosecond laser pulses minimize thermal diffusion, enabling highly precise and reproducible microstructures. As illustrated in Figure 16, femtosecond laser processing has become the preferred method for manufacturing nitinol biliary stents, which are particularly sensitive to thermal effects (Gafner et al., 2021). For temporary or degradable stents, biocompatible polymer blends such as poly-(hydroxybutyrate) (PHB) and poly-(L-lactide) (PLLA) can be used. The high peak intensities of femtosecond lasers allow nonlinear absorption in these transparent, thermally sensitive materials, producing clean, precise structures with minimal heat-affected zones, avoiding melting, charring, or microcracking (Figure 16b).

Figure 16. Stents cut from (a) L605 steel

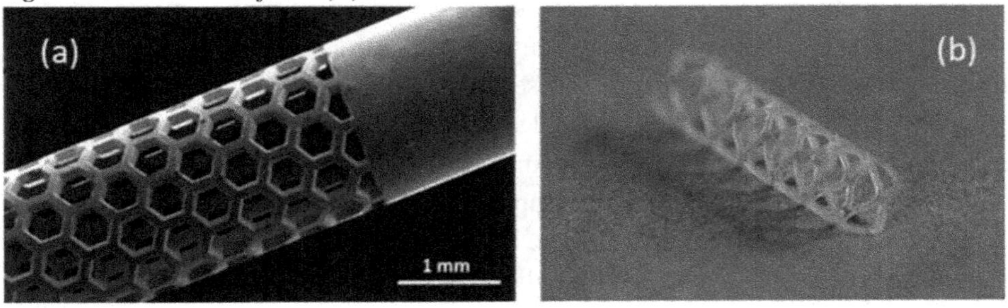

(Reprinted by permission from Laser Institute of America: Journal of Laser Applications 33:012053 (Ultrafast pulsed laser high precision micromachining of rotational symmetric parts, Gafner, M. et al.), Copyright (2021)); and (b) Stents cut made from biodegradable polymers (photo courtesy: Manutech-USD) (Gafner et al., 2021)

IV.1. Improvement of Technologies for Large-Scale Production, Compatible With Industrial Requirements

In this section, we discuss the transfer of laser processes from a laboratory setting to an industrial environment. The transition begins with simple cylindrical surfaces and progresses to more complex geometries featuring helical topologies, such as screws. For a perfectly smooth cylindrical surface, this transfer of laser patterning is relatively straightforward. A schematic representation of the process is shown in Figure 17. In the figure, the cylindrical object is defined by its diameter D and total length L. It is horizontally mounted along its central axis on a rotation stage, which in turn is supported by a linear translation stage (not shown in the figure).

As the rotation stage turns the cylindrical object at an angular velocity Ω (revolutions/s), the linear stage translates along the axis of rotation at a speed υ (m/s). The laser pulses impinge on the surface in the previously described radial configuration and remain fixed in space. The combined rotational and translational motions determine the spacing and distribution of laser pulses across the cylindrical surface. The rotation speed Ω and translation speed υ are process parameters selected based on the geometry of the cylindrical object and the optimized laser conditions established in earlier planar-processing steps.

In practice, for transferring a planar surface laser patterning process to a patterning process on a cylindrical surface, special care should be given to the speed limit; additional constraints may arise from the hardware specifications of the motorized stages. In other words, the values given by Ω and υ may not be achievable by the stages. Therefore, in some cases, the laser repetition rate ν has to be reduced to accommodate the stage's speed limits; concrete examples are to be given in the next section (Di Maio & Sedao, 2026).

Another classical strategy consists in moving the beam along the linear translation axis L at speed υ with interspot distance d_{il} followed by an incremental rotation corresponding to a fixed distance $d_{it.}$ These two steps are performed repeatedly until the entire cylindrical surface is fully laser treated.

The discussion above on cylindrical surface patterning assumes that all motion is achieved solely through translation and rotation stages. However, incorporating additional beam motion can further improve the process, particularly in terms of reducing processing time. For example, in a second scenario, the linear movement of the laser pulses relative to the cylindrical surface can be achieved either by the linear translation stage or by a galvanometer (galvo) scanner, provided its scanning direction is aligned with the intended linear motion axis.

Figure 17. Transfer from a planar to a cylindrical surface: a cylindrical object in a motorized laser surface patterning process. Key operation parameters are highlighted in the schematic

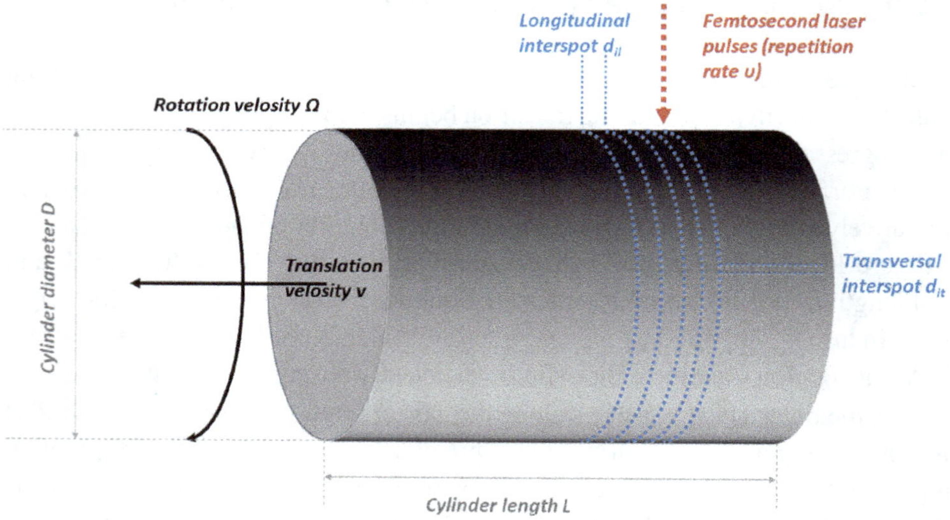

(Di Maio, Y. & Sedao, X. (2026) Ultrafast Laser Micro- and Nano-processing of Cylindrical Objects, Scaling of Laser Processing - Making Light Matter. Springer Nature) (Di Maio & Sedao, 2026)

Laser texturing of screws presents several technical micromachining constraints that accumulate issues of classical cylindrical surface while integrating more complex shapes such as threads and windings. The screw diameters can also vary depending on the position on the screw, leading to complex adaptations over the processed area to keep the same texturing properties. Such screws can be fully characterized by their inner/outer diameter, windings height/periodicity or tip length/conicity (see Figure 18). In the case of medical implants such as dental or orthopedic implants, diameters are in the order of millimeters while lengths range from a few millimeters to tens of centimeters. The variation of the diameter along the length of the workpiece, such as in screws, dental implants, or orthopedic devices, requires also a constant adaptation of the laser system to maintain consistent texture quality and alignment throughout the machining process. Configurations with great depth of field or adaptive focusing systems help maintain uniform texturing across the surface without expense at process time (Maalouf et al., 2025).

Figure 18. fs-L patterning of orthopaedic screw with technical adaptations to the screw geometry

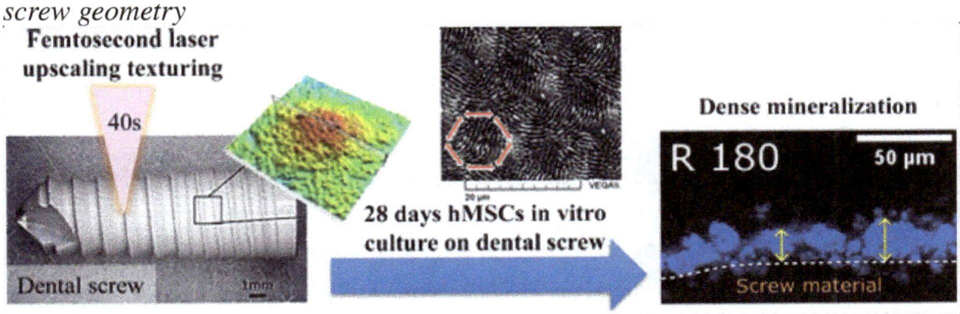

(Adapted and reprinted from, Mathieu Maalouf, et al., Femtosecond laser upscaling strategy and biological validation for dental screws with improved osteogenic performance, Surfaces and Interfaces, 2025, Surfaces and Interfaces (ScienceDirect), Copyright 2025 under Creative Commons BY 4.0 license. Retrieved from https://doi .org/10.1016/j.surfin.2025.106546)

Automating laser texturing for cylindrical components provides substantial advantages in precision, efficiency, and repeatability for industrial applications. The integration of advanced automation technologies such as vision systems, robotic handling, and real-time quality control makes the process more reliable while limiting manual intervention. High-resolution vision systems, for example, greatly facilitate the pre-inspection and alignment stages: they accurately identify the position of the part, automatically compensate for geometric variations, and thus reduce setup time and defects related to poor positioning.

Robotic handling systems, particularly multi-axis articulated arms, offer great flexibility in handling cylindrical parts. They perform pick-and-place operations autonomously and ensure smooth transfer between different stations, while maintaining the ideal orientation for laser texturing. This automation promotes continuous production and a steady flow of parts, from loading to unloading.

Real-time control further enhances process control. Tools such as optical detection, confocal microscopy, and laser interferometry make it possible to immediately assess roughness, texture uniformity, and the presence of any defects. Adaptive control algorithms can then automatically adjust laser parameters (power, pulse duration, scanning speed) to ensure consistent quality from batch to batch. Closed-loop systems also synchronize texturing with part rotation to ensure complete 360° coverage.

The most advanced motion controllers can even adapt the speed to variations in diameter or more complex shapes, ensuring perfectly uniform pattern depth and spacing.

By combining these different automation tools, laser texturing becomes a more efficient, scalable, and accurate process. The reduction in manual tasks and the implementation of intelligent quality control result in increased productivity, but also in better functional performance of cylindrical parts, whether medical implants, aerospace components, or high-engineering elements.

CONCLUSION

Femtosecond laser surface texturing represents a versatile and powerful tool for engineering multifunctional medical implant surfaces. By generating micro- and nanoscale patterns, including LIPSS, Fs-L processing improves mesenchymal stem cell adhesion, and accelerates osseointegration. The geometry, periodicity, and spatial orientation of these structures can be tuned to modulate mechanotransduction, cytoskeletal organization, and extracellular matrix deposition, providing a highly controllable platform for bone tissue integration.Beyond hard tissue interfaces, Fs-L texturing improves gingival fibroblast adhesion.. This stable interface not only can contribute to mechanical stability but also can act as a biological barrier against bacterial infiltration. Ultrafast laser technologies further extend these capabilities by enabling the design of mechano-bactericidal surfaces. By mimicking natural nanostructures, it is possible to create surfaces that mechanically disrupt bacterial cells, reducing the risk of biofilm formation and potentially eliminating the need for antibiotics or chemical coatings. To fully exploit this potential, interdisciplinary research into nanoscale bacteria-surface interactions and long-term validation is needed.

Complementing antibacterial properties, femtosecond laser texturing can be combined with advanced biofunctionalization strategies, such as immobilization or selective capture of extracellular vesicles (EVs). These EV-functionalized surfaces actively modulate tissue responses, promoting osteogenesis, angiogenesis, and soft tissue regeneration. The synergistic integration of antibacterial, osteogenic, and fibroblast-supportive functionalities positions Fs-L-textured implants as a new generation of multifunctional medical devices.

Fs-L processing is also adaptable to diverse medical devices beyond dental implants, including orthopedic screws, pacemakers, and stents. Hierarchical micro- and nano-scale patterns allow precise control of cell adhesion according to clinical needs, whether promoting integration or enabling later device removal. Automation, high-resolution motion control, and real-time monitoring make these processes scalable for industrial production while ensuring reproducibility and precision.

In conclusion, femtosecond laser surface engineering offers a comprehensive approach to the design of next-generation implants. By combining antibacterial mechano-bactericidal effects, enhanced osseointegration, improved soft tissue adhesion, and bioactive EV functionalities, Fs-L–textured surfaces can improve long-term implant performance, reduce complications, and actively contribute to tissue regeneration. Continued interdisciplinary research at the interface of photonics, surface science, microbiology, and regenerative medicine will be essential to realize the full clinical potential of these multifunctional, infection-resistant, and regenerative implants.

ACKNOWLEDGMENT

Parts of this work have been supported by the EUR MANUTECH SLEIGHT for financial support and the French ANR for the Equipex MANUTECH-USD support, through projects Lasimp, Texlid, Nanotrap and BLAST (ANR-25-CE51-2008-01). Moreover, some parts have been supported by the European Commission LaserImplant project, grant agreement number: 951730

Correction Notice: A correction to this manuscript has been published

Reason: Incorrect Figures

The publisher was contacted by the authors with copyright concerns for two images in their chapter. These figures have been replaced at the request by Dr. Sedao.

Correction Date: 25th June 2026

REFERENCES

Andreu, Z., & Yáñez-Mó, M. (2014). Tetraspanins in extracellular vesicle formation and function. *Frontiers in Immunology*, *5*, 442. DOI: 10.3389/fimmu.2014.00442 PMID: 25278937

Antich-Rosselló, M., Forteza-Genestra, M. A., Calvo, J., Gayà, A., Monjo, M., & Ramis, J. M. (2022). Customizing the extracellular vesicles release and effect by strategizing surface functionalization of titanium. *Scientific Reports*, *12*(1), 7399. DOI: 10.1038/s41598-022-11475-3 PMID: 35513419

Brandao, E., Nakhoul, A., Duffner, S., Emonet, R., Garrelie, F., Habrard, A., Jacquenet, F., Pigeon, F., Sebban, M., & Colombier, J.-P. (2023). Learning complexity to guide light-induced self-organized nanopatterns. *Physical Review Letters*, *130*(22), 226201. DOI: 10.1103/PhysRevLett.130.226201 PMID: 37327436

Chen, R., Kang, Z., Li, W., Xu, T., Wang, Y., Jiang, Q., Wang, Y., Huang, Z., Xu, X., & Huang, Z. (2024). Extracellular vesicle surface display of αPD-L1 and αCD3 antibodies via engineered late domain-based scaffold to activate T-cell anti-tumor immunity. *Journal of Extracellular Vesicles*, *13*(7), e12490. DOI: 10.1002/jev2.12490 PMID: 39051742

Chug, M. K., & Brisbois, E. J. (2022). Recent developments in multifunctional antimicrobial surfaces and applications toward advanced nitric oxide-based biomaterials. *ACS Materials Au*, *2*(4), 525–551. DOI: 10.1021/acsmaterialsau.2c00040 PMID: 36124001

Collaborators, A. R. (2022). Global burden of bacterial antimicrobial resistance in 2019: A systematic analysis. *Lancet*, *399*(10325), 629–655. DOI: 10.1016/S0140-6736(21)02724-0 PMID: 35065702

Derks, J., & Tomasi, C. (2015). Peri-implant health and disease: A systematic review of current epidemiology. *Journal of Clinical Periodontology*, *42*(S16, Suppl. 16), S158–S171. DOI: 10.1111/jcpe.12334 PMID: 25495683

Di Maio, Y., & Sedao, X. (2026). Ultrafast laser micro- and nano-processing of cylindrical objects. In *Scaling of laser processing - Making light matter*. Springer Nature.

Epperlein, N., Menzel, F., Schwibbert, K., Koter, R., Bonse, J., Sameith, J., Krüger, J., & Toepel, J. (2017). Influence of femtosecond laser produced nanostructures on biofilm growth on steel. *Applied Surface Science*, *418*, 420–424. DOI: 10.1016/j.apsusc.2017.02.174

Gafner, M., Kramer, T., Remund, S. M., Holtz, R., & Neuenschwander, B. (2021). Ultrafast pulsed laser high precision micromachining of rotational symmetric parts. *Journal of Laser Applications*, *33*(1), 012053. DOI: 10.2351/7.0000324

Heitz, J., Plamadeala, C., Muck, M., Armbruster, O., Baumgartner, W., Weth, A., Steinwender, C., Blessberger, H., Kellermair, J., Kirner, S. V., Krüger, J., Bonse, J., Guntner, A. S., & Hassel, A. W. (2017). Femtosecond laser-induced microstructures on Ti substrates for reduced cell adhesion. *Applied Physics. A, Materials Science & Processing*, *123*(12), 734. DOI: 10.1007/s00339-017-1352-0

Hoffmann, M., & Schwarz, U. S. (2013). A kinetic model for RNA-interference of focal adhesions. *BMC Systems Biology*, *7*(1), 2. DOI: 10.1186/1752-0509-7-2 PMID: 23311633

Jalil, S. A., Akram, M., Bhat, J. A., Hayes, J. J., Singh, S. C., ElKabbash, M., & Guo, C. (2020). Creating superhydrophobic and antibacterial surfaces on gold by femtosecond laser pulses. *Applied Surface Science*, *506*, 144952. DOI: 10.1016/j.apsusc.2019.144952 PMID: 32184533

Kelleher, S. M., Habimana, O., Lawler, J., O' Reilly, B., Daniels, S., Casey, E., & Cowley, A. (2016). Cicada wing surface topography: An investigation into the bactericidal properties of nanostructural features. *ACS Applied Materials & Interfaces*, *8*(24), 14966–14974. DOI: 10.1021/acsami.5b08309 PMID: 26551558

Khatoon, Z., McTiernan, C. D., Suuronen, E. J., Mah, T.-F., & Alarcon, E. I. (2018). Bacterial biofilm formation on implantable devices and approaches to its treatment and prevention. *Heliyon, 4*(4), e01067. DOI: 10.1016/j.heliyon.2018.e01067 PMID: 30619958

Konoshenko, M. Y., Lekchnov, E. A., Vlassov, A. V., & Laktionov, P. P. (2018). Isolation of extracellular vesicles: General methodologies and latest trends. *BioMed Research International, 2018*, 8545347. DOI: 10.1155/2018/8545347 PMID: 29662902

Kowal, J., Arras, G., Colombo, M., Jouve, M., Morath, J. P., Primdal-Bengtson, B., Dingli, F., Loew, D., Tkach, M., & Théry, C. (2016). Proteomic comparison defines novel markers to characterize heterogeneous populations of extracellular vesicle subtypes. *Proceedings of the National Academy of Sciences of the United States of America, 113*(8), E968–E977. DOI: 10.1073/pnas.1521230113 PMID: 26858453

Lee, S. W., Phillips, K. S., Gu, H., Kazemzadeh-Narbat, M., & Ren, D. (2021). How microbes read the map: Effects of implant topography on bacterial adhesion and biofilm formation. *Biomaterials, 268*, 120595. DOI: 10.1016/j.biomaterials.2020.120595 PMID: 33360301

Lötvall, J., Hill, A. F., Hochberg, F., Buzás, E. I., Di Vizio, D., Gardiner, C., Gho, Y. S., Kurochkin, I. V., Mathivanan, S., Quesenberry, P., Sahoo, S., Tahara, H., Wauben, M. H., Witwer, K. W., & Théry, C. (2014). Minimal experimental requirements for definition of extracellular vesicles and their functions: A position statement from the International Society for Extracellular Vesicles. *Journal of Extracellular Vesicles, 3*(1), 26913. DOI: 10.3402/jev.v3.26913 PMID: 25536934

Maalouf, M., Abou Khalil, A., Di Maio, Y., Papa, S., Sedao, X., Dalix, E., Peyroche, S., Guignandon, A., & Dumas, V. (2022). Polarization of femtosecond laser for titanium alloy nanopatterning influences osteoblastic differentiation. *Nanomaterials (Basel, Switzerland), 12*(10), 1619. DOI: 10.3390/nano12101619 PMID: 35630841

Maalouf, M., Maio, Y. D., Pallarés-Aldeiturriaga, D., Sedao, X., Hivert, L., Papa, S., Dalix, E., Thomas, M., Guignandon, A., & Dumas, V. (2025). Femtosecond laser upscaling strategy and biological validation for dental screws with improved osteogenic performance. *Surfaces and Interfaces, 66*, 106546. DOI: 10.1016/j.surfin.2025.106546

Muck, M., Wolfsjäger, B., Seibert, K., Maier, C., Lone, S. A., Hassel, A. W., Baumgartner, W., & Heitz, J. (2021). Femtosecond laser-processing of pre-anodized Ti-based bone implants for cell-repellent functionalization. *Nanomaterials (Basel, Switzerland), 11*(5), 1342. DOI: 10.3390/nano11051342 PMID: 34065199

Müller, F. A., Kunz, C., & Gräf, S. (2016). Bio-inspired functional surfaces based on laser-induced periodic surface structures. *Materials (Basel)*, *9*(6), 476. DOI: 10.3390/ma9060476 PMID: 28773596

Nabhan, J. F., Hu, R., Oh, R. S., Cohen, S. N., & Lu, Q. (2012). Formation and release of arrestin domain-containing protein 1-mediated microvesicles (ARMMs) at plasma membrane by recruitment of TSG101 protein. *Proceedings of the National Academy of Sciences of the United States of America*, *109*(11), 4146–4151. DOI: 10.1073/pnas.1200448109 PMID: 22315426

Nakhoul, A., Maurice, C., Faure, N., Garrelie, F., Pigeon, F., & Colombier, J.-P. (2022). Tailoring the surface morphology of Ni at the nanometric scale by ultrashort laser pulses. *Applied Physics. A, Materials Science & Processing*, *128*(10), 933. DOI: 10.1007/s00339-022-06046-2

Nakhoul, A., Rudenko, A., Maurice, C., Reynaud, S., Garrelie, F., Pigeon, F., & Colombier, J.-P. (2022). Boosted spontaneous formation of high-aspect ratio nanopeaks on ultrafast laser-irradiated Ni surface. *Advancement of Science*, *9*(21), e2200761. DOI: 10.1002/advs.202200761 PMID: 35618474

Papa, S. (2023). Evaluation de l'adhérence gingivale et du potentiel antibactérien de surfaces structurées par laser femtoseconde pour l'implantologie orale [Doctoral dissertation, Université de Lyon]. HAL.

Papa, S., Maalouf, M., Claudel, P., Sedao, X., Di Maio, Y., Hamzeh-Cognasse, H., Thomas, M., Guignandon, A., & Dumas, V. (2023). Key topographic parameters driving surface adhesion of Porphyromonas gingivalis. *Scientific Reports*, *13*(1), 15893. DOI: 10.1038/s41598-023-42387-5 PMID: 37741851

Pfeiffer, I., Petronis, S., Köper, I., Kasemo, B., & Zäch, M. (2010). Vesicle adsorption and phospholipid bilayer formation on topographically and chemically nanostructured surfaces. *The Journal of Physical Chemistry B*, *114*(14), 4623–4631. DOI: 10.1021/jp908283g PMID: 20232804

Richter, A. M., Buchberger, G., Stifter, D., Duchoslav, J., Hertwig, A., Bonse, J., Heitz, J., & Schwibbert, K. (2021). Spatial period of laser-induced surface nanoripples on PET determines Escherichia coli repellence. *Nanomaterials (Basel, Switzerland)*, *11*(11), 3000. DOI: 10.3390/nano11113000 PMID: 34835763

Takeda, M., Ankamatsu, S., Kita, Y., Goto, T., & Kobayashi, T. (2023). The roles of extracellular vesicles in the progression of renal cell carcinoma and their potential for future clinical application. *Nanomaterials (Basel, Switzerland)*, *13*(10), 1611. https://www.mdpi.com/2079-4991/13/10/1611. DOI: 10.3390/nano13101611 PMID: 37242027

Tao, W., Ivanovska, V., Schweickert, B., & Müller, A. (2019). Proxy indicators for antibiotic consumption; surveillance needed to control antimicrobial resistance. *Bulletin of the World Health Organization*, *97*(1), 3–3A. DOI: 10.2471/BLT.18.227348 PMID: 30618457

Théry, C., Witwer, K. W., Aikawa, E., Alcaraz, M. J., Anderson, J. D., Andriantsitohaina, R., Antoniou, A., Arab, T., Archer, F., Atkin-Smith, G. K., Ayre, D. C., Bach, J.-M., Bachurski, D., Baharvand, H., Balaj, L., Baldacchino, S., Bauer, N. N., Baxter, A. A., Bebawy, M., & Zuba-Surma, E. K. (2018). Minimal information for studies of extracellular vesicles 2018 (MISEV2018): A position statement of the International Society for Extracellular Vesicles and update of the MISEV2014 guidelines. *Journal of Extracellular Vesicles*, *7*(1), 1535750. DOI: 10.1080/20013078.2018.1535750 PMID: 30637094

Tripathy, A., Sen, P., Su, B., & Briscoe, W. H. (2017). Natural and bioinspired nanostructured bactericidal surfaces. *Advances in Colloid and Interface Science*, *248*, 85–104. DOI: 10.1016/j.cis.2017.07.030 PMID: 28780961

Willms, E., Cabañas, C., Mäger, I., Wood, M. J. A., & Vader, P. (2018). Extracellular vesicle heterogeneity: Subpopulations, isolation techniques, and diverse functions in cancer progression. *Frontiers in Immunology*, *9*, 738. DOI: 10.3389/fimmu.2018.00738 PMID: 29760691

Xie, H., Wang, Z., Zhang, L., Lei, Q., Zhao, A., Wang, H., Li, Q., Chen, Z., & Zhang, W. (2016). Development of an angiogenesis-promoting microvesicle-alginate-polycaprolactone composite graft for bone tissue engineering applications. *PeerJ*, *4*, e2040. DOI: 10.7717/peerj.2040 PMID: 27231660

Yu, J., Sane, S., Kim, J. E., Yun, S., Kim, H. J., Jo, K. B., Wright, J. P., Khoshdoozmasouleh, N., Lee, K., Oh, H. T., Thiel, K., Parvin, A., Williams, X., Hannon, C., Lee, H., & Kim, D. K. (2024). Biogenesis and delivery of extracellular vesicles: Harnessing the power of EVs for diagnostics and therapeutics. *Frontiers in Molecular Biosciences*, *10*, 1330400. DOI: 10.3389/fmolb.2023.1330400 PMID: 38234582

Zhao, J., Schlößer, H. A., Wang, Z., Qin, J., Li, J., Popp, F., Popp, M. C., Alakus, H., Chon, S.-H., Hansen, H. P., Neiss, W. F., Jauch, K.-W., Bruns, C. J., & Zhao, Y. (2019). Tumor-derived extracellular vesicles inhibit natural killer cell function in pancreatic cancer. *Cancers (Basel)*, *11*(6), 874. DOI: 10.3390/cancers11060874 PMID: 31234517

Zhao, J., Zhu, Z., Xu, Y., Song, X., Wang, Y., Peng, H., Wang, Y., Zuo, J., Shu, X., & Yin, A. (2020). Nanosecond laser ablation of Ti–6Al–4V under different temperature. *Applied Sciences (Basel, Switzerland)*, *10*(13), 4657. DOI: 10.3390/app10134657

Compilation of References

Abbaspour, N., Hurrell, R., & Kelishadi, R. (2014). Review on iron and its importance for human health. *Journal of Research in Medical Sciences : the Official Journal of Isfahan University of Medical Sciences, 19*(2), 164. PMID: 24778671

Abdul Khalil, H. P. S., Bhat, A. H., Abu Bakar, A., Tahir, P. M., Zaidul, I. S. M., & Jawaid, M. (2014). *Cellulosic nanocomposites from natural fibers for medical applications: A review. Handbook of Polymer Nanocomposites. Processing, Performance and Application* (Vol. C). Polymer Nanocomposites of Cellulose Nanoparticles.

Abdullah, F. R., Mafri, T. A., & Shen, Y. K. (2026). Principles, current applications, and future perspectives of selective laser melting in prosthodontics: A review. *Metals, 16*(4), 415. DOI: 10.3390/met16040415

Abou-Khalil, L., da Cruz, K. S., Reinhart, G., Mangelinck-Noël, N., & Nguyen-Thi, H. (2022). Influence of growth velocity on fragmentation during directional solidification of Al–14 wt.% Sn alloy studied by in-situ synchrotron X-radiography. *Acta Materialia, 241*, 118370. DOI: 10.1016/j.actamat.2022.118370

Abraham, A. M., & Subramani, V. (2023). Effect of magnesium as biomaterial in biodegradation. *Materials Today: Proceedings.*

AbuAlia, M., Fullam, S., Cinotti, F., Manninen, N., & Wimmer, M. A. (2024). Titanium nitride coatings on CoCrMo and Ti6Al4V alloys: Effects on wear and ion release. *Lubricants (Basel, Switzerland), 12*(3), 96. Advance online publication. DOI: 10.3390/lubricants12030096 PMID: 41743082

Adamovic, D., Ristic, B., & Zivic, F. (2017). Review of existing biomaterials—method of material selection for specific applications in orthopedics. Biomaterials in clinical practice: advances in clinical research and medical devices, 47-99.

Afonso, C. R. M., Vidilli, A. L., Spinelli, J. E., Riva, R., Amigó, V., & Kiminami, C. S. (2017). An assessment of microstructure and properties of laser clad coatings of ultrafine eutectic β Ti-Fe-Nb-Sn composite for implants. *Surface and Coatings Technology*, *328*, 161–171. DOI: 10.1016/j.surfcoat.2017.08.035

Ahmad Khalili, A., & Ahmad, M. R. (2015). A review of cell adhesion studies for biomedical and biological applications. *International Journal of Molecular Sciences*, *16*(8), 18149–18184. DOI: 10.3390/ijms160818149 PMID: 26251901

Ahmed, L. O., & Omer, R. A. (2024). Hydroxyapatite biomaterials: A comprehensive review of their properties, structures, clinical applications, and producing techniques. *Reviews in Inorganic Chemistry*, *44*(4), 599–618. DOI: 10.1515/revic-2024-0018

Ahmed, Y., Ankah, N. K., Ogunlakin, N., Toor, I. U., & Farooq, W. (2025). Exploring the future of metallic implants: A review of biodegradable and non-biodegradable solutions. *Corrosion Reviews*, *43*(5), 495–521. DOI: 10.1515/corrrev-2024-0125

Ahmed, Z., Wang, Z., Adil, M., Bhatti, I. A., & Cao, H. (2025). Bright antimicrobial surfaces based on Schottky interfaces: From light illumination to bacterial charging. *Smart Materials in Medicine*, *6*(1), 67–94. Advance online publication. DOI: 10.1016/j.smaim.2025.02.001

Ahsan, S., Ahmed, F., Kim, Y. G., Lee, M. S., & Jun, M. B. (2011). Colorizing stainless steel surface by femtosecond laser induced micro/nano-structures. *Applied Surface Science*, *257*(17), 7771–7777. DOI: 10.1016/j.apsusc.2011.04.027

Ajayan, A., Joseph, V., Paul, A., & Joseph, A. (2024). Coating characterization with electrochemical techniques. In *Encapsulated corrosion inhibitors for eco-benign smart coatings* (pp. 205–224). CRC Press. DOI: 10.1201/9781032677255-12

Akter, S., Azim, A. Y. M. A., & Al Faruque, M. A. (2014). Medical textiles: Significance and future prospect in Bangladesh. *European Scientific Journal, 10*(12).

Al Ali, M. (2024). Biomechanics of the Human Body. In *Fundamentals of Orthopedic Design with Non-parametric Optimization* (pp. 91–103). Springer Nature Singapore. DOI: 10.1007/978-981-97-1040-9_4

Al Jabbari, Y. S., Fehrman, J., Barnes, A. C., Zapf, A. M., Zinelis, S., & Berzins, D. W. (2012). Titanium nitride and nitrogen ion implanted coated dental materials. *Coatings*, *2*(3), 160–178. DOI: 10.3390/coatings2030160

Al Javed, M. O., & Bin Rashid, A. (2024). Laser-assisted micromachining techniques: An overview of principles, processes, and applications. *Advances in Materials and Processing Technologies*, 1–44.

Al Juhaiman, L. A. (2016). Polyvinyl pyrrolidone as a Corrosion Inhibitor for Carbon Steel in HCl. *International Journal of Electrochemical Science*, *11*(3), 2247–2262. DOI: 10.1016/S1452-3981(23)16098-6

Alam, F., Ashfaq Ahmed, M., Jalal, A. H., Siddiquee, I., Adury, R. Z., Hossain, G. M., & Pala, N. (2024). Recent progress and challenges of implantable biodegradable biosensors. *Micromachines*, *15*(4), 475. DOI: 10.3390/mi15040475 PMID: 38675286

Al-Amin, K., Kawsar, M., Mamun, M., & Sahadat Hossain, M. (2025). Fourier transform infrared spectroscopic technique for analysis of inorganic materials: A review. *Nanoscale Advances*, *7*(21), 6677–6702. DOI: 10.1039/D5NA00522A PMID: 40949700

Albrektsson, T., Chrcanovic, B., Mölne, J., & Wennerberg, A. (2017). Foreign body reactions, marginal bone loss, and allergies in relation to titanium implants. *European Journal of Oral Implantology*, *10*(1), 7–19. PMID: 30109298

Albu, D. F., Caraman, M., Cotorobai, V., Radu, A., Cobianu, C., & Gurlui, S. (2022). Thin film fabrication by pulsed laser deposition from TiO_2: Growth atmosphere effects on structure and optoelectronic properties. *Coatings*, *12*(3), 293. DOI: 10.3390/coatings12030293

AlGhamdi, K. M., Kumar, A., & Moussa, N. A. (2012). Low-level laser therapy: A useful technique for enhancing the proliferation of various cultured cells. *Lasers in Medical Science*, *27*(2), 237–249. DOI: 10.1007/s10103-011-0885-2 PMID: 21274733

Ali, M. (2014). Study of a novel configuration of laser Assisted Bioprinting (Doctoral dissertation, Université de Bordeaux).

Alontseva, D., Azamatov, B., Safarova, Y., Voinarovych, S., & Nazenova, G. (2023). A brief review of current trends in the additive manufacturing of orthopedic implants with thermal plasma-sprayed coatings to improve the implant surface biocompatibility. *Coatings*, *13*(7), 1175. DOI: 10.3390/coatings13071175

Alrmali, A. E., Misch, J., Gomez-Meda, R., Saleh, M. H., & Wang, H. L. (2026). The MICH concept for long-term implant stability. *The International Journal of Oral Implantology*, *19*(1), 13–29. PMID: 41823002

Altman, G. H., Diaz, F., Jakuba, C., Calabro, T., Horan, R. L., Chen, J., ... & Kaplan, D. L. (2003). Silk-based biomaterials. biomaterials, 24(3), 401-416.

Alvarez-Pampliega, A., Hauffman, T., Petrova, M., Breugelmans, T., Muselle, T., Van den Bergh, K., De Strycker, J., Terryn, H., & Hubin, A. (2014). Corrosion study on Al-rich metal-coated steel by odd random phase multisine electrochemical impedance spectroscopy. *Electrochimica Acta, 124*, 165–175. DOI: 10.1016/j.electacta.2013.09.159

Al-Zyoud, W., Haddadin, D., Hasan, S. A., Jaradat, H., & Kanoun, O. (2023). Biocompatibility testing for implants: A novel tool for selection and characterization. *Materials (Basel), 16*(21), 6881. DOI: 10.3390/ma16216881 PMID: 37959478

Aman, J., & Margadant, C. (2023). Integrin-dependent cell–matrix adhesion in endothelial health and disease. *Circulation Research, 132*(3), 355–378. DOI: 10.1161/CIRCRESAHA.122.322332 PMID: 36730379

Amaral, C., Paiva, M., Rodrigues, A. R., Veiga, F., & Bell, V. (2024). Global regulatory challenges for medical devices: Impact on innovation and market access. *Applied Sciences (Basel, Switzerland), 14*(20), 9304. Advance online publication. DOI: 10.3390/app14209304

Amaral, M. N., Kumar, P., Faísca, P., Ferreira, H. A., Coelho, J. M. P., Gaspar, M. M., & Reis, C. P. (2025). Gold nanoparticle-mediated photothermal therapy: Expanding the frontiers of cancer treatment and theragnostics. *Biomedicine and Pharmacotherapy, 190*, 118399. DOI: 10.1016/j.biopha.2025.118399 PMID: 40784123

Ameku, W. A., Negahdary, M., Lima, I. S., Santos, B. G., Oliveira, T. G., Paixão, T. R., & Angnes, L. (2022). Laser-scribed graphene-based electrochemical sensors: A review. *Chemosensors (Basel, Switzerland), 10*(12), 505. DOI: 10.3390/chemosensors10120505

Amirtharaj Mosas, K. K., Chandrasekar, A. R., Dasan, A., Pakseresht, A., & Galusek, D. (2022). Recent advancements in materials and coatings for biomedical implants. *Gels (Basel, Switzerland), 8*(5), 323. DOI: 10.3390/gels8050323 PMID: 35621621

Amirudin, A., & Thieny, D. (1995). Application of electrochemical impedance spectroscopy to study the degradation of polymer-coated metals. *Progress in Organic Coatings, 26*(1), 1–28. DOI: 10.1016/0300-9440(95)00581-1

Ananthakumar, B., Nagaraja, B. G., Umesh, S. M., & Raghavan, R. (2025). Predictive Medicine. *Predictive Algorithms for Rehabilitation and Assistive Systems*, 291–314. DOI: 10.4018/979-8-3373-0194-5.ch011

Ananthakumar, B., Nanjundeshwara, N., Lakshmaiah, A., & Raghavan, R. (2024). Synthetic Gene Circuits as a Promising Approach in Cancer Immunotherapy. *Critical Developments in Cancer Immunotherapy*, 391–418. DOI: 10.4018/979-8-3693-3976-3.ch011

Ananthakumar, B., Nanjundeshwara, N., Lakshmaiah, A., & Raghavan, R. (2025). Nanocarriers for Delivering Nucleic Acids in Plants. *Nanocarriers in Plant Science and Agriculture*, 207–234. DOI: 10.4018/979-8-3693-9869-2.ch008

Anders, J. J., Lanzafame, R. J., & Arany, P. R. (2015). Low-level light/laser therapy versus photobiomodulation therapy. *Photomedicine and Laser Surgery*, *33*(4), 183–184. DOI: 10.1089/pho.2015.9848 PMID: 25844681

Anderson, J. M., Rodriguez, A., & Chang, D. T. (2008). Foreign body reaction to biomaterials. In *Seminars in Immunology* (Vol. 20, Number 2, pp. 86–100). DOI: 10.1016/j.smim.2007.11.004

Andreu, Z., & Yáñez-Mó, M. (2014). Tetraspanins in extracellular vesicle formation and function. *Frontiers in Immunology*, *5*, 442. DOI: 10.3389/fimmu.2014.00442 PMID: 25278937

Angelova, L., Daskalova, A., Mincheva, R., Filipov, E., Dikovska, A., Fernandes, M. H., Vig, S., & Buchvarov, I. (2024). Design of femtosecond microstructured poly lactic acid temporal scaffolds coated with hydroxyapatite by pulse laser deposition method for bone tissue regeneration. *Optical and Quantum Electronics*, *56*(7), 1195. DOI: 10.1007/s11082-024-06500-w

Anita, N., Joany, R. M., Dorothy, R., Aslam, J., Rajendran, S., Subramania, A., Singh, G., & Verma, C. (2023). Linear polarization resistance (LPR) technique for corrosion measurements. In *Electrochemical and Analytical Techniques for Sustainable Corrosion Monitoring* (pp. 59–80). Elsevier. DOI: 10.1016/B978-0-443-15783-7.00005-0

Anselme, K., Davidson, P., Popa, A. M., Giazzon, M., Liley, M., & Ploux, L. (2010). The interaction of cells and bacteria with surfaces structured at the nanometre scale. *Acta Biomaterialia*, *6*(10), 3824–3846. DOI: 10.1016/j.actbio.2010.04.001 PMID: 20371386

Antich-Rosselló, M., Forteza-Genestra, M. A., Calvo, J., Gayà, A., Monjo, M., & Ramis, J. M. (2022). Customizing the extracellular vesicles release and effect by strategizing surface functionalization of titanium. *Scientific Reports*, *12*(1), 7399. DOI: 10.1038/s41598-022-11475-3 PMID: 35513419

Ariffin, S. H. Z., Lim, K. W., Wahab, R. M. A., Ariffin, Z. Z., Din, R. D. R., Shahidan, M. A., & Abidin, I. Z. Z. (2022). Gene expression profiles for in vitro human stem cell differentiation into osteoblasts and osteoclasts: A systematic review. *PeerJ*, *10*, e14174. DOI: 10.7717/peerj.14174 PMID: 36275474

Artemchuk, A., Osmanov, B., & Chepurnyi, Y. (2025). Evaluation of bone-to-implant contact area in patients with reduced residual bone height in the posterior maxilla: A virtual simulation study. *Journal of Oral and Maxillofacial Surgery, Medicine, and Pathology*.

Arthanari, S., Zhang, J., Xue, X., Li, Y., & Guan, Y. (2021). *Laser Induced Micro/Nano Functional Surfaces on Metals for Biomedical Applications* (pp. 439–464). DOI: 10.1007/978-981-33-6252-9_15

Arulvel, S., Dsilva Winfred Rufuss, D., Akshat, J., Kandasamy, J., & Singhal, M. (2023). Laser processing techniques for surface property enhancement: Focus on material advancement. *Surfaces and Interfaces*, *42*, 103293. DOI: 10.1016/j.surfin.2023.103293

Arun, K. L., Udhayakumar, M., & Radhika, N. (2023). A comprehensive review on various ceramic nanomaterial coatings over metallic substrates: Applications, challenges and future trends. *Journal of Bio- and Tribo-Corrosion*, *9*(1), 11. DOI: 10.1007/s40735-022-00717-6

Ashjari, M., Khoee, S., Mahdavian, A. R., & Rahmatolahzadeh, R. (2012). Self-assembled nanomicelles using PLGA–PEG amphiphilic block copolymer for insulin delivery: A physicochemical investigation and determination of CMC values. *Journal of Materials Science. Materials in Medicine*, *23*(4), 943–953. DOI: 10.1007/s10856-012-4562-1 PMID: 22354326

Aslam Khan, M. U., Aslam, M. A., Bin Abdullah, M. F., & Stojanović, G. M. (2024). Current perspectives of protein in bone tissue engineering: Bone structure, ideal scaffolds, fabrication techniques, applications, scopes, and future advances. *ACS Applied Bio Materials*, *7*(8), 5082–5106. DOI: 10.1021/acsabm.4c00362 PMID: 39007509

ASTM. (2013). *Standard Test Method for Analysis of Hemolytic Properties of Nanoparticles*. ASTM.

Atanasov, P. A., Stankova, N. E., Nedyalkov, N. N., Fukata, N., Hirsch, D., Rauschenbach, B., Amoruso, S., Wang, X., Kolev, K. N., Valova, E. I., Georgieva, J. S., & Armyanov, St. A. (2015). Fs-laser processing of medical grade polydimethylsiloxane (PDMS). *Applied Surface Science*, *374*, 229–234. Advance online publication. DOI: 10.1016/j.apsusc.2015.11.175

Atta, S., Dash, S., Mathew, A. T., & NarendraKumar, U. (2025). Recent developments on titanium based mono and multilayer nitride films deposited through HiPIMS. *Next Materials*, *9*, 100983. Advance online publication. DOI: 10.1016/j.nxmate.2025.100983

Azadmanjiri, J., Kumar, P., Srivastava, V. K., & Sofer, Z. (2020). Surface functionalization of 2D transition metal oxides and dichalcogenides via covalent and non-covalent bonding for sustainable energy and biomedical applications. *ACS Applied Nano Materials*, *3*(4), 3116–3143. DOI: 10.1021/acsanm.0c00120

Aziz, U., McAfee, M., Manolakis, I., Timmons, N., & Tormey, D. (2025). A review of optimization of additively manufactured 316/316L stainless steel process parameters, post-processing strategies, and defect mitigation. *Materials (Basel)*, *18*(12), 2870. DOI: 10.3390/ma18122870 PMID: 40573001

Badiceanu, M., Anghel, S., Mihailescu, N., Visan, A. I., Mihailescu, C. N., & Mihailescu, I. N. (2022). Coatings Functionalization via Laser versus Other Deposition Techniques for Medical Applications: A Comparative Review. *Coatings*, *12*(1), 71. DOI: 10.3390/coatings12010071

Bähr, S. H. (2025). Process gas influences during the powder bed fusion of metals using a laser beam [Doctoral dissertation, Technische Universität München].

Bai, L., Song, P., & Su, J. (2023). Bioactive elements manipulate bone regeneration. *Biomaterials Translational*, *4*(4), 248. PMID: 38282709

Baino, F., & Kargozar, S. (Eds.). (2022). *Bioactive glasses and glass-ceramics: Fundamentals and applications*. John Wiley & Sons. DOI: 10.1002/9781119724193

Balcerak-Woźniak, A., Dzwonkowska-Zarzycka, M., & Kabatc-Borcz, J. (2024). A comprehensive review of stimuli-responsive smart polymer materials—Recent advances and future perspectives. *Materials (Basel)*, *17*(17), 4255. Advance online publication. DOI: 10.3390/ma17174255 PMID: 39274645

Balena, A., Bianco, M., Pisanello, F., & De Vittorio, M. (2023). Recent advances on high-speed and holographic two-photon direct laser writing. *Advanced Functional Materials*, *33*(39), 2211773. DOI: 10.1002/adfm.202211773

Bandzerewicz, A., & Gadomska-Gajadhur, A. (2022). Into the tissues: Extracellular matrix and its artificial substitutes: Cell signalling mechanisms. *Cells*, *11*(5), 914. DOI: 10.3390/cells11050914 PMID: 35269536

Barbhuiya, N. H., Kumar, A., & Singh, S. P. (2021). A journey of laser-induced graphene in water treatment. *Transactions of the Indian National Academy of Engineering : an International Journal of Engineering and Technology*, *6*(2), 159–171. DOI: 10.1007/s41403-021-00205-2

Baskerville, S. J. J. (2022). Effects of orthopedic implant and host bone properties on stress-shielding induced osteopenia [Unpublished data or report].

Bäuerle, D. (2011). Thermal, photophysical, and photochemical processes. In *Laser Processing and Chemistry* (pp. 13–38). Springer Berlin Heidelberg. DOI: 10.1007/978-3-642-17613-5_2

Bäuerle, D. (2013). *Laser processing and chemistry*. Springer Science & Business Media.

Baylón, K., Rodríguez-Camarillo, P., Elías-Zúñiga, A., Díaz-Elizondo, J. A., Gilkerson, R., & Lozano, K. (2017). Past, present and future of surgical meshes: A review. *Membranes*, *7*(3), 47. DOI: 10.3390/membranes7030047 PMID: 28829367

Bejaoui, M., Galai, H., Touati, F., & Kouass, S. (2023). Multifunctional Roles of PVP as a Versatile Biomaterial in Solid State. In *Dosage Forms - Innovation and Future Perspectives*. IntechOpen., DOI: 10.5772/intechopen.99431

Bender, R., Féron, D., Mills, D., Ritter, S., Bäßler, R., Bettge, D., De Graeve, I., Dugstad, A., Grassini, S., Hack, T., Halama, M., Han, E.-H., Harder, T., Hinds, G., Kittel, J., Krieg, R., Leygraf, C., Martinelli, L., Mol, A., & Zheludkevich, M. (2022). Corrosion challenges towards a sustainable society. *Materials and Corrosion*, *73*(11), 1730–1751. DOI: 10.1002/maco.202213140

Beqain, M. D. Y. S. (2023). Novel substituted hydroxyapatites [Doctoral dissertation, The University of Waikato].

Berradja, A. (2019). Electrochemical Techniques for Corrosion and Tribocorrosion Monitoring: Methods for the Assessment of Corrosion Rates. In *Corrosion Inhibitors*. IntechOpen. DOI: 10.5772/intechopen.86743

Bhaduri, S. B., Narayan, R., & Balani, K. (2021). *Engineering Biomaterials for Regenerative Medicine: Novel Technologies for Clinical Applications*. Springer.

Bhattacharya, M., & Horswill, A. R. (2024). The role of human extracellular matrix proteins in defining Staphylococcus aureus biofilm infections. *FEMS Microbiology Reviews*, *48*(2), fuae002. Advance online publication. DOI: 10.1093/femsre/fuae002 PMID: 38337187

Bhuvana, G. N., Umesh, S. M., & Raghavan, R. (2025). Expanding the frontiers of Schistosomiasis Treatment: The promise of Nanoparticle-Based Therapies. *Journal of Vector Borne Diseases, 10*(4103/jvbd.jvbd_180_23). DOI: 10.4103/jvbd.jvbd_180_23

Bierwagen, G., Tallman, D., Li, J., He, L., & Jeffcoate, C. (2003). EIS studies of coated metals in accelerated exposure. *Progress in Organic Coatings, 46*(2), 149–158. DOI: 10.1016/S0300-9440(02)00222-9

Bigerelle, M., Anselme, K., Noel, B., Ruderman, I., Hardouin, P., & Iost, A. (2002). Improvement in the morphology of Ti-based surfaces: A new process to increase in vitro human osteoblast response. *Biomaterials, 23*(7), 1563–1577. DOI: 10.1016/S0142-9612(01)00271-X PMID: 11922462

Bi, J., Li, H., Ren, W., Wang, Z., & Li, C. (2023). Beam shaping technology and its application in metal laser processing. *Journal of Manufacturing Processes, 93*, 88–108. DOI: 10.1016/j.jmapro.2023.08.038

Bijapur, K., Molahalli, V., Shetty, A., Toghan, A., De Padova, P., & Hegde, G. (2023). Recent trends and progress in corrosion inhibitors and electrochemical evaluation. *Applied Sciences (Basel, Switzerland), 13*(18), 10107. DOI: 10.3390/app131810107

Bingtong, L., Yongliang, Z., & Liping, S. (2020). Identification and characterization of the peptides with calcium-binding capacity from tilapia (Oreochromis niloticus) skin gelatin enzymatic hydrolysates. *Journal of Food Science, 85*(1), 114–122. DOI: 10.1111/1750-3841.14975 PMID: 31869867

Biradar, S., Korrapati, C., Krishna, R. J., & Nanjundeshwara, N. (2025). Spatial Omics and Gene Circuits: The Future in Cancer Biology. *Spatially Variable Genes in Cancer: Development, Progression, and Treatment Response*, 383–408. DOI: 10.4018/979-8-3693-7728-4.ch014

Biradar, S., Suresh, C., Nanjundeshwara, N., & Raghavan, R. (2025). Spatially Variable Genes. *Spatially Variable Genes in Cancer*, 1–28. DOI: 10.4018/979-8-3693-7728-4.ch001

Bîrsan, D. C., & Susac, F. (2026). Recent advances and future perspectives in friction stir welding and processing. *RE:view*.

Biswas, D. J. (2026). Shaping with light: The many faces of laser material processing. In A beginner’s guide to lasers and their applications, Part 2: The laser's impact on science and humanity (pp. 297-327). Springer Nature Switzerland.

Biswas, R. K. (2023). Carbonization of Polyimide using CO2 laser and Femtosecond laser for Sensor Applications (Doctoral dissertation, NUI Galway).

Bixler, G. D. (2013). Bioinspired surface for low drag, self-cleaning, and antifouling: Shark skin, butterfly and rice leaf effects [Doctoral dissertation, The Ohio State University]. OhioLINK Electronic Theses and Dissertations Center.

Blutinger, J. D. (2022). Digital cuisine: Food printing and laser cooking [Doctoral dissertation, Columbia University].

Boccaccini, A. R., Stamboulis, A. G., Rashid, A., & Roether, J. A. (2003). Composite surgical sutures with bioactive glass coating. *Journal of Biomedical Materials Research. Part B, Applied Biomaterials*, *67*(1), 618–626. Advance online publication. DOI: 10.1002/jbm.b.10047 PMID: 14528459

Bogdan, M., & Peter, I. (2024). A comprehensive understanding of thermal barrier coatings (TBCs): Applications, materials, coating design and failure mechanisms. *Metals*, *14*(5), 575. DOI: 10.3390/met14050575

Boi, M., Marchiori, G., Berni, M., Gambardella, A., Salamanna, F., Visani, A., Bianchi, M., Fini, M., & Filardo, G. (2019). Nanoindentation: An advanced procedure to investigate osteochondral engineered tissues. *Journal of the Mechanical Behavior of Biomedical Materials*, *96*, 79–87. DOI: 10.1016/j.jmbbm.2019.04.042 PMID: 31029997

Boning, R. (2024). Review on thermal properties and reaction kinetics of Ca (OH) 2/CaO thermochemical energy storage materials. *Electrical Materials and Applications*, *1*(1), e12007. DOI: 10.1049/ema3.12007

Bonse, J., Kirner, S. V., & Krüger, J. (2020). Laser-induced periodic surface structures (LIPSS). In Handbook of Laser Micro- and Nano-Engineering (pp. 1-59). Springer. DOI: 10.1007/978-3-319-69537-2_17-2

Bonse, J., Kirner, S. V., Höhm, S., Epperlein, N., Spaltmann, D., Rosenfeld, A., & Krüger, J. (2017, February). Applications of laser-induced periodic surface structures (LIPSS). In Laser-based Micro-and Nanoprocessing XI (Vol. 10092, pp. 114-122). SPIE.

Bonse, J., Höhm, S., Kirner, S. V., Rosenfeld, A., & Krüger, J. (2012). Laser-induced periodic surface structures—A scientific evergreen. *IEEE Journal of Selected Topics in Quantum Electronics*, *23*(3), 9000615. DOI: 10.1109/JSTQE.2016.2579013

Bose, S., Dahiya, A., & Bandyopadhyay, A. (2025). Of biomaterials. *Additive Manufacturing*, 169.

Bösiger, P., Tegl, G., Richard, I. M., Le Gat, L., Huber, L., Stagl, V., & Fortunato, G. (2018). Enzyme functionalized electrospun chitosan mats for antimicrobial treatment. *Carbohydrate Polymers*, *181*, 551–559. DOI: 10.1016/j.carbpol.2017.12.002 PMID: 29254006

Boulos, A., DeFroda, S. F., Kleiner, J. E., Thomas, N., Gil, J. A., & Cruz, A. I.Jr. (2017). Inpatient orthopaedic hardware removal in children: A cross-sectional study. *Journal of Clinical Orthopaedics and Trauma*, *8*(3), 270–275. DOI: 10.1016/j.jcot.2017.06.020 PMID: 28951646

Brandao, E., Nakhoul, A., Duffner, S., Emonet, R., Garrelie, F., Habrard, A., Jacquenet, F., Pigeon, F., Sebban, M., & Colombier, J.-P. (2023). Learning complexity to guide light-induced self-organized nanopatterns. *Physical Review Letters*, *130*(22), 226201. DOI: 10.1103/PhysRevLett.130.226201 PMID: 37327436

Bredar, A. R. C., Chown, A. L., Burton, A. R., & Farnum, B. H. (2020). Electrochemical Impedance Spectroscopy of Metal Oxide Electrodes for Energy Applications. *ACS Applied Energy Materials*, *3*(1), 66–98. DOI: 10.1021/acsaem.9b01965

Breugelmans, T., Tourwé, E., Van Ingelgem, Y., Wielant, J., Hauffman, T., Hausbrand, R., Pintelon, R., & Hubin, A. (2010). Odd random phase multisine EIS as a detection method for the onset of corrosion of coated steel. *Electrochemistry Communications*, *12*(1), 2–5. DOI: 10.1016/j.elecom.2009.10.008

Bril, M., Fredrich, S., & Kurniawan, N. A. (2022). Stimuli-responsive materials: A smart way to study dynamic cell responses. *Smart Materials in Medicine*, *3*, 257–273. DOI: 10.1016/j.smaim.2022.01.010

Brown, S., Kim, H., & Camata, R. P. (2004). A Hybrid Laser/Aerosol Method for the Synthesis of Porous Nanostructured Calcium Phosphate Materials for Bone Tissue Engineering Applications. *MRS Proceedings*, *845*. DOI: 10.1557/PROC-845-AA5.35

Brown, M. S., & Arnold, C. B. (2010). Fundamentals of laser-material interaction and application to multiscale surface modification. In *Laser precision microfabrication* (pp. 91–120). Springer Berlin Heidelberg. DOI: 10.1007/978-3-642-10523-4_4

Brunello, G., Elsayed, H., & Biasetto, L. (2019). Bioactive Glass and Silicate-Based Ceramic Coatings on Metallic Implants: Open Challenge or Outdated Topic? *Materials (Basel)*, *12*(18), 2929. DOI: 10.3390/ma12182929 PMID: 31510062

Buchberger, G., Muck, M., Plamadeala, C., & Heitz, J. (2023). Laser Structuring for Biomedical Applications. In *Ultrafast Laser Nanostructuring: The Pursuit of Extreme Scales* (pp. 1105–1165). Springer International Publishing. DOI: 10.1007/978-3-031-14752-4_31

Buchman, A. (2025). Failure cases in adhesive joints and coatings. *Progress in Adhesion and Adhesives*, *9*, 317–367. DOI: 10.1002/9781394315086.ch9

Bülbül Tatbul, B. E. S. N. A. (2022). 3D-microstructuring of silicon induced by nanosecond pulsed infrared fiber laser for potential solar cell applications [Master's thesis, Middle East Technical University].

Bulgakova, N. M. (2013). Fundamentals of ultrafast laser processing. Ultrafast laser processing: from micro-to nanoscale, 99-182.

Bunaziv, I., Ren, X., Hagen, A. B., Hovig, E. W., Jevremovic, I., & Gulbrandsen-Dahl, S. (2023). Laser beam remelting of stainless steel plate for cladding and comparison with conventional CMT process. *International Journal of Advanced Manufacturing Technology*, *127*(1), 911–934. DOI: 10.1007/s00170-023-11567-y

Buser, D., Schenk, R. K., Steinemann, S., Fiorellini, J. P., Fox, C. H., & Stich, H. (1991). Influence of surface characteristics on bone integration of titanium implants. A histomorphometric study in miniature pigs. *Journal of Biomedical Materials Research*, *25*(7), 889–902. Advance online publication. DOI: 10.1002/jbm.820250708 PMID: 1918105

Butler, M. S., Henderson, I. R., Capon, R. J., & Blaskovich, M. A. T. (2023). Antibiotics in the clinical pipeline as of December 2022. *The Journal of Antibiotics*, *76*(8), 431–473. DOI: 10.1038/s41429-023-00629-8 PMID: 37291465

Butt, M. A. (2022). Thin-film coating methods: A successful marriage of high-quality and cost-effectiveness—a brief exploration. *Coatings*, *12*(8), 1115. DOI: 10.3390/coatings12081115

Byalahunasishetter, B. I., N., V., P., S., C., M., S., S., & Raghavan, R. (2025). Spatial Omics and Bioprinting. *Innovations in Precision Medicine and Genomics*, 85–110. DOI: 10.4018/979-8-3693-5787-3.ch003

Cacciotti, I. (2017). Bivalent cationic ions doped bioactive glasses: The influence of magnesium, zinc, strontium and copper on the physical and biological properties. *Journal of Materials Science*, *52*(15), 8812–8831. DOI: 10.1007/s10853-017-1010-0

Cai, A., Zhu, Y., & Qi, C. (2020). Biodegradable Inorganic Nanostructured Biomaterials for Drug Delivery. *Advanced Materials Interfaces*, *7*(20), 2000819. Advance online publication. DOI: 10.1002/admi.202000819

Cai, J., Li, F., Zhang, X., Wang, J., Yu, Z., Feng, B., & Li, Y. (2025). Application of pulsed laser deposition (PLD) technology in the preparation of two-dimensional (2D) film materials. *Materials (Basel)*, *18*(13), 2999. Advance online publication. DOI: 10.3390/ma18132999 PMID: 40649486

Calado, L. M., Taryba, M. G., Carmezim, M. J., & Montemor, M. F. (2018). Self-healing ceria-modified coating for corrosion protection of AZ31 magnesium alloy. *Corrosion Science*, *142*, 12–21. DOI: 10.1016/j.corsci.2018.06.013

Cao, J., Sun, J., Li, H., Wu, Z., & An, X. (2026). Mastering complexity toward high-performance multi-principal element alloy-based films and coatings: A review on microstructural regulation and property optimization. *Advancement of Science*, *13*(24), e14258. DOI: 10.1002/advs.202514258 PMID: 41870026

Cao, W., & Hench, L. L. (1996). Bioactive materials. *Ceramics International*, *22*(6), 493–507. DOI: 10.1016/0272-8842(95)00126-3

Cao, X., Meng, X., Zhang, L., Li, X., & Wang, C. (2020). Femtosecond laser micro/nanostructuring of zirconia ceramics for improved osteoblast responses. *Applied Surface Science*, *518*, 146178. DOI: 10.1016/j.apsusc.2020.146178

Cao, Z., Yang, X., Yang, W., Chen, F., Jiang, W., Zhan, S., Jiang, F., Li, J., Ye, C., Lang, L., Zhang, S., Feng, Z., Lai, X., Liu, Y., Mao, L., Cai, H., Teng, Y., & Xie, J. (2024). Modulation of dendritic cell function via nanoparticle-induced cytosolic calcium changes. *ACS Nano*, *18*(10), 7618–7632. DOI: 10.1021/acsnano.4c00550 PMID: 38422984

Caruso, S., Saffioti, M. R., Del Prete, A., & Rotella, G. (2025). Laser surface treatment of Ti6Al4V alloy: Finite element analysis for predicting molten pool geometry and microstructure modifications. *International Journal of Advanced Manufacturing Technology*, *137*(7–8), 3967–3976. DOI: 10.1007/s00170-025-15403-3

Catauro, M., & Vecchio Ciprioti, S. (2019). Sol-gel synthesis and characterization of hybrid materials for biomedical applications. In *Thermodynamics and Biophysics of Biomedical Nanosystems: Applications and Practical Considerations* (pp. 445–475). Springer Singapore. DOI: 10.1007/978-981-13-0989-2_13

Chandra, G. V. S. A., Biradar, S., & Raghavan, R. (2025). Recent Trends in Soft Robotics for Assistive Technologies. *Assistive Technology Solutions for Aging Adults and Individuals With Disabilities*, 119–144. DOI: 10.4018/979-8-3693-6308-9.ch006

Chang, J., & Sun, X. (2023). Laser-induced forward transfer based laser bioprinting in biomedical applications. *Frontiers in Bioengineering and Biotechnology*, *11*, 1255782. DOI: 10.3389/fbioe.2023.1255782 PMID: 37671193

Chasse, B., Xu, H., & Budhlall, B. M. (2019). In-vitro biodegradation study of poly (ε-caprolactone) films using a 3D printed helical flow prototype to simulate the physiological conditions for cardiovascular implanted devices. *Biomedical Physics & Engineering Express*, *5*(6), 065021. DOI: 10.1088/2057-1976/ab4e2b

Cheng, G. J., & Ye, C. (2009). Experiment, thermal simulation, and characterizations on transmission laser coating of hydroxyapatite on metal implant. *Journal of Biomedical Materials Research. Part A*, *92A*(1), 70–79. Advance online publication. DOI: 10.1002/jbm.a.32325 PMID: 19165793

Cheng, J., Xing, Y., Dong, E., Zhao, L., Liu, H., Chang, T., Chen, M., Wang, J., Lu, J., & Wan, J. (2022). An overview of laser metal deposition for cladding: Defect formation mechanisms, defect suppression methods and performance improvements of laser-cladded layers. *Materials (Basel)*, *15*(16), 5522. DOI: 10.3390/ma15165522 PMID: 36013658

Cheng, L., Guo, W., Cao, X., Dou, Y., Huang, L., Song, Y., Su, J., Zeng, Z., & Ye, R. (2021). Laser-induced graphene for environmental applications: Progress and opportunities. *Materials Chemistry Frontiers*, *5*(13), 4874–4891. DOI: 10.1039/D1QM00437A

Cheng, L., Wu, H., Li, J., Zhao, H., & Wang, L. (2021). Polydopamine modified ultrathin hydroxyapatite nanosheets for anti-corrosion reinforcement in polymeric coatings. *Corrosion Science*, *178*, 109064. DOI: 10.1016/j.corsci.2020.109064

Cheng, Q., Lee, B. L. P., Komvopoulos, K., Yan, Z., & Li, S. (2013). Plasma surface chemical treatment of electrospun poly (L-lactide) microfibrous scaffolds for enhanced cell adhesion, growth, and infiltration. *Tissue Engineering. Part A*, *19*(9-10), 1188–1198. DOI: 10.1089/ten.tea.2011.0725 PMID: 23281641

Chen, H., Shi, Q., & Zheng, K. (2024). Mathematical modeling of bioactive glass degradation. *Journal of Non-Crystalline Solids*, *646*, 123265. DOI: 10.1016/j.jnoncrysol.2024.123265

Chen, L., Wang, Z., Wang, L., & Li, B. (2024). Hydrodynamic expansion and plume splitting of the ultrafast laser-induced plasma of an Al-Cu alloy. *Journal of Applied Physics*, *135*(8), 083101. Advance online publication. DOI: 10.1063/1.3565061

Chen, L., Xu, P., Xu, N., Li, L., & Pang, C. (2026). Femtosecond laser-induced gradient architecture of micro-nano structures on Ti-Nb-Zr biomedical claddings: Improving the cytocompatibility of bone implants. *Materials Characterization*, *237*, 116494. DOI: 10.1016/j.matchar.2026.116494

Chen, M., Zhang, J., Qu, Q., Zhang, S., & Liu, D. (2025). Cavity-induced bubbles and pore formation of laser direct energy deposited titanium alloy: Influence of powder melting degree. *Additive Manufacturing*, *105*, 104792. DOI: 10.1016/j.addma.2025.104792

Chen, N., Tian, L., Patil, A. C., Peng, S., Yang, I. H., Thakor, N. V., & Ramakrishna, S. (2017). Neural interfaces engineered via micro- and nanostructured coatings. *Nano Today*, *14*, 59–83. DOI: 10.1016/j.nantod.2017.04.007

Chen, R., Kang, Z., Li, W., Xu, T., Wang, Y., Jiang, Q., Wang, Y., Huang, Z., Xu, X., & Huang, Z. (2024). Extracellular vesicle surface display of αPD-L1 and αCD3 antibodies via engineered late domain-based scaffold to activate T-cell anti-tumor immunity. *Journal of Extracellular Vesicles*, *13*(7), e12490. DOI: 10.1002/jev2.12490 PMID: 39051742

Chen, Z., Xing, X., Yang, Y., & Hong, M. (2024). Laser processing for medical devices. In Teoh, S.-h. (Ed.), *Designing materials for medical devices: Fundamentals* (p. 464). World Scientific., DOI: 10.1142/9789811295898_0013

Chen, Z., Zhen, Y.-H., Huang, Y., Gao, Z.-C., Sheng, H.-P., Mayrhofer, P. H., & Zhang, Z.-L. (2022). Atomic-scale understanding of the structural evolution of TiN/AlN superlattices during nanoindentation—Part I: Deformation. *Acta Materialia*, *234*, 118008. DOI: 10.1016/j.actamat.2022.118008

Cherqaoui, A., Paternoster, C., & Mantovani, D. (2026). Biodegradable metals and corrosion control: Challenges, limits and new opportunities for innovating in orthopedic fixations. *Materials (Basel)*, *19*(9), 1789. DOI: 10.3390/ma19091789 PMID: 42124196

Chichkov, B. N., Momma, C., Nolte, S., von Alvensleben, F., & Tünnermann, A. (1996). Femtosecond, picosecond and nanosecond laser ablation of solids. *Applied Physics. A, Materials Science & Processing*, *63*(2), 109–115. DOI: 10.1007/BF01567637

Chowdhury, N., Faisal, S., Kumar, A., Choudhari, A., Gupta, A. K., Kumar, A., & Kumar, A. (2024). Characterization of Hydrogel Properties in the Advancement of Bio-Tribology. In *Applications of biotribology in biomedical systems* (pp. 27–78). Springer Nature Switzerland. DOI: 10.1007/978-3-031-58327-8_2

Chowdhury, S., Yadaiah, N., Prakash, C., Ramakrishna, S., Dixit, S., Gupta, L. R., & Buddhi, D. (2022). Laser powder bed fusion: A state-of-the-art review of the technology, materials, properties & defects, and numerical modelling. *Journal of Materials Research and Technology*, *20*, 2109–2172. DOI: 10.1016/j.jmrt.2022.07.121

Chow, Y. L., Chan, C. K., & Kan, C. W. (2011). Effect of CO2 laser treatment on cotton surface. *Cellulose (London, England)*, *18*(6), 1635–1641. DOI: 10.1007/s10570-011-9603-2

Chrisey, D. B., Hubler, G. K., Perry, S. S., & Calvert, J. M. (1998). Matrix-assisted pulsed laser evaporation: A new method for thin film deposition. *Applied Physics. A, Materials Science & Processing*, *66*(2), 159–166. DOI: 10.1007/s003390050603

Chug, M. K., & Brisbois, E. J. (2022). Recent developments in multifunctional antimicrobial surfaces and applications toward advanced nitric oxide-based biomaterials. *ACS Materials Au*, *2*(4), 525–551. DOI: 10.1021/acsmaterialsau.2c00040 PMID: 36124001

Cimpoeşu, N., Trincă, L. C., Dascălu, G., Stanciu, S., Gurlui, S. O., & Mareci, D. (2016). Electrochemical Characterization of a New Biodegradable FeMnSi Alloy Coated with Hydroxyapatite-Zirconia by PLD Technique. *Journal of Chemistry*, *2016*, 1–9. DOI: 10.1155/2016/9520972

Claro, C, P. I., Pinheiro, T., Silvestre, S. L., Marques, A. C., Coelho, J., Marconcini, J. M., ... & Martins, R. (. (2022). Sustainable carbon sources for green laser-induced graphene: A perspective on fundamental principles, applications, and challenges. *Applied Physics Reviews*, *9*(4). Advance online publication. DOI: 10.1063/5.0100785

Clèries, L., Fernández-Pradas, J. M., & Morenza, J. L. (2000). Behavior in simulated body fluid of calcium phosphate coatings obtained by laser ablation. *Biomaterials*, *21*(18), 1861–1865. DOI: 10.1016/S0142-9612(00)00060-0 PMID: 10919689

Codescu, M. M., Vladescu, A., Geanta, V., Voiculescu, I., Pana, I., Dinu, M., Kiss, A. E., Braic, V., Patroi, D., Marinescu, V. E., & Iordoc, M. (2022). Zn based hydroxyapatite based coatings deposited on a novel FeMoTaTiZr high entropy alloy used for bone implants. *Surfaces and Interfaces*, *28*, 101591. DOI: 10.1016/j.surfin.2021.101591

Collaborators, A. R. (2022). Global burden of bacterial antimicrobial resistance in 2019: A systematic analysis. *Lancet*, *399*(10325), 629–655. DOI: 10.1016/S0140-6736(21)02724-0 PMID: 35065702

Constantinescu, S., Gherasim, O., Dorcioman, G., Grumezescu, V., Iosub, G., Niculescu, A.-G., Moldoveanu, E.-T., Rădulescu, D. M., & Grumezescu, A. M. (2024). MAPLE-prepared graphene oxide-based coatings for improved orthopedic screws used in knee interventions. *Romanian Journal of Morphology and Embryology = Revue Roumaine de Morphologie et Embryologie*, *65*(3), 433–442. DOI: 10.47162/RJME.65.3.05 PMID: 39529336

Córdoba-Torres, P. (2017). Relationship between constant-phase element (CPE) parameters and physical properties of films with a distributed resistivity. *Electrochimica Acta*, *225*, 592–604. DOI: 10.1016/j.electacta.2016.12.087

Cortese, Y. J., Wagner, V. E., Tierney, M., Devine, D., & Fogarty, A. (2018). Review of catheter-associated urinary tract infections and in vitro urinary tract models. *Journal of Healthcare Engineering*, *2986742*, 1–16. Advance online publication. DOI: 10.1155/2018/2986742 PMID: 30405898

Costa, B. C., Ponciano Gomes, J. A., & Soares, P. (2020). Surface texturing of titanium for biomedical applications: From mechanical to biological functionality. *Applied Surface Science*, *532*, 147464.

Costache, F., & Reif, J. (2004). Femtosecond laser induced Coulomb explosion from calcium fluoride. *Thin Solid Films*, *453-454*, 334–339. DOI: 10.1016/j.tsf.2003.11.096

Cotrut, C. M., Blidisel, A., & Vranceanu, D. M., Vladescu (Dragomir), A., Ungureanu, E., Pana, I., Dinu, M., Vitelaru, C., Parau, A. C., Pruna, V., Magurean, M. S., & Titorencu, I. (. (2024). Evaluation of the In Vitro Behavior of Electrochemically Deposited Plate-like Crystal Hydroxyapatite Coatings. *Biomimetics*, *9*(11), 704. DOI: 10.3390/biomimetics9110704 PMID: 39590276

Cui, B., Zhou, P., & Lv, Y. (2025). Research progress in and defect improvement measures for laser cladding. *Materials (Basel)*, *18*(13), 3206. DOI: 10.3390/ma18133206 PMID: 40649694

da Rocha Silva, I., da Silva Barreto, A. T., Seixas, R. S., Paes, P. N. G., do Nascimento Lunz, J., Thiré, R. M. D. S. M., & Jardim, P. M. (2023). Novel Strategy for Surface Modification of Titanium Implants towards the Improvement of Osseointegration Property and Antibiotic Local Delivery.

da Silva, M. J., Alves, W., Graeff, C. F. de O., & D'Alpino, P. H. P. (2021). Modified Synthesis and Physicochemical Characterization of a Bioglass-Based Composite for Guided Bone Regeneration. *TheScientificWorldJournal*, *2021*, 1–9. DOI: 10.1155/2021/4295433 PMID: 34899084

Da Silva, V. D. (2006). *Mechanics and strength of materials*. Springer Berlin Heidelberg. DOI: 10.1007/3-540-30813-X

Dai, S., Zhang, D., Shi, Q., Han, X., Wang, S., & Du, Z. (2013). Biomimetic fabrication and tunable wetting properties of three-dimensional hierarchical ZnO structures by combining soft lithography templated with lotus leaf and hydrothermal growth. *CrystEngComm*, *15*(25), 5040–5047. DOI: 10.1039/c3ce40238j

Dalton, J. E., & Cook, S. D. (1995). In vivo mechanical and histological characteristics of HA-coated implants vary with coating vendor. *Journal of Biomedical Materials Research*, *29*(2), 239–245. DOI: 10.1002/jbm.820290215 PMID: 7738072

Daniyan, A. A., Akpomejero, T. L., Ige, O. O., & Olubambi, P. A. (2023). Corrosion prevention of biomedical implants: Surface coating techniques perspective. *ChemistrySelect*, *8*(40), e202300223. DOI: 10.1002/slct.202300223

Das, A., & Shukla, M. (2016). Surface morphology, bioactivity, and antibacterial studies of pulsed laser deposited hydroxyapatite coatings on Stainless Steel 254 for orthopedic implant applications. *Proceedings of the Institution of Mechanical Engineers, Part L: Journal of Materials: Design and Applications, 233*. DOI: 10.1177/1464420716663029

Das, A., Majumder, S. D., Kozak, D., & Ding, C. F. (2024). Laser-Induced Forward Transfer for Biosensor Application. Mechanical Engineering in Biomedical Applications: Bio-3D Printing, Biofluid Mechanics, Implant Design, Biomaterials, Computational Biomechanics, Tissue Mechanics, 77-117.

Daskalova, A., Bliznakova, I., & Kochev, V. (2022). Surface engineering of biomedical implants using femtosecond lasers: Advances and perspectives. *Journal of Biomedical Materials Research. Part B, Applied Biomaterials*, *110*(3), 451–468. DOI: 10.1002/jbm.b.35027

Daskalova, A., Ostrowska, B., Zhelyazkova, A., Święszkowski, W., Trifonov, A., Declercq, H., Nathala, C., Szlazak, K., Lojkowski, M., Husinsky, W., & Buchvarov, I. (2018). Improving osteoblasts cells proliferation via femtosecond laser surface modification of 3D-printed poly-ε-caprolactone scaffolds for bone tissue engineering applications. *Applied Physics. A, Materials Science & Processing*, *124*(6), 413. DOI: 10.1007/s00339-018-1831-y

Das, P., Kumar, T. S., Sahu, K. K., & Gollapudi, S. (2022). Corrosion, stress corrosion cracking and corrosion fatigue behavior of magnesium alloy bioimplants. *Corrosion Reviews*, *40*(4), 289–333. DOI: 10.1515/corrrev-2021-0088

de Lange, G. L., & Donath, K. (1989). Interface between bone tissue and implants of solid hydroxyapatite or hydroxyapatite-coated titanium implants. *Biomaterials*, *10*(2), 121–125. Advance online publication. DOI: 10.1016/0142-9612(89)90044-6 PMID: 2706299

De Mey, K., De Decker, I., & Hoeksema, H.. (2025). Validity of laser speckle contrast imaging for predicting wound healing potential in burns. *Burns*, *51*(4), 107449. Advance online publication. DOI: 10.1016/j.burns.2025.107449 PMID: 40112657

De Micco, S., Bellucci, D., & Cannillo, V. (2025). The Assessment of Bioactivity and Biological Responsiveness in Bioactive Glasses and Ceramics: A Review of Available Techniques. *Materials (Basel)*, *18*(18), 4393. DOI: 10.3390/ma18184393 PMID: 41010235

De Riccardis, M. F. (2023). Green coatings: Materials, deposition processes, and applications. In *Handbook of green and sustainable nanotechnology: Fundamentals, developments and applications* (pp. 1627–1654). Springer International Publishing. DOI: 10.1007/978-3-031-16101-8_41

de Sá Guimarães, A. M. (2021). Developing an antimicrobial coating to prevent abdominal aortic graft infections (AGIs).

Degli Esposti, L., Zheng, K., Piancastelli, A., Ionescu, A. C., Adamiano, A., Boccaccini, A. R., & Iafisco, M. (2024). Composite materials of amorphous calcium phosphate and bioactive glass nanoparticles for preventive dentistry. *Ceramics International*, *50*(1), 593–602. DOI: 10.1016/j.ceramint.2023.10.137

Deng, X., Qasim, M., & Ali, A. (2021). Engineering and polymeric composition of drug-eluting suture: A review. *Journal of Biomedical Materials Research. Part A*, *109*(10), 2065–2081. DOI: 10.1002/jbm.a.37194 PMID: 33830631

Deng, Z., Liu, D., Liu, G., Xiong, Y., Xin, S., Li, S., Li, C., & Chen, T. (2023). Study on the corrosion resistance of SiC particle-reinforced hydroxyapatite-silver gradient bioactive ceramic coatings prepared by laser cladding. *Surface and Coatings Technology*, *467*, 129734. DOI: 10.1016/j.surfcoat.2023.129734

Depablos-Rivera, O., Pedraza-Hernández, R., Martínez, E., Valdivia-Barrientos, R., & Rangel, R. H.Jr. (2019). Hybrid magnetron sputtering and pulsed laser ablation for the synthesis of ZnO–Au composite thin films. *Thin Solid Films*, *687*, 137518. Advance online publication. DOI: 10.1016/j.tsf.2019.04.032

Derks, J., & Tomasi, C. (2015). Peri-implant health and disease: A systematic review of current epidemiology. *Journal of Clinical Periodontology*, *42*(S16, Suppl. 16), S158–S171. DOI: 10.1111/jcpe.12334 PMID: 25495683

Despres, L., Costil, S., Cormier, J., Villechaise, P., & Cariou, R. (2021). Impact of Laser Texturing on Ni-Based Single Crystal Superalloys. *Metals*, *11*(11), 1737. DOI: 10.3390/met11111737

Devine, D. M. (Ed.). (2017). *Bioresorbable polymers and their biomedical applications*. Smithers Rapra.

Dhafer, G., Al-Shroofy, M. N., & Al-Kaisy, H. A. (2022). Electrostatic Deposition of Bio-Composite Polymer/Hydroxyapatite Coatings on 316L Stainless Steel. *Key Engineering Materials*, *911*, 40–45. Advance online publication. DOI: 10.4028/p-40btz7

Dhinasekaran, D., Kaliaraj, G. S., Jagannathan, M., Rajendran, A. R., Prakasarao, A., Ganesan, S., & Subramanian, B. (2021). Pulsed laser deposition of nanostructured bioactive glass and hydroxyapatite coatings: Microstructural and electrochemical characterization. *Materials Science and Engineering C, 130*(April), 112459. DOI: 10.1016/j.msec.2021.112459 PMID: 34702534

Di Maio, Y., & Sedao, X. (2026). Ultrafast laser micro- and nano-processing of cylindrical objects. In *Scaling of laser processing - Making light matter.* Springer Nature.

Di Siena, M., Genna, S., Moretti, P., Ponticelli, G. S., Venettacci, S., & Russo, P. (2023). Study of the laser-material interaction for innovative hybrid structures: Thermo-mechanical characterization of polyethylene-based polymers. *Polymer Testing, 120*, 107947. DOI: 10.1016/j.polymertesting.2023.107947

Di, H., Yu, Z., Ma, Y., Zhang, C., Li, F., Lv, L., Pan, Y., Shi, H., & He, Y. (2016). Corrosion-resistant hybrid coatings based on graphene oxide–zirconia dioxide/epoxy system. *Journal of the Taiwan Institute of Chemical Engineers, 67*, 511–520. DOI: 10.1016/j.jtice.2016.08.008

Dikova, T., Vasilev, T., & Dolgov, N. (2019). Failure of ceramic coatings on cast and selective laser melted Co-Cr dental alloys under tensile test: Experiment and finite element analysis. *Engineering Failure Analysis, 105*, 1045–1054. DOI: 10.1016/j.engfailanal.2019.07.018

Dilova, T., Atanasova, G., Dikovska, A., & Nedyalkov, N. (2019). The effect of light irradiation on the gas-sensing properties of nanocomposites based on ZnO and Ag nanoparticles. *Applied Surface Science, 505*, 144625. Advance online publication. DOI: 10.1016/j.apsusc.2019.144625

Dinu, M., Bita, B., Parau, A. C., Ristoscu, C., & Negut, I. (2025). Bioactive Glass and Melittin Thin Films Deposited by MAPLE for Titanium Implant Functionalization. *Materials (Basel), 18*(10), 2410. DOI: 10.3390/ma18102410 PMID: 40429147

Dinu, M., Cotrut, C. M., Petreus, T., Hauffman, T., & Baciu, F., & Vladescu (Dragomir), A. (. (2024). Structural, electrochemical, biological, and mechanical assessment of functionally graded Cr-based multilayers for enhanced metal-ceramic bond strength in dental restorations. *Materials Chemistry and Physics, 312*, 128655. DOI: 10.1016/j.matchemphys.2023.128655

Dinu, M., Hauffman, T., Cordioli, C., Vladescu, A., Braic, M., Hubin, A., & Cotrut, C. M. (2017). Protective performance of Zr and Cr based silico-oxynitrides used for dental applications by means of potentiodynamic polarization and odd random phase multisine electrochemical impedance spectroscopy. *Corrosion Science, 115*, 118–128. DOI: 10.1016/j.corsci.2016.11.018

Dinu, M., Kiss, A. E., Parau, A. C., Braic, V., Vitelaru, C., Braic, M., Pana, I., Balaceanu, M., & Vladescu, A. (2013). Influence of Thermal Treatment on the Roughness, Corrosion Resistance and Wettability of Hydroxyapatite Films Deposited by RF Magnetron Sputtering. *Key Engineering Materials, 587*, 297–302. Advance online publication. . DOI: 10.4028/www.scientific.net/KEM.587.297

Dinu, M., Pana, I., Scripca, P., Sandu, I. G., Vitelaru, C., & Vladescu, A. (2020). Improvement of CoCr Alloy Characteristics by Ti-Based Carbonitride Coatings Used in Orthopedic Applications. *Coatings, 10*(5), 495. DOI: 10.3390/coatings10050495

Diwakar, V., Kumar Yadav, S., Yusufzai, M. Z. K., & Vashista, M. (2024). Advances in laser cladding: Material trends, applications, and future perspectives. In *Hybrid materials and advanced manufacturing* (pp. 121–142). CRC Press.

Doi, K., Nakayama, Y., & Matsuda, T. (1996). Novel compliant and tissue-permeable microporous polyurethane vascular prosthesis fabricated using an excimer laser ablation technique. *Journal of Biomedical Materials Research, 31*(1), 27–33. DOI: 10.1002/(SICI)1097-4636(199605)31:1<27::AID-JBM4>3.0.CO;2-S PMID: 8731146

Dorishetty, P., Balu, R., Athukoralalage, S. S., Greaves, T. L., Mata, J., De Campo, L., Saha, N., Zannettino, A. C. W., Dutta, N. K., & Choudhury, N. R. (2020). Tunable biomimetic hydrogels from silk fibroin and nanocellulose. *ACS Sustainable Chemistry & Engineering, 8*(6), 2375–2389. DOI: 10.1021/acssuschemeng.9b05317

Dorozhkin, S. V. (2022). Calcium orthophosphate (CaPO4)-based bioceramics: Preparation, properties, and applications. *Coatings, 12*(10), 1380. DOI: 10.3390/coatings12101380

Dorozhkin, S. V. (2023). There are over 60 ways to produce biocompatible calcium orthophosphate (CaPO4) deposits on various substrates. *Journal of Composites Science, 7*(7), 273. DOI: 10.3390/jcs7070273

Drevet, R., Fauré, J., & Benhayoune, H. (2023). Bioactive calcium phosphate coatings for bone implant applications: A review. *Coatings, 13*(6), 1091. DOI: 10.3390/coatings13061091

Drevet, R., Velard, F., Potiron, S., Laurent-Maquin, D., & Benhayoune, H. (2011). In vitro dissolution and corrosion study of calcium phosphate coatings elaborated by pulsed electrodeposition current on Ti6Al4V substrate. *Journal of Materials Science. Materials in Medicine, 22*(4), 753–761. Advance online publication. DOI: 10.1007/s10856-011-4251-5 PMID: 21290169

Drobný, P., Mercier, D., Koula, V., Škrobáková, S. I., Čaplovič, Ľ., & Sahul, M. (2021). Evaluation of Adhesion Properties of Hard Coatings by Means of Indentation and Acoustic Emission. *Coatings, 11*(8), 919. DOI: 10.3390/coatings11080919

Duan, X., Yang, Y., Zhang, T., Zhu, B., Wei, G., & Li, H. (2024). Research progress of metal biomaterials with potential applications as cardiovascular stents and their surface treatment methods to improve biocompatibility. *Heliyon, 10*(4), e25515. DOI: 10.1016/j.heliyon.2024.e25515 PMID: 38375258

Du, T., Liu, J., Dong, J., Xie, H., Wang, X., Yang, X., & Yang, Y. (2024). Multifunctional coatings of nickel-titanium implant toward promote osseointegration after operation of bone tumor and clinical application: A review. *Frontiers in Bioengineering and Biotechnology, 12*, 1325707. DOI: 10.3389/fbioe.2024.1325707 PMID: 38444648

Duta, L. (2021). In Vivo Assessment of Synthetic and Biological-Derived Calcium Phosphate-Based Coatings Fabricated by Pulsed Laser Deposition: A Review. *Coatings, 11*(1), 99. DOI: 10.3390/coatings11010099

Duta, L., Stan, G. E., Popa, A. C., Husanu, M. A., Moga, S., Socol, M., Zgura, I., Miculescu, F., Urzica, I., Popescu, A. C., & Mihailescu, I. (2016). Thickness influence on in vitro biocompatibility of titanium nitride thin films synthesized by pulsed laser deposition. *Materials (Basel), 9*(1), 38. Advance online publication. DOI: 10.3390/ma9010038 PMID: 28787846

Dutta Majumdar, J., & Manna, I. (2011). Laser material processing. *International Materials Reviews, 56*(5-6), 341–388. DOI: 10.1179/1743280411Y.0000000003

Eason, R. (Ed.). (2007). *Pulsed laser deposition of thin films: Applications-led growth of functional materials*. Wiley.

Easton, Z. H. W., Essink, M. A. J., Rodriguez Comas, L., Wurm, F. R., & Gojzewski, H. (2023). Acceleration of Biodegradation Using Polymer Blends and Composites. *Macromolecular Chemistry and Physics, 224*(6), 2200421. Advance online publication. DOI: 10.1002/macp.202200421

Elashnikov, R., Ulbrich, P., Vokatá, B., Pavlíčková, V. S., Švorčík, V., Lyutakov, O., & Rimpelová, S. (2021). Physically switchable antimicrobial surfaces and coatings: General concept and recent achievements. *Nanomaterials (Basel, Switzerland)*, *11*(11), 3083. DOI: 10.3390/nano11113083 PMID: 34835852

Eldessoky, A. E., Khalefa, M. M., & Abu-Seida, A. M. (2023). Regenerative endodontic therapy in mature teeth with necrotic pulp and apical periodontitis using two disinfection protocols. *BMC Oral Health*, *23*(1), 386. Advance online publication. DOI: 10.1186/s12903-023-02863-w PMID: 36949460

Elmowafy, E. M., Tiboni, M., & Soliman, M. E. (2019). Biocompatibility, biodegradation and biomedical applications of poly (lactic acid)/poly (lactic-co-glycolic acid) micro and nanoparticles. *Journal of Pharmaceutical Investigation*, *49*(4), 347–380. DOI: 10.1007/s40005-019-00439-x

Epperlein, N., Menzel, F., Schwibbert, K., Koter, R., Bonse, J., Sameith, J., Krüger, J., & Toepel, J. (2017). Influence of femtosecond laser produced nanostructures on biofilm growth on steel. *Applied Surface Science*, *418*, 420–424. DOI: 10.1016/j.apsusc.2017.02.174

Eraković, S., Janković, A., Ristoscu, C., Duta, L., Serban, N., Visan, A., Mihailescu, I. N., Stan, G. E., Socol, M., Iordache, O., Dumitrescu, I., Luculescu, C. R., Janaćković, Dj., & Miškovic-Stanković, V. (2014). Antifungal activity of Ag:hydroxyapatite thin films synthesized by pulsed laser deposition on Ti and Ti modified by TiO2 nanotubes substrates. *Applied Surface Science*, *293*, 37–45. DOI: 10.1016/j.apsusc.2013.12.029

Esmailzadeh, S., Aliofkhazraei, M., & Sarlak, H. (2018). Interpretation of Cyclic Potentiodynamic Polarization Test Results for Study of Corrosion Behavior of Metals: A Review. *Protection of Metals and Physical Chemistry of Surfaces*, *54*(5), 976–989. DOI: 10.1134/S207020511805026X

Espiritu, J. W. (2025). Imaging magnesium-based implants and their biological surroundings.

Estevam-Alves, R., Günther, D., Dani, S., Eckhardt, S., Roch, T., Mendonca, C. R., Cestari, I. N., & Lasagni, A. F. (2016). UV direct laser interference patterning of polyurethane substrates as tool for tuning its surface wettability. *Applied Surface Science*, *374*, 222–228. DOI: 10.1016/j.apsusc.2015.11.119

Fader, D. J., & Ratner, D. (2000). Principles of CO2/erbium laser safety. *Dermatologic Surgery*, *26*(3), 235–239. DOI: 10.1046/j.1524-4725.2000.09187.x PMID: 10759800

Fan, L., Xue, X., Wu, L., Zhang, S., Zhao, T., Wang, T., Qian, H., Xie, B., Tofil, S., & Yao, J. (2024). Regulation of High-Index Crystal Facets with Laser-Induced Periodic Surface Structures on CoFe2O4 Epitaxial Films for Ethanol Gas Sensing. *Advanced Materials Interfaces*, *11*(27), 2400217. DOI: 10.1002/admi.202400217

Farjaminejad, S., Farjaminejad, R., Hasani, M., Garcia-Godoy, F., Abdouss, M., Marya, A., Harsoputranto, A., & Jamilian, A. (2024). Advances and challenges in polymer-based scaffolds for bone tissue engineering: A path towards personalized regenerative medicine. *Polymers*, *16*(23), 3303. DOI: 10.3390/polym16233303 PMID: 39684048

Farooq, R. S., Yaqoob, K., Liaqat, U., Ahmed, R., & Baig, M. A. (2026). Pulsed laser deposition of hydroxyapatite coatings on titanium substrates with varying surface topographies. *Optics & Laser Technology*, *195*, 114557. DOI: 10.1016/j.optlastec.2025.114557

Fattahi, F. S. (2020). Poly (lactic acid) nano-structures for cartilage regeneration and joint repair: Strategies and ideas. *Recent Trends Nanosci. Technol*, *1*, 1–19.

Faure, J., Balamurugan, A., Benhayoune, H., Torres, P., Balossier, G., & Ferreira, J. M. F. (2009). Morphological and chemical characterisation of biomimetic bone like apatite formation on alkali treated Ti6Al4V titanium alloy. *Materials Science and Engineering C*, *29*(4), 1252–1257. DOI: 10.1016/j.msec.2008.09.047

Ferguson, J. I. (2023). Residual stress and microstructural evolution of composites and coatings for extreme environments [Doctoral dissertation, Purdue University].

Fernandes, C., & Taurino, I. (2022). Biodegradable molybdenum (Mo) and tungsten (W) devices: One step closer towards fully-transient biomedical implants. *Sensors (Basel)*, *22*(8), 3062. DOI: 10.3390/s22083062 PMID: 35459047

Fielding, G. A., Roy, M., Bandyopadhyay, A., & Bose, S. (2012). Antibacterial and biological characteristics of plasma sprayed silver and strontium doped hydroxyapatite coatings. *Acta Biomaterialia*, *8*(8), 3144–3152. DOI: 10.1016/j.actbio.2012.04.004 PMID: 22487928

Fitz-Gerald, J. M., Chrisey, D. B., Piqu, A., Auyeung, R. C. Y., Mohdi, R., Young, H. D., Wu, H. D., Lakeou, S., & Chung, R. (2000). Matrix assisted pulsed laser evaporation direct write (MAPLE DW): A new method to rapidly prototype active and passive electronic circuit elements. [OPL]. *Proceedings of the Materials Research Society*, *624*, 143. DOI: 10.1557/PROC-624-143

Flores, J. F., Olaya, J. J., Colás, R., Rodil, S. E., Valdez, B. S., & Fuente, I. G. (2006). Corrosion behaviour of TaN thin PVD films on steels. *Corrosion Engineering, Science and Technology*, *41*(2), 168–176. DOI: 10.1179/174327806X107941

Florian, C., Kirner, S. V., Krüger, J., & Bonse, J. (2020). Surface functionalization by laser-induced periodic surface structures. *Journal of Laser Applications*, *32*(2), 022063. DOI: 10.2351/7.0000103

Floroian, L., Florescu, M., Sima, F., Popescu-Pelin, G., Ristoscu, C., & Mihailescu, I. N. (2012). Synthesis of biomaterial thin films by pulsed laser technologies: Electrochemical evaluation of bioactive glass-based nanocomposite coatings for biomedical applications. *Materials Science and Engineering C*, *32*(5), 1152–1157. DOI: 10.1016/j.msec.2012.03.001

Floroian, L., Ristoscu, C., Candiani, G., Pastori, N., Moscatelli, M., Mihailescu, N., Negut, I., Badea, M., Gilca, M., Chiesa, R., & Mihailescu, I. N. (2017). Antimicrobial thin films based on ayurvedic plants extracts embedded in a bioactive glass matrix. *Applied Surface Science*, *417*, 224–233. DOI: 10.1016/j.apsusc.2017.02.197

Floroian, L., Ristoscu, C., Mihailescu, N., Negut, I., Badea, M., Ursutiu, D., Chifiriuc, M., Urzica, I., Dyia, H., Bleotu, C., & Mihailescu, I. (2016). Functionalized Antimicrobial Composite Thin Films Printing for Stainless Steel Implant Coatings. *Molecules (Basel, Switzerland)*, *21*(6), 740. DOI: 10.3390/molecules21060740 PMID: 27294895

Floroian, L., Samoila, C., Badea, M., Munteanu, D., Ristoscu, C., Sima, F., Negut, I., Chifiriuc, M. C., & Mihailescu, I. N. (2015). Stainless steel surface biofunctionalization with PMMA-bioglass coatings: Compositional, electrochemical corrosion studies and microbiological assay. *Journal of Materials Science. Materials in Medicine*, *26*(6), 195. DOI: 10.1007/s10856-015-5527-y PMID: 26085116

Francolini, I., & Donelli, G. (2010). Prevention and control of biofilm-based medical-device-related infections. *FEMS Immunology and Medical Microbiology*, *59*(3), 227–238. DOI: 10.1111/j.1574-695X.2010.00665.x PMID: 20412300

Frayssinet, P., Tourenne, F., Rouquet, N., Giammara, B., & Bonel, G. (1992). Biological influence of crystallinity of hydroxyapatite coatings for orthopaedic use on cell growth. In B. C (Ed.), *Thermal Spray: International Advances in Coatings Technology Conference Proceedings* (pp. 477–486). ASM International, Material Park, Ohio.

Fröjd, V., Wennerberg, A., & Franke Stenport, V. (2010). Importance of Ca2+ Modifications for Osseointegration of Smooth and Moderately Rough Anodized Titanium Implants – A Removal Torque and Histological Evaluation in Rabbit. *Clinical Implant Dentistry and Related Research*, *14*(5), 737–745. Advance online publication. DOI: 10.1111/j.1708-8208.2010.00315.x PMID: 20977616

Fu, W., Yan, X., Gurumukhi, Y., Garimella, V. S., King, W. P., & Miljkovic, N. (2022). High power and energy density dynamic phase change materials using pressure-enhanced close contact melting. *Nature Energy*, *7*(3), 270–280. DOI: 10.1038/s41560-022-00986-y

Gafner, M., Kramer, T., Remund, S. M., Holtz, R., & Neuenschwander, B. (2021). Ultrafast pulsed laser high precision micromachining of rotational symmetric parts. *Journal of Laser Applications*, *33*(1), 012053. DOI: 10.2351/7.0000324

Gajjeraman, S., Narayanan, K., Hao, J., Qin, C., & George, A. (2007). Matrix Macromolecules in Hard Tissues Control the Nucleation and Hierarchical Assembly of Hydroxyapatite. *The Journal of Biological Chemistry*, *282*(2), 1193–1204. Advance online publication. DOI: 10.1074/jbc.M604732200 PMID: 17052984

Gao, B., Sun, Z., Tong, Y., Yu, H., & Wang, F. (2025). Research progress on personalized bone implants based on additive manufacturing. *Micromachines*, *16*(12), 1339. DOI: 10.3390/mi16121339 PMID: 41470504

Gao, B., Zhao, H., Peng, L., & Sun, Z. (2022). A review of research progress in selective laser melting (SLM). *Micromachines*, *14*(1), 57. DOI: 10.3390/mi14010057 PMID: 36677118

Gao, H., Wang, Y., Ye, J., Du, B., Wang, D., Li, S., Cui, Q., Wang, S., & Zhang, T. (2025). Study on the ablation behavior of high-intensity lasers in vacuum. *Applied Sciences (Basel, Switzerland)*, *15*(2), 848. Advance online publication. DOI: 10.3390/app15020848

Garcia-Ortega, D. Y. (2024). Comprehensive treatment strategy for improving surgical resection rate of retroperitoneal sarcomas: A histology-specific approach narrative review. *Frontiers in Oncology*, *14*, 1432900. DOI: 10.3389/fonc.2024.1432900 PMID: 39435281

Gatto, M. L. (2022). Biomaterials for tissue engineering by additive manufacturing.

Gaudiuso, R. (2023). Pulsed laser deposition of carbon-based materials: A focused review of methods and results. *Processes (Basel, Switzerland)*, *11*(8), 2373. Advance online publication. DOI: 10.3390/pr11082373

Geetha, M., Singh, A. K., Asokamani, R., & Gogia, A. K. (2009). Ti based biomaterials, the ultimate choice for orthopaedic implants – A review. *Progress in Materials Science, 54*(3), 397–425. DOI: 10.1016/j.pmatsci.2008.06.004

GERING. (2023). *C.* Design Strategies for Polysaccharide Hydrogels Used in Soft Tissue Engineering.

Gerits, E., Kucharíková, S., Van Dijck, P., Erdtmann, M., Krona, A., Lövenklev, M., Fröhlich, M., Dovgan, B., Impellizzeri, F., Braem, A., Vleugels, J., Robijns, S. C. A., Steenackers, H. P., Vanderleyden, J., De Brucker, K., Thevissen, K., Cammue, B. P. A., Fauvart, M., Verstraeten, N., & Michiels, J. (2016). Antibacterial activity of a new broad-spectrum antibiotic covalently bound to titanium surfaces. *Journal of Orthopaedic Research, 34*(12), 2191–2198. DOI: 10.1002/jor.23238 PMID: 27003909

Ghanbari, H., Viatge, H., Kidane, A. G., Burriesci, G., Tavakoli, M., & Seifalian, A. M. (2009). Polymeric heart valves: New materials, emerging hopes. *Trends in Biotechnology, 27*(6), 359–367. DOI: 10.1016/j.tibtech.2009.03.002 PMID: 19406497

Gherasim, O., Grumezescu, A. M., Grumezescu, V., Andronescu, E., Negut, I., Bîrcă, A. C., Gălă eanu, B., & Hudi ă, A. (2021). Bioactive Coatings Loaded with Osteogenic Protein for Metallic Implants. *Polymers, 13*(24), 4303. DOI: 10.3390/polym13244303 PMID: 34960852

Gherasim, O., Grumezescu, A. M., Grumezescu, V., Negut, I., Dumitrescu, M. F., Stan, M. S., Nica, I. C., Holban, A. M., Socol, G., & Andronescu, E. (2021). Bioactive Coatings Based on Hydroxyapatite, Kanamycin, and Growth Factor for Biofilm Modulation. *Antibiotics (Basel, Switzerland), 10*(2), 160. DOI: 10.3390/antibiotics10020160 PMID: 33562515

Gholamzadeh, J., Fatemi, S. M., Mollaei, N., & Abedi, A. (2024). Processing of ultrafine/nano-grained microstructures through additive manufacturing techniques: A critical review. *Journal of Ultrafine Grained and Nanostructured Materials, 57*(2), 203–221.

Ghorbani, F., Li, D., Zhong, Z., Sahranavard, M., Qian, Z., Ni, S., Zhang, Z., Zamanian, A., & Yu, B. (2021). Bioprinting a cell-laden matrix for bone regeneration: A focused review. *Journal of Applied Polymer Science, 138*(8), 49888. DOI: 10.1002/app.49888

Ghosh, J., Rupanty, N. S., Khan, F., Noor, T., Jahangir, R., Mirmohammadsadeghi, S., & Islam, T. (2025). Discover Applied Sciences. *Discover, 7*, 49.

Gnilitskyi, I.. (2022). Femtosecond laser-induced surface modifications of Ti-6Al-4V alloys: Toward improved bioactivity and antibacterial performance. *Applied Surface Science*, *573*, 151511.

Gomah, M. E., Li, G., Jiahui, X., Omar, A. A., Haoran, H., & Zaki, M. M. (2024). Micro to macro-cracking mechanism in thermally treated granodiorite followed by different cooling techniques. *International Journal of Fracture*, *246*(2), 161–180. DOI: 10.1007/s10704-023-00740-9

Gonzalez-Garcia, Y., Garcia, S. J., & Mol, J. M. C. (2016). *Electrochemical Techniques for the Study of Self Healing Coatings*., DOI: 10.1007/978-94-017-7540-3_9

Goriainov, V., Cook, R., Latham, J. M., Dunlop, D. G., & Oreffo, R. O. C. (2014). Bone and metal: An orthopaedic perspective on osseointegration of metals. *Acta Biomaterialia*, *10*(10), 4043–4057. DOI: 10.1016/j.actbio.2014.06.004 PMID: 24932769

Gotfredson, K., Wennerberg, A., Johansson, C., Skovgaard, L. T., & Hjørting-Hansen, E. (1995). Anchorage of TiO2-blasted, HA-coated, and machined implants: An experimental study with rabbits. *Journal of Biomedical Materials Research*, *29*(10), 1223–1231. Advance online publication. DOI: 10.1002/jbm.820291009 PMID: 8557724

Gouyau, J., Duval, R. E., Boudier, A., & Lamouroux, E. (2021). Investigation of nanoparticle metallic core antibacterial activity: Gold and silver nanoparticles against Escherichia coli and Staphylococcus aureus. *International Journal of Molecular Sciences*, *22*(4), 1905. DOI: 10.3390/ijms22041905 PMID: 33672995

Gracheva, A., Polozov, I., & Popovich, A. (2025). Additive manufacturing of biodegradable metallic implants by selective laser melting: Current research status and application perspectives. *Metals*, *15*(7), 754. DOI: 10.3390/met15070754

Gradi teanu-Pircalabioru, G., Negut, I., Dinu, M., Parau, A. C., Bita, B., Duta, L., Ristoscu, C., & Sava, B. (2025). Enhancing orthopaedic implant efficacy: The development of cerium-doped bioactive glass and polyvinylpyrrolidone composite coatings via MAPLE technique. *Biomedical Materials (Bristol, England)*, *20*(1), 015019. DOI: 10.1088/1748-605X/ad98d5 PMID: 39612575

Graeser, H. (2023). *Still on the way: A father's journey to hope*. Wipf and Stock Publishers.

Griffis, J. C., Shahed, K., Meinert, K., Yilmaz, B., Lear, M., & Manogharan, G. (2025). Multi-material laser powder bed fusion: Effects of build orientation on defects, material structure and mechanical properties. npj. *Advances in Manufacturing*, *2*(1), 5.

Gristina, A. G. (1987). Biomaterial-Centered Infection: Microbial Adhesion Versus Tissue Integration. *Science, 237*(4822), 1588–1595. DOI: 10.1126/science.3629258 PMID: 3629258

Grohol, C. M. (2024). Laser cladding of aluminum alloys and high-fidelity modeling of the molten pool dynamics in laser melting of metals [Doctoral dissertation, Purdue University Graduate School].

Grubova, I., Priamushko, T., Chudinova, E., Surmeneva, M., Korneva, O., Epple, M., Prymak, O., Shulepov, I., & Surmenev, R. (2016). Formation and Characterization of Crystalline Hydroxyapatite Coating with the (002) Texture. *IOP Conference Series. Materials Science and Engineering, 116*, 012016. Advance online publication. DOI: 10.1088/1757-899X/116/1/012016

Grumezescu, V., & Duta, L. (2026). Collagen Type I as a biological barrier interface in biomimetic microfluidic devices: Properties, applications, and challenges. *Biomimetics, 11*(1), 66. DOI: 10.3390/biomimetics11010066 PMID: 41589983

Grumezescu, V., Grumezescu, A., Ficai, A., Negut, I., Vasile, B., Gălă eanu, B., & Hudi ă, A. (2022). Composite Coatings for Osteoblast Growth Attachment Fabricated by Matrix-Assisted Pulsed Laser Evaporation. *Polymers, 14*(14), 2934. DOI: 10.3390/polym14142934 PMID: 35890714

Grundmeier, G., Schmidt, W., & Stratmann, M. (2000). Corrosion protection by organic coatings: Electrochemical mechanism and novel methods of investigation. *Electrochimica Acta, 45*(15–16), 2515–2533. DOI: 10.1016/S0013-4686(00)00348-0

Grunenwald, A., Keyser, C., Sautereau, A. M., Crubézy, E., Ludes, B., & Drouet, C. (2014). Revisiting carbonate quantification in apatite (bio)minerals: A validated FTIR methodology. *Journal of Archaeological Science, 49*, 134–141. DOI: 10.1016/j.jas.2014.05.004

Gul, I. A., Abdul-Rani, A. M., Ahmad, A., Al-Amin, M., Aliyu, A. A., & Garba, E. (2025). Surface Wettability Response of Hydroxyapatite-Doped Coatings on Metallic Biomaterials: A Concise Review. *Diffusion and Defect Data, Solid State Data. Part B, Solid State Phenomena, 370*, 61–72. Advance online publication. DOI: 10.4028/p-c4DEVS

Guo, J., Gao, F., Zhang, F., Zhao, X., & Zhao, R. (2025). Biomineralization-Driven Advances in Materials Science and Biomedical Engineering. *JACS Au, 5*(9), 4134–4154. DOI: 10.1021/jacsau.5c00669 PMID: 41001633

Gupta, A., Holoidovsky, L., Thamaraiselvan, C., Thakur, A. K., Singh, S. P., Meijler, M. M., & Arnusch, C. J. (2019). Silver-doped laser-induced graphene for potent surface antibacterial activity and anti-biofilm action. *Chemical Communications*, *55*(48), 6890–6893. DOI: 10.1039/C9CC02415H PMID: 31134255

Gupta, N. K., Ganesan, G., Siddhartha, S., Karade, S. R., Paul, A. K., Dubey, S., Ely, R. H., & Karunakaran, K. P. (2025). In situ pre-heating in wire arc additive manufacturing: Design, development, and experimental investigation on residual stresses and metallurgical and mechanical properties. *Journal of Materials Engineering and Performance*, *34*(12), 12256–12267. DOI: 10.1007/s11665-024-10011-w

Gu, Y., Yuan, J., & Chen, L. (2023). Switching of control mechanisms during the rapid solidification of a melt pool. *Physical Review Materials*, *7*(10), 103401. DOI: 10.1103/PhysRevMaterials.7.103401

GVS, A. C., Ananthakumar, B., & Raghavan, R. (2025). Enhancing Assistive Technologies With Neuromorphic Computing. *Revolutionizing AI with Brain-Inspired Technology: Neuromorphic Computing*, 207–232. DOI: 10.4018/979-8-3693-6303-4.ch009

Habelitz, S., Marshall, G. W.Jr, Balooch, M., & Marshall, S. J. (2002). Nanoindentation and storage of teeth. *Journal of Biomechanics*, *35*(7), 995–998. DOI: 10.1016/S0021-9290(02)00039-8 PMID: 12052404

Hahn, J., Gögele, C., & Schulze-Tanzil, G. (2023). Could an Anterior Cruciate Ligament Be Tissue-Engineered from Silk? *Cells*, *12*(19), 2350. DOI: 10.3390/cells12192350 PMID: 37830564

Haider, A. J., Alawsi, T., Haider, M. J., Taha, B. A., & Marhoon, H. A. (2022). A comprehensive review on pulsed laser deposition technique to effective nanostructure production: Trends and challenges. *Optical and Quantum Electronics*, *54*(8), 488. DOI: 10.1007/s11082-022-03786-6

Hajjar, S., & Zhou, X. (2023). pH sensing at the intersection of tissue homeostasis and inflammation. *Trends in Immunology*, *44*(10), 807–825. DOI: 10.1016/j.it.2023.08.008 PMID: 37714775

Hamblin, M. R. (2017). Mechanisms and applications of the anti-inflammatory effects of photobiomodulation. *AIMS Biophysics*, *4*(3), 337–361. DOI: 10.3934/biophy.2017.3.337 PMID: 28748217

Hanawa, T. (2011). A comprehensive review of techniques for biofunctionalization of titanium. *Journal of Periodontal & Implant Science*, *41*(6), 263. DOI: 10.5051/jpis.2011.41.6.263 PMID: 22324003

Han, Y., Chen, H., Sun, Y., Liu, J., Wei, S., Xie, B., Zhang, Z., Zhu, Y., Li, M., Yang, J., Chen, W., Cao, P., & Yang, Y. (2024). Ubiquitous short-range order in multi-principal element alloys. *Nature Communications*, *15*(1), 6486. DOI: 10.1038/s41467-024-49606-1 PMID: 39090088

Haque, F. (2003). Application of Nanoindentation Development of Biomedical to Materials. *Surface Engineering*, *19*(4), 255–268. DOI: 10.1179/026708403322499173

Harilal, S. S., Brumfield, B. E., LaHaye, N. L., Hartig, K. C., & Phillips, M. C. (2018). Optical spectroscopy of laser-produced plasmas for standoff isotopic analysis. *Applied Physics Reviews*, *5*(2), 021301. Advance online publication. DOI: 10.1063/1.5016053

Harilal, S. S., Freeman, J. R., Diwakar, P. K., & Hassanein, A. (2014). Femtosecond laser ablation: Fundamentals and applications. In *Laser-Induced Breakdown Spectroscopy* (pp. 143–166). Springer., DOI: 10.1007/978-3-642-45085-3_6

Hassan, H. H., Abdelghani, E., & Amin, M. A. (2007). Inhibition of mild steel corrosion in hydrochloric acid solution by triazole derivatives. *Electrochimica Acta*, *52*(22), 6359–6366. DOI: 10.1016/j.electacta.2007.04.046

Hegemann, D., Brunner, H., & Oehr, C. (2003). Plasma treatment of polymers for surface and adhesion improvement. *Nuclear Instruments & Methods in Physics Research. Section B, Beam Interactions with Materials and Atoms*, *208*, 281–286. DOI: 10.1016/S0168-583X(03)00644-X

Heimann, R. B. (2024). Plasma-sprayed osseoconductive hydroxylapatite coatings for endoprosthetic hip implants: Phase composition, microstructure, properties, and biomedical functions. *Coatings*, *14*(7), 787. DOI: 10.3390/coatings14070787

Heitz, J., Plamadeala, C., Muck, M., Armbruster, O., Baumgartner, W., Weth, A., Steinwender, C., Blessberger, H., Kellermair, J., Kirner, S. V., Krüger, J., Bonse, J., Guntner, A. S., & Hassel, A. W. (2017). Femtosecond laser-induced microstructures on Ti substrates for reduced cell adhesion. *Applied Physics. A, Materials Science & Processing*, *123*(12), 734. DOI: 10.1007/s00339-017-1352-0

Hench, L. L. (1991). Bioceramics: From Concept to Clinic. *Journal of the American Ceramic Society*, *74*(7), 1487–1510. DOI: 10.1111/j.1151-2916.1991.tb07132.x

Hench, L. L., & Polak, J. M. (2002). Third-Generation Biomedical Materials. *Science*, *295*(5557), 1014–1017. DOI: 10.1126/science.1067404 PMID: 11834817

Herrera Hernández, H., & Ruiz Reynoso, M. A., C. Trinidad González, J., O. González Morán, C., G. Miranda Hernández, J., Mandujano Ruiz, A., Morales Hernández, J., & Orozco Cruz, R. (2020). Electrochemical Impedance Spectroscopy (EIS): A Review Study of Basic Aspects of the Corrosion Mechanism Applied to Steels. In *Electrochemical Impedance Spectroscopy*. IntechOpen. DOI: 10.5772/intechopen.94470

Hinojosa-Ventura, G., Acosta-Cuevas, J. M., Velázquez-Carriles, C. A., Navarro-López, D. E., López-Alvarez, M. Á., Ortega-de la Rosa, N. D., & Silva-Jara, J. M. (2025). From Basic to Breakthroughs: The Journey of Microfluidic Devices in Hydrogel Droplet Generation. *Gels (Basel, Switzerland)*, *11*(5), 309. DOI: 10.3390/gels11050309 PMID: 40422329

Hirschorn, B., Orazem, M. E., Tribollet, B., Vivier, V., Frateur, I., & Musiani, M. (2010). Determination of effective capacitance and film thickness from constant-phase-element parameters. *Electrochimica Acta*, *55*(21), 6218–6227. DOI: 10.1016/j.electacta.2009.10.065

Hoffmann, M., & Schwarz, U. S. (2013). A kinetic model for RNA-interference of focal adhesions. *BMC Systems Biology*, *7*(1), 2. DOI: 10.1186/1752-0509-7-2 PMID: 23311633

Hofmann, S. (2025). Segregation at grain boundaries. In *Surface segregation phenomena* (pp. 107–134). CRC Press.

Høiby, N., Bjarnsholt, T., Givskov, M., Molin, S., & Ciofu, O. (2010). Antibiotic resistance of bacterial biofilms. *International Journal of Antimicrobial Agents*, *35*(4), 322–332. DOI: 10.1016/j.ijantimicag.2009.12.011 PMID: 20149602

Homa, K., Zakrzewski, W., Dobrzyński, W., Piszko, P. J., Piszko, A., Matys, J., Wiglusz, R. J., & Dobrzyński, M. (2024). Surface functionalization of titanium-based implants with a nanohydroxyapatite layer and its impact on osteoblasts: A systematic review. *Journal of Functional Biomaterials*, *15*(2), 45. DOI: 10.3390/jfb15020045 PMID: 38391898

Hong, M. H., Lee, J. H., Jung, H. S., Shin, H., & Shin, H. (2022). Biomineralization of bone tissue: Calcium phosphate-based inorganics in collagen fibrillar organic matrices. *Biomaterials Research*, *26*(1), 42. DOI: 10.1186/s40824-022-00288-0 PMID: 36068587

Hoppe, A., Güldal, N. S., & Boccaccini, A. R. (2011). A review of the biological response to ionic dissolution products from bioactive glasses and glass-ceramics. In *Biomaterials* (Vol. 32, Number 11, pp. 2757–2774). DOI: 10.1016/j.biomaterials.2011.01.004

Hoque, M. A., Yao, C.-W., Khanal, M., & Lian, I. (2022). Tribocorrosion Behavior of Micro/Nanoscale Surface Coatings. *Sensors (Basel), 22*(24), 9974. DOI: 10.3390/s22249974 PMID: 36560343

Horbett, T. A. (2018). Fibrinogen adsorption to biomaterials. *Journal of Biomedical Materials Research. Part A, 106*(10), 2777–2788. DOI: 10.1002/jbm.a.36460 PMID: 29896846

Hsu, K., Yang, L., Medina, F., Mireles, J., Menon, M., Meyer, M., & Godfrey, D. (2026). Laser powder-bed fusion for metals. In *Additive manufacturing of metals* (pp. 139–199). Springer Nature. DOI: 10.1007/978-3-032-04961-2_4

Huang, A. G., Liu, Y., Hu, S. H., Wang, Q. H., & Zhen, L. L. (2013). Laser Cladding of Fluoridated Hydroxyapatite Coatings on Titanium Alloy for Bone Replacement Structures. *Advanced Materials Research, 815*, 540–546. Advance online publication. . DOI: 10.4028/www.scientific.net/AMR.815.540

Huang, J., Liu, Y., Jin, S., Wang, Z., Qi, Y., Zhang, J., Wang, K., & Qiu, R. (2022). Uniformity control of laser-induced periodic surface structures. *Frontiers in Physics (Lausanne), 10*, 932284. DOI: 10.3389/fphy.2022.932284

Huang, V. M., Wu, S.-L., Orazem, M. E., Pébère, N., Tribollet, B., & Vivier, V. (2011). Local electrochemical impedance spectroscopy: A review and some recent developments. *Electrochimica Acta*. Advance online publication. DOI: 10.1016/j.electacta.2011.03.018

Huang, Z., Tsui, G. C.-P., Deng, Y., & Tang, C.-Y. (2020). Two-photon polymerization nanolithography technology for fabrication of stimulus-responsive micro/nanostructures for biomedical applications. *Nanotechnology Reviews, 9*(1), 1118–1136. DOI: 10.1515/ntrev-2020-0073

Hudson, M. C., Ramp, W. K., Nicholson, N. C., Williams, A. S., & Nousiainen, M. T. (1999). Staphylococcus aureus adhesion to bone matrix and bone-associated biomaterials. *FEMS Microbiology Letters, 173*(2), 279–284. DOI: 10.1111/j.1574-6968.1999.tb13514.x PMID: 10227156

Huerta Murillo, D. (2019). Development of Laser Induced Hierarchical Surface Structures on Ti-6Al-4V using short and ultra-short pulsed lasers for modified wettability applications (Doctoral dissertation, Industriales).

Hussein, B. M. A., Safi, I. N., Al Shammari, A. M., & Tawfiq, T. A. (2019). Implementation and characterization of coating pure titanium dental implant with sintered β-TCP using Nd:YAG laser. *The Saudi Dental Journal, 31*(2), 242–250. DOI: 10.1016/j.sdentj.2018.12.004 PMID: 30983835

Hussein, M. A., Kumar, A. M., Ankah, N., & Azeem, M. A. (2021). Thermal treatment effect on the surface and in vitro corrosion characteristics of arc deposited TiN coating on Ti alloy for orthopedic applications. *Ceramics International, 47*(16), 23203–23213. DOI: 10.1016/j.ceramint.2021.05.032

Ibrahim, M. S., Seresht, H. B., Kum, C. H., Cho, J. H., Jin, G., An, S. H., Ye, S., Kim, S., Wagner, W. R., & Chun, Y. (2025). Novel laser-textured grooves extended to the sidewall edges of CoCr surfaces for rapid and selective endothelialization following coronary artery stenting. *Biomaterials, 321*, 123299. DOI: 10.1016/j.biomaterials.2025.123299 PMID: 40188719

Ikeuba, A. I., Njoku, C. N., Ekerenam, O. O., Njoku, D. I., Udoh, I. I., Daniel, E. F., Uzoma, P. C., Etim, I.-I. N., & Okonkwo, B. O. (2024). A review of the electrochemical and galvanic corrosion behavior of important intermetallic compounds in the context of aluminum alloys. *RSC Advances, 14*(43), 31921–31953. DOI: 10.1039/D4RA06070A PMID: 39385762

Imani, A. (2024). Corrosion behaviour of a biodegradable rare-earth magnesium alloy in modified simulated body fluid [Doctoral dissertation, University of British Columbia].

Irimiciuc, S. A., Nica, P., Agop, M., & Focsa, C. (2019). Target properties – Plasma dynamics relationship in laser ablation of metals: Common trends for fs, ps and ns irradiation regimes. *Applied Surface Science, 506*, 144926. Advance online publication. DOI: 10.1016/j.apsusc.2019.144926

Iskandar, M. E., Lock, J. Y., Aslani, A., & Liu, H. (2011). Controlling the Biodegradation of Magnesium Implants Through Nanostructured Coatings. *Volume 2: Biomedical and Biotechnology Engineering; Nanoengineering for Medicine and Biology*. DOI: 10.1115/IMECE2011-65901

Ivanova, E. P., Hasan, J., Webb, H. K., Truong, V. K., Watson, G. S., Watson, J. A., Baulin, V. A., Pogodin, S., Wang, J. Y., Tobin, M. J., Löbbe, C., & Crawford, R. J. (2012). Natural bactericidal surfaces: Mechanical rupture of Pseudomonas aeruginosa cells by cicada wings. *Small, 8*(16), 2489–2494. DOI: 10.1002/smll.201200528 PMID: 22674670

Ivanovska, A., Petrović, A., Lazarević-Pašti, T., Ilic-Tomic, T., Dimić-Mišić, K., Lađarević, J., & Bradić, J. (2025). Development of Bioactive Cotton, Wool, and Silk Fabrics Functionalized with Origanum vulgare L. for Healthcare and Medical Applications: An In Vivo Study. *Pharmaceutics, 17*(7), 856. DOI: 10.3390/pharmaceutics17070856 PMID: 40733065

Jadav, H., Badheka, V., Upadhyay, G., & Mehta, K. (2022). Dissimilar welding of magnesium alloy to aluminium alloy: A review. *Advances in Materials and Processing Technologies*, *8*(4), 3794–3821. DOI: 10.1080/2374068X.2021.2016564

Jadhav, N., & Gelling, V. J. (2019). Review—The Use of Localized Electrochemical Techniques for Corrosion Studies. *Journal of the Electrochemical Society*, *166*(11), C3461–C3476. DOI: 10.1149/2.0541911jes

Jalil, S. A., Akram, M., Bhat, J. A., Hayes, J. J., Singh, S. C., ElKabbash, M., & Guo, C. (2020). Creating superhydrophobic and antibacterial surfaces on gold by femtosecond laser pulses. *Applied Surface Science*, *506*, 144952. DOI: 10.1016/j.apsusc.2019.144952 PMID: 32184533

Jamesh, M., Kumar, S., & Sankara Narayanan, T. S. N. (2011). Corrosion behavior of commercially pure Mg and ZM21 Mg alloy in Ringer's solution – Long term evaluation by EIS. *Corrosion Science*, *53*(2), 645–654. DOI: 10.1016/j.corsci.2010.10.011

Jażdżewska, M., & Majkowska-Marzec, B. (2017). Hydroxyapatite deposition on the laser modified Ti13Nb13Zr alloy. *Advances in Materials Science*, *17*(4), 5–13. Advance online publication. DOI: 10.1515/adms-2017-0017

Jażdżewska, M., Majkowska-Marzec, B., Zieliński, A., Ostrowski, R., Frączek, A., Karwowska, G., & Olive, J.-M. (2023). Mechanical Properties and Wear Susceptibility Determined by Nanoindentation Technique of Ti13Nb13Zr Titanium Alloy after "Direct Laser Writing.". *Materials (Basel)*, *16*(13), 4834. DOI: 10.3390/ma16134834 PMID: 37445148

Jelinek, M., Kocourek, T., Remsa, J., Weiserová, M., Jurek, K., Miksovsky, J., Strnad, J., Galandáková, A., & Ulrichová, J. (2013). Antibacterial, cytotoxicity and physical properties of laser—Silver doped hydroxyapatite layers. *Materials Science and Engineering C*, *33*(3), 1242–1246. DOI: 10.1016/j.msec.2012.12.018 PMID: 23827567

Jeyaprakash, N., Prabu, G., & Yang, C. H. (Eds.). (2025). *Tailoring of engineering material properties through laser cladding: Metallurgical, mechanical and tribological properties*. CRC Press. DOI: 10.1201/9781032693576

Jiang, J., Wan, F., Yang, J., Hao, W., Wang, Y., Yao, J., Shao, Z., Zhang, P., Chen, J., Zhou, L., & Chen, S. (2014). Enhancement of osseointegration of polyethylene terephthalate artificial ligament by coating of silk fibroin and depositing of hydroxyapatite. *International Journal of Nanomedicine*, ●●●, 4569–4580. DOI: 10.2147/IJN.S69137 PMID: 25302023

Jiménez-Boland, D., Robles-Fernández, A., Martín-Rodríguez, A., Cuadros, M. Á., Traverso, J. Á., Sánchez-Moreno, P., & Bramini, M. (2025). Breaking Barriers in Glioblastoma Targeting through Advanced Nanoparticle Cell Membrane Coating. *ACS Applied Materials & Interfaces*, *17*(24), 35288–35303. DOI: 10.1021/acsami.5c07306 PMID: 40488690

Jin, H. W., Noumbissi, S., & Wiedemann, T. G. (2024). Comparison of zirconia implant surface modifications for optimal osseointegration. *Journal of Functional Biomaterials*, *15*(4), 91. DOI: 10.3390/jfb15040091 PMID: 38667548

John, M., Kuruveri, U. B., & Menezes, P. L. (2022). Laser cladding-based surface modification of carbon steel and high-alloy steel for extreme condition applications. *Coatings*, *12*(10), 1444. DOI: 10.3390/coatings12101444

Johns, D. A., Shivashankar, V. Y., Krishnamma, S., & Johns, M. (2014). Use of photoactivated disinfection and platelet-rich fibrin in regenerative endodontics. *Journal of Conservative Dentistry : JCD*, *17*(5), 487–490. DOI: 10.4103/0972-0707.139850 PMID: 25298655

Jonas, R. A., Schoen, F. J., Levy, R. J., & Castaneda, A. R. (1986). Biological sealants and knitted Dacron: Porosity and histological comparisons of vascular graft materials with and without collagen and fibrin glue pretreatments. *The Annals of Thoracic Surgery*, *41*(6), 657–663. DOI: 10.1016/S0003-4975(10)63083-4 PMID: 2940983

Kanaginahal, G. M., Kiran, M. C., Shahapurkar, K., Mahale, R. S., & Kakkamari, P. (2024). Automobile applications of mechanically alloyed magnesium and titanium material. In *Mechanically alloyed novel materials: Processing, applications, and properties* (pp. 379–406). Springer Nature. DOI: 10.1007/978-981-97-6504-1_16

Kan, C. W. (2008). Effects of laser irradiation on polyester textile properties. *Journal of Applied Polymer Science*, *107*(3), 1584–1589. DOI: 10.1002/app.26632

Kang, C., & Sakaguchi, M. (2025). Interfacial bonding and residual stress of single splats on solid substrates: A literature review. *Coatings*, *15*(11), 1259. DOI: 10.3390/coatings15111259

Kannan, S., Sivashanmugam, P., Shivalingam, C., & Thayumanavan, P. (2025). Tailored bioactive glass coating on titanium implants to elevate bioactivity and longevity. *Journal of Sol-Gel Science and Technology*, *115*(1), 63–72. DOI: 10.1007/s10971-025-06780-y

Kantaros, A., Ganetsos, T., Pallis, E., & Papoutsidakis, M. (2025). Biomimetic Additive Manufacturing: Engineering Complexity Inspired by Nature's Simplicity. *Biomimetics*, *10*(7), 453. DOI: 10.3390/biomimetics10070453 PMID: 40710266

Kapadia, B. H., Berg, R. A., Daley, J. A., Fritz, J., Bhave, A., & Mont, M. A. (2016). Periprosthetic joint infection. *Lancet*, *387*(10016), 386–394. DOI: 10.1016/S0140-6736(14)61798-0 PMID: 26135702

Karimi, F., O'Connor, A. J., Qiao, G. G., & Heath, D. E. (2018). Integrin clustering matters: A review of biomaterials functionalized with multivalent integrin-binding ligands to improve cell adhesion, migration, differentiation, angiogenesis, and biomedical device integration. *Advanced Healthcare Materials*, *7*(12), 1701324. DOI: 10.1002/adhm.201701324 PMID: 29577678

Kartsonakis, I. A., Athanasopoulou, E., Snihirova, D., Martins, B., Koklioti, M. A., Montemor, M. F., Kordas, G., & Charitidis, C. A. (2014). Multifunctional epoxy coatings combining a mixture of traps and inhibitor loaded nanocontainers for corrosion protection of AA2024-T3. *Corrosion Science*, *85*, 147–159. DOI: 10.1016/j.corsci.2014.04.009

Karu, T. I. (2010). Multiple roles of cytochrome c oxidase in mammalian cells under red and near-IR radiation. *Photomedicine and Laser Surgery*, *28*(2), 173–180. DOI: 10.1089/pho.2009.2541 PMID: 19743963

Kasman, Ş., & Ozan, S. (2025). Marine Shell-Inspired Laser Surface Texturing: Characterizing the Surface Properties of Co-Based Alloy. *Journal of Bionic Engineering*, *23*(1), 341–353. Advance online publication. DOI: 10.1007/s42235-025-00712-w

Kazemi, M., Ahangarani, S., Esmailian, M., & Shanaghi, A. (2020). Investigation on the corrosion behavior and biocompatibility of Ti-6Al-4V implant coated with HA/TiN dual layer for medical applications. *Surface and Coatings Technology*, *397*, 126044. DOI: 10.1016/j.surfcoat.2020.126044

Kelleher, S. M., Habimana, O., Lawler, J., O' Reilly, B., Daniels, S., Casey, E., & Cowley, A. (2016). Cicada wing surface topography: An investigation into the bactericidal properties of nanostructural features. *ACS Applied Materials & Interfaces*, *8*(24), 14966–14974. DOI: 10.1021/acsami.5b08309 PMID: 26551558

Kendig, M., & Scully, J. (1990). Basic Aspects of Electrochemical Impedance Application for the Life Prediction of Organic Coatings on Metals. *Corrosion*, *46*(1), 22–29. DOI: 10.5006/1.3585061

Khan, M. S., Ali, S., Westerbaan, D., Duley, W., Biro, E., & Zhou, Y. N. (2022). The effect of laser impingement angle on the optimization of melt pool geometry to improve process stability during high-speed laser welding of thin-gauge automotive steels. *Journal of Manufacturing Processes*, *78*, 242–253. DOI: 10.1016/j.jmapro.2022.04.022

Khatoon, Z., McTiernan, C. D., Suuronen, E. J., Mah, T.-F., & Alarcon, E. I. (2018). Bacterial biofilm formation on implantable devices and approaches to its treatment and prevention. *Heliyon*, *4*(4), e01067. DOI: 10.1016/j.heliyon.2018.e01067 PMID: 30619958

Khobragade, S. S., Deshmukh, M., Vyas, U., & Ingle, R. G. (2025). Innovative Approaches in Bone Tissue Engineering: Strategies for Cancer Treatment and Recovery. *International Journal of Molecular Sciences*, *26*(9), 3937. DOI: 10.3390/ijms26093937 PMID: 40362178

Khudair, N. Y., Al-Amir, Q. R., & Rashid, H. K. (n.d.). Simulation of the thermal Marangoni effect in a free surface cavity: Heatline and entropy visualizations.

Kim, C. S., Jung, K. H., Kim, H., Kim, C. B., & Kang, I. K. (2016). Collagen-grafted porous HDPE/PEAA scaffolds for bone reconstruction. *Biomaterials Research*, *20*(1), 23. DOI: 10.1186/s40824-016-0071-5 PMID: 27468356

Kim, H.-M., Himeno, T., Kawashita, M., Kokubo, T., & Nakamura, T. (2004). The mechanism of biomineralization of bone-like apatite on synthetic hydroxyapatite: An in vitro assessment. *Journal of the Royal Society, Interface*, *1*(1), 17–22. DOI: 10.1098/rsif.2004.0003 PMID: 16849149

Kim, M. S., Chang, H., Zheng, L., Yan, Q., Pfleger, B. F., Klier, J., Nelson, K., Majumder, E. L.-W., & Huber, G. W. (2023). A review of biodegradable plastics: Chemistry, applications, properties, and future research needs. *Chemical Reviews*, *123*(16), 9915–9939. DOI: 10.1021/acs.chemrev.2c00876 PMID: 37470246

Kim, Y., Choi, Y., Lee, S. A., Choi, W. S., & Kang, K. T. (2024). Complex oxide thin films: A review on pulsed laser epitaxy growth. *Current Applied Physics*, *68*, 113–130. DOI: 10.1016/j.cap.2024.09.001

Kirchhof, S., Strasser, A., Wittmann, H. J., Messmann, V., Hammer, N., Goepferich, A. M., & Brandl, F. P. (2015). New insights into the cross-linking and degradation mechanism of Diels–Alder hydrogels. *Journal of Materials Chemistry. B*, *3*(3), 449–457. DOI: 10.1039/C4TB01680G PMID: 32262047

Kisała, J., Gnilitskyi, I., Cieniek, B., Krzemiński, P., Marchewka, M., Barylyak, A., & Bobitski, Y. (2021). Synthesis of micro-spikes and herringbones structures by femtosecond laser pulses on a titanium plate—A new material for water organic pollutants degradation. *Materials (Basel)*, *14*(19), 5556. DOI: 10.3390/ma14195556 PMID: 34639953

Klee, D., & Höcker, H. (1999). Polymers for biomedical applications: improvement of the interface compatibility. In *Biomedical Applications Polymer Blends* (pp. 1–57). Springer Berlin Heidelberg. DOI: 10.1007/3-540-48838-3_1

Kligman, S., Ren, Z., Chung, C. H., Perillo, M. A., Chang, Y. C., Koo, H., Zheng, Z., & Li, C. (2021). The impact of dental implant surface modifications on osseointegration and biofilm formation. *Journal of Clinical Medicine, 10*(8), 1641. DOI: 10.3390/jcm10081641 PMID: 33921531

Knabe, C. (2011). Bioactivity : Mechanisms. In *Comprehensive Biomaterials*. Elsevier Ltd., DOI: 10.1016/B978-0-08-055294-1.00026-X

Kobielarz, M., Gazińska, M., Tomanik, M., Stępak, B., Szustakiewicz, K., Filipiak, J., Antończak, A., & Pezowicz, C. (2019). Physicochemical and mechanical properties of CO2 laser-modified biodegradable polymers for medical applications. *Polymer Degradation & Stability, 165*, 182–195. DOI: 10.1016/j.polymdegradstab.2019.05.010

Kokubo, T., & Takadama, H. (2008). Simulated Body Fluid (SBF) as a Standard Tool to Test the Bioactivity of Implants. *Handbook of Biomineralization: Biological Aspects and Structure Formation, 3*, 97–109. DOI: 10.1002/9783527619443.ch51

Kokubo, T., Kushitani, H., Sakka, S., Kitsugi, T., & Yamamuro, T. (1990). Solutions able to reproduce in vivo surface-structure changes in bioactive glass-ceramic A-W3. *Journal of Biomedical Materials Research, 24*(6), 721–734. DOI: 10.1002/jbm.820240607 PMID: 2361964

Kokubo, T., & Takadama, H. (2006). How useful is SBF in predicting in vivo bone bioactivity? *Biomaterials, 27*(15), 2907–2915. DOI: 10.1016/j.biomaterials.2006.01.017 PMID: 16448693

Kolaklieva, L., Grozdanov, I., Zafirova, K., Boev, V., Srebrenkoska, V., Mitrovic, M., & Krumme, A. (2019). Pulsed laser deposition of aluminum nitride films: Correlation between mechanical, optical and structural properties. *Coatings, 9*(3), 195. DOI: 10.3390/coatings9030195

Kong, L., Heydari, Z., Lami, G. H., Saberi, A., Baltatu, M. S., & Vizureanu, P. (2023). A comprehensive review of the current research status of biodegradable zinc alloys and composites for biomedical applications. *Materials (Basel), 16*(13), 4797. DOI: 10.3390/ma16134797 PMID: 37445111

König, K. (2023). Medical femtosecond laser. *Journal of the European Optical Society. Rapid Publications, 19*(2), 36. DOI: 10.1051/jeos/2023032

Konoshenko, M. Y., Lekchnov, E. A., Vlassov, A. V., & Laktionov, P. P. (2018). Isolation of extracellular vesicles: General methodologies and latest trends. *BioMed Research International, 2018*, 8545347. DOI: 10.1155/2018/8545347 PMID: 29662902

Kőrösi, L., Rodio, M., Dömötör, D., Kovács, T., Papp, S., Diaspro, A., Intartaglia, R., & Dékány, I. (2016). Ultrasmall, ligand-free Ag nanoparticles with high antibacterial activity prepared by pulsed laser ablation in liquid. *Journal of Chemistry, 4143560*, 1–8. Advance online publication. DOI: 10.1155/2016/4143560

Korrapati, C., Chowdareddy, S., Suresh, C., & Raghavan, R. (2026). Cancer Cell Membrane-Coated Nanoparticles: Therapeutic Advancements and Hybrid Systems. *Applications of Nanomaterials for Oncology*, 131–160. DOI: 10.4018/979-8-3373-3765-4.ch005

Korrapati, C., Yaraladdi, A. B., Vaddi, H., & Raghavan, R. (2024). Applications of the Saffron European Kitchen Spice Crocus sativus. *Therapeutic and Pharmacological Applications of Ethnobotany*, 98–127. DOI: 10.4018/979-8-3693-1986-4.ch005

Korshed, P., Li, L., Liu, Z., Mironov, A., & Wang, T. (2018). Antibacterial mechanisms of a novel type picosecond laser-generated silver-titanium nanoparticles and their toxicity to human cells. *International Journal of Nanomedicine, 13*, 1007–1019. DOI: 10.2147/IJN.S140222 PMID: 29317818

Kosińska, E., Sadlik, J., Tomala, A., & Sobczak-Kupiec, A. (2024). Thermal stability of titanium/hydroxyapatite systems. In *Surface modifications of titanium and its alloys for biomedical applications* (pp. 179–196). CRC Press.

Kowal, J., Arras, G., Colombo, M., Jouve, M., Morath, J. P., Primdal-Bengtson, B., Dingli, F., Loew, D., Tkach, M., & Théry, C. (2016). Proteomic comparison defines novel markers to characterize heterogeneous populations of extracellular vesicle subtypes. *Proceedings of the National Academy of Sciences of the United States of America, 113*(8), E968–E977. DOI: 10.1073/pnas.1521230113 PMID: 26858453

Kryou, C., & Zergioti, I. (2023). Laser-induced forward transfer on regenerative medicine applications. *Biomedical Materials & Devices, 1*(1), 5–20. DOI: 10.1007/s44174-022-00040-1

Kumar Yadav, Y., Kumar Singh, A., & Siddhartha. (2023). Insights to improve the tribo-performance of materials used under slurry erosion applications: A review. Proceedings of the Institution of Mechanical Engineers, Part L: Journal of Materials: Design and Applications, 237(1), 3-32.

Kumar, D., & Ansari, N. (Eds.). (2024). Magnesium alloys for biomedical applications: Advances and challenges.

Kumar, A., Byadwal, M., Kumar, A., & Kumar, A. (2023). Laser micromachining in biomedical industry. In *Laser-based technologies for sustainable manufacturing* (pp. 169–206). CRC Press. DOI: 10.1201/9781003402398-8

Kumari, R., Kumar, S., Das, A. K., Murmu, A. M., & Kanchan, K. (2024). Microstructural characterization and corrosion analysis of HA/TiO2 and HA/ZrO2 composite coating on Ti- alloy by laser cladding. *Applied Surface Science Advances, 24*, 100655. DOI: 10.1016/j.apsadv.2024.100655

Kumar, P., Rajak, D. K., & Hussain, M. (Eds.). (2026). *Advanced coating and cladding technologies*. CRC Press. DOI: 10.1201/9781003653226

Kumar, S., Nasna, P., & Ghosh, G. (2025). Recent advancement in biocompatible materials, hybrid bioactive coating, surface modification and post-processing techniques for the fabrication of biomedical implant: Critical review and future prospects. *Proceedings of the Institution of Mechanical Engineers. Part C, Journal of Mechanical Engineering Science, 239*(9), 3247–3286. DOI: 10.1177/09544062241307455

Kunz, C., Müller, F. A., & Gräf, S. (2018). Multifunctional hierarchical surface structures by femtosecond laser processing. *Materials (Basel), 11*(5), 789. DOI: 10.3390/ma11050789 PMID: 29757240

Kurakula, M., & Koteswara Rao, G. S. N. (2020). Moving polyvinyl pyrrolidone electrospun nanofibers and bioprinted scaffolds toward multidisciplinary biomedical applications. *European Polymer Journal, 136*, 109919. DOI: 10.1016/j.eurpolymj.2020.109919

Kurdi, A., Tabbakh, T., & Basak, A. K. (2023). Microstructural and Nanoindentation Investigation on the Laser Powder Bed Fusion Stainless Steel 316L. *Materials (Basel), 16*(17), 5933. DOI: 10.3390/ma16175933 PMID: 37687627

Kwak, H., Gao, J., An, Y., Kasule, M., Doyle, P., Kleinhammes, A., & Wu, Y. (2022, February 21). Connection Between Wettability and NMR-Detected Isotherms. *International Petroleum Technology Conference*. DOI: 10.2523/IPTC-22408-EA

Kwiecień, K., Pina, S., Oliveira, J. M., Reis, R. L., & Pamuła, E. (2026). Bioceramics for osteochondral tissue engineering and regeneration. In *Osteochondral tissue engineering* (pp. 1–23). Springer Nature. DOI: 10.1007/978-3-031-80976-7_21-1

Kylychbekov, S., Abdisatarov, B., Allamyradov, Y., Annamuradov, B., Duta, L., & Er, A. O. (2026). The mechanism of hydroxyapatite coatings degradation at high substrate temperatures. *Journal of Applied Physics, 139*(6), 065304. Advance online publication. DOI: 10.1063/5.0297660

Kyriakides, T. R., Raj, A., Tseng, T. H., Xiao, H., Nguyen, R., Mohammed, F. S., Halder, S., Xu, M., Wu, M. J., Bao, S., & Sheu, W. C. (2021). Biocompatibility of nanomaterials and their immunological properties. *Biomedical Materials (Bristol, England), 16*(4), 042005. DOI: 10.1088/1748-605X/abe5fa PMID: 33578402

Lai, W.-F. (2025). Design and Applications of Polymersomes for Oral Drug Administration. *ACS Applied Materials & Interfaces, 17*(21), 30423–30435. DOI: 10.1021/acsami.5c04658 PMID: 40370090

Lakshmaiah, A., & Raghavan, R. (2026). Cardiovascular Protector Activity of Medicinal Mushrooms. *Medicinal Mushrooms: Conservation and Bioprospecting*, 209–218. DOI: 10.1007/978-1-0716-4993-0_28

Lakshmaiah, A., Korrapati, C., Suresh, C., & Raghavan, R. (2025). Synthetic Biology and AI. *Spatially Variable Genes in Cancer*, 265–290. DOI: 10.4018/979-8-3693-7728-4.ch010

Lampin, M., Warocquier-Clérout, R., Legris, C., Degrange, M., & Sigot-Luizard, M. F. (1997). Correlation between substratum roughness and wettability, cell adhesion, and cell migration. *Journal of Biomedical Materials Research: An Official Journal of The Society for Biomaterials and The Japanese Society for Biomaterials, 36*(1), 99–108. DOI: 10.1002/(SICI)1097-4636(199707)36:1<99::AID-JBM12>3.0.CO;2-E PMID: 9212394

Lam, T. N., Lee, A., Chiu, Y. R., Kuo, H. F., Kawasaki, T., Harjo, S., Jain, J., Lee, S. Y., & Huang, E. W. (2022). Estimating fine melt pool, coarse melt pool, and heat affected zone effects on the strengths of additive manufactured AlSi10Mg alloys. *Materials Science and Engineering A, 856*, 143961. DOI: 10.1016/j.msea.2022.143961

Lanzafame, R. J., Brondon, P., Stadler, I., DeVore, D. P., Soltz, R., & Soltz, B. A. (2005). Histologic assessment of mesh fixation following laser-assisted tissue soldering in a lapine model. *Lasers in Surgery and Medicine, 37*(2), 130–137. DOI: 10.1002/lsm.20201 PMID: 16044423

Lapitskaya, V., Nikolaev, A., Khabarava, A., Sadyrin, E., Antipov, P., Abdulvakhidov, K., Aizikovich, S., & Chizhik, S. (2023). The influence of nitrogen flow on the stoichiometric composition, structure, mechanical, and microtribological properties of TiN coatings. *Materials (Basel), 17*(1), 120. Advance online publication. DOI: 10.3390/ma17010120 PMID: 38203974

Lasagni, A. F., Gachot, C., Trinh, K. E., Hans, M., Rosenkranz, A., Roch, T., & Mücklich, F. (2017, March). Direct laser interference patterning, 20 years of development: from the basics to industrial applications. In *Laser-based micro-and nanoprocessing XI* (Vol. 10092, pp. 186–196). SPIE.

Lata, S., Mondal, S., Mishra, R., Priya, S., Deshwal, R. K., Thapliyal, S., Talukdar, N., Dhasmana, A., Rustagi, S., Bora, J., & Malik, S. (2025). Advancements in biodegradable implants: Evaluation in large animal surgical models. *Annals of Medicine and Surgery (London), 87*(10), 6480–6495. DOI: 10.1097/MS9.0000000000003578 PMID: 41181418

Lawrence, B. J., & Madihally, S. V. (2008). Cell colonization in degradable 3D porous matrices. *Cell Adhesion & Migration, 2*(1), 9–16. DOI: 10.4161/cam.2.1.5884 PMID: 19262124

Lawrence, M., & Jiang, Y. (2017). Porosity, pore size distribution, micro-structure. In *Bio-aggregates based building materials: state-of-the-art report of the RILEM technical committee 236-BBM* (pp. 39–71). Springer Netherlands. DOI: 10.1007/978-94-024-1031-0_2

Lee, B. L. P., Jeon, H., Wang, A., Yan, Z., Yu, J., Grigoropoulos, C., & Li, S. (2012). Femtosecond laser ablation enhances cell infiltration into three-dimensional electrospun scaffolds. *Acta Biomaterialia, 8*(7), 2648–2658. DOI: 10.1016/j.actbio.2012.04.023 PMID: 22522128

Lee, S. J., Lee, D., Yoon, T. R., Kim, H. K., Jo, H. H., Park, J. S., Lee, J. H., Kim, W. D., Kwon, I. K., & Park, S. A. (2016). Surface modification of 3D-printed porous scaffolds via mussel-inspired polydopamine and effective immobilization of rhBMP-2 to promote osteogenic differentiation for bone tissue engineering. *Acta Biomaterialia, 40*, 182–191. DOI: 10.1016/j.actbio.2016.02.006 PMID: 26868173

Lee, S. W., Phillips, K. S., Gu, H., Kazemzadeh-Narbat, M., & Ren, D. (2021). How microbes read the map: Effects of implant topography on bacterial adhesion and biofilm formation. *Biomaterials, 268*, 120595. DOI: 10.1016/j.biomaterials.2020.120595 PMID: 33360301

Lee, S. Y., Shim, E. S., Kang, H. S., Pang, S. S., & Kang, J. S. (2005). Fabrication of ZnO thin film diode using laser annealing. *Thin Solid Films, 473*(1), 31–34. DOI: 10.1016/j.tsf.2004.06.194

Lei, B., Wang, L., Chen, X., & Chae, S.-K. (2013). Biomimetic and molecular level-based silicate bioactive glass-gelatin hybrid implants for loading-bearing bone fixation and repair. *Journal of Materials Chemistry. B, 1*(38), 5153. Advance online publication. DOI: 10.1039/c3tb20941e PMID: 32261107

Leśniak, D., Zasadziński, J., Libura, W., Gronostajski, Z., Śliwa, R., Leszczyńska-Madej, B., Kaszuba, M., Woźnicki, A., Płonka, B., Widomski, P., & Madura, J. (2024). Latest advances in extrusion processes of light metals. *Archives of Civil and Mechanical Engineering*, *24*(3), 184. DOI: 10.1007/s43452-024-00988-5

Le, T. S. D., Phan, H. P., Kwon, S., Park, S., Jung, Y., Min, J., Chun, B. J., Yoon, H., Ko, S. H., Kim, S.-W., & Kim, Y. J. (2022). Recent advances in laser-induced graphene: Mechanism, fabrication, properties, and applications in flexible electronics. *Advanced Functional Materials*, *32*(48), 2205158. DOI: 10.1002/adfm.202205158

Levard, C., Hotze, E. M., Lowry, G. V., & Brown, G. E. (2012). Environmental transformations of silver nanoparticles: Impact on stability and toxicity. In *Environmental Science and Technology*. DOI: 10.1021/es2037405

Liang, J., Cai, Z., He, X., Luo, Y., Zheng, D., Sun, S., Liu, Q., Li, L., Chu, W., Alfaifi, S., Luo, F., Yao, Y., Tang, B., & Sun, X. (2024). Electroreduction of alkaline/natural seawater: Self-cleaning Pt/carbon cathode and on-site co-synthesis of H2 and Mg hydroxide nanoflakes. *Chem*, *10*(10), 3067–3087. DOI: 10.1016/j.chempr.2024.05.018

Li, C. J., Luo, X. T., Dong, X. Y., Zhang, L., & Li, C. X. (2022). Recent research advances in plasma spraying of bulk-like dense metal coatings with metallurgically bonded lamellae. *Journal of Thermal Spray Technology*, *31*(1), 5–27. DOI: 10.1007/s11666-022-01327-x PMID: 37520913

Li, C. J., Luo, X. T., Yao, S. W., Li, G. R., Li, C. X., & Yang, G. J. (2022). The bonding formation during thermal spraying of ceramic coatings: A review. *Journal of Thermal Spray Technology*, *31*(4), 780–817. DOI: 10.1007/s11666-022-01379-z

Li, C., Wang, F., Douglas, G., Zhang, Z., Guidoin, R., & Wang, L. (2017). Comprehensive mechanical characterization of PLA fabric combined with PCL to form a composite structure vascular graft. *Journal of the Mechanical Behavior of Biomedical Materials*, *69*, 39–49. DOI: 10.1016/j.jmbbm.2016.11.005 PMID: 28038405

Li, D., Dai, D., Xiong, G., Lan, S., & Zhang, C. (2023). Composite nanocoatings of biomedical magnesium alloy implants: Advantages, mechanisms, and design strategies. *Advancement of Science*, *10*(18), 2300658. DOI: 10.1002/advs.202300658 PMID: 37097626

Li, F. M., Huang, L., Zaman, S., Guo, W., Liu, H., Guo, X., & Xia, B. Y. (2022). Corrosion chemistry of electrocatalysts. *Advanced Materials*, *34*(52), 2200840. DOI: 10.1002/adma.202200840 PMID: 35334145

Liguori, A., Capasso, L., & Di Benedetto, A.. (2019). The pulsed electron deposition technique for biomedical applications: A review. *Coatings*, *9*(1), 16. Advance online publication. DOI: 10.3390/coatings10010016

Li, J., Ding, J., Liu, T., Liu, J. F., Yan, L., & Chen, X. (2017). Poly (lactic acid) controlled drug delivery. In *Industrial Applications of Poly (lactic acid)* (pp. 109–138). Springer International Publishing. DOI: 10.1007/12_2017_11

Li, J., Wang, C., Wu, Y., Fan, E., Zhu, G., Zhang, X., Zou, Y., & Shi, S. (2025). Advances in inside-laser material feeding additive manufacturing technology. *International Journal of Advanced Manufacturing Technology*, *140*(11), 5713–5736. DOI: 10.1007/s00170-025-16549-w

Li, J., Zhang, F., Lyu, H., Yin, P., Shi, L., Li, Z., Zhang, L., Di, C., & Tang, P. (2024). Evolution of musculoskeletal electronics. *Advanced Materials*, *36*(28), 2303311. DOI: 10.1002/adma.202303311 PMID: 38561020

Li, J., Zhou, P., Attarilar, S., & Shi, H. (2021). Innovative surface modification procedures to achieve micro/nano-graded Ti-based biomedical alloys and implants. *Coatings*, *11*(6), 647. Advance online publication. DOI: 10.3390/coatings11060647

Li, K., Zhang, M., Zhang, Z., Jin, P., Wang, Y., Yan, W., Zhu, L., Zhang, D. Z., & Murr, L. E. (2025). High performance realization of functionally graded materials based on integrated optimal design and additive manufacturing: A review. *International Materials Reviews*, *70*(6), 497–547. DOI: 10.1177/09506608251354889

Li, M., Huang, K., & Yi, X. (2023). Crack formation mechanisms and control methods of laser cladding coatings: A review. *Coatings*, *13*(6), 1117. DOI: 10.3390/coatings13061117

Li, M., Zhao, Z., Yang, Q., Wei, Y., & Li, J. (2023). Investigation of the effect of laser energy density on selective laser sintering of PA12 using a multi-physics field simulation method. *International Journal of Advanced Manufacturing Technology*, *129*(5), 1987–1998. DOI: 10.1007/s00170-023-12434-6

Lin, J.-S., Tsai, T.-B., Say, W.-C., Chiu, C., & Chen, S.-H. (2017). In vitro study of electrodeposited fluoridated hydroxyapatite coating on G-II titanium with a nanostructured TiO2 interlayer. *Biomedical Materials (Bristol, England)*, *12*(2), 025018. Advance online publication. DOI: 10.1088/1748-605X/aa6264 PMID: 28374679

Lin, Y., Zhou, Z., & Li, K. Y. (2019). Improved wear resistance at high contact stresses of hydrogen-free diamond-like carbon coatings by carbon/carbon multilayer architecture. *Applied Surface Science*, *477*, 137–146. DOI: 10.1016/j.apsusc.2017.11.118

Lin, Z., & Hong, M. (2021). *Femtosecond laser precision engineering: from micron, submicron, to nanoscale*. Ultrafast Science.

Lin, Z., Huang, T., Ye, X., Zhong, M., Li, L., Jiang, J., Zhang, W., Fan, L., & Zhu, H. (2013). Thinning of large-area graphene film from multilayer to bilayer with a low-power CO2 laser. *Nanotechnology*, *24*(27), 275302. DOI: 10.1088/0957-4484/24/27/275302 PMID: 23764487

Lipovka, A., Fatkullin, M., Shchadenko, S., Petrov, I., Chernova, A., Plotnikov, E., Menzelintsev, V., Li, S., Qiu, L., Cheng, C., Rodriguez, R. D., & Sheremet, E. (2023). Textile electronics with laser-induced graphene/polymer hybrid fibers. *ACS Applied Materials & Interfaces*, *15*(32), 38946–38955. DOI: 10.1021/acsami.3c06968 PMID: 37466067

Li, T., Zhang, L., Qu, X., & Lei, B. (2025). Advanced thermoactive nanomaterials for thermomedical tissue regeneration: Opportunities and challenges. *Small Methods*, *9*(1), 2400510. DOI: 10.1002/smtd.202400510 PMID: 39588862

Li, T., Zhi, D. D., Guo, Z. H., Li, J. Z., Chen, Y., & Meng, F. B. (2022). 3D porous biomass-derived carbon materials: Biomass sources, controllable transformation and microwave absorption application. *Green Chemistry*, *24*(2), 647–674. DOI: 10.1039/D1GC02566J

Liu, H., Sun, Z., Chen, Y., Zhang, W., Chen, X., & Wong, C. P. (2022). Laser processing of flexible in-plane micro-supercapacitors: Progresses in advanced manufacturing of nanostructured electrodes. *ACS Nano*, *16*(7), 10088–10129. DOI: 10.1021/acsnano.2c02812 PMID: 35786945

Liu, R., Suárez, J. M., Weisblum, B., Gellman, S. H., & McBride, S. M. (2014). Synthetic polymers active against Clostridium difficile vegetative cell growth and spore outgrowth. *Journal of the American Chemical Society*, *136*(41), 14498–14504. DOI: 10.1021/ja506798e PMID: 25279431

Liu, S., Zeng, T. H., Hofmann, M., Burcombe, E., Wei, J., Jiang, R., Kong, J., & Chen, Y. (2011). Antibacterial activity of graphite, graphite oxide, graphene oxide, and reduced graphene oxide: Membrane and oxidative stress. *ACS Nano*, *5*(9), 6971–6980. DOI: 10.1021/nn202451x PMID: 21851105

Liu, X. (2025). Review of protective coatings for corrosion mitigation in chemical machinery: Performance and mechanical aspects. *Journal of Adhesion Science and Technology*, •••, 1–50.

Liu, X., Chu, P. K., & Ding, C. (2019). Surface modification of titanium, titanium alloys, and related materials for biomedical applications. *Materials Science and Engineering R Reports*, *47*(3-4), 49–121. DOI: 10.1016/j.mser.2004.11.001

Liu, X., Xiong, J., Lv, Y., & Zuo, Y. (2009). Study on corrosion electrochemical behavior of several different coating systems by EIS. *Progress in Organic Coatings*, *64*(4), 497–503. DOI: 10.1016/j.porgcoat.2008.08.012

Liu, Y., Rui, Z., Cheng, W., Song, L., Xu, Y., Li, R., & Zhang, X. (2021). Characterization and evaluation of a femtosecond laser-induced osseointegration and an anti-inflammatory structure generated on a titanium alloy. *Regenerative Biomaterials*, *8*(2), rbab006. DOI: 10.1093/rb/rbab006 PMID: 33738120

Liu, Y., Wang, S., Huo, J., Zhang, X., Wen, H., Zhang, D., Zhao, Y., Kang, D., Guo, W., & Ngo, H. H. (2024). Adsorption recovery of phosphorus in contaminated water by calcium modified biochar derived from spent coffee grounds. *The Science of the Total Environment*, *909*, 168426. DOI: 10.1016/j.scitotenv.2023.168426 PMID: 37944608

Liu, Z., Ma, S., Duan, S., Xuliang, D., Sun, Y., Zhang, X., Xu, X., Guan, B., Wang, C., Hu, M., Qi, X., Zhang, X., & Gao, P. (2016). Modification of Titanium Substrates with Chimeric Peptides Comprising Antimicrobial and Titanium-Binding Motifs Connected by Linkers To Inhibit Biofilm Formation. *ACS Applied Materials & Interfaces*, *8*(8), 5124–5136. DOI: 10.1021/acsami.5b11949 PMID: 26863404

Li, W., Yang, L., Huang, J., Zheng, C., Chen, Y., Li, Y., Yang, D., Li, S., Chen, Z., Cai, W., & Lai, Y. (2024). Progress on fiber engineering for fabric innovation in ecological hydrophobic design and multifunctional applications. *Industrial Chemistry & Materials*, *2*(3), 393–423. DOI: 10.1039/D4IM00048J

Li, X., Wang, M., Xu, L., Xu, T., Wu, W., Pan, S., Wang, C., Zhang, W., & Cai, X. (2024). Progress in indentation test for material characterization: A systematic review. *Results in Surfaces and Interfaces*, *17*, 100358. DOI: 10.1016/j.rsurfi.2024.100358

Li, Y., Fischer, R., Zboray, R., Boillat, P., Camenzind, M., Toncelli, C., & Rossi, R. M. (2020). Laser-engraved textiles for engineering capillary flow and application in microfluidics. *ACS Applied Materials & Interfaces*, *12*(26), 29908–29916. DOI: 10.1021/acsami.0c03988 PMID: 32506905

Logesh, M., Toan, N. K., Ahn, S.-G., & Choe, H.-C. (2024). Engineering bioactive MWCNT-reinforced hydroxyapatite coatings on Ti29Nb5Zr alloy by plasma electrolytic oxidation method. *Journal of Alloys and Compounds*, *1006*, 176339. DOI: 10.1016/j.jallcom.2024.176339

Lopera, H. A. C., & Godoy, M. A. M. (2025). Orthophosphates synthetized from calcium and calcium magnesium phosphates, types of resources and applications: A review. *ES Materials and Manufacturing*, *31*, 1997.

Lorusso, A., Campanella, D., Di Fonzo, F., Loconsole, A., Girolami, M., & Rea, M.. (2024). The potential of spot size control in shaping the thickness distribution of PLD films. *Materials (Basel)*, *17*(11), 2712. Advance online publication. DOI: 10.3390/ma17112712 PMID: 38893975

Lötvall, J., Hill, A. F., Hochberg, F., Buzás, E. I., Di Vizio, D., Gardiner, C., Gho, Y. S., Kurochkin, I. V., Mathivanan, S., Quesenberry, P., Sahoo, S., Tahara, H., Wauben, M. H., Witwer, K. W., & Théry, C. (2014). Minimal experimental requirements for definition of extracellular vesicles and their functions: A position statement from the International Society for Extracellular Vesicles. *Journal of Extracellular Vesicles*, *3*(1), 26913. DOI: 10.3402/jev.v3.26913 PMID: 25536934

Low, Y. J., Andriyana, A., Ang, B. C., & Zainal Abidin, N. I. (2020). Bioresorbable and degradable behaviors of PGA: Current state and future prospects. *Polymer Engineering and Science*, *60*(11), 2657–2675. DOI: 10.1002/pen.25508

Lu, G., Ni, E., Jiang, Y., Wu, W., & Li, H. (2024). Room-temperature liquid metals for flexible electronic devices. *Small*, *20*(9), 2304147. DOI: 10.1002/smll.202304147 PMID: 37875665

Lu, J., Wu, X., Wang, Z., Zhang, C., Zeng, Q., Wang, C., & Wen, W. (2025). Study on cooling category and heat transfer mechanisms of piston oscillating cooling in a dual gallery structure. *Applied Thermal Engineering*, *276*, 126862. DOI: 10.1016/j.applthermaleng.2025.126862

Lu, K., Xu, Y., Gao, S., Chen, B., Zhu, D., Zhang, Y., Chen, W., Liu, D., Dong, L., Wang, J., Chen, S., & Li, B. (2025). Biomimetic design of advanced ceramics for hard tissue repair. *Journal of the American Ceramic Society*, *108*(11), e20642. DOI: 10.1111/jace.20642

Lu, K., Zhu, J., Guo, D., Yang, M., Sun, H., Wang, Z., Hui, X., & Wu, Y. (2022). Microstructures, corrosion resistance and wear resistance of high-entropy alloys coatings with various compositions prepared by laser cladding: A review. *Coatings*, *12*(7), 1023. DOI: 10.3390/coatings12071023

Lutz, A., van den Berg, O., Wielant, J., De Graeve, I., & Terryn, H. (2016). A Multiple-Action Self-Healing Coating. *Frontiers in Materials*, *2*. Advance online publication. DOI: 10.3389/fmats.2015.00073

Luu, C. H., Nguyen, N. T., & Ta, H. T. (2024). Unravelling surface modification strategies for preventing medical device-induced thrombosis. *Advanced Healthcare Materials*, *13*(1), 2301039. DOI: 10.1002/adhm.202301039 PMID: 37725037

Lu, X., Chang, L., Tran, M. A., Komljenovic, T., Bowers, J. E., & Srinivasan, K. (2024). Emerging integrated laser technologies in the visible and short near-infrared regimes. *Nature Photonics*, *18*(10), 1010–1023. DOI: 10.1038/s41566-024-01529-5

Lu, X., Feng, X., Werber, J. R., Chu, C., Zucker, I., Kim, J. H., Osuji, C. O., & Elimelech, M. (2017). Enhanced antibacterial activity through the controlled alignment of graphene oxide nanosheets. *Proceedings of the National Academy of Sciences of the United States of America*, *114*(46), E9793–E9801. DOI: 10.1073/pnas.1710996114 PMID: 29078354

Lu, X., Zhao, Y., Peng, X., Lu, C., Wu, Z., Xu, H., Qin, Y., Xu, Y., Wang, Q., Hao, Y., & Geng, D. (2025). Comprehensive overview of interface strategies in implant osseointegration. *Advanced Functional Materials*, *35*(16), 2418849. DOI: 10.1002/adfm.202418849

Lu, Y., Chen, L., Wu, Z., Zhou, P., Dai, J., Li, J., Wen, Q., Fan, Y., Zeng, F., Chen, Y., & Fu, S. (2023). Self-driven bioactive hybrids co-deliver doxorubicin and indocyanine green nanoparticles for chemo/photothermal therapy of breast cancer. *Biomedicine and Pharmacotherapy*, *169*, 115846. DOI: 10.1016/j.biopha.2023.115846 PMID: 37944443

M, M. K., P, D. A., J, S. P., Moger, S. S., & Raghavan, R. (2026). Spatial Multi-Omics in Cancers: Integrating Molecular Landscapes for Precision Oncology. *Spatially Resolved Single-Cell and Multiple Omics in Human Diseases*, 217–248. DOI: 10.4018/979-8-3373-4849-0.ch007

Maalouf, M., Abou Khalil, A., Di Maio, Y., Papa, S., Sedao, X., Dalix, E., Peyroche, S., Guignandon, A., & Dumas, V. (2022). Polarization of femtosecond laser for titanium alloy nanopatterning influences osteoblastic differentiation. *Nanomaterials (Basel, Switzerland)*, *12*(10), 1619. DOI: 10.3390/nano12101619 PMID: 35630841

Maalouf, M., Maio, Y. D., Pallarés-Aldeiturriaga, D., Sedao, X., Hivert, L., Papa, S., Dalix, E., Thomas, M., Guignandon, A., & Dumas, V. (2025). Femtosecond laser upscaling strategy and biological validation for dental screws with improved osteogenic performance. *Surfaces and Interfaces*, *66*, 106546. DOI: 10.1016/j.surfin.2025.106546

Madelat, N., Wouters, B., Jiryaeisharahi, Z., Hubin, A., Terryn, H., & Hauffman, T. (2023). An ORP-EIS approach to distinguish the contribution of the buried interface to the electrochemical behaviour of coated aluminium. *Electrochimica Acta*, *455*, 142400. DOI: 10.1016/j.electacta.2023.142400

Madhuri, A., Parthasarathy, S., & Swain, B. P. (2025). Mechanism of the wettability of coating materials. In *Advances in mechanical coating: Processing, properties and applications* (pp. 137–163). Springer Nature. DOI: 10.1007/978-981-96-7484-8_5

Magyari, K., Baia, L., Vulpoi, A., Simon, S., Popescu, O., & Simon, V. (2014). Bioactivity evolution of the surface functionalized bioactive glasses. *Journal of Biomedical Materials Research. Part B, Applied Biomaterials*, *103*(2), 261–272. Advance online publication. DOI: 10.1002/jbm.b.33203 PMID: 24820252

Mahajan, U., Bhatia, H., Singh, A., Gupta, R., & Mahapatra, P. S. (2024). Titanium nitride as a promising alternative to plasmonic metals: Properties, processing, and applications. *Materials Advances*, *5*(7), 3642–3667.

Mahamood, R. M. (2017). *Laser metal deposition process of metals, alloys, and composite materials*. Springer., DOI: 10.1007/978-3-319-64985-6

Maheshwaran, S., Kannan, A. R., & Shanmugam, N. S. (2026). Microstructure and mechanical performance of Inconel 617 thin wall fabricated via wire arc additive manufacturing. *Welding in the World*, *70*(2), 503–519. DOI: 10.1007/s40194-025-02242-9

Mahmoud, Z. H., AL-Salman, H. N. K., & Kianfar, E. (2024). Nanoindentation: Introduction and applications of a non-destructive analysis. *Nano TransMed*, *3*, 100057. DOI: 10.1016/j.ntm.2024.100057

Majkowska-Marzec, B., Tęczar, P., Bartmański, M., Bartosewicz, B., & Jankiewicz, B. J. (2020). Mechanical and Corrosion Properties of Laser Surface-Treated Ti13N-b13Zr Alloy with MWCNTs Coatings. *Materials (Basel)*, *13*(18), 3991. DOI: 10.3390/ma13183991 PMID: 32916961

Ma, K., Cheng, Y., Jeyaprakash, N., Zhou, J., Wan, Y., & Yang, W. (2023). Temperature gradient and solidification rate simulation model of the microstructure of laser-cladded 27SiMn. *Metals*, *13*(10), 1682. DOI: 10.3390/met13101682

Maleki, H., Azimi, B., Ismaeilimoghadam, S., & Danti, S. (2022). Poly (lactic acid)-based electrospun fibrous structures for biomedical applications. *Applied Sciences (Basel, Switzerland)*, *12*(6), 3192. DOI: 10.3390/app12063192

Malinauskas, M., Farsari, M., Piskarskas, A., & Juodkazis, S. (2013). Ultrafast laser nanostructuring of photopolymers: A decade of advances. *Physics Reports*, *533*(1), 1–31. DOI: 10.1016/j.physrep.2013.07.005

Ma, M.-G., & Su, R.-C. (2011). Biomineralization and Biomimetic Synthesis of Biomineral and Nanomaterials. In *Advances in Biomimetics*. InTech., DOI: 10.5772/13838

Mandel, S., & Tas, A. C. (2010). Brushite (CaHPO4 ·2H2O) to octacalcium phosphate (Ca 8 (HPO4) 2 (PO4)4·5H2O) transformation in DMEM solutions at 36.5 °C. *Materials Science and Engineering C*, *30*(2), 245–254. DOI: 10.1016/j.msec.2009.10.009 PMID: 30011614

Manne, B., Thiruvayapati, H., Bontha, S., Rangarasaiah, R. M., Das, M., & Balla, V. K. (2018). Surface design of Mg-Zn alloy temporary orthopaedic implants: Tailoring wettability and biodegradability using laser surface melting. *Surface and Coatings Technology*, *347*, 337–349. DOI: 10.1016/j.surfcoat.2018.05.017

Mao, J., Kim, J. H., & Seo, S. (2024). Current status and outlook of low-melting-point metals in biomedical applications. *Advanced Functional Materials*, *34*(31), 2307708. DOI: 10.1002/adfm.202307708

Mareci, D., Cimpoesu, N., & Popa, M. I. (2012). Electrochemical and SEM characterization of NiTi alloy coated with chitosan by PLD technique. *Materials and Corrosion*, *63*(11), 985–991. DOI: 10.1002/maco.201206501

Marin, E., Boschetto, F., & Pezzotti, G. (2020). Biomaterials and biocompatibility: An historical overview. *Journal of Biomedical Materials Research. Part A*, *108*(8), 1617–1633. DOI: 10.1002/jbm.a.36930 PMID: 32196949

Marin, E., & Lanzutti, A. (2025). History of metallic orthopedic materials. *Metals*, *15*(4), 378. DOI: 10.3390/met15040378

Martínez-Calderon, M., Martín-Palma, R. J., Rodríguez, A., Gómez-Aranzadi, M., García-Ruiz, J. P., Olaizola, S. M., & Manso-Silván, M. (2020). Biomimetic hierarchical micro/nano texturing of TiAlV alloys by femtosecond laser processing for the control of cell adhesion and migration. *Physical Review Materials*, *4*(5), 056008. DOI: 10.1103/PhysRevMaterials.4.056008

Maruo, S., Nakamura, O., & Kawata, S. (1997). Three-dimensional microfabrication with two-photon-absorbed photopolymerization. *Optics Letters*, *22*(2), 132–134. DOI: 10.1364/OL.22.000132 PMID: 18183126

Maskowicz, D., Maroszek, K., Jendrzejewski, R., & Sawczak, M. (2024). Hydroxyapatite Nanocoatings Deposited by Means of Resonant Matrix-Assisted Pulsed Laser Evaporation. *Materials (Basel)*, *17*(23), 5778. DOI: 10.3390/ma17235778 PMID: 39685214

Massa, M. A., Covarrubias, C., Bittner, M., Fuentevilla, I. A., Capetillo, P., Von Marttens, A., & Carvajal, J. C. (2014). Synthesis of new antibacterial composite coating for titanium based on highly ordered nanoporous silica and silver nanoparticles. *Materials Science and Engineering C*, *45*, 146–153. DOI: 10.1016/j.msec.2014.08.057 PMID: 25491813

Matei, A. T., & Visan, A. I. (2026). Biocompatible Thin Films Deposited by Laser Techniques. *Materials (Basel)*, *19*(5), 925. DOI: 10.3390/ma19050925 PMID: 41828192

Matei, A. T., Visan, A. I., & Negut, I. (2025). Laser-fabricated micro/nanostructures: Mechanisms, fabrication techniques, and applications. *Micromachines*, *16*(5), 573. Advance online publication. DOI: 10.3390/mi16050573 PMID: 40428699

Matos, R. J. R., Silva, J. C., Soares, P. I. P., & Borges, J. P. (2023). Polyvinylpyrrolidone Nanofibers Incorporating Mesoporous Bioactive Glass for Bone Tissue Engineering. *Biomimetics*, *8*(2), 206. DOI: 10.3390/biomimetics8020206 PMID: 37218792

Matsugaki, A., Aramoto, G., Ninomiya, T., Sawada, H., Hata, S., & Nakano, T. (2015). Abnormal arrangement of a collagen/apatite extracellular matrix orthogonal to osteoblast alignment is constructed by a nanoscale periodic surface structure. *Biomaterials*, *37*, 134–143. DOI: 10.1016/j.biomaterials.2014.10.025 PMID: 25453944

Matsui, H., Takao, S., Higashi, K., Kaneko, T., Samjeské, G., Uruga, T., Tada, M., & Iwasawa, Y. (2022). Operando imaging of Ce radical scavengers in a practical polymer electrolyte fuel cell by 3D fluorescence CT–XAFS and depth-profiling nano-XAFS–SEM/EDS techniques. *ACS Applied Materials & Interfaces*, *14*(5), 6762–6776. DOI: 10.1021/acsami.1c22336 PMID: 35077130

Matusiewicz, H. (2014). Potential release of in vivo trace metals from metallic medical implants in the human body: From ions to nanoparticles – A systematic analytical review. *Acta Biomaterialia*, *10*(6), 2379–2403. DOI: 10.1016/j.actbio.2014.02.027 PMID: 24565531

Maxian, S. H., Zawadsky, J. P., & Dunn, M. (1993). In vitro evaluation of amorphous calcium phosphate and poorly crystallized hydroxyapatite coatings on titanium implants. *Journal of Biomedical Materials Research*, *27*(1), 111–117. DOI: 10.1002/jbm.820270114 PMID: 8380594

Mei, X., Yang, Z., Wang, X., Shi, A., Blanchard, J., Elahi, F., Kang, H., Orive, G., & Zhang, Y. S. (2025). Integrating microfluidic and bioprinting technologies: Advanced strategies for tissue vascularization. *Lab on a Chip*, *25*(5), 764–786. DOI: 10.1039/D4LC00280F PMID: 39775452

Mellon, S. J., & Tanner, K. E. (2012). Bone and its adaptation to mechanical loading: A review. *International Materials Reviews*, *57*(5), 235–255. DOI: 10.1179/1743280412Y.0000000008

Meng, D., Shi, X., Jiang, J., Liu, F., Li, H., & Liang, X. (2026). Experimental study on thermal runaway propagation characteristics and combustion behaviors of LFP and NCM lithium-ion batteries induced by dual heat sources. *Applied Thermal Engineering*, *294*, 130589. DOI: 10.1016/j.applthermaleng.2026.130589

Meng, F., & Du, Y. (2024). Research progress on laser powder bed fusion additive manufacturing of zinc alloys. *Materials (Basel)*, *17*(17), 4309. DOI: 10.3390/ma17174309 PMID: 39274701

Mettler, J. J. H. (2024). Investigation of high and low spatial frequency laser induced periodic surface structures on metals (Doctoral dissertation, University of Illinois at Urbana-Champaign).

Milošev, I., Kosec, T., & Strehblow, H.-H. (2008). XPS and EIS study of the passive film formed on orthopaedic Ti–6Al–7Nb alloy in Hank's physiological solution. *Electrochimica Acta*, *53*(9), 3547–3558. DOI: 10.1016/j.electacta.2007.12.041

Milošev, I., Metikoš-Huković, M., & Strehblow, H.-H. (2000). Passive film on orthopaedic TiAlV alloy formed in physiological solution investigated by X-ray photoelectron spectroscopy. *Biomaterials*, *21*(20), 2103–2113. DOI: 10.1016/S0142-9612(00)00145-9 PMID: 10966021

Mirkani, A., Nabid, M. R., & Pakian, S. (2025). Microfluidics-Facilitated Synthesis of Methacrylated Chitosan-Coated Liposomes: An Innovative Platform for Prolonged Ocular Retention and Controlled Drug Release. *Biomacromolecules*, *26*(8), 5336–5352. DOI: 10.1021/acs.biomac.5c00849 PMID: 40665681

Miu, E. V., McKone, J. R., & Mpourmpakis, G. (2022). The sensitivity of metal oxide electrocatalysis to bulk hydrogen intercalation: Hydrogen evolution on tungsten oxide. *Journal of the American Chemical Society*, *144*(14), 6420–6433. DOI: 10.1021/jacs.2c00825 PMID: 35289172

Mndlovu, H., Kumar, P., du Toit, L. C., & Choonara, Y. E. (2024). A review of bio material degradation assessment approaches employed in the biomedical field. *NPJ Materials Degradation*, *8*(1), 66. DOI: 10.1038/s41529-024-00487-1

Mocanu, A. C., Miculescu, F., Stan, G. E., Pasuk, I., Tite, T., Pascu, A., Butte, T. M., & Ciocan, L. T. (2022). Modulated laser cladding of implant-type coatings by bovine-bone-derived hydroxyapatite powder injection on Ti6Al4V substrates—Part I: Fabrication and physico-chemical characterization. *Materials (Basel), 15*(22), 7971. DOI: 10.3390/ma15227971 PMID: 36431457

Mogosanu, D. E., Giol, E. D., Vandenhaute, M., Dragusin, D. M., Samal, S. K., & Dubruel, P. (2014). Polyester biomaterials for regenerative medicine. In The Design, Synthetic Strategies and Biocompatibility of Polymer Scaffolds for Biomedical Application (pp. 155-197). Bentham Science Publishers. DOI: 10.2174/9781608058761114010009

Mohammadi, Z., Ziaei-Moayyed, A. A., & Sheikh-Mehdi Mesgar, A. (2008). In vitro dissolution of plasma-sprayed hydroxyapatite coatings with different characteristics: Experimental study and modeling. *Biomedical Materials (Bristol, England), 3*(1), 015006. Advance online publication. DOI: 10.1088/1748-6041/3/1/015006 PMID: 18458493

Mohammed, A. O. A., Hamdan, Y. M., & Ali, H. O. (2026). Additive manufacturing of ceramic components: A review of 3D printing technologies for industrial applications. *Saudi Journal of Engineering and Technology, 11*(1), 1–22. DOI: 10.36348/sjet.2026.v11i01.001

Mohammed, A., Kannan, S., Pervaiz, S., Ali, S., Thomas, K. K., & Karthikeyan, R. (2022). Sustainable machining of Mg-9Al-1.4 Zn foam used for temporary biomedical implant using cryogenic cooling. *Materials (Basel), 15*(19), 6678. DOI: 10.3390/ma15196678 PMID: 36234019

Mohammed-Sadhakathullah, A. H., Paulo-Mirasol, S., Torras, J., & Armelin, E. (2023). Advances in functionalization of bioresorbable nanomembranes and nanoparticles for their use in biomedicine. *International Journal of Molecular Sciences, 24*(12), 10312. DOI: 10.3390/ijms241210312 PMID: 37373461

Mohazzab, B. F., Jaleh, B., Fattah-alhosseini, A., Mahmoudi, F., & Momeni, A. (2020). Laser surface treatment of pure titanium: Microstructural analysis, wear properties, and corrosion behavior of titanium carbide coatings in Hank's physiological solution. *Surfaces and Interfaces, 20*, 100597. DOI: 10.1016/j.surfin.2020.100597

Mokhtari, F., Symes, R. W., Simon, Ž., Dharmasiri, B., Henderson, L. C., Joosten, M. W., & Varley, R. J. (2025). An energy efficient and sustainable approach to structural health monitoring in carbon fiber composites: Harnessing sound-induced vibration with Ti3C2Tx MXene/AgNPs modified P(VDF-TrFE) sensors. *Journal of Materials Chemistry. A, Materials for Energy and Sustainability*, *13*(9), 6482–6492. DOI: 10.1039/D4TA07797K

Molaei, M., Al-Mangour, B., Alhawiti, A., Alidrisi, H., & Manladan, S. M. (2025). Plasma electrolytic oxidation (PEO) coatings for biomedical implants: A review on enhancing antibacterial efficacy through controlled antibiotic release. *Coatings*, *15*(8), 925. Advance online publication. DOI: 10.3390/coatings15080925

Molla, S., Abedin, M. M., & Siddique, I. M. (2024). Exploring the versatility of medical textiles: Applications in implantable and non-implantable medical textiles. *World Journal of Advanced Research and Reviews*, *21*(1), 603–615. DOI: 10.30574/wjarr.2024.21.1.0058

Montesissa, M., Tommasini, V., Rubini, K., Boi, M., Baldini, N., & Boanini, E. (2025). State of Art and Perspective of Calcium Phosphate-Based Coatings Coupled with Bioactive Compounds for Orthopedic Applications. *Nanomaterials (Basel, Switzerland)*, *15*(15), 1199. DOI: 10.3390/nano15151199 PMID: 40801737

Moore, M. J., Jabbari, E., Ritman, E. L., Lu, L., Currier, B. L., Windebank, A. J., & Yaszemski, M. J. (2004). Quantitative analysis of interconnectivity of porous biodegradable scaffolds with micro-computed tomography. *Journal of Biomedical Materials Research. Part A*, *71*(2), 258–267. DOI: 10.1002/jbm.a.30138 PMID: 15376269

Moriarty, T. F., Metsemakers, W. J., Morgenstern, M., Hofstee, M. I., Vallejo Diaz, A., Cassat, J. E., Wildemann, B., Depypere, M., Schwarz, E. M., & Richards, R. G. (2022). Fracture-related infection. *Nature Reviews. Disease Primers*, *8*(1), 67. DOI: 10.1038/s41572-022-00396-0 PMID: 36266296

Moro, T., Yoshio Takatori, M. D., & Kawaguchi, H. (2011). Cartilage-mimicking, High-density Brush Structure Improves Wear Resistance of Crosslinked Polyethylene. *Clinical Orthopaedics and Related Research*, *469*(8), 2327–2336. DOI: 10.1007/s11999-010-1718-5 PMID: 21132412

Morris, H., & Murray, R. (2020). Medical textiles. Textile progress, 52(1-2), 1-127.

Morris, H., & Murray, R. (2021). Textile Materials for Healthcare and Medical Applications. In Medical Textiles (pp. 35-62). CRC Press. DOI: 10.1201/9781003170570-3

Mostafa, A. M. (2025). The influence of various parameters on the ablation and deposition mechanisms in pulsed laser deposition. *Plasmonics*, *20*(7), 5627–5645. DOI: 10.1007/s11468-024-02706-6

Mousavian, S. M. H., & Bautin, V. A. (2026). Microbiologically influenced corrosion in magnesium implants: Mechanisms, challenges, and opportunities for biomedical innovation. *Journal of Bio- and Tribo-Corrosion*, *12*(1), 5. DOI: 10.1007/s40735-025-01065-x

Mubeen, M., Khalid, S., Tabish, M., Mahmood, M., Jawad, M., Malik, M. U., & Yasin, G. (2025). Advancements in magnesium applications, biocompatible implants to nanoparticles in medicine and biomedical innovations. *MetalMat*, *2*(2), e70003. DOI: 10.1002/metm.70003

Muck, M., Wolfsjäger, B., Seibert, K., Maier, C., Lone, S. A., Hassel, A. W., Baumgartner, W., & Heitz, J. (2021). Femtosecond laser-processing of pre-anodized Ti-based bone implants for cell-repellent functionalization. *Nanomaterials (Basel, Switzerland)*, *11*(5), 1342. DOI: 10.3390/nano11051342 PMID: 34065199

Mukhametkaliyev, T. M., Surmeneva, M. A., Vladescu, A., Cotrut, C. M., Braic, M., Dinu, M., Vranceanu, M. D., Pana, I., Mueller, M., & Surmenev, R. A. (2017). A biodegradable AZ91 magnesium alloy coated with a thin nanostructured hydroxyapatite for improving the corrosion resistance. *Materials Science and Engineering C*, *75*, 95–103. DOI: 10.1016/j.msec.2017.02.033 PMID: 28415551

Müller, F. A., Kunz, C., & Gräf, S. (2016). Bio-inspired functional surfaces based on laser-induced periodic surface structures. *Materials (Basel)*, *9*(6), 476. DOI: 10.3390/ma9060476 PMID: 28773596

Mundekkad, D., & Mallya, A. R. (2025). Biomimicry at the nanoscale—A review of nanomaterials inspired by nature. *Nano Trends*, *10*, 100119. DOI: 10.1016/j.nwnano.2025.100119

Municoy, S., Echazú, M. I. Á., Antezana, P. E., Galdopórpora, J. M., Olivetti, C., Mebert, A. M., Foglia, M. L., Tuttolomondo, M. V., Alvarez, G. S., Hardy, J. G., & Desimone, M. F. (2020). Stimuli-responsive materials for tissue engineering and drug delivery. *International Journal of Molecular Sciences*, *21*(13), 4724. Advance online publication. DOI: 10.3390/ijms21134724 PMID: 32630690

Muresan, L. M. (2023). Nanocomposite coatings for anti-corrosion properties of metallic substrates. *Materials (Basel)*, *16*(14), 5092. DOI: 10.3390/ma16145092 PMID: 37512373

Murzin, S. P. (2024). Artificial intelligence-driven innovations in laser processing of metallic materials. *Metals*, *14*(12), 1458. DOI: 10.3390/met14121458

Murzin, S. P. (2024). Digital engineering in photonics: Optimizing laser processing. *Photonics*, *11*(10), 935. DOI: 10.3390/photonics11100935

Nabhan, J. F., Hu, R., Oh, R. S., Cohen, S. N., & Lu, Q. (2012). Formation and release of arrestin domain-containing protein 1-mediated microvesicles (ARMMs) at plasma membrane by recruitment of TSG101 protein. *Proceedings of the National Academy of Sciences of the United States of America*, *109*(11), 4146–4151. DOI: 10.1073/pnas.1200448109 PMID: 22315426

Nadi, A., Boukhriss, A., Bentis, A., Jabrane, E., & Gmouh, S. (2018). Evolution in the surface modification of textiles: A review. *Textile Progress*, *50*(2), 67–108. DOI: 10.1080/00405167.2018.1533659

Nagashri, N., Archana, L., & Raghavan, R. (2024). Microbial Volatile Compounds (MVOCs) in Food Industries and Food Safety Applications. *Innovations, Securities, and Case Studies Across Healthcare, Business, and Technology*, 75–96. DOI: 10.4018/979-8-3693-1906-2.ch005

Nagel, M., & Lippert, T. (2012). Laser-induced forward transfer for the fabrication of devices. Nanomaterials: processing and characterization with lasers, 255-306.

Nakhoul, A., Maurice, C., Faure, N., Garrelie, F., Pigeon, F., & Colombier, J.-P. (2022). Tailoring the surface morphology of Ni at the nanometric scale by ultrashort laser pulses. *Applied Physics. A, Materials Science & Processing*, *128*(10), 933. DOI: 10.1007/s00339-022-06046-2

Nakhoul, A., Rudenko, A., Maurice, C., Reynaud, S., Garrelie, F., Pigeon, F., & Colombier, J.-P. (2022). Boosted spontaneous formation of high-aspect ratio nanopeaks on ultrafast laser-irradiated Ni surface. *Advancement of Science*, *9*(21), e2200761. DOI: 10.1002/advs.202200761 PMID: 35618474

Nanjundeshwara, N., Lakshmaiah, A., & Raghavan, R. (2025). Aflatoxigenic Fungi. *Aflatoxigenic Fungi*, 225–239. DOI: 10.1201/9781003487043-29

Narayan, R. J. (2010). The next generation of biomaterial development. Philosophical Transactions of the Royal Society A: Mathematical, Physical and Engineering Sciences, 368(1917), 1831-1837.

Narayanan, S., Singh, M., & Kandasubramanian, B. (2025). Innovative Approaches in Regenerative Medicine for the Development of Sustainable and Bioactive Implants. *Biomedical Materials & Devices*, 1–25.

Narazaki, A. (2021). Characterization of Laser-Processed Samples. In *Handbook of Laser Micro- and Nano-Engineering* (pp. 1–38). Springer International Publishing., DOI: 10.1007/978-3-319-69537-2_64-1

Nascimento, M. D., Brito, T. O., Lima, A. M., & Elias, C. N. (2023). Protein interactions with osseointegrable titanium implants. *Brazilian Journal of Oral Sciences, 22*, e238749. DOI: 10.20396/bjos.v22i00.8668749

Nasef, M. M., Gupta, B., Shameli, K., Verma, C., Ali, R. R., & Ting, T. M. (2021). Engineered bioactive polymeric surfaces by radiation induced graft copolymerization: Strategies and applications. *Polymers, 13*(18), 3102. DOI: 10.3390/polym13183102 PMID: 34578003

Negut, I., Bita, B., Dinu, M., Parau, A. C., Tozar, T., Gherasim, O., Grumezescu, V., Ristoscu, C., Marutescu, L., & Gradisteanu Pircalabioru, G. (2025). Multifunctional Bioactive Coatings Based on Bioglass, Vitamin D3, and Melittin Deposited via MAPLE for Enhanced Osseointegration, Antibacterial Activity, and Corrosion Resistance of Titanium Implants. *Advanced Materials Interfaces, 12*(22), e00505. Advance online publication. DOI: 10.1002/admi.202500505

Negut, I., Bita, B., Dorcioman, G., Dinu, M., Parau, A. C., Ristoscu, C., & Gradisteanu-Pircalabioru, G. (2026). Laser-Engineered Multilayer Coatings Based on Zinc Oxide and Lovastatin-Functionalized Bioactive Glasses for Corrosion-Resistant and Antimicrobial Stainless Steel Implants. *Biomimetics, 11*(4), 227. DOI: 10.3390/biomimetics11040227 PMID: 42041457

Negut, I., Chiriac, A. P., & Grumezescu, A. M. (2024). Advances in antimicrobial coatings for preventing infections in implantable and wearable medical devices. *Coatings, 14*(3), 256. DOI: 10.3390/coatings14030256

Negut, I., Floroian, L., Ristoscu, C., Mihailescu, C. N., Mirza Rosca, J. C., Tozar, T., Badea, M., Grumezescu, V., Hapenciuc, C., & Mihailescu, I. N. (2020). Functional Bioglass—Biopolymer Double Nanostructure for Natural Antimicrobial Drug Extracts Delivery. *Nanomaterials (Basel, Switzerland), 10*(2), 385. DOI: 10.3390/nano10020385 PMID: 32098412

Negut, I., Gradisteanu-Pircalabioru, G., Dinu, M., Bita, B., Parau, A. C., Grumezescu, V., Ristoscu, C., & Chifiriuc, M. C. (2023). Bioglass and Vitamin D3 Coatings for Titanium Implants: Osseointegration and Corrosion Protection. *Biomedicines, 11*(10), 2772. DOI: 10.3390/biomedicines11102772 PMID: 37893145

Negut, I., Ristoscu, C., Tozar, T., Dinu, M., Parau, A. C., Grumezescu, V., Hapenciuc, C., Popa, M., Stan, M. S., Marutescu, L., Mihailescu, I. N., & Chifiriuc, M. C. (2022). Implant Surfaces Containing Bioglasses and Ciprofloxacin as Platforms for Bone Repair and Improved Resistance to Microbial Colonization. *Pharmaceutics, 14*(6), 1175. DOI: 10.3390/pharmaceutics14061175 PMID: 35745748

Nemani, S. K., Annavarapu, R. K., Mohammadian, B., Raiyan, A., Heil, J., Haque, M. A., Abdelaal, A., & Sojoudi, H. (2018). Surface modification of polymers: Methods and applications. *Advanced Materials Interfaces, 5*(24), 1801247. DOI: 10.1002/admi.201801247

Nikam, S., Sawant, M., & McFadden, S. (Eds.). (2026). *Additive manufacturing of thin-walled structures*. CRC Press. DOI: 10.1201/9781003571209

Nikolova, M. P., & Apostolova, M. D. (2022). Advances in multifunctional bioactive coatings for metallic bone implants. *Materials (Basel), 16*(1), 183. Advance online publication. DOI: 10.3390/ma16010183 PMID: 36614523

Ni, X., Feng, J., Liang, M., Zhou, F., Xia, Y., Dong, Z., Xue, Q., Li, Z., Pu, F., & Xia, P. (2025). Enhancing bone repair with β-TCP-based composite scaffolds: A review of design strategies and biological mechanisms. *Orthopedic Research and Reviews, 17*, 313–340. DOI: 10.2147/ORR.S525959 PMID: 40688750

Notingher, I., Jones, J. R., Verrier, S., Bisson, I., Embanga, P., Edwards, P., Polak, J. M., & Hench, L. L. (2003). Application of FTIR and Raman Spectroscopy to Characterisation of Bioactive Materials and Living Cells. *Journal of Spectroscopy, 17*(2–3), 275–288. DOI: 10.1155/2003/893584

Nsakabwebwe, C., Makhatha, M. E., Tsoeunyane, G. M., & Baruwa, A. D. (2022). Corrosion Inhibition Efficiency of Polyvinylpyrrolidone-Cysteine on Mild Steel in 1.0 M HCl Solution. *Journal of Bio- and Tribo-Corrosion, 8*(2), 48. DOI: 10.1007/s40735-022-00647-3

Obilor, A. F., Pacella, M., Wilson, A., & Silberschmidt, V. V. (2022). Micro-texturing of polymer surfaces using lasers: A review. *International Journal of Advanced Manufacturing Technology, 120*(1), 103–135. DOI: 10.1007/s00170-022-08731-1

Odugbemi, M. (2018). Biopolymers for bioartificial pancreas (Doctoral dissertation, University of Westminster).

Oladele, I. O., Adekola, S. A., Agbeboh, N. I., Isola-Makinde, B. A., & Adewuyi, B. O. (2024). Synthesis and application of sustainable tricalcium phosphate based biomaterials from agro-based materials: A review. *Biomedical Engineering and Computational Biology*, *15*, 11795972241293525. DOI: 10.1177/11795972241293525 PMID: 39524096

Oliviero, O., Ventre, M., & Netti, P. A. (2012). Functional porous hydrogels to study angiogenesis under the effect of controlled release of vascular endothelial growth factor. *Acta Biomaterialia*, *8*(9), 3294–3301. DOI: 10.1016/j.actbio.2012.05.019 PMID: 22641106

Omenogor, C. E., & Adeniran, A. A. (2024). Advancing Precision Healthcare: The Integration of Nanotechnology, Millimeter Wave Sensing, Laser Technology, Fibre Bragg Grating, and Deep Learning Models. International Journal of Research Publication and Reviews.

Omigbodun, F. T., Oladapo, B. I., & Osa-uwagboe, N. (2026). Exploring the frontier of polylactic acid/hydroxyapatite composites in bone regeneration and their revolutionary biomedical applications: A review. *Journal of Reinforced Plastics and Composites*, *45*(5-6), 1230–1250. DOI: 10.1177/07316844241278045

Ong, J. L., & Lucas, L. C. (1994). Post-deposition heat treatments for ion beam sputter deposited calcium phosphate coatings. *Biomaterials*, *15*(5), 337–341. DOI: 10.1016/0142-9612(94)90245-3 PMID: 8061124

Örnek, C., Leygraf, C., & Pan, J. (2019). On the Volta potential measured by SKPFM – fundamental and practical aspects with relevance to corrosion science. *Corrosion Engineering, Science and Technology*, *54*(3), 185–198. DOI: 10.1080/1478422X.2019.1583436

Oyane, A. (2019). Calcium Phosphate Coating Using Supersaturated Solutions for Biomedical Applications. *Vacuum and Surface Science*, *62*(4), 217–222. Advance online publication. DOI: 10.1380/vss.62.217

P., A., D., K. M., M., J., P., S., & Raghavan, R. (2025). The Cancer Exposome. *Cancer Exposomics and Environmental Influences on Carcinogenesis*, 161–192. DOI: 10.4018/979-8-3373-2165-3.ch004

Pagar, R. R., Musale, S. R., Pawar, G., Kulkarni, D., & Giram, P. S. (2022). Comprehensive review on the degradation chemistry and toxicity studies of functional materials. *ACS Biomaterials Science & Engineering*, *8*(6), 2161–2195. DOI: 10.1021/acsbiomaterials.1c01304 PMID: 35522605

Paital, S. R., Balani, K., Agarwal, A., & Dahotre, N. B. (2009). Fabrication and evaluation of a pulse laser-induced Ca–P coating on a Ti alloy for bioapplication. *Biomedical Materials (Bristol, England)*, *4*(1), 015009. DOI: 10.1088/1748-6041/4/1/015009 PMID: 19020344

Paital, S. R., Cao, Z., He, W., & Dahotre, N. B. (2010). Wetting effects on in vitro bioactivity and in vitro biocompatibility of laser micro-textured Ca-P coating. *Biofabrication*, *2*(2), 025001. Advance online publication. DOI: 10.1088/1758-5082/2/2/025001 PMID: 20811129

Paital, S. R., He, W., & Dahotre, N. B. (2010). Laser pulse dependent micro textured calcium phosphate coatings for improved wettability and cell compatibility. *Journal of Materials Science. Materials in Medicine*, *21*(7), 2187–2200. Advance online publication. DOI: 10.1007/s10856-010-4085-6 PMID: 20464459

Pakuła, D., Staszuk, M., Dziekońska, M., Kožmín, P., & Čermák, A. (2019). Laser Micro-Texturing of Sintered Tool Materials Surface. *Materials (Basel)*, *12*(19), 3152. DOI: 10.3390/ma12193152 PMID: 31561577

Pallares, R. M., Barmin, R. A., Wang, A., Kiessling, F., & Lammers, T. (2025). Clinical cancer nanomedicines. *Journal of Controlled Release*, *385*, 113991. DOI: 10.1016/j.jconrel.2025.113991 PMID: 40582644

Pallavi, V. N., Chowdareddy, S., Byalahunasishetter, B. I., & Raghavan, R. (2026). CRISPR-Cas9 Technology for Precision Cancer Medicine: Target Discovery, Disease Modeling, and Therapeutics. *CRISPR-Cas9 Technology for Precision Medicine*, 35–72. DOI: 10.4018/979-8-3373-4857-5.ch002

Pana, I., Vitelaru, C., Kiss, A., Zoita, N. C., Dinu, M., & Braic, M. (2017). Design, fabrication and characterization of TiO 2 -SiO 2 multilayer with tailored color glazing for thermal solar collectors. *Materials & Design*, *130*, 275–284. DOI: 10.1016/j.matdes.2017.05.063

Pandey, C., Rokaya, D., & Bhattarai, B. P. (2022). Contemporary concepts in osseointegration of dental implants: A review. *BioMed Research International*, *2022*(1), 6170452. DOI: 10.1155/2022/6170452 PMID: 35747499

Pandit, S., Gaska, K., Kádár, R., & Mijakovic, I. (2021). Graphene-based antimicrobial biomedical surfaces. *ChemPhysChem*, *22*(3), 250–263. DOI: 10.1002/cphc.202000769 PMID: 33244859

Pan, J., Thierry, D., & Leygraf, C. (1996). Electrochemical impedance spectroscopy study of the passive oxide film on titanium for implant application. *Electrochimica Acta*, *41*(7–8), 1143–1153. DOI: 10.1016/0013-4686(95)00465-3

Pan, Q., Cao, Y., Xue, W., Zhu, D., & Liu, W. (2019). Picosecond laser-textured stainless steel superhydrophobic surface with an antibacterial adhesion property. *Langmuir*, *35*(33), 11414–11421. DOI: 10.1021/acs.langmuir.9b01333 PMID: 31397568

Papa, S. (2023). Evaluation de l'adhérence gingivale et du potentiel antibactérien de surfaces structurées par laser femtoseconde pour l'implantologie orale [Doctoral dissertation, Université de Lyon]. HAL.

Papa, S., Maalouf, M., Claudel, P., Sedao, X., Di Maio, Y., Hamzeh-Cognasse, H., Thomas, M., Guignandon, A., & Dumas, V. (2023). Key topographic parameters driving surface adhesion of Porphyromonas gingivalis. *Scientific Reports*, *13*(1), 15893. DOI: 10.1038/s41598-023-42387-5 PMID: 37741851

Papavlu, A. P., Dinca, V., Filipescu, M., & Dinescu, M. (2017). Matrix-assisted pulsed laser evaporation of organic thin films: Applications in biology and chemical sensors. In InTech eBooks. DOI: 10.5772/intechopen.70676

Papavlu, A. P., Dinca, V., Filipescu, M., & Dinescu, M. (2017). Matrix-Assisted Pulsed Laser Evaporation of Organic Thin Films: Applications in Biology and Chemical. Laser Ablation: From Fundamentals to Applications, 171.

Parangusan, H., Bhadra, J., & Al-Thani, N. (2021). A review of passivity breakdown on metal surfaces: Influence of chloride- and sulfide-ion concentrations, temperature, and pH. *Emergent Materials*, *4*(5), 1187–1203. DOI: 10.1007/s42247-021-00194-6

Park, H., Park, J. J., Bui, P. D., Yoon, H., Grigoropoulos, C. P., Lee, D., & Ko, S. H. (2024). Laser-based selective material processing for next-generation additive manufacturing. *Advanced Materials*, *36*(34), 2307586. DOI: 10.1002/adma.202307586 PMID: 37740699

Park, J. H., Pattipaka, S., Hwang, G., Park, M., Woo, Y. M., Kim, Y. B., Lee, H. E., Jeong, C. K., Zhang, T., Min, Y., Park, K., Lee, K. J., & Ryu, J. (2024). Light–material interactions using laser and flash sources for energy conversion and storage applications. *Nano-Micro Letters*, *16*(1), 314. Advance online publication. DOI: 10.1007/s40820-024-01483-5 PMID: 39186184

Park, J., Um, S. H., Seo, Y., Lee, J., Kim, Y. C., Ok, M. R., Hwang, S.-W., Sun, J.-Y., Han, H.-S., & Jeon, H. (2023). Improving hydroxyapatite coating ability on biodegradable metal through laser-induced hydrothermal coating in liquid precursor: Application in orthopedic implants. *Bioactive Materials*, *25*, 796–806. DOI: 10.1016/j.bioactmat.2022.06.020 PMID: 37056265

Pastrafidou, M., Binas, V., & Kartsonakis, I. A. (2025). Designing the next generation: A physical chemistry approach to surface coating materials. *Applied Sciences (Basel, Switzerland), 15*(19), 10817. DOI: 10.3390/app151910817

Patel, H., Ostergard, D. R., & Sternschuss, G. (2012). Polypropylene mesh and the host response. *International Urogynecology Journal, 23*(6), 669–679. DOI: 10.1007/s00192-012-1718-y PMID: 22430945

Patenge, N., Arndt, K., Eggert, T., Zietz, C., Kreikemeyer, B., Bader, R., Nebe, B., Stranak, V., Hippler, R., & Podbielski, A. (2012). Evaluation of antimicrobial effects of novel implant materials by testing the prevention of biofilm formation using a simple small scale medium-throughput growth inhibition assay. *Biofouling, 28*(3), 267–277. DOI: 10.1080/08927014.2012.671305 PMID: 22435853

Patrucco, A. (2017). KERATIN-BASED 3D SCAFFOLD DESIGN FOR BONE TISSUE ENGINEERING.

Patrucco, A., Cristofaro, F., Simionati, M., Zoccola, M., Bruni, G., Fassina, L., Visai, L., Magenes, G., Mossotti, R., Montarsolo, A., & Tonin, C. (2016). Wool fibril sponges with perspective biomedical applications. *Materials Science and Engineering C, 61*, 42–50. DOI: 10.1016/j.msec.2015.11.073 PMID: 26838822

Pearson, I. (2025). Persistent proton corridors as a hidden driver of cervical cancer: A signal-based framework for prevention and intervention.

Peethan, A., Unnikrishnan, V. K., Chidangil, S., & George, S. D. (2020). Laser-Assisted Tailoring of Surface Wettability-Fundamentals and Applications: A Critical Review. *Progress in Adhesion and Adhesives, 5*, 331–365. DOI: 10.1002/9781119749882.ch11

Pei, X., Wang, J., Cong, Y., & Fu, J. (2021). Recent progress in polymer hydrogel bioadhesives. *Journal of Polymer Science, 59*(13), 1312–1337. DOI: 10.1002/pol.20210249

Pelletier, H., Nelea, V., Mille, P., & Muller, D. (2004). Determination of mechanical properties of pulsed laser-deposited hydroxyapatite thin film implanted at high energy with N+ and Ar+ ions, using nanoindentation. *Journal of Materials Science, 39*(11), 3605–3611. DOI: 10.1023/B:JMSC.0000030712.81704.b0

Pelling, H., Nzakizwanayo, J., Milo, S., Denissen, J., Smits, W. P., Thet, N. T., & Jenkins, A. T. A. (2019). Bacterial biofilm formation on indwelling urethral catheters. *Letters in Applied Microbiology, 68*(4), 277–293. DOI: 10.1111/lam.13144 PMID: 30811615

Pereira, L. M., Nawaz, Q., Ferreira, N. M., Pereira, A. B., Detsch, R., Boccaccini, A. R., & Mesquita-Guimarães, J. (2025). Effects of S53P4 bioactive glass coated on double-size laser-textured zirconia substrates: Cell adhesion and mineralisation behaviour. *Ceramics International.*

Perrakis, G., Tsilipakos, O., Tsibidis, G. D., & Stratakis, E. (2024). Impact of hybrid electromagnetic surface modes on the formation of low spatial frequency LIPSS: A universal approach. *Laser & Photonics Reviews*, *18*(7), 2301090. DOI: 10.1002/lpor.202301090

Pesode, P., & Barve, S. (2023). Additive manufacturing of metallic biomaterials: Sustainability aspect, opportunity, and challenges. *Journal of Industrial and Production Engineering*, *40*(6), 464–505. DOI: 10.1080/21681015.2023.2229341

Pethő, Á. G., Fülöp, T., Orosz, P., & Tapolyai, M. (2024). Magnesium is a vital ion in the body—It is time to consider its supplementation on a routine basis. *Clinics and Practice*, *14*(2), 521–535. DOI: 10.3390/clinpract14020040 PMID: 38525719

Petrović, S., Peruško, D., Skoulas, E., Kovač, J., Mitrić, M., Potočnik, J., Rakočević, Z., & Stratakis, E. (2019). Laser-assisted surface texturing of Ti/Zr multilayers for mesenchymal stem cell response. *Coatings*, *9*(12), 854. DOI: 10.3390/coatings9120854

Pfeiffer, I., Petronis, S., Köper, I., Kasemo, B., & Zäch, M. (2010). Vesicle adsorption and phospholipid bilayer formation on topographically and chemically nanostructured surfaces. *The Journal of Physical Chemistry B*, *114*(14), 4623–4631. DOI: 10.1021/jp908283g PMID: 20232804

Phillips, K. C., Gandhi, H. H., Mazur, E., & Sundaram, S. K. (2015). Ultrafast laser processing of materials: A review. *Advances in Optics and Photonics*, *7*(4), 684–712. DOI: 10.1364/AOP.7.000684

Phogat, D., Rani, P., Biswas, A., Balani, K., & Awasthi, S. (2025). Architecturally Refined Cerium-Integrated Hydroxyapatite/CNT Nanocomposite Coatings: Enhanced Mechanics and Biofunction for Orthopaedic Implantation. *Macromolecular Bioscience*, *25*(12), 70106. Advance online publication. DOI: 10.1002/mabi.70106 PMID: 41059707

Piqué, A., Chrisey, D. B., & Fitz-Gerald, J. M. (2003). Laser-induced forward transfer: Fundamentals and applications. *Applied Physics. A, Materials Science & Processing*, *77*(2), 167–173. DOI: 10.1007/s00339-003-2143-5

Pohl, R. (2015). Laser-induced forward transfer of pure metals//Towards 3D printing.

Polshettiwar, S. (2025). Polymers for Vascular Applications. Materials Research Foundations, 172.

Pramesti, D., Winarto, W., Suharno, B., & Ismadi, S. (2025). With two types of buffer layer on Q55. In Proceedings of the International Conference on Smart and Advanced Manufacturing: ICSAM 2024 (p. 41). Springer Nature.

Prasadh, S., Raguraman, S., Wong, R., & Gupta, M. (2022). Current status and outlook of temporary implants (magnesium/zinc) in cardiovascular applications. *Metals*, *12*(6), 999. DOI: 10.3390/met12060999

Pratiwi, L., Elisa, E., & Sutanto, H. (2024). Probing the protrusions: Lamellipodia and filopodia in cancer invasion and beyond. *Mechanobiology in Medicine*, *2*(2), 100064. DOI: 10.1016/j.mbm.2024.100064 PMID: 40395858

Prefac, G.-A., Milea, M.-L., Vadureanu, A.-M., Muraru, S., Dobrin, D.-I., Isopencu, G.-O., Jinga, S.-I., Raileanu, M., Bacalum, M., & Busuioc, C. (2020). CeO2 Containing Thin Films as Bioactive Coatings for Orthopaedic Implants. *Coatings*, *10*(7), 642. DOI: 10.3390/coatings10070642

Prodana, M., Stoian, A. B., Ionita, D., Brajnicov, S., Boerasu, I., Enachescu, M., & Burnei, C. (2024). In-Depth Characterization of Two Bioactive Coatings Obtained Using MAPLE on TiTaZrAg. *Materials (Basel)*, *17*(12), 2989. DOI: 10.3390/ma17122989 PMID: 38930358

Pshyk, O., Nakonechna, O., & Coy, E. (2025). Current View of Nanoindentation: Recent Developments and Application in Material Characterization. *ACS Applied Materials & Interfaces*, *17*(25), 36275–36300. DOI: 10.1021/acsami.5c05434 PMID: 40499095

Puleo, D. A., & Nanci, A. (1999). Understanding and controlling the bone-implant interface. In *Biomaterials*. DOI: 10.1016/S0142-9612(99)00160-X

Pulido, L., Ghanem, E., Joshi, A., Purtill, J. J., & Parvizi, J. (2008). Periprosthetic joint infection: The incidence, timing, and predisposing factors. *Clinical Orthopaedics and Related Research*, *466*(7), 1710–1715. DOI: 10.1007/s11999-008-0209-4 PMID: 18421542

Pu, X., Li, G., & Huang, H. (2016). Preparation, anti-biofouling and drag-reduction properties of a biomimetic shark skin surface. *Biology Open*, *5*(4), 389–396. DOI: 10.1242/bio.016899 PMID: 26941105

Qazi, M. A. (2025). Zr-induced phase-strengthened high entropy alloy (HEA) coatings for hydrogen embrittlement resistance developed via thermal spraying.

Qin, S. (2026). An advanced modeling framework with well-designed experiments to unveil molten pool dynamics and particle spattering behavior [Doctoral dissertation, University of Dublin].

Raabe, D. (2023). The materials science behind sustainable metals and alloys. *Chemical Reviews, 123*(5), 2436–2608. DOI: 10.1021/acs.chemrev.2c00799 PMID: 36848879

Radhi, A. B., Khashan, K. S., & Sulaiman, G. M. (2024). Antibacterial Activity of Gold Nanoparticles Produced by One-Step Pulsed Laser Ablation in Liquid. *Plasmonics, 19*(3), 1173–1185. DOI: 10.1007/s11468-023-02081-8

Raghavan, R., Moger, S. S., Umesh, S., & Bhuvana, G. N. (2025). Revolutionizing Pain Management. *Unveiling Technological Advancements and Interdisciplinary Solutions for Pain Care*, 129–164. DOI: 10.4018/979-8-3693-9501-1.ch005

Raghavendra, G., Prasad, P. S., & Gara, D. K. (Eds.). (2025). *Bioactive glass materials for biological applications*. CRC Press. DOI: 10.1201/9781003660781

Rahaman, M. N., Xiao, W., & Bal, B. S. (2014). Development of Implants Composed of Hollow Hydroxyapatite Microspheres for Bone Regeneration. *Biomaterials Science: Processing. Properties and Applications IV: Ceramic Transactions, 251*, 45–56.

Rahman, M. M., Ahammad, R., Neon, A. K., Rhaman, M., Ansari, M. J., Uddin, M. N., Islam, M. M., & Nazir, M. A. (2026). Progresses and challenges in additive manufacturing of bulk metallic glasses. *Journal of Manufacturing and Materials Processing, 10*(4), 121. DOI: 10.3390/jmmp10040121

Ramakrishna, T. R., Nalder, T. D., Yang, W., Marshall, S. N., & Barrow, C. J. (2018). Controlling enzyme function through immobilisation on graphene, graphene derivatives and other two dimensional nanomaterials. *Journal of Materials Chemistry. B, 6*(20), 3200–3218. DOI: 10.1039/C8TB00313K PMID: 32254379

Ramasubramanian, B., Sundarrajan, S., Rao, R. P., Reddy, M. V., Chellappan, V., & Ramakrishna, S. (2022). Novel low-carbon energy solutions for powering emerging wearables, smart textiles, and medical devices. *Energy & Environmental Science, 15*(12), 4928–4981. DOI: 10.1039/D2EE02695C

Ramirez, G. (2023). *Cytotoxic Effects of Dental Implant Debris After Decontamination Treatment Using Er: YAG Laser*. Temple University.

Rammelt, U., & Reinhard, G. (1992). Application of electrochemical impedance spectroscopy (EIS) for characterizing the corrosion-protective performance of organic coatings on metals. *Progress in Organic Coatings, 21*(2–3), 205–226. DOI: 10.1016/0033-0655(92)87005-U

Ranakoti, L., Gangil, B., Bhandari, P., Singh, T., Sharma, S., Singh, J., & Singh, S. (2023). Promising role of polylactic acid as an ingenious biomaterial in scaffolds, drug delivery, tissue engineering, and medical implants: Research developments, and prospective applications. *Molecules (Basel, Switzerland)*, *28*(2), 485. DOI: 10.3390/molecules28020485 PMID: 36677545

Ranjan, A., Jindal, V., & Tyagi, R. (2022). Effect of load on tribological properties of Ti–TiB–Fe composites processed via spark plasma sintering (SPS). *Transactions of the Indian Institute of Metals*, *75*(11), 2847–2856. DOI: 10.1007/s12666-022-02651-0

Ranjan, A., Shetty, P., Despande, R., Biradar, A., Khan, W., & Kulshrestha, R. (2023). A comparative surface evaluation of orthodontic mini-implants before and after en masse retraction-A SEM study. *Journal of Orthodontic Science*, *12*(1). Advance online publication. DOI: 10.4103/jos.jos_166_21 PMID: 37351404

Ranjan, A., Tyagi, R., & Jindal, V. (2022). Reciprocating wear of Ti–TiB in situ composites synthesised via vacuum arc melting. *Journal of Materials Engineering and Performance*, *31*(12), 9985–9996. DOI: 10.1007/s11665-022-07002-0

RASLAN, W. M., EL-HALWAGY, A. A., & ELSAYAD, H. (2020). Recent advances in plasma/nanoparticles treatments of textile fibers. Journal of Textiles. *Coloration and Polymer Science*, *17*(2), 87–105.

Rau, J. V., Antoniac, I., Cama, G., Komlev, V. S., & Ravaglioli, A. (2016). Bioactive Materials for Bone Tissue Engineering. *BioMed Research International*, *1–3*, 1–3. Advance online publication. DOI: 10.1155/2016/3741428 PMID: 26942191

Rau, J. V., Antoniac, I., Fosca, M., De Bonis, A., Blajan, A. I., Cotrut, C., Graziani, V., Curcio, M., Cricenti, A., Niculescu, M., Ortenzi, M., & Teghil, R. (2016). Glass-ceramic coated Mg-Ca alloys for biomedical implant applications. *Materials Science and Engineering C*, *64*, 362–369. Advance online publication. DOI: 10.1016/j.msec.2016.03.100 PMID: 27127065

Rau, J. V., De Bonis, A., Curcio, M., Schuhladen, K., Barbaro, K., De Bellis, G., Teghil, R., & Boccaccini, A. R. (2020). Borate and Silicate Bioactive Glass Coatings Prepared by Nanosecond Pulsed Laser Deposition. *Coatings*, *10*(11), 1105. DOI: 10.3390/coatings10111105

Rethfeld, B., Ivanov, D. S., Garcia, M. E., & Anisimov, S. I. (2017). Modelling ultrafast laser ablation. *Journal of Physics. D, Applied Physics*, *50*(19), 193001. DOI: 10.1088/1361-6463/50/19/193001

Rezwan, K., Chen, Q. Z., Blaker, J. J., & Boccaccini, A. R. (2006). Biodegradable and bioactive porous polymer/inorganic composite scaffolds for bone tissue engineering. *Biomaterials*, *27*(18), 3413–3431. DOI: 10.1016/j.biomaterials.2006.01.039 PMID: 16504284

Ribba, L., Parisi, M., D'accorso, N. B., & Goyanes, S. (2014). Electrospun nanofibrous mats: From vascular repair to osteointegration. *Journal of Biomedical Nanotechnology*, *10*(12), 3508–3535. DOI: 10.1166/jbn.2014.2046 PMID: 26000368

Richter, A. M., Buchberger, G., Stifter, D., Duchoslav, J., Hertwig, A., Bonse, J., Heitz, J., & Schwibbert, K. (2021). Spatial period of laser-induced surface nanoripples on PET determines Escherichia coli repellence. *Nanomaterials (Basel, Switzerland)*, *11*(11), 3000. DOI: 10.3390/nano11113000 PMID: 34835763

Ristoscu, C., & Mihailescu, I. N. (2014). Biomimetic coatings by pulsed laser deposition. In Laser Technology in Biomimetics: Basics and Applications (pp. 179-208). Springer. DOI: 10.1007/978-3-642-41341-4_7

Riveiro, A., Maçon, A. L., del Val, J., Comesaña, R., & Pou, J. (2018). Laser surface texturing of polymers for biomedical applications. *Frontiers in Physics (Lausanne)*, *6*, 16. DOI: 10.3389/fphy.2018.00016

Robau-Porrua, A., González, J. E., Arancibia-Castillo, R., Picardo, A., Araneda-Hernández, E., & Torres, Y. (2025). Design, fabrication, and characterization of novel dental implants with porosity gradient obtained by selective laser melting. *Materials & Design*, *251*, 113660. DOI: 10.1016/j.matdes.2025.113660

Rohanová, D., Boccaccini, A. R., Horkavcová, D., Bozděchová, P., Bezdička, P., & Častorálová, M. (2014). Is non-buffered DMEM solution a suitable medium for *in vitro* bioactivity tests? *Journal of Materials Chemistry. B*, *2*(31), 5068–5076. DOI: 10.1039/C4TB00187G PMID: 32261840

Rohwerder, M. (2009). Conducting polymers for corrosion protection: A review. *International Journal of Materials Research*, *100*(10), 1331–1342. DOI: 10.3139/146.110205

Ruff, C., Holt, B., & Trinkaus, E. (2006). Who's afraid of the big bad Wolff?: "Wolff's law" and bone functional adaptation. *American Journal of Physical Anthropology*, *129*(4), 484–498. DOI: 10.1002/ajpa.20371 PMID: 16425178

Rupp, F., Gittens, R. A., Scheideler, L., Marmur, A., Boyan, B. D., Schwartz, Z., & Geis-Gerstorfer, J. (2014). A review on the wettability of dental implant surfaces I: Theoretical and experimental aspects. *Acta Biomaterialia*, *10*(7), 2894–2906. DOI: 10.1016/j.actbio.2014.02.040 PMID: 24590162

Ryou, S., Xie, M., & Lee, J.. (2025). Machine-learning-enabled on-the-fly analysis of RHEED during thin-film growth. *Journal of Vacuum Science & Technology. A, Vacuum, Surfaces, and Films*, *43*(3), 032702. Advance online publication. DOI: 10.1116/6.0004493

S., G. V., C., A., Biradar, S., & Raghavan, R. (2025). Neuroscience-Driven Solutions for Speech Impairment. *Advancing Medical Research Through Neuroscience*, 119–146. DOI: 10.4018/979-8-3693-5464-3.ch006

Saba, T., Saad, K. S. K., & Rashid, A. B. (2024). Precise surface engineering: Leveraging chemical vapor deposition for enhanced biocompatibility and durability in biomedical implants. *Heliyon*, *10*(18), e37976. DOI: 10.1016/j.heliyon.2024.e37976 PMID: 39328539

Sachan, R., Jaipan, P., Zhang, J. Y., Degan, S., Erdmann, D., Tedesco, J., Vanderwal, L., Stafslien, S. J., Negut, I., Visan, A., Dorcioman, G., Socol, G., Cristescu, R., Chrisey, D. B., & Narayan, R. J. (2025). Printing amphotericin B on microneedles using matrixassisted pulsed laser evaporation. *International Journal of Bioprinting*, *3*(2), 147–157. DOI: 10.18063/IJB.2017.02.004 PMID: 33094188

Saha, S. (2023). Evaluating the potential of wrought and additively manufactured Co-Cr alloy for prosthodontic applications [Master's thesis, The University of North Dakota].

Sahadat Hossain, M., & Ahmed, S. (2023). FTIR spectrum analysis to predict the crystalline and amorphous phases of hydroxyapatite: A comparison of vibrational motion to reflection. *RSC Advances*, *13*(21), 14625–14630. DOI: 10.1039/D3RA02580B PMID: 37197675

Sahani, S. (2025). Plasma Innovations: Revolutionizing Sustainable Textile Treatments. In: Dutta, S., Bansal, P., Kapoor, A., Reddy, C.C., Venkatraman, P. (eds) Selected Articles from the International Conference on Sustainable Fashion and Technical Textiles. ICSFT 2024. Proceedings in Technology Transfer. Springer, pp 78–94. Singapore. DOI: 10.1007/978-981-96-3169-8_7

Saha, S., & Roy, S. (2022). Metallic dental implants wear mechanisms, materials, and manufacturing processes: A literature review. *Materials (Basel)*, *16*(1), 161. DOI: 10.3390/ma16010161 PMID: 36614500

Sai Charan, M., Naik, A. K., Kota, N., Laha, T., & Roy, S. (2022). Review on developments of bulk functionally graded composite materials. *International Materials Reviews*, *67*(8), 797–863. DOI: 10.1080/09506608.2022.2026863

Saikia, N. (2024). Inorganic-based nanoparticles and biomaterials as biocompatible scaffolds for regenerative medicine and tissue engineering: Current advances and trends of development. *Inorganics*, *12*(11), 292. DOI: 10.3390/inorganics12110292

Salama, M., Vaz, M. F., Colaço, R., Santos, C., & Carmezim, M. (2022). Biodegradable iron and porous iron: Mechanical properties, degradation behaviour, manufacturing routes and biomedical applications. *Journal of Functional Biomaterials*, *13*(2), 72. DOI: 10.3390/jfb13020072 PMID: 35735927

Samipour, S., Taghvaei, H., Mohebbi-Kalhori, D., & Rahimpour, M. R. (2020). Plasma treatment and chitosan coating: A combination for improving PET surface properties. *Surface Innovations*, *8*(1-2), 76–88. DOI: 10.1680/jsuin.19.00030

Sanjari, S., Etemadi Haghighi, S., Saraeian, P., & Alinia-ziazi, A. (2025). Modeling Degradation Behavior of (Bio) Degradable Polymers for Medical Devices: A Comparative Review of Physio-chemical Approaches. *Journal of Polymers and the Environment*, *33*(2), 660–697. DOI: 10.1007/s10924-024-03376-5

Sankar, M., Suwas, S., Balasubramanian, S., & Manivasagam, G. (2017). Comparison of electrochemical behavior of hydroxyapatite coated onto WE43 Mg alloy by electrophoretic and pulsed laser deposition. *Surface and Coatings Technology*, *309*, 840–848. DOI: 10.1016/j.surfcoat.2016.10.077

Sarantopoulou, E., Petrou, P. S., Kollia, Z., Palles, D., Spyropoulos-Antonakakis, N., Kakabakos, S., & Cefalas, A.-C. (2011). Protein immobilization and detection on laser processed polystyrene surfaces. *Journal of Applied Physics*, *110*(6), 064309. Advance online publication. DOI: 10.1063/1.3627160

Sarkar, S. K. (2022). Lasers in materials processing and synthesis. *Handbook on Synthesis Strategies for Advanced Materials*, *II*, 791–831. DOI: 10.1007/978-981-16-1803-1_19

Sattar, H., Hu, Z., Zheng, W., Zhou, J., Shahzad, M. K., Rehman, F., Guo, L., & Luo, W. (2023). Exploring the potential and recent advancement in laser opto-ultrasonic detection for material characterization: A state-of-the-art review. *Optics & Laser Technology*, *171*, 110316. Advance online publication. DOI: 10.1016/j.optlastec.2023.110316

Saxena, P., & Shukla, P. (2023). A review on recent developments and advances in environmental gas sensors to monitor toxic gas pollutants. *Environmental Progress & Sustainable Energy*, *42*(5), e14126. DOI: 10.1002/ep.14126

Scarlat, E. V. R., & Mitu, B. (2020, June). the Biocompatibility of the Textile Medical. In Advances in Human Factors and Ergonomics in Healthcare and Medical Devices: Proceedings of the AHFE 2020 Virtual Conference on Human Factors and Ergonomics in Healthcare and Medical Devices, July 16-20, 2020, USA (Vol. 1205, p. 217). Springer Nature.

Schille, J., Schneider, L., Mauersberger, S., Szokup, S., Höhn, S., Pötschke, J., Reiß, F., Leidich, E., & Löschner, U. (2020). High-Rate Laser Surface Texturing for Advanced Tribological Functionality. *Lubricants (Basel, Switzerland)*, *8*(3), 33. DOI: 10.3390/lubricants8030033

Schmidt, M., Hassel, T., & Vollertsen, F.. (2024). Dynamic beam shaping—Improving laser materials processing by tailored energy deposition. *CIRP Annals*, *73*(2), 501–524. DOI: 10.1016/j.cirp.2024.05.005

Schneider, C. W., Lippert, T., & Wokaun, A. (2023). PLD plasma plume analysis: A summary of the PSI contributions. *Applied Sciences (Basel, Switzerland)*, *13*(2), 1186. Advance online publication. DOI: 10.3390/app13021186

Schwarz, S., Rung, S., & Hellmann, R. (2017). Two-dimensional low spatial frequency laser-induced periodic surface structuring of sapphire. *Journal of Laser Micro Nanoengineering*, *12*(2), 67–71. DOI: 10.2961/jlmn.2017.02.0003

Schwibbert, K., Richter, A. M., Krüger, J., & Bonse, J. (2024). Laser-textured surfaces: A way to control biofilm formation? *Laser & Photonics Reviews*, *18*(5), 2300753. DOI: 10.1002/lpor.202300753

Sedghamiz, E., Liu, M., & Wenzel, W. (2022). Challenges and limits of mechanical stability in 3D direct laser writing. *Nature Communications*, *13*(1), 2115. DOI: 10.1038/s41467-022-29749-9 PMID: 35440637

Seetharaman, S., Sankaranarayanan, D., Wong, E., & Gupta, M. (2025). Magnesium-based syntactic, metastable and nano-composites.

Sekine, I. (1997). Recent evaluation of corrosion protective paint films by electrochemical methods. *Progress in Organic Coatings*, *31*(1–2), 73–80. DOI: 10.1016/S0300-9440(97)00020-9

Sergi, R., Bellucci, D., & Cannillo, V. (2020). A Comprehensive Review of Bioactive Glass Coatings: State of the Art, Challenges and Future Perspectives. *Coatings*, *10*(8), 757. DOI: 10.3390/coatings10080757

Serra, P., & Piqué, A. (2019). Laser-induced forward transfer: Fundamentals and applications. *Advanced Materials Technologies*, *4*(1), 1800099. DOI: 10.1002/admt.201800099

Shabanloo, R., Montazer, M., Farahani, A., & Karimi, N. (2025). A review on surface modification of nanofibrous textiles for diverse applications: Focus on medical uses. *Heliyon*, *11*(2), e41863. DOI: 10.1016/j.heliyon.2025.e41863 PMID: 39897825

Shabir, S., Khan, N. Z., & Siddiquee, A. N. (2025). Laser-induced surface modifications in biomedical materials: A comprehensive review. *Surface Topography : Metrology and Properties*, *13*(2), 023002. DOI: 10.1088/2051-672X/addc85

Shakiba, M., Rezvani Ghomi, E., Khosravi, F., Jouybar, S., Bigham, A., Zare, M., Abdouss, M., Moaref, R., & Ramakrishna, S. (2021). Nylon—A material introduction and overview for biomedical applications. *Polymers for Advanced Technologies*, *32*(9), 3368–3383. DOI: 10.1002/pat.5372

Shamim, Z., Awan, M. T., Jamal, A., & Shahid, M. N. (2025). Structural and Mechanical Features of Biomedical Textiles. In Biomedical Textiles (pp. 69-99). Jenny Stanford Publishing. DOI: 10.1201/9781003651697-3

Shapagina, N. A., & Dushik, V. V. (2025). Bilayer coating systems: Functional interlayers and top layers for enhanced performance. *Materials (Basel)*, *18*(22), 5217. DOI: 10.3390/ma18225217 PMID: 41304061

Sharma, C. P., Gupta, A., Khadka, M., Ben-Yoav, H., Ronen, A., & Arnusch, C. J. (2024). Silver nanoparticle doped laser-induced graphene fabrication methodology affects silver nanoparticle size, distribution, biological and electrochemical properties. *Environmental Science. Nano*, *11*(4), 1582–1596. DOI: 10.1039/D3EN00371J

Sharma, S., Sudhakara, P., Singh, J., Ilyas, R. A., Asyraf, M. R. M., & Razman, M. R. (2021). Critical review of biodegradable and bioactive polymer composites for bone tissue engineering and drug delivery applications. *Polymers*, *13*(16), 2623. DOI: 10.3390/polym13162623 PMID: 34451161

Shein, P., Cilip, C. M., Quinto, G., Behrens, A., & Fried, N. M. (2008, August). Selective laser suture lysis with a compact, low-cost, red diode laser. In 2008 30th Annual International Conference of the IEEE Engineering in Medicine and Biology Society (pp. 4358-4360). IEEE. DOI: 10.1109/IEMBS.2008.4650175

Shen, D., Zhang, X., & Zhu, L. (2023). Laser processing for electricity generators: Physics, methods and applications. *Nano Energy*, *120*, 109182. Advance online publication. DOI: 10.1016/j.nanoen.2023.109182

Shen, S., Cai, S., Xu, G., Zhao, H., Niu, S., & Zhang, R. (2015). Influence of heat treatment on bond strength and corrosion resistance of sol–gel derived bioglass–ceramic coatings on magnesium alloy. *Journal of the Mechanical Behavior of Biomedical Materials*, *45*, 166–174. DOI: 10.1016/j.jmbbm.2015.02.005 PMID: 25728582

Shepelin, N. A., Tehrani, Z. P., Ohannessian, N., Schneider, C. W., Pergolesi, D., & Lippert, T. (2023). A practical guide to pulsed laser deposition: From fundamentals to advanced functional thin films. *Chemical Society Reviews*, *52*(7), 2294–2321. DOI: 10.1039/D2CS00938B PMID: 36916771

Sherazi, T. A., Rehman, T., Naqvi, S. A. R., Shaikh, A. J., Shahzad, S. A., Abbas, G., Raza, R., & Waseem, A. (2015). Surface functionalization of solid state ultra-high molecular weight polyethylene through chemical grafting. *Applied Surface Science*, *359*, 593–601. DOI: 10.1016/j.apsusc.2015.10.080

Sheriff, P. (2026). A proposed archaeological pedestrian survey on Mt. Rainier–Attempting recovery of R5C Curtis Commando and her crew. Sciences, 1880, 146th.

Shi, G., Zhang, R., Cao, Y., & Yang, G. (2024). A review of the vaporization behavior of some metal elements in the LPBF process. *Micromachines*, *15*(7), 846. DOI: 10.3390/mi15070846 PMID: 39064359

Shivakoti, I., Kibria, G., Cep, R., Pradhan, B. B., & Sharma, A. (2021). Laser Surface Texturing for Biomedical Applications: A Review. *Coatings*, *11*(2), 124. DOI: 10.3390/coatings11020124

Shi, X., Li, X., Jiang, L., Qu, L., Zhao, Y., Ran, P., Wang, Q., Cao, Q., Ma, T., & Lu, Y. (2015). Femtosecond laser rapid fabrication of large-area rose-like micropatterns on freestanding flexible graphene films. *Scientific Reports*, *5*(1), 17557. DOI: 10.1038/srep17557 PMID: 26615800

Shrivastava, A., Kumar, A., Aggarwal, L. M., Pradhan, S., Choudhary, S., Ashish, A., Kashyap, K., & Mishra, S. (2024). Evolution of bioelectric membrane potentials: Implications in cancer pathogenesis and therapeutic strategies. *The Journal of Membrane Biology*, *257*(5/6), 281–305. DOI: 10.1007/s00232-024-00323-2 PMID: 39183198

Shuai, C., Li, D., Xie, H., Yao, X., Peng, S., & Gao, C. (2025). Programmable lamellar eutectic Zn-2Al-Mg biodegradable implants manufactured by laser powder bed fusion for synergistic strength-ductility and osteogenesis. *Advanced Healthcare Materials*, *14*(32), e01917. DOI: 10.1002/adhm.202501917 PMID: 40817637

Sidhu, M. S., & Dhingra, N. (2022). Ablation of materials using femtosecond lasers and electron beams. In *Terahertz*. Ultrafast Lasers and Their Medical and Industrial Applications. IntechOpen. DOI: 10.5772/intechopen.106198

Sikka, M., Garg, S., & Kumari, N. (2024). Bandages (compression, retention, and support). In *Textile Materials for Good Health and Wellbeing: Design and Applications* (pp. 155–170). Springer Nature Singapore.

Singh, A. V., Chandrasekar, V., Prabhu, V. M., Bhadra, J., Laux, P., Bhardwaj, P., Al-Ansari, A. A., Aboumarzouk, O. M., Luch, A., & Dakua, S. P. (2024). Sustainable bioinspired materials for regenerative medicine: Balancing toxicology, environmental impact, and ethical considerations. *Biomedical Materials (Bristol, England)*, *19*(6), 060501. DOI: 10.1088/1748-605X/ad85bb PMID: 39389102

Singha, K., Pandit, P., & Maity, S. (2022). Wool composites for hygienic/medical applications. In *Wool Fiber Reinforced Polymer Composites* (pp. 387–406). Woodhead Publishing. DOI: 10.1016/B978-0-12-824056-4.00007-8

Singhal, J. P., Singh, H., & Ray, A. R. (1988). Absorbable suture materials: Preparation and properties. *Polymer Reviews (Philadelphia, Pa.)*, *28*(3-4), 475–502.

Singh, C., Wong, C. S., & Wang, X. (2015). Medical textiles as vascular implants and their success to mimic natural arteries. *Journal of Functional Biomaterials*, *6*(3), 500–525. DOI: 10.3390/jfb6030500 PMID: 26133386

Sirdeshmukh, N., & Dongre, G. (2021). Laser micro & nano surface texturing for enhancing osseointegration and antimicrobial effect of biomaterials: A review. *Materials Today: Proceedings*, *44*, 2348–2355. DOI: 10.1016/j.matpr.2020.12.433

Smith, D. S., Lefeuvre, P., Renaux, M., Nait-Ali, B., & Leriche, A. (2023). Study of neck formation and densification in porous hydroxyapatite ceramics using thermal conductivity measurements. *Open Ceramics*, *13*, 100329. DOI: 10.1016/j.oceram.2023.100329

Sobolev, A., Sviridov, A., & Romanov, D. (2021). Surface functionalization of zirconia and titanium implants by laser microstructuring: A review. *Optics & Laser Technology*, *140*, 106992. DOI: 10.1016/j.optlastec.2021.106992

Soloviev, A., Burdonov, K., Chen, S. N., Eremeev, A., Korzhimanov, A., Pokrovskiy, G. V., Pikuz, T. A., Revet, G., Sladkov, A., Ginzburg, V., Khazanov, E., Kuzmin, A., Osmanov, R., Shaikin, I., Shaykin, A., Yakovlev, I., Pikuz, S., Starodubtsev, M., & Fuchs, J. (2017). Experimental evidence for short-pulse laser heating of solid-density target to high bulk temperatures. *Scientific Reports*, *7*(1), 11675. Advance online publication. DOI: 10.1038/s41598-017-11675-2 PMID: 28939883

Somasekhar, L. (2021). Bioink Optimization and Effects of Microgravity on 3D Bioprinted Cell Laden Constructs.

Song, S. H., Lee, S.-W., Kim, Y.-K., Lee, Y., Ryu, J.-J., Heo, S.-J., & Lee, J.-H. (2020). Application of a novel CVD TiN coating on a biomedical Co–Cr alloy: An evaluation of coating layer and substrate characteristics. *Materials (Basel)*, *13*(5), 1145. Advance online publication. DOI: 10.3390/ma13051145 PMID: 32150803

Song, Z., & Li, H. (2022). Plasma spraying with wire feeding: A facile route to enhance the coating/substrate interfacial metallurgical bonding. *Coatings*, *12*(5), 615. DOI: 10.3390/coatings12050615

Statha, D., Sfiniadakis, I., Rallis, M., Anastassopoulou, J., & Alexandratou, E. (2025). Investigating the wound healing potential of low-power 661 nm laser light in a pigmented hairless murine model. *Free Radical Biology & Medicine*, *233*, S1–S102. DOI: 10.1016/j.freeradbiomed.2025.05.062 PMID: 40338500

Stępak, B. D., Antończak, A. J., Szustakiewicz, K., Kozioł, P. E., & Abramski, K. M. (2014). Degradation of poly (L-lactide) under KrF excimer laser treatment. *Polymer Degradation & Stability*, *110*, 156–164. DOI: 10.1016/j.polymdegradstab.2014.08.028

Stewart, P. S. (2015). Antimicrobial tolerance in biofilms. *Microbiology Spectrum*, *3*(3), MB-0010–MB-2014. DOI: 10.1128/microbiolspec.MB-0010-2014 PMID: 26185072

Stuart, B. W., Gimeno-Fabra, M., Segal, J., Ahmed, I., & Grant, D. M. (2015). Degradation and Characterization of Resorbable Phosphate-Based Glass Thin-Film Coatings Applied by Radio-Frequency Magnetron Sputtering. *ACS Applied Materials & Interfaces*, *7*(49), 27362–27372. DOI: 10.1021/acsami.5b08957 PMID: 26523618

Subraveti, V., Richter, B., Yeratapally, S. R., & Oskay, C. (2024). Three-dimensional prediction of lack-of-fusion porosity volume fraction and morphology for powder bed fusion additively manufactured Ti–6Al–4V. *Integrating Materials and Manufacturing Innovation*, *13*(2), 511–525. DOI: 10.1007/s40192-024-00347-5

Suhag, D., Kaushik, S., & Taxak, V. B. (2024). *Handbook of Biomaterials for Medical Applications* (Vol. 2).

Sun, L., Berndt, C. C., Gross, K. A., & Kucuk, A. (2001). Material fundamentals and clinical performance of plasma-sprayed hydroxyapatite coatings: A review. *Journal of Biomedical Materials Research*, *58*(5), 570–592. DOI: 10.1002/jbm.1056 PMID: 11505433

Sun, L., Guo, J., Chen, H., Zhang, D., Shang, L., Zhang, B., & Zhao, Y. (2021). Tailoring materials with specific wettability in biomedical engineering. *Advancement of Science*, *8*(19), 2100126. DOI: 10.1002/advs.202100126 PMID: 34369090

Surmenev, M., Grubova, I., Glukhova, N., Khrapov, D., Koptyug, A., Volkova, A., Ivanov, Y., Cotrut, C. M., Vladescu, A., Teresov, A., Koval, N., Tyurin, A., & Surmenev, R. (2021). New Ti–35Nb–7Zr–5Ta alloy manufacturing by electron beam melting for medical application followed by high current pulsed electron beam treatment. *Metals*, *11*(7), 1066. DOI: 10.3390/met11071066

Surmenev, R. A., Surmeneva, M. A., & Ivanova, A. A. (2014). Significance of calcium phosphate coatings for the enhancement of new bone osteogenesis – A review. *Acta Biomaterialia*, *10*(2), 557–579. DOI: 10.1016/j.actbio.2013.10.036 PMID: 24211734

Su, Y., Li, D., Su, Y., Lu, C., Niu, L., Lian, J., & Li, G. (2016). Improvement of the Biodegradation Property and Biomineralization Ability of Magnesium-Hydroxyapatite Composites with Dicalcium Phosphate Dihydrate and Hydroxyapatite Coatings. *ACS Biomaterials Science &amp. ACS Biomaterials Science & Engineering*, *2*(5), 818–828. Advance online publication. DOI: 10.1021/acsbiomaterials.6b00013 PMID: 33440579

Su, Y., Lu, C., Hu, X., Guo, Y., Xun, X., Zhang, Z., Li, G., Lian, J., & Ren, L. (2018). Improving the Degradation Resistance and Surface Biomineralization Ability of Calcium Phosphate Coatings on a Biodegradable Magnesium Alloy via a Sol-Gel Spin Coating Method. *Journal of The Electrochemical Society, 165*. https://doi.org/ DOI: 10.1149/2.0901803jes

Sypniewska, J., & Szkodo, M. (2022). Influence of Laser Modification on the Surface Character of Biomaterials: Titanium and Its Alloys—A Review. *Coatings*, *12*(10), 1371. DOI: 10.3390/coatings12101371

Szczęsny, G., Kopec, M., & Kowalewski, Z. L. (2025). Toxicity, irritation, and allergy of metal implants: Historical perspective and modern solutions. *Coatings*, *15*(3), 361. DOI: 10.3390/coatings15030361

Tabatabaeian, A., Ghasemi, A. R., Shokrieh, M. M., Marzbanrad, B., Baraheni, M., & Fotouhi, M. (2022). Residual stress in engineering materials: A review. *Advanced Engineering Materials*, *24*(3), 2100786. DOI: 10.1002/adem.202100786

Tajabadi, M., Barghamadi, H., Lahiji, Z. P., & Momeninasab, E. (2026). Scaffolds, grafts, and coatings. In Musculoskeletal repair and regeneration: New tools, technology and techniques (pp. 171-294). DOI: 10.1142/9789819819645_0004

Takeda, M., Ankamatsu, S., Kita, Y., Goto, T., & Kobayashi, T. (2023). The roles of extracellular vesicles in the progression of renal cell carcinoma and their potential for future clinical application. *Nanomaterials (Basel, Switzerland)*, *13*(10), 1611. https://www.mdpi.com/2079-4991/13/10/1611. DOI: 10.3390/nano13101611 PMID: 37242027

Tanaka, Y., Xiao, L., & Miwa, N. (2022). Hydrogen-rich bath with nano-sized bubbles improves antioxidant capacity based on oxygen radical absorbing and inflammation levels in human serum. *Medical Gas Research*, *12*(3), 91–99. DOI: 10.4103/2045-9912.330692 PMID: 34854419

Tang, W., Fischer, N. G., Kong, X., Sang, T., & Ye, Z. (2024). Hybrid coatings on dental and orthopedic titanium implants: Current advances and challenges. *BMEmat*, *2*(4), e12105. DOI: 10.1002/bmm2.12105

Tao, W., Ivanovska, V., Schweickert, B., & Müller, A. (2019). Proxy indicators for antibiotic consumption; surveillance needed to control antimicrobial resistance. *Bulletin of the World Health Organization*, *97*(1), 3–3A. DOI: 10.2471/BLT.18.227348 PMID: 30618457

Tavafoghi, M., & Cerruti, M. (2016). The role of amino acids in hydroxyapatite mineralization. In *Journal of the Royal Society Interface*. DOI: 10.1098/rsif.2016.0462

Thakur, A. J. (2022). *The elements of fracture fixation (E-book)*. Elsevier Health Sciences.

Thakur, A. K., Lin, B., Nowrin, F. H., & Malmali, M. (2021). Comparing structure and sorption characteristics of laser-induced graphene (LIG) from various polymeric substrates. *ACS ES&T Water*, *2*(1), 75–87. DOI: 10.1021/acsestwater.1c00259

Thakur, A., Kumar, A., Kaya, S., Marzouki, R., Zhang, F., & Guo, L. (2022). Recent advancements in surface modification, characterization and functionalization for enhancing the biocompatibility and corrosion resistance of biomedical implants. *Coatings*, *12*(10), 1459. DOI: 10.3390/coatings12101459

Thapliyal, D., Kumar, P., & Saxena, A. (2025). Nanoparticle-doped antibacterial and antifungal coatings: A review. *Polymers*, *17*(2), 247. Advance online publication. DOI: 10.3390/polym17020247 PMID: 39861318

Theerthagiri, J., Karuppasamy, K., Lee, S. J., Shwetharani, R., Kim, H., Pasha, S. K. K., Ashokkumar, M., & Choi, M. Y. (2022). Fundamentals and comprehensive insights on pulsed laser synthesis of advanced materials for diverse photo- and electrocatalytic applications. *Light, Science & Applications*, *11*(1), 210. Advance online publication. DOI: 10.1038/s41377-022-00904-7 PMID: 35945216

Theodorou, G., Goudouri, O. M., Kontonasaki, E., Chatzistavrou, X., Papadopoulou, L., Kantiranis, N., & Paraskevopoulos, K. M. (2011). Comparative Bioactivity Study of 45S5 and 58S Bioglasses in Organic and Inorganic Environment. *Bioceramics Development and Applications*, *1*, 1–4. DOI: 10.4303/bda/D110154

Théry, C., Witwer, K. W., Aikawa, E., Alcaraz, M. J., Anderson, J. D., Andriantsitohaina, R., Antoniou, A., Arab, T., Archer, F., Atkin-Smith, G. K., Ayre, D. C., Bach, J.-M., Bachurski, D., Baharvand, H., Balaj, L., Baldacchino, S., Bauer, N. N., Baxter, A. A., Bebawy, M., & Zuba-Surma, E. K. (2018). Minimal information for studies of extracellular vesicles 2018 (MISEV2018): A position statement of the International Society for Extracellular Vesicles and update of the MISEV2014 guidelines. *Journal of Extracellular Vesicles*, *7*(1), 1535750. DOI: 10.1080/20013078.2018.1535750 PMID: 30637094

Thiévenaz, V., Meijer, J. G., Lohse, D., & Sauret, A. (2025). On the shape of air bubbles trapped in ice. *Proceedings of the National Academy of Sciences of the United States of America*, *122*(10), e2415027122. DOI: 10.1073/pnas.2415027122 PMID: 40035763

Todorov, D. N. (2020, November). Practical research of marking and cutting of textiles with increased resistance, using CO2 laser. []. IOP Publishing.]. *Journal of Physics: Conference Series*, *1681*(1), 012014. DOI: 10.1088/1742-6596/1681/1/012014

Trentin, A., de L. Gasparini, A., Faria, F. A., Harb, S. V., dos Santos, F. C., Pulcinelli, S. H., Santilli, C. V., & Hammer, P. (2020). Barrier properties of high performance PMMA-silica anticorrosion coatings. *Progress in Organic Coatings*, *138*, 105398. DOI: 10.1016/j.porgcoat.2019.105398

Trentin, A., Pakseresht, A., Duran, A., Castro, Y., & Galusek, D. (2022). Electrochemical Characterization of Polymeric Coatings for Corrosion Protection: A Review of Advances and Perspectives. *Polymers*, *14*(12), 2306. DOI: 10.3390/polym14122306 PMID: 35745882

Tripathy, A., Sen, P., Su, B., & Briscoe, W. H. (2017). Natural and bioinspired nanostructured bactericidal surfaces. *Advances in Colloid and Interface Science*, *248*, 85–104. DOI: 10.1016/j.cis.2017.07.030 PMID: 28780961

Tsang, C. Y., & Zhang, Y. (2024). Nanomaterials for light-mediated therapeutics in deep tissue. *Chemical Society Reviews*, *53*(6), 2898–2931. DOI: 10.1039/D3CS00862B PMID: 38265834

Tuikampee, T., Chaijareenont, P., Rungsiyakull, P., & Yavirach, A. (2024). Titanium surface modification techniques to enhance osteoblasts and bone formation. *Metals*, *14*(5), 515. Advance online publication. DOI: 10.3390/met14050515

Tu, K. T., & Chung, C. K. (2016). Rapid prototyping of biodegradable microneedle arrays by integrating CO2 laser processing and polymer molding. *Journal of Micromechanics and Microengineering*, *26*(6), 065015. DOI: 10.1088/0960-1317/26/6/065015

Tynan, J., Ward, P., Byrne, G., & Dowling, D. P. (2009). Deposition of Biodegradable Polycaprolactone Coatings Using an In-line Atmospheric Pressure Plasma System. *Plasma Processes and Polymers*, *6*(S1). Advance online publication. DOI: 10.1002/ppap.200930310

Tzagiollari, A. (2024). Development of a biomimetic phosphoserine modified calcium phosphate adhesive for the regeneration of bone defects [Doctoral dissertation].

Vahabzadeh, S., Roy, M., Bandyopadhyay, A., & Bose, S. (2015). Phase stability and biological property evaluation of plasma sprayed hydroxyapatite coatings for orthopedic and dental applications. *Acta Biomaterialia*, *17*, 47–55. Advance online publication. DOI: 10.1016/j.actbio.2015.01.022 PMID: 25638672

Vakulov, Z., Khakhulin, D., Zamburg, E., Mikhaylichenko, A., Smirnov, V. A., Tominov, R., Klimin, V. S., & Ageev, O. A. (2021). Towards scalable large-area pulsed laser deposition. *Materials (Basel)*, *14*(17), 4854. Advance online publication. DOI: 10.3390/ma14174854 PMID: 34500943

Van den Borre, C. E., Zigterman, B. G., Mommaerts, M. Y., & Braem, A. (2022). How surface coatings on titanium implants affect keratinized tissue: A systematic review. *Journal of Biomedical Materials Research. Part B, Applied Biomaterials*, *110*(7), 1713–1723. DOI: 10.1002/jbm.b.35025 PMID: 35103386

van Ede, M. C., & Angst, U. (2024). Tafel slopes and exchange current densities of oxygen reduction and hydrogen evolution on steel. *Corrosion Engineering, Science and Technology*, *59*(1), 39–55. DOI: 10.1177/1478422X241227829 PMID: 42007378

van Gaalen, K. (2023). An experimental and computational investigation of the corrosion behaviour and mechanical performance of a bioabsorbable medical grade magnesium alloy.

van Raemdonck, W., Ducheyne, P., & de Meester, P. (1984). Auger Electron Spectroscopic Analysis of Hydroxylapatite Coatings on Titanium. *Journal of the American Ceramic Society*, *67*(6), 381–384. Advance online publication. DOI: 10.1111/j.1151-2916.1984.tb19721.x

Vardaki, M., Prodana, M., Stoian, A. B., & Ionita, D. (2018). Corrosion and bioactivity of a bioinspired coating on tizr alloys. *UPB Scientific Bulletin, Series B. Chemistry and Materials Science*, *80*(1), 209–218.

Varut, R. M., Trasca, D. M., Stoica, G. A., Sirbulet, C., Arsenie, C. C., & Popescu, C. (2025). Animal models as foundational tools in preclinical orthopedic implant research. *Biomedicines*, *13*(10), 2468. DOI: 10.3390/biomedicines13102468 PMID: 41153751

Vasile, E., Serafim, A., Dragusin, D.-M., Petrea, C., Iovu, H., & Stancu, I.-C. (2012). Apatite formation on active nanostructured coating based on functionalized gold nanoparticles. *Journal of Nanoparticle Research*, *14*(6), 918. Advance online publication. DOI: 10.1007/s11051-012-0918-1

Vázquez-López, A., de la Vega, J., Collado, I., Carmona, F. J., Pradanos, P., Prolongo, S. G., & Wang, D. Y. (2024). Graphene Oxide/Polylactic Acid-Based Face Mask to Combat H3N2: A Strategy against Influenza. *ACS Applied Nano Materials*, *7*(6), 6460–6470. DOI: 10.1021/acsanm.4c00183

Vedani, M. (2025). *Metal science in modern manufacturing technologies*. Springer Nature. DOI: 10.1007/978-3-031-97125-9

Veerapandian, M., & Yun, K. (2011). Functionalization of biomolecules on nanoparticles: Specialized for antibacterial applications. *Applied Microbiology and Biotechnology*, *90*(5), 1655–1667. DOI: 10.1007/s00253-011-3291-6 PMID: 21523475

Verdoia, M., Kedhi, E., Suryapranata, H., Galasso, G., Dudek, D., & De Luca, G. (2020). Poly (L-lactic acid) bioresorbable scaffolds versus metallic drug-eluting stents for the treatment of coronary artery disease: A meta-analysis of 11 randomized trials. *Catheterization and Cardiovascular Interventions*, *96*(4), 813–824. DOI: 10.1002/ccd.28594 PMID: 31730255

Verma, P. K., Vasudev, H., & Kumar, V. (2025). Effect of electro-corrosion behavior of different grades of stainless steel used in bio-implants: A review. Advanced Manufacturing Processes, 269-283.

Verma, P. K., Vasudev, H., & Singh, B. (2025). Environmental and economic considerations in the fabrication of coatings and claddings. In *Microwave processing of metallic materials* (pp. 216–236). CRC Press. DOI: 10.1201/9781003614081-12

Visan, A. I., & Negut, I. (2025). Polymeric Composite Thin Films Deposited by Laser Techniques for Antimicrobial Applications—A Short Overview. *Polymers*, *17*(15), 2020. DOI: 10.3390/polym17152020 PMID: 40808071

Visan, A. I., & Popescu-Pelin, G. F. (2024). Advanced Laser Techniques for the Development of Nature-Inspired Biomimetic Surfaces Applied in the Medical Field. *Coatings*, *14*(10), 1290. DOI: 10.3390/coatings14101290

Vivier, V., & Orazem, M. E. (2022). Impedance analysis of electrochemical systems. *Chemical Reviews*, *122*(12), 11131–11168. DOI: 10.1021/acs.chemrev.1c00876 PMID: 35687869

Vladescu, A., Pârâu, A., Pana, I., Cotrut, C. M., Constantin, L. R., Braic, V., & Vrânceanu, D. M. (2019). In vitro activity assays of sputtered HAp coatings with SiC addition in various simulated biological fluids. *Coatings*, *9*(6), 389. Advance online publication. DOI: 10.3390/coatings9060389

Vorobyev, A. Y., & Guo, C. (2013). Direct femtosecond laser surface nano/microstructuring and its applications. *Laser & Photonics Reviews*, *7*(3), 385–407. DOI: 10.1002/lpor.201200017

Vranceanu, D. M., Parau, A. C., Cotrut, C. M., Kiss, A. E., Constantin, L. R., Braic, V., & Vladescu, A. (2019). In vitro evaluation of Ag doped hydroxyapatite coatings in acellular media. *Ceramics International*, *45*(8), 11050–11061. DOI: 10.1016/j.ceramint.2019.02.191

Walsh, F. C., Ponce de León, C., Kerr, C., Court, S., & Barker, B. D. (2008). Electrochemical characterisation of the porosity and corrosion resistance of electrochemically deposited metal coatings. *Surface and Coatings Technology*, *202*(21), 5092–5102. DOI: 10.1016/j.surfcoat.2008.05.008

Wang, W., & Yeung, K. W. K. (2017). Bone grafts and biomaterials substitutes for bone defect repair: A review. In *Bioactive Materials*. DOI: 10.1016/j.bioactmat.2017.05.007

Wang, B., Zhang, Z., & Pan, H. (2023). Bone apatite nanocrystal: Crystalline structure, chemical composition, and architecture. *Biomimetics*, *8*(1), 90. DOI: 10.3390/biomimetics8010090 PMID: 36975320

Wang, H., Deng, D., Zhai, Z., & Yao, Y. (2024). Laser-processed functional surface structures for multi-functional applications-a review. *Journal of Manufacturing Processes*, *116*, 247–283. DOI: 10.1016/j.jmapro.2024.02.062

Wang, H., Dong, B. S., Chen, Z. B., Liu, J. Q., Haghdadi, N., Lu, R. Q., Primig, S., Wang, Z. Y., Pan, Z. X., Li, H. J., Ringer, S. P., & Liao, X. Z. (2024). Effect of compositional heterogeneity on the mechanical properties of a single-phase Cu-9Al alloy with different grain sizes. *Acta Materialia*, *263*, 119531. DOI: 10.1016/j.actamat.2023.119531

Wang, H., Lee, J.-K., Moursi, A., & Lannutti, J. (2003). Ca/P ratio effects on the degradation of hydroxyapatite in vitro. *Journal of Biomedical Materials Research. Part A*, *67*(2), 599–608. DOI: 10.1002/jbm.a.10538 PMID: 14566803

Wang, H., Zhao, Z., Liu, P., & Guo, X. (2022). Laser-induced graphene based flexible electronic devices. *Biosensors (Basel)*, *12*(2), 55. DOI: 10.3390/bios12020055 PMID: 35200316

Wang, J. (2023). *Influence of solution temperature on chemical deposited magnesium and corrosion behavior of bare and coated magnesium in simulated body fluid.* Washington State University.

Wang, K., Liu, W., Hong, Y., Sohan, H. S., Tong, Y., Hu, Y., Zhang, M., Zhang, J., Xiang, D., Fu, H., & Ju, J. (2023). An overview of technological parameter optimization in the case of laser cladding. *Coatings*, *13*(3), 496. DOI: 10.3390/coatings13030496

Wang, L., Feng, S., Wang, Y., Zhao, X., Ge, J., Gao, T., & Di, F. (2026). Porosity defects in additively manufactured metal materials: Formation mechanisms, impact on performance and regulation. *International Materials Reviews*, *71*(2), 97–128. DOI: 10.1177/09506608251371459

Wang, L., Liu, K., Li, J., & Geng, S. (2025). A review on corrosion behavior and surface modification technology of nickel aluminum bronze alloys: Current research and prospects. *Advanced Engineering Materials*, *27*(1), 2401779. DOI: 10.1002/adem.202401779

Wang, M. F., Yan, T., Gao, M. C., Han, C. W., Yan, Z. Q., Gao, Y. Z., Zhang, W., & Yi, Z. (2025). A review of the advances in implant technology: Accomplishments and challenges for the design of functionalized surface structures. *Biomedical Materials (Bristol, England)*, *20*(3), 032003. DOI: 10.1088/1748-605X/adca7c PMID: 40199334

Wang, Q., Xu, S., Xing, X., & Wang, N. (2022). Progress in fabrication and applications of micro/nanostructured superhydrophobic surfaces. *Surface Innovations*, *10*(2), 89–109. DOI: 10.1680/jsuin.21.00031

Wang, T., Li, Q., Huo, F., Chen, H., & Xu, T. (2025). Comparison of Microstructure and Properties of CoCrMo Coatings Prepared by High-Speed and Conventional Laser Cladding. *Coatings*, *15*(10), 1200. DOI: 10.3390/coatings15101200

Wang, Y., Wang, H., Peng, S., Xia, B., & Zhu, H. (2024). The dominant role of recrystallization and grain growth behaviors in the simulated welding heat-affected zone of high-Mn steel. *Materials (Basel)*, *17*(10), 2218. DOI: 10.3390/ma17102218 PMID: 38793285

Wang, Z., Qi, F., Luo, H., Xu, G., & Wang, D. (2022). Inflammatory microenvironment of skin wounds. *Frontiers in Immunology*, *13*, 789274. DOI: 10.3389/fimmu.2022.789274 PMID: 35300324

Wang, Z., Tong, S., Niu, J., Cao, C., Gao, A., Jiao, Y., Fu, Y., Li, D., Pan, X., Cui, D., Sheng, N., Yan, L., Cui, S., Lin, S., & Liu, Y. (2025). Microneedles: Multifunctional devices for drug delivery, body fluid extraction, and bio-sensing. *Nanoscale*, *17*(2), 740–773. DOI: 10.1039/D4NR03538K PMID: 39606819

Wang, Z., Xu, Z., Zhao, W., & Sahai, N. (2015). A potential mechanism for amino acid-controlled crystal growth of hydroxyapatite. *Journal of Materials Chemistry. B*, *3*(47), 9157–9167. Advance online publication. DOI: 10.1039/C5TB01036E PMID: 32263130

Watson, J. L., Seinkmane, E., Styles, C. T., Mihut, A., Krüger, L. K., McNally, K. E., Planelles-Herrero, V. J., Dudek, M., McCall, P. M., Barbiero, S., Vanden Oever, M., Peak-Chew, S. Y., Porebski, B. T., Zeng, A., Rzechorzek, N. M., Wong, D. C. S., Beale, A. D., Stangherlin, A., Riggi, M., & Derivery, E. (2023). Macromolecular condensation buffers intracellular water potential. *Nature*, *623*(7988), 842–852. DOI: 10.1038/s41586-023-06626-z PMID: 37853127

Webster, T. J., Massa-Schlueter, E. A., Smith, J. L., & Slamovich, E. B. (2004). Osteoblast response to hydroxyapatite doped with divalent and trivalent cations. *Biomaterials*, *25*(11), 2111–2121. DOI: 10.1016/j.biomaterials.2003.09.001 PMID: 14741626

Wei, F., Wei, Y., Yao, X., Li, X., Wei, Z., Zhang, S., Luo, X., Guo, Z., Liu, Q., & Zhu, Q. (2025). Enhancing bond strength of heterogeneous metal-polymer components the perspective of surface micro-nano morphology construction. *Journal of Materials Science*, *60*(14), 6023–6058. DOI: 10.1007/s10853-025-10795-9

Wei, J., Igarashi, T., Okumori, N., Igarashi, T., Maetani, T., Liu, B., & Yoshinari, M. (2009). Influence of surface wettability on competitive protein adsorption and initial attachment of osteoblasts. *Biomedical Materials (Bristol, England)*, *4*(4), 045002. Advance online publication. DOI: 10.1088/1748-6041/4/4/045002 PMID: 19525576

Wei, L., & Gao, Z. (2023). Recent research advances on corrosion mechanism and protection, and novel coating materials of magnesium alloys: A review. *RSC Advances*, *13*(12), 8427–8463. DOI: 10.1039/D2RA07829E PMID: 36926015

Wei, X., Zheng, S., An, Y., Yu, S., Liu, G., & Pu, J. (2024). Microstructure and tribocorrosion properties of Fe-based amorphous composite coatings fabricated by laser cladding. *Journal of Materials Science*, *59*(9), 4039–4058. DOI: 10.1007/s10853-024-09436-4

Wen, D., Yuan, R., Cao, K., Yang, F., & Chen, R. (2024). Advancements in atomic-scale interface engineering for flexible electronics: Enhancing flexibility and durability. *Nanotechnology*, *35*(41), 412501. DOI: 10.1088/1361-6528/ad64db PMID: 39025081

Wennerberg, A., & Albrektsson, T. (2009). On implant surfaces: A review of current knowledge and opinions. *The International Journal of Oral & Maxillofacial Implants*, *25*(1), 63–74. DOI: 10.1111/j.1600-051X.2008.01321.x PMID: 20209188

Williams, D. F., & Williams, R. L. (2013). Degradative Effects of the Biological Environment on Metals and Ceramics. *Biomaterials Science*, (1998), 728–738. DOI: 10.1016/B978-0-08-087780-8.00062-0

Williams, D. F. (2008). On the mechanisms of biocompatibility. *Biomaterials*, *29*(20), 2941–2953. DOI: 10.1016/j.biomaterials.2008.04.023 PMID: 18440630

Williams, D. F. (2022). Biocompatibility pathways and mechanisms for bioactive materials: The bioactivity zone. *Bioactive Materials*, *10*, 306–322. DOI: 10.1016/j.bioactmat.2021.08.014 PMID: 34901548

Willms, E., Cabañas, C., Mäger, I., Wood, M. J. A., & Vader, P. (2018). Extracellular vesicle heterogeneity: Subpopulations, isolation techniques, and diverse functions in cancer progression. *Frontiers in Immunology*, *9*, 738. DOI: 10.3389/fimmu.2018.00738 PMID: 29760691

Wolstenholme, J. (1973). Electrochemical methods of assessing the corrosion of painted metals—A review. *Corrosion Science*, *13*(7), 521–530. DOI: 10.1016/S0010-938X(73)80002-2

World Health Organization. (2011). Report on the burden of endemic health care-associated infection worldwide. https://iris.who.int/bitstream/handle/10665/80135/

Wu, C. H., Chang, F. Y., Lin, C. J., & Teng, P. T. (2026). Development of magnesium alloy stents with layered double hydroxide coating for improved corrosion resistance and biochemical stability in AVF applications. *Journal of Functional Biomaterials*, *17*(2), 76. DOI: 10.3390/jfb17020076 PMID: 41745538

Wu, C. S. (2015). Influence of modified polyester on the material properties of collagen-based biocomposites and in vitro evaluation of cytocompatibility. *Materials Science and Engineering C*, *48*, 310–319. DOI: 10.1016/j.msec.2014.12.013 PMID: 25579928

Wu, J., Huang, Y., Tan, Y., Zhou, J., Yao, J., Tan, L., & Liu, F. (2025). Particle size-dependent solidification heterogeneity: Simulation and experimental decoupling of gas-atomized Ni-based superalloy powders. *Powder Technology*, •••, 121658.

Wu, L., Wei, S., Cheng, X., He, N., Kang, X., Zhou, H., Yang, S., & Pan, X. (2024). Release of ions enhanced the antibacterial performance of laser-generated, uncoated Ag nanoparticles. *Colloids and Surfaces. B, Biointerfaces*, *238*, 113894. DOI: 10.1016/j.colsurfb.2024.114131 PMID: 39094211

Wu, T., Li, H., Xue, J., Mo, X., & Xia, Y. (2019). Photothermal welding, melting, and patterned expansion of nonwoven mats of polymer nanofibers for biomedical and printing applications. *Angewandte Chemie International Edition*, *58*(46), 16416–16421. DOI: 10.1002/anie.201907876 PMID: 31373102

Wu, Y., Heikal, L., Ferns, G., Ghezzi, P., Nokhodchi, A., & Maniruzzaman, M. (2019). 3D bioprinting of novel biocompatible scaffolds for endothelial cell repair. *Polymers*, *11*(12), 1924. DOI: 10.3390/polym11121924 PMID: 31766610

Xia, B., Liu, Y., Xing, Y., Shi, Z., & Pan, X. (2025). Biodegradable medical implants: Reshaping future medical practice. *Advancement of Science*, *12*(35), e08014. DOI: 10.1002/advs.202508014 PMID: 40782070

Xia, J., Zhang, J., Jiao, H., Niu, X., Ji, X., He, T., & Wang, Z. (2026). Research progress and prospects of laser coating technology. Light. *Advances in Manufacturing*, *6*(4), 870–895.

Xie, H., Hong, G., Zhou, Q., Wu, Y., Chen, Y., You, R., & Han, F. (2023). Nanocomposite surface modification. In *Innovative Development in Micromanufacturing Processes* (pp. 296–371). CRC Press. DOI: 10.1201/9781003364948-15

Xie, H., Wang, Z., Zhang, L., Lei, Q., Zhao, A., Wang, H., Li, Q., Chen, Z., & Zhang, W. (2016). Development of an angiogenesis-promoting microvesicle-alginate-polycaprolactone composite graft for bone tissue engineering applications. *PeerJ*, *4*, e2040. DOI: 10.7717/peerj.2040 PMID: 27231660

Xie, W., Wei, X., Kang, H., Jiang, H., Chu, Z., Lin, Y., Hou, Y., & Wei, Q. (2023). Static and dynamic: Evolving biomaterial mechanical properties to control cellular mechanotransduction. *Advancement of Science*, *10*(9), 2204594. DOI: 10.1002/advs.202204594 PMID: 36658771

Xie, Y., Liu, F., Zhang, X., Jin, Y., Li, Q., Shen, H., Fu, H., & Mao, J. (2022). Benefits and risks of essential trace elements in chronic kidney disease: A narrative review. *Annals of Translational Medicine*, *10*(24), 1400. DOI: 10.21037/atm-22-5969 PMID: 36660676

Xing, Y., Xiong, W., Li, X., & Zhang, J. (2021). Biofunctionalization of zirconia implant surfaces via femtosecond laser-induced microstructures. *Journal of Biomedical Materials Research. Part A*, *109*(10), 1875–1886.

Xu, L.. (2025). L-arginine-induced versatile hydrogel with skin adaptiveness for wound healing. *Chemical Engineering Journal*. Advance online publication. DOI: 10.1016/j.cej.2025.152376

Xu, L., Liu, Y., Zhou, W., & Yu, D. G. (2022). Electrospun Medical Sutures for Wound Healing: A Review. *Polymers*, *2022*(14), 1637. DOI: 10.3390/polym14091637 PMID: 35566807

Xu, T. Z., Zhang, S., Wang, L., Du, Y., Wu, C. L., Zhang, C. H., Sun, X. Y., Chen, H. T., & Chen, J. (2023). Influence of scanning speed on the microstructure, nanoindentation characteristics and tribological behavior of novel maraging steel coatings by laser cladding. *Materials Characterization*, *205*, 113335. DOI: 10.1016/j.matchar.2023.113335

Xu, Z., Wang, Y. A., Ng, V., Yin, H., & Xu, S. (2025). Advancements in Laser-Processed Functional Surfaces for Medical Devices: A Current Review. *Nanomaterials (Basel, Switzerland)*, *15*(13), 999. DOI: 10.3390/nano15130999 PMID: 40648706

Yadav, B., Mohan Muralikrishna, G., Vaidya, M., Wilde, G., & Divinski, S. V. (2026). Grain boundary diffusion in compositionally complex alloys: A comprehensive review. *International Materials Reviews*, *71*(2), 129–177. DOI: 10.1177/09506608251390882

Yadroitsev, I., Krakhmalev, P., & Yadroitsava, I. (2014). Selective laser melting of Ti6Al4V alloy for biomedical applications: Temperature monitoring and microstructural evolution. *Journal of Alloys and Compounds*, *583*, 404–409. DOI: 10.1016/j.jallcom.2013.08.183

Yandrapalli, N. (2025). Bottom-up development of lipid-based synthetic cells for practical applications. *Trends in Biotechnology*, *0*(0), 2150–2169. Advance online publication. DOI: 10.1016/j.tibtech.2025.03.009 PMID: 40263003

Yang, C.-J., Mei, X.-S., Tian, Y.-L., Zhang, D.-W., Li, Y., & Liu, X.-P. (2016). Modification of wettability property of titanium by laser texturing. *International Journal of Advanced Manufacturing Technology*, *87*(5-8), 1663–1670. Advance online publication. DOI: 10.1007/s00170-016-8601-9

Yang, J., Zhang, W., Lin, B., Mao, S., Liu, G., Tan, K., & Tang, J. (2024). Enhancement of Local Osseointegration and Implant Stability of Titanium Implant in Osteoporotic Rats by Biomimetic Multilayered Structures Containing Catalpol. *ACS Omega*, *9*(27), 29544–29556. DOI: 10.1021/acsomega.4c02322 PMID: 39005760

Yang, L., Wei, J., Ma, Z., Song, P., Ma, J., Zhao, Y., Huang, Z., Zhang, M., Yang, F., & Wang, X. (2019). The fabrication of micro/nano structures by laser machining. *Nanomaterials (Basel, Switzerland)*, *9*(12), 1789. DOI: 10.3390/nano9121789 PMID: 31888222

Yang, S., & Zhang, J. (2017). Matrix-Assisted Pulsed Laser Evaporation (MAPLE) technique for deposition of hybrid nanostructures. *Front. Nanosci. Nanotechnol*, *3*(1). Advance online publication. DOI: 10.15761/FNN.1000146

Yang, Y. L., Paital, S. R., & Dahotre, N. B. (2010). Wetting and in vitro bioactivity of laser processed CaP coating with presence and variation of SiO2 on Ti–6Al–4V. *Materials Technology*, *25*(3-4), 137–142. Advance online publication. DOI: 10.1179/175355510X12723642365287

Yang, Y., Paital, S. R., & Dahotre, N. B. (2010). Effects of SiO2 substitution on wettability of laser deposited Ca-P biocoating on Ti-6Al-4V. *Journal of Materials Science. Materials in Medicine*, *21*(9), 2511–2521. Advance online publication. DOI: 10.1007/s10856-010-4105-6 PMID: 20549311

Yang, Y., Wen, C., & Hodgson, P. D. (2020). Laser surface modification of titanium alloys for biomedical applications. *Materials & Design*, *191*, 108626.

Yang, Z., Wu, C., Shi, H., Luo, X., Sun, H., Wang, Q., & Zhang, D. (2022). Advances in barrier membranes for guided bone regeneration techniques. *Frontiers in Bioengineering and Biotechnology*, *10*, 921576. DOI: 10.3389/fbioe.2022.921576 PMID: 35814003

Yan, L., Yuan, Y., Ouyang, L., Li, H., Mirzasadeghi, A., & Li, L. (2016). Improved mechanical properties of the new Ti-15Ta-xZr alloys fabricated by selective laser melting for biomedical application. *Journal of Alloys and Compounds*, *688*, 156–162. DOI: 10.1016/j.jallcom.2016.07.002

Yao, Q., Li, W., Yu, S., Ma, L., Jin, D., Boccaccini, A. R., & Liu, Y. (2015). Multifunctional chitosan/polyvinyl pyrrolidone/45S5 Bioglass® scaffolds for MC3T3-E1 cell stimulation and drug release. *Materials Science and Engineering C*, *56*, 473–480. DOI: 10.1016/j.msec.2015.06.046 PMID: 26249617

Yeo, T., Ko, Y. G., Kim, E. J., Kwon, O. K., Chung, H. Y., & Kwon, O. H. (2021). Promoting bone regeneration by 3D-printed poly (glycolic acid)/hydroxyapatite composite scaffolds. *Journal of Industrial and Engineering Chemistry*, *94*, 343–351. DOI: 10.1016/j.jiec.2020.11.004

Yi, G., Teong, S. P., Zhang, Y., & Ying, J. Y. (2026). Inorganic Antimicrobial Materials. *ACS Nano*, *20*(15), 11495–11523. Advance online publication. DOI: 10.1021/acsnano.5c21363 PMID: 41945705

Yin, Y., Prabhakar, M., Ebbinghaus, P., Corrêa da Silva, C., & Rohwerder, M. (2022). Neutral inhibitor molecules entrapped into polypyrrole network for corrosion protection. *Chemical Engineering Journal*, *440*, 135739. DOI: 10.1016/j.cej.2022.135739

Yong, J., Gröger, S., Wu, Z., Ruf, S., Ye, Y., & Chen, X. (2023). Photobiomodulation therapy and pulp-regenerative endodontics: A narrative review. *Bioengineering (Basel, Switzerland)*, *10*(3), 371. Advance online publication. DOI: 10.3390/bioengineering10030371 PMID: 36978762

Young, O. M., Xu, X., Sarker, S., & Sochol, R. D. (2024). Direct laser writing-enabled 3D printing strategies for microfluidic applications. *Lab on a Chip*, *24*(9), 2371–2396. DOI: 10.1039/D3LC00743J PMID: 38576361

Yu, C., Ye, J., Chang, H., Zhou, W., Han, X., Li, M., & Gao, H. (2023). Experimental research on characteristics of impulse coupling and plasma plume generated by laser irradiating copper target with nanosecond pulsed laser propulsion. *Aerospace (Basel, Switzerland)*, *10*(6), 544. Advance online publication. DOI: 10.3390/aerospace10060544

Yue, W., Zhang, Y., Zheng, Z., & Lai, Y. (2024). Hybrid laser additive manufacturing of metals: A review. *Coatings*, *14*(3), 315. DOI: 10.3390/coatings14030315

Yu, J., Huang, W., & Li, J.. (2024). Recent advances on pulsed laser deposition of large-area 2D materials. *Small Methods*, *8*(10), 2301282. Advance online publication. DOI: 10.1002/smtd.202301282 PMID: 39082065

Yu, J., Sane, S., Kim, J. E., Yun, S., Kim, H. J., Jo, K. B., Wright, J. P., Khoshdoozmasouleh, N., Lee, K., Oh, H. T., Thiel, K., Parvin, A., Williams, X., Hannon, C., Lee, H., & Kim, D. K. (2024). Biogenesis and delivery of extracellular vesicles: Harnessing the power of EVs for diagnostics and therapeutics. *Frontiers in Molecular Biosciences*, *10*, 1330400. DOI: 10.3389/fmolb.2023.1330400 PMID: 38234582

Zaludin, M. A. F., Jamal, Z. A. Z., Derman, M. N., & Kasmuin, M. Z. (2019). Corrosion analysis of hydroxyapatite coated magnesium in simulated body fluid (SBF), phosphate buffered saline (PBS), and Ringer's solution. *Materials Today: Proceedings*, *16*, 1686–1691. DOI: 10.1016/j.matpr.2019.06.037

Zarif, M. E., Bita, B., Yehia-Alexe, S. A., Negut, I., Gradisteanu Pircalabioru, G., Andronescu, E., & Groza, A. (2024). Biological and Physicochemical Analysis of Sr-Doped Hydroxyapatite/Chitosan Composite Layers. *Polymers, 16*(13), 1922. DOI: 10.3390/polym16131922 PMID: 39000777

Zhang, D., Wang, M., Zhao, Y., Yang, C., & Yang, Z. (2019). Modification of Wettability Property of NITI Alloy by Laser Texturing and Carbon Ion Implantation. *2019 IEEE International Conference on Manipulation, Manufacturing and Measurement on the Nanoscale (3M-NANO)*, 224–227. DOI: 10.1109/3M-NANO46308.2019.8947423

Zhang, F., Cai, S., Xu, G., Shen, S., Li, Y., Zhang, M., & Wu, X. (2016). Corrosion behavior of mesoporous bioglass-ceramic coated magnesium alloy under applied forces. *Journal of the Mechanical Behavior of Biomedical Materials, 56*, 146–155. DOI: 10.1016/j.jmbbm.2015.11.029 PMID: 26703229

Zhang, H., Ijaz, M., & Blaikie, R. J. (2023). Recent review of surface plasmons and plasmonic hot electron effects in metallic nanostructures. *Frontiers of Physics, 18*(6), 63602. DOI: 10.1007/s11467-023-1328-9

Zhang, J. M., Lin, C. J., Feng, Z. D., & Tian, Z. W. (1998). Mechanistic studies of electrodeposition for bioceramic coatings of calcium phosphates by an in situ pH-microsensor technique. *Journal of Electroanalytical Chemistry (Lausanne, Switzerland), 452*(2), 235–240. DOI: 10.1016/S0022-0728(98)00107-7

Zhang, J., Yin, C., Farbiz, F., Jafary-Zadeh, M., & Sing, S. L. (2025). Advancing machine learning applications for in-situ monitoring and control in laser-based metal additive manufacturing: A state-of-the-art review. *Virtual and Physical Prototyping, 20*(1), e2592732. DOI: 10.1080/17452759.2025.2592732

Zhang, L.. (2021). Laser-induced periodic surface structures (LIPSS) on titanium with simultaneous antibacterial and biocompatible effects. *Journal of Biomedical Materials Research. Part A, 109*(3), 362–373. PMID: 32506802

Zhang, L., Xiong, N., Liu, Y., & Gan, L. (2021). Biomimetic cell-adhesive ligand-functionalized peptide composite hydrogels maintain stemness of human amniotic mesenchymal stem cells. *Regenerative Biomaterials, 8*(2), rbaa057. DOI: 10.1093/rb/rbaa057 PMID: 33738111

Zhang, Q., Chen, J., Feng, J., Cao, Y., Deng, C., & Zhang, X. (2003). Dissolution and mineralization behaviors of HA coatings. *Biomaterials, 24*(26), 4741–4748. Advance online publication. DOI: 10.1016/S0142-9612(03)00371-5 PMID: 14530071

Zhang, Q., & Guan, Y. (2023). Laser induced periodic surface structure on titanium alloy and its effect on microstructure refinement and corrosion behavior. *Journal of Alloys and Compounds*, *936*, 168188. DOI: 10.1016/j.jallcom.2022.168188

Zhang, S., Wang, J., Zheng, Z., Yan, J., Zhang, L., Li, Y., Zhang, J., Li, G., Wang, X., & Kaplan, D. (2021). Porous nerve guidance conduits reinforced with braided composite structures of silk/magnesium filaments for peripheral nerve repair. *Acta Biomaterialia*, *134*, 116–130. DOI: 10.1016/j.actbio.2021.07.028 PMID: 34289421

Zhang, T., Zhang, X., Mao, M., Li, J., Wei, T., & Sun, H. (2020). Chitosan/hydroxyapatite composite coatings on porous Ti6Al4V titanium implants: In vitro and in vivo studies. *Journal of Periodontal & Implant Science*, *50*(6), 392. DOI: 10.5051/jpis.1905680284 PMID: 33350179

Zhang, X. D. (2024). Flow dynamics of liquid metal. In *Handbook of liquid metals* (pp. 89–134). Springer Nature. DOI: 10.1007/978-981-97-1614-2_7

Zhang, Y., & Chen, X.. (2025). Near-infrared light-responsive on-demand puerarin releasing intelligent hydrogel for chronic wound healing. *Materials Letters*. Advance online publication. DOI: 10.1016/j.matlet.2025.135678

Zhang, Y., Wang, X., Yan, K., Zhu, H., Wang, B., & Zou, B. (2023). Laser micro/nano-structuring pushes forward smart sensing: Opportunities and challenges. *Advanced Functional Materials*, *33*(8), 2211272. DOI: 10.1002/adfm.202211272

Zhang, Y., Wu, C., & Ding, J. (2021). Laser-induced hierarchical surface structures for guiding cell behavior and tissue integration. *Bioactive Materials*, *6*(7), 2091–2105. DOI: 10.1016/j.bioactmat.2021.01.033

Zhang, Y., Xia, X., Ma, K., Xia, G., Wu, M., Cheung, Y. H., Yu, H., Zou, B., Zhang, X., Farha, O. K., & Xin, J. H. (2023). Functional textiles with smart properties: Their fabrications and sustainable applications. *Advanced Functional Materials*, *33*(33), 2301607. DOI: 10.1002/adfm.202301607

Zhao, C., Shi, B., Chen, S., Du, D., Sun, T., Simonds, B. J., Fezzaa, K., & Rollett, A. D. (2022). Laser melting modes in metal powder bed fusion additive manufacturing. *Reviews of Modern Physics*, *94*(4), 045002. DOI: 10.1103/RevModPhys.94.045002

Zhao, C., Yu, Z., Du, W., Zhang, Z., Ma, L., Zhang, T., Bai, Y., & Chen, S. (2025). Laser cladding of high-temperature alloy: Research status and development. *Advanced Engineering Materials*, *27*(10), 2402956. DOI: 10.1002/adem.202402956

Zhao, J., Schlößer, H. A., Wang, Z., Qin, J., Li, J., Popp, F., Popp, M. C., Alakus, H., Chon, S.-H., Hansen, H. P., Neiss, W. F., Jauch, K.-W., Bruns, C. J., & Zhao, Y. (2019). Tumor-derived extracellular vesicles inhibit natural killer cell function in pancreatic cancer. *Cancers (Basel)*, *11*(6), 874. DOI: 10.3390/cancers11060874 PMID: 31234517

Zhao, J., Zhu, Z., Xu, Y., Song, X., Wang, Y., Peng, H., Wang, Y., Zuo, J., Shu, X., & Yin, A. (2020). Nanosecond laser ablation of Ti–6Al–4V under different temperature. *Applied Sciences (Basel, Switzerland)*, *10*(13), 4657. DOI: 10.3390/app10134657

Zhao, W., Zhu, J., Hang, J., & Zeng, W. (2022). Biomaterials to promote vascularization in tissue engineering organs and ischemic fibrotic diseases. *MedComm - Biomaterials and Applications*, *1*(1), e16. DOI: 10.1002/mba2.16

Zhou, X., Yu, W., & Huang, Q. (2020). Synergistic effect of laser microtexturing and hydroxyapatite coating on titanium implants: Enhancing osseointegration. *Ceramics International*, *46*(15), 23677–23685.

Zhou, Y., Liu, K., & Zhang, H. (2023). Biomimetic Mineralization: From Microscopic to Macroscopic Materials and Their Biomedical Applications. *ACS Applied Bio Materials*, *6*(9), 3516–3531. DOI: 10.1021/acsabm.3c00109 PMID: 36944024

Zhou, Y., Wang, J., Yang, Y., Yang, M., Zheng, H., Xie, D., Wang, D., & Shen, L. (2022). Laser additive manufacturing of zinc targeting for biomedical application. *International Journal of Bioprinting*, *8*(1), 501. DOI: 10.18063/ijb.v8i1.501 PMID: 35187283

Zhou, Z., Sun, W., Lu, W., Zhang, S., & Wang, J. (2025). Advancements in understanding the characteristics of gas atomized powders: A review. *Advanced Engineering Materials*, *27*(16), 2500270. DOI: 10.1002/adem.202500270

Zhu, D., Zuo, P., Li, F., Tian, H., Liu, T., Hu, L., Huang, H., Liu, J., & Qian, X. (2024). Fabrication and applications of surface micro/nanostructures by femtosecond laser. *Colloid and Interface Science Communications*, *59*, 100700. DOI: 10.1016/j.colcom.2024.100770

Zhuikov, V. A., Bonartsev, A. P., Bagrov, D. V., Yakovlev, S. G., Myshkina, V. L., Makhina, T. K., Bessonov, I. V., Kopitsyna, M. N., Morozov, A. S., Rusakov, A. A., Useinov, A. S., Shaitan, K. V., & Bonartseva, G. A. (2017). Mechanics and surface ultrastructure changes of poly(3-hydroxybutyrate) films during enzymatic degradation in pancreatic lipase solution. *Molecular Crystals and Liquid Crystals (Philadelphia, Pa.)*, *648*(1), 236–243. Advance online publication. DOI: 10.1080/15421406.2017.1302580

Zhu, J., Huang, X., & Song, W. (2021). Physical and chemical sensors on the basis of laser-induced graphene: Mechanisms, applications, and perspectives. *ACS Nano*, *15*(12), 18708–18741. DOI: 10.1021/acsnano.1c05806 PMID: 34881870

Zhu, K., & Fuh, Y. H. J. (2026). Modeling and simulation of laser-based additive manufacturing process. In *Monitoring and control of laser additive manufacturing process: Theory and applications* (pp. 93–145). Springer Nature. DOI: 10.1007/978-981-95-5226-9_3

Zia, A. W., Liu, R., & Wu, X. (2022). Structural design and mechanical performance of composite vascular grafts. *Bio-Design and Manufacturing*, *5*(4), 757–785. DOI: 10.1007/s42242-022-00201-7

Zimbone, M., Buccheri, M. A., Cacciato, G., Sanz, R., Rappazzo, G., Boninelli, S., Reitano, R., Romano, L., Privitera, V., & Grimaldi, M. G. (2015). Photocatalytical and antibacterial activity of TiO2 nanoparticles obtained by laser ablation in water. *Applied Catalysis B: Environmental*, *165*, 487–494. DOI: 10.1016/j.apcatb.2014.10.031

Zuo, S., Wang, T., & Wang, Y.. (2025). High-performance NIR laser-beam shaping and materials processing at 350 W with a spatial light modulator. *Photonics*, *12*(6), 544. Advance online publication. DOI: 10.3390/photonics12060544

Zurita-Méndez, N. N., Carbajal-De la Torre, G., Flores-Merino, M. V., & Espinosa-Medina, M. A. (2022). Development of Bioactive Glass-Collagen-Hyaluronic Acid-Polycaprolactone Scaffolds for Tissue Engineering Applications. *Frontiers in Bioengineering and Biotechnology*, *10*, 825903. Advance online publication. DOI: 10.3389/fbioe.2022.825903 PMID: 35252134

About the Contributors

Irina Negut is a Senior Researcher at the National Institute for Laser, Plasma, and Radiation Physics in the Lasers Department, specialized in laser processing of thin films. With a strong background in surface physics and engineering, her research focuses on developing advanced coatings and materials for biomedical, electronic, and other functional applications. Her expertise lies in laser-based thin film processing, employing techniques such as Pulsed Laser Deposition and Matrix-Assisted Pulsed Laser Evaporation to fabricate and modify nanostructured coatings. She also works on the integration of nanoparticles and hybrid inorganic-organic materials to enhance biocompatibility and impart antimicrobial properties to different surfaces.

Carmen Ristoscu received a PhD in Physics from the University of Bucuresti in 2003. Dr. Ristoscu is Scientific Researcher 1st degree at National Institute for Lasers, Plasma and Radiation Physics (INFLPR), Romania. The main topics of interest are: laser processing using (reactive) Pulsed Laser Deposition and Matrix-Assisted Pulsed Laser Evaporation of various materials with potential applications in bio-nano-medicine; laser interactions; lasers and plasma physics; nanostructured thin films technology; surface physics and engineering.

***

Sercan Basit is a faculty member at Kırşehir Ahi Evran University. He received his Bachelor's degree in Metallurgical and Materials Engineering from Yıldız Technical University (YTU) in 2012, Master's (2015) and PhD (2020) degrees from YTU, Metallurgical and Materials Engineering, Department of Production. After his work in the fields of solidification, alloy development and green recovery, he received the title of Ph. During this time, he developed his project experience by taking part in various tasks in many TUBITAK and university-based projects. He is currently involved in national and international projects in the field of biodegradable metallic materials development and is an active member of BIOMATAM research group at Yıldız Technical University. He has been teaching since 2020 at Kırşehir Ahi Evran University, Department of Mechanical Engineering with the title of Dr. Lecturer in the field of materials.

Virginie Dumas, research engineer at Centrale Lyon ENISE (LTDS), has been working for 13 years on cell–surface interactions, with a particular focus on laser-structured materials. She was awarded the CNRS Crystal Medal in 2024 and has (co-)authored more than 20 international publications in this field. She leads the national ANR BLAST project (2026–2030) dedicated to optimizing laser-processed surfaces for antibacterial applications.

Alain Guignandon, Associate Professor at the Faculty of Medicine, University Jean Monnet (UJM). He has been an Associate Professor in Cell Biology at the Department of Bio-Engineering for Health, Medical School, since 2008, and at the EUR Manutech-Sleight Graduate School since 2017. He completed his PhD in bone physiology at Jean Monnet University in 1997, followed by postdoctoral research in cell signaling in space at the University of Liège's GIGA Center, supported by an ESA fellowship. As a mechanobiologist, his current research at the INSERM SAINBIOSE unit focuses on how biomaterial surfaces modulate cell mechanics and explores the design of nature-inspired topographies created by ultrashort lasers in collaboration with the Hubert Curien Laboratory.

Toiwiya Hassane is a PhD student in cellular biology at Jean Monnet University, where she conducts her research at the SAINBIOSE and Hubert Curien laboratories and at the LTDS laboratory of École Centrale de Lyon. She holds a Master's degree in Life Therapies and Technologies from Sorbonne Paris Nord University, during which she worked on the role of extracellular vesicles in the resistance of immunological cancers to therapies. Her thesis work focuses on the use of extracellular vesicles as a therapeutic means of optimizing the performance and biocompatibility of metal implants.

Ramyakrishna Jangama, graduated from department of Life Sciences, Sri Sathya Sai University for Human Excellence, Kalaburagi, KA 585313, India.

Rupesh Kumar is an Assistant Professor in the Department of Automation and Robotics at Dr. D. Y. Patil Institute of Technology, Pimpri, Pune, India.

Iulian Pana is a scientific researcher at the National Institute of Research and Development in Optoelectronics – INOE 2000 (Research Center for Advanced

Surface Processing and Analysis by Vacuum Technologies - ReCAST) and has worked in the field of research for the last 15 years. His research interests focus on modeling and simulation of optical and physical systems, as well as on the deposition and investigation of thin-film properties produced by Physical Vapor Deposition (PVD) techniques. Dr. Iulian Pana has received his Bachelor's (2011) and Master's (2013) degrees in Physics-Informatics from the Faculty of Physics, University of Bucharest (Romania). In 2018, Dr. Iulian Pana obtained a Ph.D. in Physics and, to date, has published more than 30 scientific articles as first or co-author in ISI journals.

Ramya Raghavan, Assistant Professor at Sri Sathya Sai University for Human Excellence, India. Dr. Raghavan research interests focus on epigenetic factors contributing to intellectual disability. Further investigated neural markers, molecular mechanisms associated with severe speech impairment. Her recent editorial work on a book dedicated to advancing cancer treatment through spatial omics approaches underscores her commitment to innovation. Her expertise extends to the critical intersection of the exposome and Social Determinants of Health (SDOH). Dr. Raghavan's work underscores how a holistic understanding of environmental exposures, from the molecular level to population-wide influences, exposomics, and their implications for public health.

Sushila Sahani is an assistant professor in the School of Fashion Design and Technology at Lovely Professional University in Phagwara, Punjab, India. She holds a doctorate in apparel and textile science. Her academic and professional journey blends expertise in textiles from material properties to sustainable innovations, with interests in fashion design, intellectual property and ethical consumerism. Her research and educational work explore the connections between textile knowledge and design thinking, highlighting the economic, cultural and sustainable impact of the fashion and textile industry.

Xxx Sedao is a Research Engineer at the Hubert Curien Laboratory, specializing in ultrafast laser–matter interactions and laser micro/nanofabrication. He has contributed to numerous European FP7 and H2020 projects, as well as French ANR programs. Dr. Sedao brings extensive experience in developing advanced laser processing technologies at the interface of multiple application domains.

Alina Vladescu has completed her PhD in Materials Science and Engineering from the University Politehnica of Bucharest, Romania. She works at National Institute for Optoelectronics, Department for Advanced Surface Processing and Analysis by Vacuum Technologies since 2002. She is also associate professor at University Politehnica of Bucharest since 2015 and research scientist at National

Research Tomsk Polytechnic University, Russia, since 2016. She has published more than 123 papers in reputed journals, 14 patents, 4 books. She was Guest Editor in Frontiers in Materials (2016-2017), Composite Interfaces (2016), Coatings (2019-2021).

Index

A

B

C

D

E

F

G

H

I

L

M

N

O

P

R

S

T

U

W

Z

www.ingramcontent.com/pod-product-compliance
Lightning Source LLC
LaVergne TN
LVHW062330300926
849161LV00007B/1644
9798337341040